Solutions Manual For

Organic Chemistry,
revised printing

T. W. Graham Solomons
Jack E. Fernandez
University of South Florida

JOHN WILEY & SONS

New York   Santa Barbara   Chichester   Brisbane   Toronto

Copyright ® 1976, 1977 by John Wiley & Sons, Inc.

ISBN 0-471-81218-8

Printed in the United States of America

10 9 8 7 6 5 4 3 2 1

TO THE STUDENT

This manual presents solutions to all the problems in the text including the end-of-chapter problems. In many instances we have given not only the answer, but also an explanation of the reasoning that leads to the solution. Although many problems have more than one solution, we have generally given only one. Thus, you should not necessarily assume that your answer is incorrect if it differs from the one given here.

The heart of organic chemistry lies in problem-solving. This is as true for the practicing organic chemist as for the beginning student. But, a problem in organic chemistry is like a riddle; once you have seen the answer, it is impossible for you to go through the process of solving it. The essential value of a problem lies in the mental exercise of the problem-solving process, and you cannot get this exercise if you already know the answer. The best way to use this manual, therefore, is to check the problems you have worked, or to find explanations for unsolved problems only after you have made serious attempts at working them.

Problems involving syntheses are always intriguing to chemists and students alike. These problems are probably unlike any that you have seen in other courses. In them you are asked to put together a series of reactions that will convert one compound into another. To do this, it is not enough to start your reasoning with the starting materials because you must also keep in mind the desired product. In some instances you can work from both ends simultaneously, and this is best done by trying to find an intermediate that will link the starting materials and products in a sequence of reactions. In many cases, however, a synthesis problem is best solved by reasoning backward, by starting your thinking with the product and by trying to discover a series of reactions that will lead back to the starting compounds. Sherlock Holmes in *A Study in Scarlet* said:

Most people if you describe a train of events to them, will tell you what the result would be. They can put these events together in their minds, and argue from them that something will come to pass. There are a few people, however, who, if you told them a result, would be able to evolve from their own inner conciousness what the steps were which led up to that result. This power is what I mean when I talk of reasoning backward, or analytically.

This power is what the organic chemistry student must develop.

1
A STUDY OF THE COMPOUNDS OF CARBON

1.1
Using the elemental symbol to denote the nucleus and inner shell electrons:

(a) $(\text{Na} \cdot) + (: \overset{\cdot\cdot}{\underset{\cdot\cdot}{Cl}} \cdot) \longrightarrow \text{Na}^+ + (: \overset{\cdot\cdot}{\underset{\cdot\cdot}{Cl}} :)^-$

(b) $(\text{Mg} :) + 2 (: \overset{\cdot\cdot}{\underset{\cdot\cdot}{F}} :) \longrightarrow \text{Mg}^{+2} + 2 (: \overset{\cdot\cdot}{\underset{\cdot\cdot}{F}} :)^-$

(c) $(\text{K} \cdot) + (: \overset{\cdot\cdot}{\underset{\cdot\cdot}{Br}} :) \longrightarrow \text{K}^+ + (: \overset{\cdot\cdot}{\underset{\cdot\cdot}{Br}} :)^-$

1.2

(a) $\text{H} : \overset{\cdot\cdot}{\underset{\cdot\cdot}{Br}} :$, $\text{H}-\overset{\cdot\cdot}{\underset{\cdot\cdot}{Br}} :$

(b) $: \overset{\cdot\cdot}{\underset{\cdot\cdot}{Br}} : \overset{\cdot\cdot}{\underset{\cdot\cdot}{Br}} :$, $: \overset{\cdot\cdot}{\underset{\cdot\cdot}{Br}}-\overset{\cdot\cdot}{\underset{\cdot\cdot}{Br}} :$

(c) $: \overset{\cdot\cdot}{O} : : C : : \overset{\cdot\cdot}{O} :$, $: \overset{\cdot\cdot}{O}=C=\overset{\cdot\cdot}{O} :$

(d) $\text{H} : \overset{H}{\underset{\overset{\cdot\cdot}{H}}{C}} : \text{H}$, $\text{H}-\overset{\overset{H}{|}}{\underset{\underset{H}{|}}{C}}-\text{H}$

(e) $\text{H} : \overset{\cdot\cdot}{O} : \overset{:\overset{\cdot\cdot}{O}:}{N} : : \overset{\cdot\cdot}{O} :$, $\text{H}-\overset{:\overset{\cdot\cdot}{O}:}{\underset{\overset{|}{}}{O}} \overset{}{N}=\overset{\cdot\cdot}{O} :$

(f) $\text{H} : \overset{\cdot\cdot}{O} : N : : \overset{\cdot\cdot}{O} :$, $\text{H}-\overset{\cdot\cdot}{O}-N=\overset{\cdot\cdot}{O} :$

(g) $\text{H} : \overset{\cdot\cdot}{\underset{\cdot\cdot}{O}} : \overset{\cdot\cdot}{\underset{\cdot\cdot}{O}} : \text{H}$, $\text{H}-\overset{\cdot\cdot}{\underset{\cdot\cdot}{O}}-\overset{\cdot\cdot}{\underset{\cdot\cdot}{O}}-\text{H}$

(h) $\text{H} : \overset{H}{\underset{\overset{\cdot\cdot}{H}}{Si}} : \text{H}$, $\text{H}-\overset{\overset{H}{|}}{\underset{\underset{H}{|}}{Si}}-\text{H}$

(i) $\text{H} : \overset{\cdot\cdot}{\underset{\overset{\cdot\cdot}{H}}{N}} : \text{H}$, $\text{H}-\overset{\cdot\cdot}{\underset{\underset{H}{|}}{N}}-\text{H}$

(j) $: \overset{\cdot\cdot}{\underset{\cdot\cdot}{Cl}} : \overset{:\overset{\cdot\cdot}{\underset{\cdot\cdot}{Cl}}:}{P} : \overset{\cdot\cdot}{\underset{\cdot\cdot}{Cl}} :$, $: \overset{\cdot\cdot}{\underset{\cdot\cdot}{Cl}}-\overset{\overset{|}{}}{\underset{\underset{:\overset{\cdot\cdot}{\underset{\cdot\cdot}{Cl}}:}{}}{P}}-\overset{\cdot\cdot}{\underset{\cdot\cdot}{Cl}} :$

(k) $: \overset{\cdot\cdot}{\underset{\cdot\cdot}{F}} : \overset{:\overset{\cdot\cdot}{F}:}{N} : \overset{\cdot\cdot}{\underset{\cdot\cdot}{F}} :$, $: \overset{\cdot\cdot}{\underset{\cdot\cdot}{F}}-\overset{\overset{|}{}}{\underset{\underset{:\overset{\cdot\cdot}{F}:}{}}{N}} \overset{\cdot\cdot}{\underset{\cdot\cdot}{F}} :$

(l) $\text{H} : \overset{H}{\underset{\overset{\cdot\cdot}{H}}{C}} : \overset{\cdot\cdot}{\underset{\cdot\cdot}{Cl}} :$, $\text{H}-\overset{\overset{H}{|}}{\underset{\underset{H}{|}}{C}}-\overset{\cdot\cdot}{\underset{\cdot\cdot}{Cl}} :$

(m) $\text{H} : \overset{\cdot\cdot}{\underset{\overset{\cdot\cdot}{H}}{O}} :$, $\text{H}-\overset{\cdot\cdot}{\underset{\underset{H}{|}}{O}} :$

(n) $: \overset{\cdot\cdot}{\underset{\cdot\cdot}{O}} : \text{H}$, $: \overset{\cdot\cdot}{\underset{\cdot\cdot}{O}}-\text{H}$

(o) $\left[:\ddot{O}:N::\ddot{O}:\right]^{-}$, $\left[:\ddot{O}-N=\ddot{O}:\right]^{-}$ (q) $\left[H:\underset{\underset{H}{\overset{H}{:}}}{\overset{H}{N}}:H\right]^{+}$ $:\ddot{Cl}:^{-}$, $\left[H-\underset{\underset{H}{|}}{\overset{\overset{H}{|}}{N}}-H\right]^{+}$ $:\ddot{Cl}:^{-}$

$\quad\quad:\ddot{O}:$

(p) $\left[:\ddot{O}:N::\ddot{O}:\right]^{-}$, $\left[:\ddot{O}-N=\ddot{O}:\right]^{-}$ (r) $Mg^{+2}\left[:\ddot{O}:\underset{\underset{:\ddot{O}:}{}}{\overset{:\ddot{O}:}{S}}:\ddot{O}:\right]^{-2}$, $Mg^{2+}\left[:\ddot{O}-\underset{\underset{:\ddot{O}:}{\|}}{\overset{\overset{:\ddot{O}:}{\|}}{S}}-\ddot{O}:\right]^{-2}$

1.3

Formal charge = group number $-$ [½ (number of shared electrons) + (number of unshared electrons)]

Charge on the ion = number of protons + number of electrons

= sum of all formal charges;

		Formal Charge	Total Charge
(a) $H-\underset{\underset{H}{\|}}{\overset{\overset{H}{\|}}{B}}-H$	H	$1-(1+0)=\quad 0$	-1
	B	$3-(4+0)=-1$	
(b) $H-\ddot{O}-\underset{\underset{:\ddot{O}}{\|}}{\overset{\overset{}{S}}{}}-\ddot{O}-H$	H	$1-(1+0)=\quad 0$	0
	O (OH)	$6-(2+4)=\quad 0$	
	O (=O)	$6-(2+4)=\quad 0$	
	S	$6-(4+2)=\quad 0$	
(c) $:\ddot{F}-\underset{\underset{:\ddot{F}:}{\|}}{\overset{\overset{:\ddot{F}:}{\|}}{B}}-\ddot{F}:$	F	$7-(1+6)=\quad 0$	-1
	B	$3-(4+0)=-1$	
(d) $H-\overset{\cdot\cdot}{\underset{\underset{H}{\|}}{O}}-H$	H	$1-(1+0)=\quad 0$	$+1$
	O	$6-(3+2)=+1$	
(e) $:\ddot{F}-\underset{\underset{:\ddot{F}:}{\|}}{\overset{\cdot\cdot}{N}}-\ddot{F}:$	F	$7-(1+6)=\quad 0$	0
	N	$5-(3+2)=\quad 0$	
(f) $H-\underset{\underset{H}{\|}}{\overset{\cdot\cdot}{C}}-H$	H	$1-(1+0)=\quad 0$	-1
	C	$4-(3+2)=-1$	
(g) $H-\underset{\underset{H}{\|}}{C}-H$	H	$1-(1+0)=\quad 0$	$+1$
	C	$4-(3+0)=+1$	
(h) $H-\underset{\underset{H}{\|}}{\overset{\cdot}{C}}-H$	H	$1-(1+0)=\quad 0$	0
	C	$4-(3+1)=\quad 0$	
(i) $H-\overset{\cdot\cdot}{C}-H$	H	$1-(1+0)=\quad 0$	0
	C	$4-(2+2)=\quad 0$	

1.4

(a) No formal charges   (d) No formal charges

(b) No formal charges   (e) $CH_3-\overset{\overset{\displaystyle CH_3}{|}}{\underset{\underset{\displaystyle :\overset{..}{O}:^-}{|}}{N^+}}-CH_3$

(c) No formal charges   (f) $CH_3-\overset{+}{\underset{\underset{\displaystyle :\overset{..}{O}:^-}{|}}{N}}=\overset{..}{O}:$

1.5

(a) H—Br   (c) H—H (dipole moment = 0)

(b) $\overset{\longrightarrow}{I-Cl}$ (d) Cl—Cl (dipole moment = 0)

1.6

(a) Tetrahedral   (c) Trigonal planar   (e) Tetrahedral

(b) Tetrahedral   (d) Tetrahedral   (f) Linear

1.7

The two C=O bond moments are opposed and cancel each other:

$$\overset{\longleftarrow\;+\;+\;\longrightarrow}{O = C = O}$$

If the bond angle were other than 180°, then the individual bond moments would not cancel. There would be a resultant dipole moment.

1.8

The direction of polarity of the N—H bond is opposite to that of the N—F bond.

In NH_3, the resultant N—H bond polarities and the polarity of the unshared electron pair are in the same direction.

In NF_3 the resultant N—F bond polarities partially cancel the polarity of the unshared electron pair.

1.9

BF_3 is trigonal planar. Its B—F bonds are all necessarily equal in polarity, and the F—B—F bond angles are all equal (120°).

Therefore, the resultant of the three bond moments is zero.

1.10

Assuming a 100-g sample, the amounts of the elements are:

	Weight	Moles (A)		B	
C	57.45	$\frac{57.45}{12.01}$ = 4.78	$\frac{4.78}{0.300}$ = 15.9	= 16
H	5.40	$\frac{5.40}{1.008}$ = 5.36	$\frac{5.36}{0.300}$ = 17.9	= 18
N	8.45	$\frac{8.45}{14.01}$ = 0.603	$\frac{0.603}{0.300}$ = 2.01	= 2
S	9.61	$\frac{9.61}{32.06}$ = 0.300	$\frac{0.300}{0.300}$ = 1.00	= 1
O*	$\frac{19.09}{100.00}$	$\frac{19.09}{16.00}$ = 1.19	$\frac{1.19}{0.300}$ = 3.97	= 4

(* by difference from 100)

The empirical formula is thus $C_{16}H_{18}N_2SO_4$. The empirical formula weight (334.4) is within the range given for the molecular weight (330 ± 10), thus the molecular formula for Penicillin G is the same as the empirical formula.

1.11

1.12

The equal length O—O bonds in ozone can be accounted for by the two equivalent resonance structures,

The hybrid structure thus has two equivalent (1½) bonds joining the middle oxygen to the other two. The unshared electron pair on the middle oxygen accounts for the fact that the molecule is bent.

1.13

(c)

$$
\begin{array}{c}
\text{H} \\
\text{H}\ \overset{\text{H}-\text{C}-\text{H}}{\underset{}{|}}\ \text{H} \quad \text{H} \\
\text{H} - \text{C} - \text{C} - \text{C} - \text{C} - \text{H} \\
\underset{\text{H}}{|}\ \ \underset{\text{H}-\text{C}-\text{H}}{|}\ \ \underset{\text{H}}{|}\ \ \underset{\text{H}}{|} \\
\text{H}
\end{array}
$$

(f)

$$
\begin{array}{c}
\text{H} \quad \text{H} \quad \text{H} \quad \text{H} \\
\text{H} - \text{C} - \text{C} - \text{C} - \text{C} - \overset{..}{\text{O}}\text{H} \\
\underset{\text{H}}{|}\ \ \underset{\text{H}}{|}\ \ \underset{\text{H}}{|}\ \ \underset{\text{H}}{|}
\end{array}
$$

(d)

$$
\begin{array}{c}
\text{H} \quad :\!\overset{..}{\text{C}}\text{l}:\ :\!\overset{..}{\text{C}}\text{l}:\ \text{H} \\
\text{H} - \text{C} - \text{C} - \text{C} - \text{C} - \text{H} \\
\underset{\text{H}}{|}\ \ \underset{\text{H}}{|}\ \ \underset{\text{H}}{|}\ \ \underset{\text{H}}{|}
\end{array}
$$

(g)

$$
\begin{array}{c}
\text{H} \quad :\!\overset{..}{\underset{}{\text{O}}} \quad \text{H} \quad \overset{\text{H}-\text{C}-\text{H}}{|} \quad \text{H} \\
\text{H} - \text{C} - \text{C} - \text{C} - \text{C} - \text{C} - \text{H} \\
\underset{\text{H}}{|}\ \ \ \ \underset{\text{H}}{|}\ \ \underset{\text{H}}{|}\ \ \underset{\text{H}}{|}
\end{array}
$$

(e)

$$
\begin{array}{c}
\text{H} \quad :\!\overset{..}{\text{O}}\text{H} \quad \text{H} \quad \text{H} \\
\text{H} - \text{C} - \text{C} - \text{C} - \text{C} - \text{H} \\
\underset{\text{H}}{|}\ \ \ \underset{\text{H}}{|}\ \ \underset{\text{H}}{|}\ \ \underset{\text{H}}{|}
\end{array}
$$

(h)

$$
\begin{array}{c}
\text{H} \quad \text{H} \quad :\!\overset{..}{\text{O}}\text{H} \quad \overset{\text{H}-\text{C}-\text{H}}{|} \quad \text{H} \\
\text{H} - \text{C} - \text{C} - \text{C} - \text{C} - \text{C} - \text{H} \\
\underset{\text{H}}{|}\ \ \underset{\text{H}}{|}\ \ \underset{\text{H}}{|}\ \ \underset{\text{H}}{|}\ \ \underset{\text{H}}{|}
\end{array}
$$

1.14
(a) and (d) are structural isomers.
(e) and (f) are structural isomers.

1.15
(a) To calculate the percentage composition from the molecular formula, first determine the weight of each element in one mole of the compound. For $C_6H_{12}O_6$,

$$C_6 = 6 \times 12.01 = 72.06 \qquad \frac{72.06}{180.2} = 0.400 = 40.0\%$$

$$H_{12} = 12 \times 1.008 = 12.10 \qquad \frac{12.10}{180.2} = 0.0671 = 6.7\%$$

$$O_6 = 6 \times 16.00 = 96.00 \qquad \frac{96.00}{180.2} = 0.533 = 53.3\%$$

Molecular Wt.   180.16

Then determine the percentage of each element using the formula.

$$\text{Percentage of A} = \frac{\text{Weight of A}}{\text{Molecular Weight}} \times 100$$

(b) $C_2 = 2 \times 12.01 = 24.02$ $\dfrac{24.02}{75.07} = 0.320 = 32.0\%$

$H_5 = 5 \times 1.008 = 5.04$ $\dfrac{5.04}{75.07} = 0.067 = 6.7\%$

$N = 1 \times 14.01 = 14.01$ $\dfrac{14.01}{75.07} = 0.187 = 18.7\%$

$O_2 = 2 \times 16.00 = 32.00$ $\dfrac{32.00}{75.07} = 0.426 = 42.6\%$

(c) $C_3 = 3 \times 12.01 = 36.03$ $\dfrac{36.03}{280.77} = 0.128 = 12.8\%$

$H_5 = 5 \times 1.008 = 5.04$ $\dfrac{5.04}{280.77} = 0.018 = 1.8\%$

$Br_3 = 3 \times 79.90 = 239.70$ $\dfrac{239.70}{280.77} = 0.854 = 85.4\%$

1.16

(a) H : C : N : : C : : S :
     with H above and below the C

(d) H : C : N : : C : : O :
     with H above and below the C

(b) H : C : C : : : N : O :
     with H above and below the C, + and − charges

(e) H : C : : C : : O :

(c) H : C : O : N : O :
     with :O: above N, H above and below C, + and − charges

(f) H : C : : N : : N :
     with + and − charges

1.17

(a)

(b)

(c)

(d)

1.18

| Electron Configuration | Orbital Arrangement |

(a) Be $1s^2 2s^2$

(b) B $1s^2 2s^2 2p_x^1$

(c) C $1s^2 2s^2 2p_x^1 2p_y^1$

(d) N $1s^2 2s^2 2p_x^1 2p_y^1 2p_z^1$

(e) O $1s^2 2s^2 2p_x^2 2p_y^1 2p_z^1$

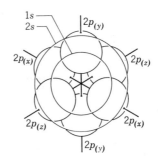

1.19

(a) NH_2 (b) CH   (c) C_2H_4O (d) C_5H_7N (e) CH_2 (f) CH

1.20

Empirical Formula	Empirical Formula Weight	$\left(\dfrac{\text{Molecular Wt.}}{\text{Emp. Form. Wt.}}\right)$	Molecular Formula
(a) CH_2O	30	$\dfrac{179}{30} \cong 6$	$C_6H_{12}O_6$
(b) CHN	27	$\dfrac{80}{27} \cong 3$	$C_3H_3N_3$
(c) CCl_2	83	$\dfrac{410}{83} \cong 5$	C_5Cl_{10}

1.21

First we must determine the empirical formula. Assuming that the difference between the percentages given and 100 percent is due to oxygen,

$$
\text{C:} \quad 40.04 \qquad \frac{40.04}{12.01} = 3.33 \qquad \frac{3.33}{3.33} = 1
$$

$$
\text{H:} \quad 6.69 \qquad \frac{6.69}{1.008} = 6.64 \qquad \frac{6.64}{3.33} \cong 2
$$

$$
\text{O:} \quad \underline{53.27} \qquad \frac{53.27}{16.00} = 3.33 \qquad \frac{3.33}{3.33} = 1
$$
$$
100.00
$$

The empirical formula is thus CH_2O.

To determine the molecular formula we must first determine the molecular weight. At standard temperature and pressure, the volume of one mole of an ideal gas is 22.4 liters. Assuming ideal behavior,

$$\frac{1.00 \text{ g}}{0.746 \text{ liter}} = \frac{M}{22.4 \text{ liters}} \quad \text{Where M = Molecular weight.}$$

$$M = \frac{(1.00)(22.4)}{0.746} = 30.0 \text{ g}$$

The empirical formula weight (30.0) equals the molecular weight, thus the molecular formula is the same as the empirical formula.

1.22

As in problem 1.21, the molecular weight is found by the equation

$$\frac{1.251 \text{ g}}{1.00 \text{ liter}} = \frac{M}{22.4 \text{ liter}}$$

$$M = (1.251)(22.4)$$

$$M = 28.02$$

To determine the empirical formula, we must determine the amount of carbon in 3.926 g of carbon dioxide, and the amount of hydrogen in 1.608 g of water.

$$C: \quad \left(3.926 \text{ g } CO_2\right)\left(\frac{12.01 \text{ g C}}{44.01 \text{ g } CO_2}\right) = 1.071 \text{ g carbon}$$

$$H: \quad \left(1.608 \text{ g } H_2O\right)\left(\frac{2.016 \text{ g H}}{18.016 \text{ g } H_2O}\right) = \frac{0.179 \text{ g hydrogen}}{1.250 \text{ g sample}}$$

The weight of C and H in a 1.250 g sample is 1.250 g. Therefore there are no other elements present in ethene.

To determine the empirical formula we proceed as in problem 1.10 except that the sample size is 1.250 instead of 100 g.

$$C: \quad \frac{1.071}{12.01} = 0.0892 \qquad \frac{0.0892}{0.0892} = 1$$

$$H: \quad \frac{0.179}{1.008} = 0.178 \qquad \frac{0.178}{0.0892} = 2$$

The empirical formula is thus CH_2. The empirical formula weight (14) is one-half the molecular weight. Thus the molecular formula is C_2H_4.

1.23

Using the procedure of problem 1.10,

$$C: \quad 59.10 \qquad \frac{59.10}{12.01} = 4.92 \qquad \frac{4.92}{0.817} = 6.02 \cong 6$$

$$H: \quad 4.92 \qquad \frac{4.92}{1.008} = 4.88 \qquad \frac{4.88}{0.817} = 5.97 \cong 6$$

$$N: \quad 22.91 \qquad \frac{22.91}{14.01} = 1.64 \qquad \frac{1.64}{0.817} = 2$$

$$O: \quad \frac{13.07}{100.00} \qquad \frac{13.07}{16.00} = 0.817 \qquad \frac{0.817}{0.817} = 1$$

The empirical formula is thus $C_6H_6N_2O$. The empirical formula weight is 123.13 which is equal to the molecular weight within experimental error. The molecular formula is thus the same as the empirical formula.

1.24

C:	40.88	$\dfrac{40.88}{12.01} = 3.40$	$\dfrac{3.40}{0.619} = 5.5$	$5.5 \times 2 = 11$
H:	3.74	$\dfrac{3.74}{1.008} = 3.71$	$\dfrac{3.71}{0.619} = 6$	$6 \times 2 = 12$
Cl:	21.95	$\dfrac{21.95}{35.45} = 0.619$	$\dfrac{0.619}{0.619} = 1$	$1 \times 2 = 2$
N:	8.67	$\dfrac{8.67}{14.01} = 0.619$	$\dfrac{0.619}{0.619} = 1$	$1 \times 2 = 2$
O:	$\dfrac{24.76}{100.00}$	$\dfrac{24.76}{16.00} = 1.55$	$\dfrac{1.55}{0.619} = 2.5$	$2.5 \times 2 = 5$

The empirical formula is thus $C_{11}H_{12}Cl_2N_2O_5$. The empirical formula weight (323) is equal to the molecular weight, therefore the molecular formula is the same as the empirical formula.

1.25

(a) Tetramethyllead, because it is a relatively low-boiling liquid, must be a molecular compound. Its bonds are covalent. Lead fluoride, a high-melting solid, is ionic. A higher temperature (greater energy) is required to separate ions of opposite charges than to separate neutral molecules.

(b) The C—Pb bonds are covalent because lead and carbon have similar electronegativities. Lead and fluorine, on the other hand, have widely differing electronegativities, and Pb—F bonds are therefore ionic.

1.26

$$H : \overset{\cdot\cdot}{\underset{\cdot\cdot}{Cl}} : \longleftrightarrow H \overset{-}{:} \overset{\cdot\cdot}{\underset{\cdot\cdot}{Cl}} : \overset{+}{} \longleftrightarrow H^+ : \overset{\cdot\cdot}{\underset{\cdot\cdot}{Cl}} : \overset{-}{}$$

$$\text{(I)} \qquad\qquad \text{(II)} \qquad\qquad \text{(III)}$$

(a) III is more reasonable than II because in III the negative charge resides on chlorine, the more electronegative atom.

(b) III would contribute more than II.

(c) The dipole moment, $\overset{\longrightarrow}{H-Cl}$, can be accounted for by a hybrid of structures I, II, and III, but in which III contributes more than II.

1.27

The three equivalent resonance structures,

when combined result in equivalent N—O bonds that are intermediate in length between single and double bonds.

1.28

$$
\begin{array}{cccc}
:\overset{..}{\underset{..}{O}} & :\overset{..}{\underset{}{O}} & :\overset{..}{\underset{..}{O}} & :\overset{..}{\underset{..}{O}}: \\
:\overset{..}{O}-\overset{||}{S}-\overset{..}{O}:^{-} & :\overset{..}{O}-\overset{||}{S}=\overset{..}{O}:^{-} & :\overset{..}{O}=\overset{||}{S}-\overset{..}{O}:^{-} & :\overset{..}{O}=\overset{|}{S}=\overset{..}{O}: \\
:\underset{..}{O}: & :\underset{..}{O}: & :\underset{..}{O}: & :\underset{..}{O}:
\end{array}
$$

Yes. Note that *each* oxygen atom in the first six structures appears three times doubly bonded to S and three times singly bonded to S. In the next four structures *each* oxygen appears doubly bonded once and singly bonded three times. Thus each oxygen has the same average type of bonding, hence the same bond length.

Unlike second period elements, sulfur has five vacant $3d$ orbitals that can accept electrons. Its valence shell can therefore expand, in principle, to as many as 18 valence electrons. Actually, structures with more than two double bonds and two single bonds to sulfur can be considered insignificant.

1.29

(a) $e = \dfrac{\mu}{d} = \dfrac{1.08 \times 10^{-18} \text{ esu-cm}}{1.27 \times 10^{-8} \text{ cm}}$

$e = 0.85 \times 10^{-10} \text{ esu}$

(b) $\dfrac{0.85 \times 10^{-10} \text{ esu}}{4.8 \times 10^{-10} \text{ esu/electron}} = 0.18 \text{ electron}$

1.30

A carbon-chlorine bond is longer than a carbon-fluorine bond because chlorine is a larger atom than fluorine. Thus in $\overset{\delta+}{C}H_3-\overset{\delta-}{C}l$ the distance, d, that separates the charges is greater than in $\overset{\delta+}{C}H_3-\overset{\delta-}{F}$. The greater value of d for CH_3Cl more than compensates for the smaller value of e and thus the dipole moment ($e \times d$) is larger.

1.31

(a) $\underset{H}{\overset{H}{>}}C=\overset{..}{\underset{..}{O}}:$

(b) Formaldehyde is actually a hybrid of the resonance structures shown below. In the second structure notice that the electronegative oxygen atom bears a negative charge.

This means that this structure is a "reasonable" structure even though it involves separated charges and even though the carbon has only six electrons in its valence shell. The contribution made by the second structure makes the carbon-oxygen bond highly polar and accounts for the fact that formaldehyde has an unusually high dipole moment.

In CH_3F no such resonance is possible and thus it has a lower dipole moment even though fluorine is more electronegative than oxygen.

1.32

If the compound contains iron, each molecule must contain at least one atom of iron. This means that one mole of the compound must contain at least 55.85 grams of iron. Therefore,

$$\text{MW of ferrocene} = 55.85 \ \frac{\text{grams Fe}}{\text{mole}} \times \frac{1.000 \ \text{gram}}{0.3002 \ \text{gram Fe}}$$

$$= 186.0 \ \frac{\text{grams}}{\text{mole}}$$

1.33

(a) While the structures differ in the position of their electrons they also differ in the positions of their nuclei and thus *they are not resonance structures*. (In cyanic acid the hydrogen nucleus is bonded to oxygen; in isocyanic acid it is bonded to nitrogen.)

(b) The anion obtained from either acid is a resonance hybrid of the following structures:

1.34

If we assume that the electrons of a multiple bond behave as though they were a single pair, we predict the following shapes and bond angles.

(a) CO_2, linear, bond angle 180°

180°

(The observed bond angle is 180°.)

(b) SO_2, bent, bond angle ~120°

~120°

(The observed bond angle is 119.5°.)

(c) SO_3, triangular, bond angles 120°

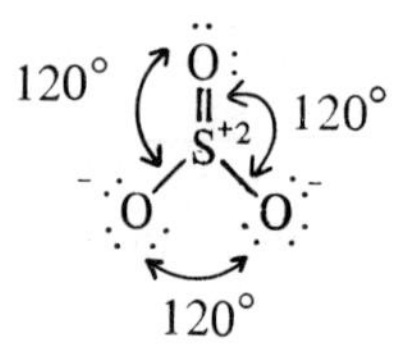

(The observed bond angles are $120°$.)

1.35

(a) Sulfur dioxide is a hybrid of the two equivalent resonance structures shown below. Thus the bonds are the same length.

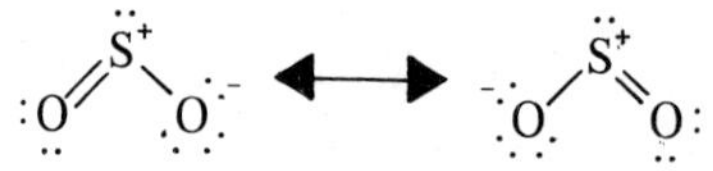

(b) Sulfur trioxide is a hybrid of three equivalent resonance structures thus all three bonds are the same length.

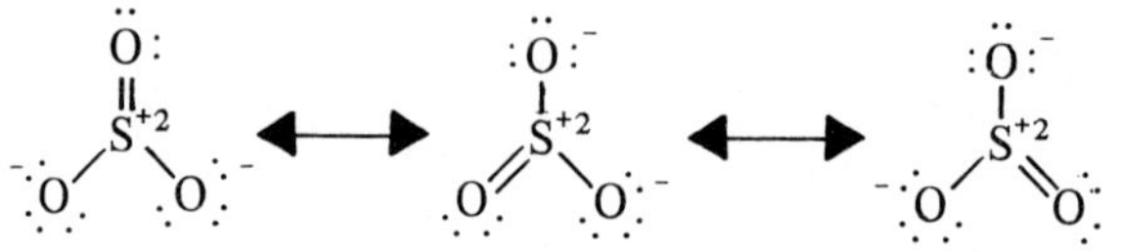

2 FUNCTIONAL GROUPS AND ORGANIC COMPOUNDS: THE MAJOR REACTION TYPES

2.1

(a)
$$Cl-\underset{\underset{Cl}{|}}{C}HCH_3 + Cl_2 \longrightarrow Cl-\underset{\underset{Cl}{|}}{\overset{\overset{Cl}{|}}{C}}CH_3 + Cl-\underset{\underset{Cl}{|}}{C}HCH_2-Cl$$

1,1-dichloroethane $\longrightarrow$ two trichloroethanes

(b)
$$Cl-CH_2CH_2-Cl + Cl_2 \longrightarrow Cl-\underset{\underset{Cl}{|}}{C}HCH_2-Cl$$

1,2-dichloroethane $\longrightarrow$ one trichloroethane (the same as the second compound above)

(c) See above   (d) Yes, the trichloroethanes are isomers.

(e)
$$Cl-\underset{\underset{Cl}{|}}{\overset{\overset{Cl}{|}}{C}}CH_3 + Cl_2 \longrightarrow Cl-\underset{\underset{Cl}{|}}{\overset{\overset{Cl}{|}}{C}}CH_2Cl$$

1,1,1-trichloroethane $\longrightarrow$ one tetrachloroethane

$$Cl-\underset{\underset{Cl}{|}}{C}HCH_2-Cl + Cl_2 \longrightarrow Cl-\underset{\underset{Cl}{|}}{\overset{\overset{Cl}{|}}{C}}CH_2-Cl + Cl-\underset{\underset{Cl}{|}}{C}H\underset{\underset{Cl}{|}}{C}H-Cl$$

1,1,2 trichloroethane $\longrightarrow$ two tetrachloroethanes, one of which is the same as the one formed from 1,1,1-trichloroethane.

Thus there are only two tetrachloroethanes

(f) See (e) above.

(g) Only one pentachloroethane is possible:

$$Cl-\underset{\underset{Cl}{|}}{\overset{\overset{Cl}{|}}{C}}-\underset{\underset{Cl}{|}}{C}H-Cl$$

2.2

(a) Counting the double bond as two electron pairs located in the region of space between the two carbon atoms, each carbon has three atoms attached to it:

11

The maximum separation of the electrons bonding the three atoms about each carbon atom occurs when they are equally spaced about that atom; i.e., when the bond angles are ~ 120° and the molecule is planar.

(b) The H−C−H bond angle is less than 120° because the repulsive force between the electron pairs of the C−H bonds is less than that between the electron pairs of the C−H bonds and the four electrons of the carbon−carbon double bond.

2.3

(a) The C−X bond moments in the *trans*-isomers point in opposite directions and therefore cancel:

X = Cl or Br

trans -   *cis-*

In the *cis-* isomers the bond moments are additive

(b) The C−Cl bond moment is larger than the C−Br bond moment because Cl is more electronegative than Br.

(c) Yes   (d) Yes

2.4

(a)

$Cl{-}CH_2CH{=}CH_2$

cis - trans isomers

(b)

cis - trans isomers   *cis - trans* isomers

(c) $CH_3CH{=}CCl_2$ $Cl{-}CHCH{=}CH_2$ $Cl{-}CH_2C{=}CH_2$
                                         | |
                                         Cl   Cl

$Cl_3CCH{=}CH_2$ $Cl{-}CHC{=}CH_2$ $Cl{-}CH_2CH{=}CCl_2$ $CH_3C{=}CCl_2$
                            | | |
                           Cl Cl   Cl

cis - trans isomers   *cis - trans* isomers

(d)　$Cl_3CC=CH_2$
$\quad\quad\quad\overset{|}{Cl}$

$$\underset{\text{cis - \textit{trans} isomers}}{\underbrace{\begin{array}{cc} \underset{H}{\overset{Cl_3C}{\diagdown}}C=C\underset{H}{\overset{Cl}{\diagup}} & \underset{H}{\overset{Cl_3C}{\diagdown}}C=C\underset{Cl}{\overset{H}{\diagup}} \end{array}}}$$

$Cl_2CHCH=CCl_2$

$$\underset{\text{cis - \textit{trans} isomers}}{\underbrace{\begin{array}{cc} \underset{Cl}{\overset{Cl_2CH}{\diagdown}}C=C\underset{Cl}{\overset{H}{\diagup}} & \underset{Cl}{\overset{Cl_2CH}{\diagdown}}C=C\underset{H}{\overset{Cl}{\diagup}} \end{array}}}$$

$ClCH_2C=CCl_2$
$\quad\quad\quad\overset{|}{Cl}$

(e)

$$\underset{\text{cis - \textit{trans} isomers}}{\underbrace{\begin{array}{cc} \underset{Cl}{\overset{Cl_3C}{\diagdown}}C=C\underset{H}{\overset{Cl}{\diagup}} & \underset{Cl}{\overset{Cl_3C}{\diagdown}}C=C\underset{Cl}{\overset{H}{\diagup}} \end{array}}}$$

$Cl_3CCH=CCl_2$　　$Cl_2CHC=CCl_2$
$\quad\quad\quad\quad\quad\quad\quad\quad\quad\quad\quad\overset{|}{Cl}$

(f)　See (a-e) above.

2.5

(a)　$CH_3\overset{\overset{\displaystyle Cl}{|}}{\underset{\underset{\displaystyle CH_3}{|}}{C}}CH_3$　(b)　$CH_3\overset{}{\underset{\underset{\displaystyle I}{|}}{C}HCH_2CH_3}$

2.6
(a)　Yes, counting the electrons in the triple bond together and as occupying the region between the two carbon atoms, the remaining electrons (in the C–H bonds) are at a maximum separation from the electrons in the triple bond when the bond angle is 180°.

(b)　$H-C\equiv C-\overset{|}{\underset{|}{C}}-$

(c)　$H-C\equiv C-C\overset{\diagup H}{\underset{\diagdown H}{}}_{H}$　or　$H-\bigcirc\!\!\!\equiv\!\!\!\bigcirc-\bigcirc\overset{\diagup H}{\underset{\diagdown H}{}}_{H}$

2.7
(a)　Yes, the halogen adds to the less hydrogenated carbon, i.e., to the one that already bears a chlorine atom.

(b)　$CH_3C\equiv CH + HCl \longrightarrow CH_3\underset{\underset{\displaystyle Cl}{|}}{C}=CH_2 \overset{HCl}{\longrightarrow} CH_3\overset{\overset{\displaystyle Cl}{|}}{\underset{\underset{\displaystyle Cl}{|}}{C}}CH_3$

2.8
(a)　$HC\equiv CH + 2HCl \longrightarrow CH_3CHCl_2$

(b) $CH_2{=}CH_2 + Cl_2 \xrightarrow[\text{dark}]{\text{CCl}_4} CH_2ClCH_2Cl$

(c) $CH_2{=}CH_2 + Br_2 \xrightarrow[\text{dark}]{\text{CCl}_4} CH_2BrCH_2Br$

(d) $HC{\equiv}CH + 2Br_2 \xrightarrow[\text{dark}]{\text{CCl}_4} CHBr_2CHBr_2$

(e) $HC{\equiv}CH + HCl \longrightarrow CH_2{=}CHCl \xrightarrow{\text{HBr}} CH_3CHClBr$

(f) $CH_3C{\equiv}CH + HCl \longrightarrow CH_3\overset{\overset{\displaystyle Cl}{|}}{C}{=}CH_2 \xrightarrow{\text{HBr}} CH_3\overset{\overset{\displaystyle Cl}{|}}{\underset{\underset{\displaystyle Br}{|}}{C}}CH_3$

(g) $CH_3C{\equiv}CH + 2HBr \longrightarrow CH_3CBr_2CH_3$

(h) $CH_3CH_3 + Br_2 \xrightarrow[\substack{\text{excess} \\ CH_3CH_3}]{\text{light}} CH_3CH_2Br + HBr$

$CH_2{=}CH_2 + HBr \longrightarrow CH_3CH_2Br$

2.9

(a) $K_a = \dfrac{[H_3O^+][CF_3COO^-]}{[CF_3COOH]} = 1$

let $[H_3O^+] = [CF_3COO^-] = X$

then $[CF_3COOH] = 0.1 - X$

$\therefore \quad \dfrac{(X)(X)}{0.1 - X} = 1 \quad$ or $\quad X^2 = 0.1 - X$

$X^2 + X - 0.1 = 0$

Using the quadratic formula, $X = \dfrac{-b \pm \sqrt{b^2 - 4ac}}{2a}$,

$X = \dfrac{-1 \pm \sqrt{1 + 0.4}}{2} = \dfrac{-1 \pm \sqrt{1.4}}{2} = \dfrac{-1 \pm 1.183}{2} = \dfrac{+0.183}{2}$

$X = 0.0915$ (We can exclude negative values of X.)

$[H_3O^+] = [CF_3COO^-] = 0.0915 \underline{M}$

(b) Percentage ionized $= \dfrac{[H_3O^+]}{0.1} \times 100 = \dfrac{(0.0915)(100)}{0.1}$

Percentage ionized $= 91.5\%$

2.10

Molecules of N-propylamine can form hydrogen bonds to each other,

$$CH_3CH_2CH_2N\begin{smallmatrix} \diagup H \quad\quad H \diagdown \\ \\ \cdots\cdots\cdots\cdots \\ \diagdown H\cdots\cdots H\diagup \end{smallmatrix}NCH_2CH_2CH_3 \ ,$$

whereas molecules of trimethylamine, because they have no hydrogens attached to nitrogen, cannot form hydrogen bonds to each other.

2.11
(a) Alkyne   (b) Carboxylic acid   (c) Alcohol

(d) Aldehyde   (e) Alkane   (f) Ketone

2.12

(a) Carbon-carbon double bonds, hydroxyl group

(b) Ketone group, hydroxyl group, carbon-carbon double bond

(c) Carbon-carbon double bond, ester group

(d) Amide groups

(e) Aldehyde group, hydroxyl groups

(f) Carbon-carbon double bond, ether linkage

(g) Carbon-carbon double bond, hydroxyl group

2.13
(a) Elimination   (b) Substitution   (c) Substitution

(d) Oxidation   (e) Condensation   (f) Condensation

2.14

(a) $CH_3CH_2CCl_3$ $CH_3\underset{\underset{Cl}{|}}{C}HCHCl_2$ $CH_3\underset{\underset{Cl}{|}}{\overset{\overset{Cl}{|}}{C}}CH_2Cl$ $ClCH_2CH_2CHCl_2$

$ClCH_2\underset{\underset{Cl}{|}}{C}HCH_2Cl$ (five trichloropropanes)

(b) $CH_3\underset{\underset{Cl}{|}}{C}HCCl_3$ $ClCH_2CH_2CCl_3$ $CH_3\underset{\underset{Cl}{|}}{\overset{\overset{Cl}{|}}{C}}CHCl_2$ $Cl_2CHCH_2CHCl_2$

$ClCH_2\underset{\underset{Cl}{|}}{C}HCHCl_2$ $ClCH_2\underset{\underset{Cl}{|}}{\overset{\overset{Cl}{|}}{C}}CH_2Cl$ (six tetrachloropropanes)

2.15

(a) $CH_3\underset{\underset{CH_3}{|}}{C}HCH_2Br$ + KOH $\xrightarrow{\text{ethanol}}$ $CH_3\overset{\overset{CH_3}{|}}{C}=CH_2$ + KBr + H_2O

(b) By dehydrobromination as shown in (a) above, and then by adding HBr to the resulting alkene, isobutyl bromide can be converted into tert-butyl bromide:

$$CH_3\overset{\underset{\displaystyle |}{CH_3}}{C}=CH_2 + HBr \xrightarrow{CCl_4} CH_3\overset{\underset{\displaystyle |}{\underset{\displaystyle Br}{\overset{\displaystyle CH_3}{\overset{\displaystyle |}{C}}}}}{}CH_3 \quad \text{(Markovnikov's rule)}$$

2.16

(a) Primary   (b) Secondary   (c) Tertiary   (d) Secondary

(e) Secondary   (f) Tertiary

2.17

(a) Secondary   (b) Primary   (c) Tertiary   (d) Primary

(e) Secondary

2.18

(a) $2H_3O^+ + CO_3^{-2} \longrightarrow [H_2CO_3] + 2H_2O \longrightarrow 3H_2O + CO_2$

(b) $H_3O^+ + CH_3\overset{\overset{\displaystyle O}{\parallel}}{C}O^- \longrightarrow H_2O + CH_3\overset{\overset{\displaystyle O}{\parallel}}{C}OH$

(c) $CO_3^{-2} + H_2O \longrightarrow HCO_3^- + OH^-$

(d) $:H^- + H_2O \longrightarrow H_2 + OH^-$

(e) $:CH_3^- + H_2O \longrightarrow CH_4 + OH^-$

(f) $:CH_3^- + HC\equiv CH \longrightarrow HC\equiv C:^- + CH_4$

(g) $H_3O^+ + NH_3 \longrightarrow NH_4^+ + H_2O$

(h) $NH_4^+ + NH_2^- \longrightarrow 2\,NH_3$

(i) $CH_3CH_2O^- + H_2O \longrightarrow CH_3CH_2OH + OH^-$

2.19

Oxygen-containing compounds contain either $=\ddot{O}:$ or $-\ddot{O}-$. Both of these are Brønsted-Lowry bases in the presence of the strong proton donor, sulfuric acid. The equation for the reaction using an ether as an example is

$$R-\ddot{O}-R + H_2SO_4 \rightleftharpoons \underbrace{R-\overset{\overset{\displaystyle H}{|}}{\underset{\displaystyle \cdot\cdot}{O}}-{}^+R}_{\text{Salt}} + HSO_4^-$$

The salt is soluble in the highly polar H_2SO_4.

2.20

(a) Ethyl alcohol because its molecules can form hydrogen bonds to each other. Methyl ether molecules have no hydrogens attached to oxygen.

(b) Ethylene glycol because its molecules have more OH groups and will therefore participate in more extensive hydrogen bonding.

(c) Heptane because it has a higher molecular weight. (Neither compound can form hydrogen bonds.)

(d) 1-Propanol because its molecules can form hydrogen bonds to each other. Acetone molecules have no hydrogens attached to oxygen.

(e) *Cis*-1,2-dichloroethane because its molecules have a higher dipole moment.

(f) Propionic acid because its molecules can form hydrogen bonds to each other.

2.21
(a) Reduction with lithium aluminum hydride ($LiAlH_4$).

(b) Oxidation with chromic acid (H_2CrO_4).

(c) Reduction with $LiAlH_4$, or preferably with the milder sodium boronydride ($NaBH_4$).

(d) Oxidation with H_2CrO_4.

(e) Condensation with methyl alcohol in the presence of an acid catalyst.

(f) Self-condensation of methyl alcohol at $150°$ (under pressure) in the presence of a strong acid catalyst.

2.22

(a) $CH_3OCH_2CH_3$ (b) $CH_3CH_2CH_2OH$ (c) $CH_3\overset{\displaystyle OH}{\underset{|}{C}}HCH_3$

(d) $CH_3\overset{O}{\overset{\|}{C}}OCH_2CH_3$ $CH_3CH_2\overset{O}{\overset{\|}{C}}OCH_3$ (e) $CH_3CH_2CH_2CH_2X$

(f) $CH_3CH_2CHXCH_3$ (g) $CH_3\overset{\displaystyle CH_3}{\underset{\underset{\displaystyle X}{|}}{\overset{|}{C}}}CH_3$ (h) $CH_3\overset{\displaystyle CH_3}{\underset{\underset{\displaystyle O}{\|}}{\overset{|}{C}}}HCH$ or $CH_3CH_2CH_2\underset{\underset{\displaystyle O}{\|}}{C}H$

(i) $CH_3\overset{O}{\overset{\|}{C}}CH_2CH_3$ (j) $CH_3CH_2\overset{\displaystyle CH_3}{\overset{|}{C}}HNH_2$ (k) $CH_3CH_2CH_2NHCH_3$

(l) $CH_3CH_2N(CH_3)_2$ (m) $CH_3CH_2CH_2\overset{O}{\overset{\|}{C}}NH_2$ (n) $CH_3\overset{O}{\overset{\|}{C}}NHCH_2CH_3$

2.23

Staggered   Eclipsed

2.24
Basic strength depends upon ability to accept a proton. In $(CF_3)_3N$:, the high electroneg-

ativity of fluorine reduces the availability of the lone electron pair on nitrogen. In $(CH_3)_3N$:, the lone pair is more available to bond with a proton. (An alternate view is that the conjugate acid is rendered less stable by the presence of the electronegative fluorines.)

$$CF_3-\underset{\underset{CF_3}{|}}{\overset{\overset{CF_3}{|}}{N}}: \ + \ HA \ \rightleftharpoons \ CF_3-\underset{\underset{CF_3}{|}}{\overset{\overset{CF_3}{|}}{\overset{+}{N}}}-H \ + \ A^-$$

2.25

An ester group.

2.26

(a) $CH_3CH_2\overset{O}{\overset{||}{C}}-OCH_2CH_3 \xrightarrow[\text{(a)}]{HO^-} CH_3CH_2\overset{O}{\overset{||}{C}}-O^- + CH_3CH_2OH$

$\xrightarrow[\text{(b)}]{H_3O^+} CH_3CH_2\overset{O}{\overset{||}{C}}-OH + H_2O$

(b) $CH_3\overset{O}{\overset{||}{C}}-O-\bigcirc \xrightarrow[\text{(a)}]{OH^-} CH_3\overset{O}{\overset{||}{C}}-O^- + HO-\bigcirc$

$\xrightarrow[\text{(b)}]{H_3O^+} CH_3\overset{O}{\overset{||}{C}}-OH$

(c) $CH_3CH_2CH_2\overset{O}{\overset{||}{C}}-NH_2 \xrightarrow[\text{(a)}]{OH^-} CH_3CH_2CH_2\overset{O}{\overset{||}{C}}-O^- + NH_3 \xrightarrow[\text{(b)}]{2H_3O^+}$

$CH_3CH_2CH_2\overset{O}{\overset{||}{C}}-OH + NH_4^+ + 2H_2O$

(d) $CH_3\overset{O}{\overset{||}{C}}-\underset{\underset{CH_3}{|}}{N}-CH_3 \xrightarrow[\text{(a)}]{OH^-} CH_3\overset{O}{\overset{||}{C}}-O^- + (CH_3)_2NH$

$\xrightarrow[\text{(b)}]{2H_3O^+} CH_3\overset{O}{\overset{||}{C}}-OH + (CH_3)_2NH_2^+ + 2H_2O$

(e) $CH_3CH_2\overset{O}{\overset{||}{C}}-NHCH_2CH_3 \xrightarrow[\text{(a)}]{OH^-} CH_3CH_2\overset{O}{\overset{||}{C}}-O^- + CH_3CH_2NH_2$

$\xrightarrow[\text{(b)}]{2H_3O^+} CH_3CH_2\overset{O}{\overset{||}{C}}-OH + CH_3CH_2NH_3^+ + 2H_2O$

2.27

The attractive forces between hydrogen fluoride molecules are the very strong dipole-dipole attractions that we call *hydrogen bonds*. (The partial positive charge of a hydrogen fluoride molecule is relatively exposed because it resides on the hydrogen nucleus. By contrast, the positive charge of an ethyl fluoride molecule is buried in the ethyl group and is shielded by the surrounding electrons. Thus the positive end of one hydrogen fluoride molecule can approach the negative end of another hydrogen fluoride molecule much more closely with the result that the attractive force between them is much stronger.)

2.28

Since both molecules are nonpolar, the only intermolecular forces that we need to consider are van der Waals forces. Tetrafluoromethane is a generally spherical, compact molecule and its outer electrons (of the fluorine atoms) are very tightly held. Consequently, it will be difficult for a temporary dipole in one molecule of CF_4 to induce a very large temporary dipole in an adjacent molecule and, as a result, the van der Waals forces acting between them will be small. Hexane is a much larger molecule and the outer electrons are more loosely held. The outer electrons of hexane are more easily distorted—hexane is said to be more *polarizable*—thus the van der Waals forces acting between hexane molecules will be much larger.

2.29

In the molecules or atoms with greater molecular weight there are not only more electrons, but the outermost electrons are further from the nucleus where they are more loosely held. Thus the temporary dipoles that occur in a given molecule can be larger. Because the outermost electrons of adjacent molecules are also more loosely held, the induced dipoles in the adjacent molecules will also be larger. Consequently the attractive forces between the heavier molecules will be larger and the boiling points will be higher.

2.30

(a)

$$H-\overset{\overset{\displaystyle H}{|}}{\underset{\underset{\displaystyle H}{|}}{C}}-O-H$$

$$3H = 3(-1)$$
$$1O = +1$$
$$\overline{\text{total} = -2} = \text{oxidation state of C}$$

$$H-\overset{\overset{\displaystyle O}{\|}}{C}-O-H$$

$$1H = -1$$
$$3O = +3$$
$$\overline{\text{total} = +2} = \text{oxidation state of C}$$

$$H-\overset{\overset{\displaystyle O}{\|}}{C}-H$$

$$2H = -2$$
$$2O = +2$$
$$\overline{\text{total} = 0} = \text{oxidation state of C}$$

(b) CH_4, CH_3OH, $H\overset{\overset{\displaystyle O}{\|}}{C}H$, $H\overset{\overset{\displaystyle O}{\|}}{C}OH$, CO_2
      −4   −2   0   +2   +4

(c) a change from −2 to 0

(d) an oxidation, since the oxidation state increases

(e) a reduction from +6 to +3

(f) Methanol undergoes an increase in oxidation state of 2 and chromium undergoes a decrease of 3. Since the amount of oxidation must equal the amount of reduction, two moles of H_2CrO_4 will be required to oxidize three moles of CH_3OH, and thus two-thirds of a mole of H_2CrO_4 will be required for one mole of CH_3OH.

2.31

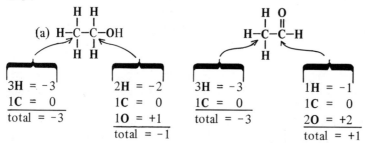

$$
\begin{array}{ll}
3H = -3 & 2H = -2 \\
1C = \ 0 & 1C = \ 0 \\
\hline
\text{total} = -3 & 1O = +1 \\
& \hline
& \text{total} = -1
\end{array}
\qquad
\begin{array}{ll}
3H = -3 & 1H = -1 \\
1C = \ 0 & 1C = \ 0 \\
\hline
\text{total} = -3 & 2O = +2 \\
& \hline
& \text{total} = +1
\end{array}
$$

(b) Only the carbon of the $-CH_2OH$ group of ethanol undergoes a change in oxidation state. The oxidation state of the CH_3- group remains unchanged.

(c) H–C–C–OH

$$
\begin{array}{ll}
3H = -3 & 1C = \ \ 0 \\
1C = \ 0 & 3O = +3 \\
\hline
\text{total} = -3 & \text{total} = +3
\end{array}
$$

(d) Each atom of silver undergoes a decrease in oxidation state of 1 when $Ag_2O \longrightarrow$ 2 Ag and each carbon undergoes an increase of 2 when $CH_3CHO \longrightarrow CH_3COOH$. Thus one mole of Ag_2O will be required to oxidize one mole of acetaldehyde.

2.32

(a) If we consider the hydrogenation of ethene as an example, we find that the oxidation state of carbon decreases. Thus, because the

$$
\text{H–C=C–H} + H_2 \xrightarrow{\text{Ni}} \text{H–C–C–H}
$$

$$
\begin{array}{ll}
2H = -2 & 3H = -3 \\
2C = \ 0 & 1C = \ 0 \\
\hline
\text{total} = -2 & \text{total} = -3
\end{array}
$$

reaction involves the *addition* of *hydrogen*, it is both an *addition reaction* and a *reduction.*

(b) The hydrogenation of acetaldehyde is not only an addition reaction, it is also a *reduction* because the carbon atom of the $>C=O$ group goes from a $+1$ to a -1 oxidation state. The reverse reaction (the *dehydrogenation* of ethyl alcohol) is not only an *elimination* reaction, it is also an *oxidation.*

3 ALKANES AND CYCLOALKANES: THEIR STRUCTURES, PROPERTIES, AND SYNTHESES

3.1

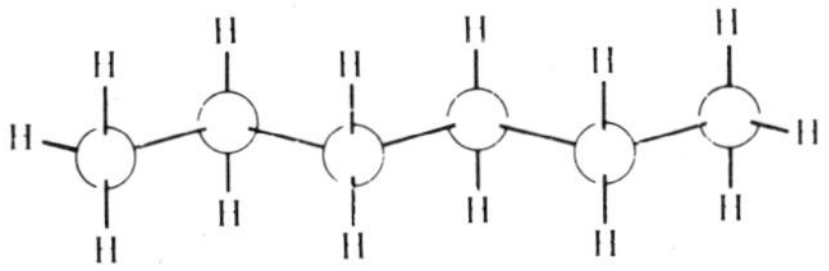

(1) CH₃CH₂CH₂CH₂CH₂CH₃

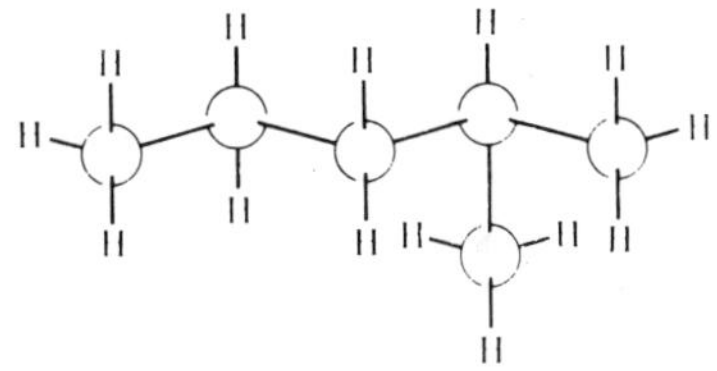

(2) CH₃CH₂CH₂CHCH₃
            |
           CH₃

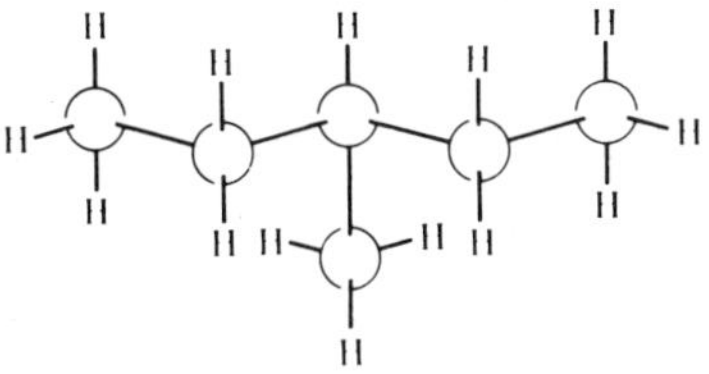

(3) CH₃CH₂CHCH₂CH₃
          |
         CH₃

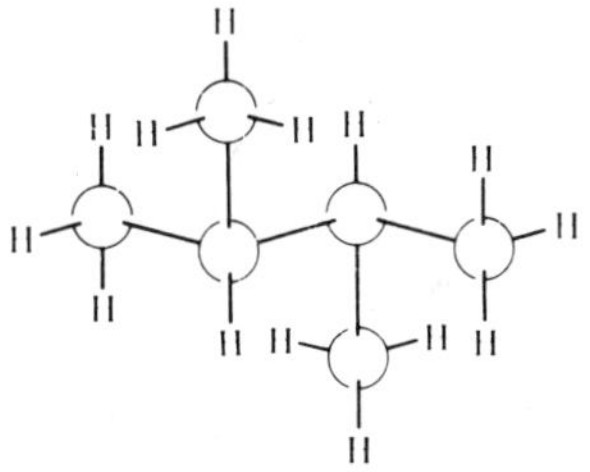

(4) CH₃CHCHCH₃
        | |
       CH₃ CH₃

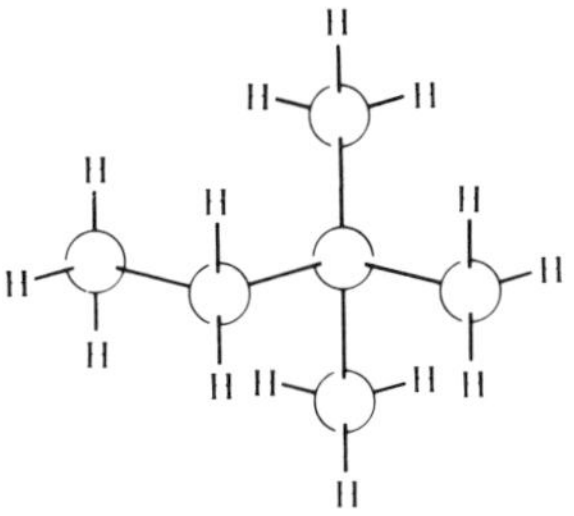

                 CH₃
                  |
(5) CH₃CH₂CCH₃
                  |
                 CH₃

3.2

(a) Refer to problem 3.1 above:

(1) Hexane, (2) 2-methylpentane, (3) 3-methylpentane, (4) 2,3-dimethylbutane, (5) 2,2-dimethylbutane.

(b) $CH_3CH_2CH_2CH_2CH_2CH_2CH_3$ heptane

$$CH_3CH_2CH_2CH_2\underset{\underset{\displaystyle CH_3}{|}}{C}HCH_3$$ 2-methylhexane

$$CH_3CH_2CH_2\underset{\underset{\displaystyle CH_3}{|}}{C}HCH_2CH_3$$ 3-methylhexane

$$CH_3CH_2CH_2\underset{\underset{\displaystyle CH_3}{|}}{\overset{\overset{\displaystyle CH_3}{|}}{C}}CH_3$$ 2, 2-dimethylpentane

$$CH_3CH_2\underset{\underset{\displaystyle CH_3}{|}}{C}H\overset{\overset{\displaystyle CH_3}{|}}{C}HCH_3$$ 2, 3-dimethylpentane

$$CH_3\underset{\underset{\displaystyle CH_3}{|}}{C}HCH_2\underset{\underset{\displaystyle CH_3}{|}}{C}HCH_3$$ 2, 4-dimethylpentane

$$CH_3CH_2\underset{\underset{\displaystyle CH_3}{|}}{\overset{\overset{\displaystyle CH_3}{|}}{C}}CH_2CH_3$$ 3, 3-dimethylpentane

$$CH_3\underset{\underset{\displaystyle CH_3}{|}}{C}H-\underset{\underset{\displaystyle CH_3}{|}}{\overset{\overset{\displaystyle CH_3}{|}}{C}}CH_3$$ 2, 2, 3-trimethylbutane

$$CH_3CH_2\underset{\underset{\displaystyle CH_2CH_3}{|}}{C}HCH_2CH_3$$ 3-ethylpentane

3.3

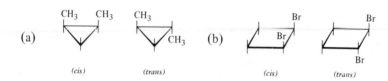

(a)   *(cis)*   *(trans)*   (b)   *(cis)*   *(trans)*

3.4

(a)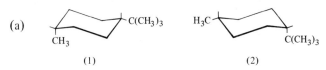

                (1)   (2)

(b) No. In (1), the methyl group is axial and the *tert*-butyl group is equatorial; in (2) the situation is reversed.

(c) The *tert*-butyl group is larger than the methyl; conformation (1) is more stable because the *tert*-butyl group is equatorial.

(d) The preferred conformation at equilibrium is (1).

3.5

(a) Conformations of *cis*-isomer are equivalent, (e, a) and (a, e).

              (a, e)   (e, a)

(b) Conformations of *trans*-isomer are not equivalent, (e, e) and (a, a).

              (e, e)   (a, a)

(c) The *trans*-(e, e) conformation is more stable than the *trans*-(a,a).

(d) The *trans*-(e, e) would be more highly populated at equilibrium.

3.6

(a) $CH_3I + 2 Li \xrightarrow{\text{ether}} CH_3Li + LiI$

      $2 CH_3Li + CuI \longrightarrow (CH_3)_2CuLi + LiI$

      $(CH_3)_2CuLi + CH_3CH_2CH_2CH_2-Cl \longrightarrow CH_3CH_2CH_2CH_2CH_3 + CH_3Cu$
                                                               $+ LiCl$

(b) $(CH_3)_2CH-Br + 2Li \xrightarrow{\text{ether}} (CH_3)_2CHLi + LiBr$

                         $\xrightarrow{\text{CuI}} [(CH_3)_2CH]_2CuLi + LiI$

                                            CH_3
                                                       |

    $[(CH_3)_2CH]_2CuLi + CH_3CH_2-Br \longrightarrow CH_3CHCH_2CH_3 + (CH_3)_2CHCu$
                                                                $+ LiBr$

(c) [cyclohexyl]—Br +2Li $\xrightarrow{\text{ether}}$ [cyclohexyl]—Li + Li Br

$\xrightarrow{\text{CuI}}$ ([cyclohexyl])$_2$—Cu Li + LiI

([cyclohexyl])$_2$—CuLi + CH$_3$—I $\longrightarrow$ [cyclohexyl]—CH$_3$ + [cyclohexyl]—Cu + LiI

(d) [cyclopentyl]—Br + 2Li $\xrightarrow{\text{ether}}$ [cyclopentyl]—Li + Li Br

$\xrightarrow{\text{CuI}}$ ([cyclopentyl])$_2$—CuLi + LiI

([cyclopentyl])$_2$—CuLi + CH$_3$CH$_2$CH$_2$—Br $\longrightarrow$ [cyclopentyl]—CH$_2$CH$_2$CH$_3$ + [cyclopentyl]—Cu

+ LiBr

3.7

(a) CH$_3$CHCHCH$_2$CH$_3$
      | |
      Cl Cl

$\quad$ CH$_3$
$\qquad$ |
(b) CH$_3$CCH$_3$
$\qquad$ |
$\qquad$ I

(c) CH$_3$CH$_2$CHCH$_2$CH$_3$
$\qquad\qquad$ |
$\qquad\qquad$ CH$_2$
$\qquad\qquad$ |
$\qquad\qquad$ CH$_3$

(d) CH$_3$CH—CH—CHCH$_2$CH$_2$CH$_2$CH$_2$CH$_2$CH$_3$
$\qquad$ |$\quad$ |$\quad$ |
$\qquad$ CH$_3$ CH$_3$ CH$_3$

(e) CH$_3$CH$_2$CH$_2$CHCH$_2$CH$_2$CH$_2$CH$_2$CH$_3$
$\qquad\qquad$ |
$\qquad\qquad$ CH
$\qquad\quad$ CH$_3$ CH$_3$

(f) [cyclopropane with CH$_3$ and CH$_3$]

(g) [cyclobutane with CH$_3$, CH$_3$ and CH$_3$]

(h) [cyclobutane with H$_3$C and CH$_3$]

(i) [cyclohexane with H and CH(CH$_3$)$_2$]

(j) [cyclohexane with H, CH$_3$, H and CH(CH$_3$)$_2$]

(k) CH$_3$CHCH$_2$CH$_2$CH$_2$—Cl
$\qquad$ |
$\qquad$ CH$_3$

$\qquad\qquad$ CH$_3$ CH$_3$
$\qquad\qquad$ |$\quad$ |
(l) CH$_3$CCH$_2$CCH$_2$CH$_2$CH$_2$CH$_3$
$\qquad\qquad$ |$\quad$ |
$\qquad\qquad$ CH$_3$ CH$_3$

$\qquad$ CH$_3$
$\qquad$ |
(m) CH$_3$CCH$_2$—Cl
$\qquad$ |
$\qquad$ CH$_3$

$\qquad\qquad$ CH$_3$
$\qquad\qquad$ |
(n) CH$_3$CHCH$_2$CH$_3$

3.8

(a) 3,4-dimethylhexane   (e) ethylcyclohexane

(b) 2-methylbutane   (f) cyclopentylcyclopentane

(c) 2,4-dimethylpentane   (g) 6-isobutyl-2-methyldecane

(d) 3-methylpentane

3.9

		Common	IUPAC
(a)	$CH_3CH_2CH_2-Cl$	*n*-propyl chloride	1-chloropropane
	CH_3CHCH_3 $\mid$ Cl	isopropyl chloride	2-chloropropane
(b)	$CH_3CH_2CH_2CH_2-Br$	*n*-butyl bromide	1-bromobutane
	$\quad\quad Br$ $\quad\quad \mid$ $CH_3CH_2CHCH_3$	sec-butyl bromide	2-bromobutane
	CH_3CHCH_2-Br $\mid$ CH_3	isobutyl bromide	1-bromo-2-methylpropane
	$\quad\quad CH_3$ $\quad\quad \mid$ CH_3C-Br $\quad\quad \mid$ $\quad\quad CH_3$	tert-butyl bromide	2-bromo-2-methylpropane

3.10

(a) $CH_3\overset{\displaystyle CH_3}{\underset{\displaystyle CH_3}{\overset{\mid}{\underset{\mid}{C}}}}CH_3$ 2,2-dimethylpropane (neopentane)

(b) $CH_3\overset{\displaystyle CH_3}{\overset{\mid}{C}H}CH_2CH_3$ 2-methylbutane (isopentane)

(c) $CH_3CH_2CH_2CH_2CH_3$ pentane

(d) cyclopentane structure   cyclopentane

(e) $CH_3\overset{\displaystyle CH_3}{\overset{\mid}{C}H}-\overset{\displaystyle CH_3}{\overset{\mid}{C}H}CH_3$ 2,3-dimethylbutane

3.11

(a) $CH_3\overset{\displaystyle CH_3}{\overset{\mid}{C}H}\overset{\displaystyle CH_3}{\overset{\mid}{C}H}CHCH_3$ 2,4-dimethyl-3-isopropylpentane

$\quad\quad\quad\quad CH_3CH\ CH_3$

(b) 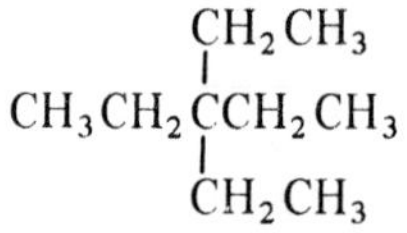
$$CH_3CH_2\underset{\underset{CH_2CH_3}{|}}{\overset{\overset{CH_2CH_3}{|}}{C}}CH_2CH_3$$
3,3-diethylpentane

(c) 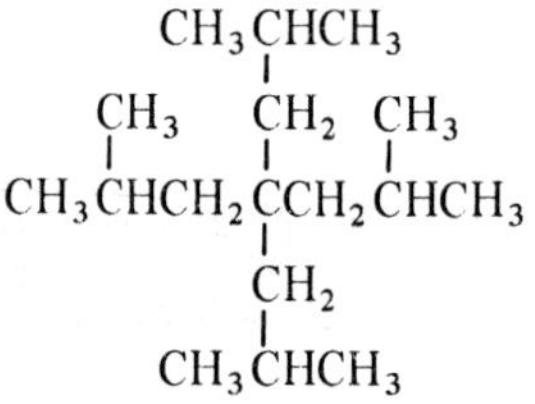
$$CH_3\underset{\underset{\underset{CH_3CHCH_3}{|}}{CH_2}}{C}HCH_2\underset{\overset{CH_2}{|}}{\underset{|}{C}}CH_2CHCH_3$$
with CH_3CHCH_3 and CH_2, CH_3 groups
4,4-diisobutyl-2,6-dimethylheptane

(d) 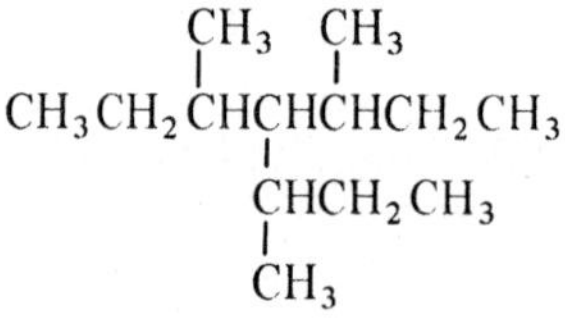
$$CH_3CH_2CHCHCHCH_2CH_3$$
with CH_3, CH_3 and $CHCH_2CH_3$, CH_3
4-*sec*-butyl-3,5-dimethylheptane

(e) 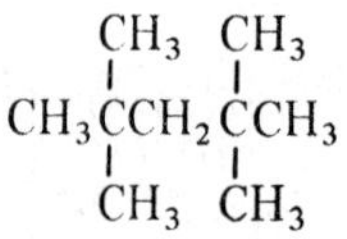
$$CH_3\underset{\underset{CH_3}{|}}{\overset{\overset{CH_3}{|}}{C}}CH_2\underset{\underset{CH_3}{|}}{\overset{\overset{CH_3}{|}}{C}}CH_3$$
2,2,4,4-tetramethylpentane

(f) 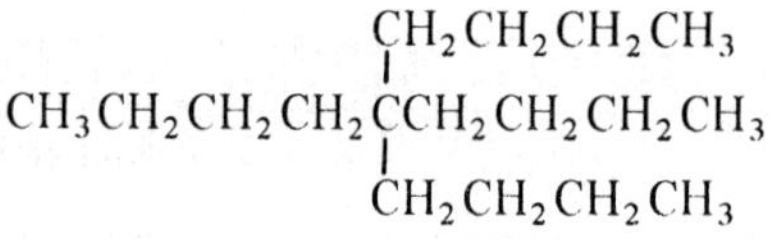
$$CH_3CH_2CH_2CH_2\underset{\underset{CH_2CH_2CH_2CH_3}{|}}{\overset{\overset{CH_2CH_2CH_2CH_3}{|}}{C}}CH_2CH_2CH_2CH_3$$
5,5-dibutylnonane

3.12

(a) neopentane $\longrightarrow$
$$CH_3\underset{\underset{CH_3}{|}}{\overset{\overset{CH_3}{|}}{C}}CH_2-Cl$$

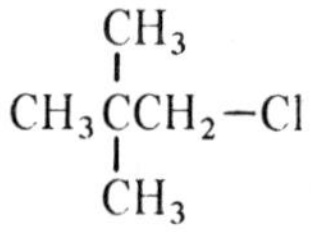

(b) pentane $\longrightarrow CH_3CH_2CH_2CH_2\underset{\underset{Cl}{|}}{C}H_2$, $CH_3CH_2CH_2\underset{\underset{Cl}{|}}{C}HCH_3$, $CH_3CH_2\underset{\underset{Cl}{|}}{C}HCH_2CH_3$

(c) isopentane $\longrightarrow CH_3\underset{\overset{CH_3}{|}}{C}HCH_2\underset{\underset{Cl}{|}}{C}H_2$, $CH_3\underset{\overset{CH_3}{|}}{C}H\underset{\underset{Cl}{|}}{C}HCH_3$, $CH_3\underset{\underset{Cl}{|}}{\overset{\overset{CH_3}{|}}{C}}CH_2CH_3$, $\underset{\underset{Cl}{|}}{C}H_2\underset{\overset{CH_3}{|}}{C}HCH_2CH_3$

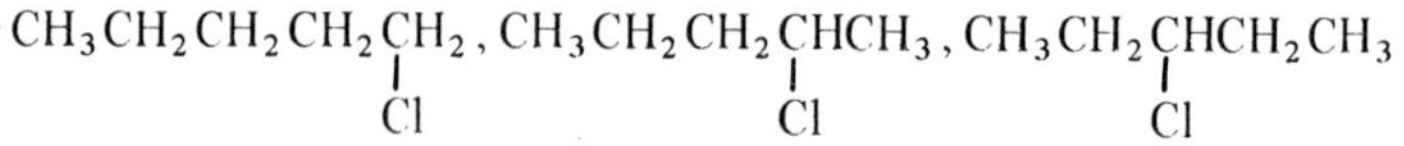

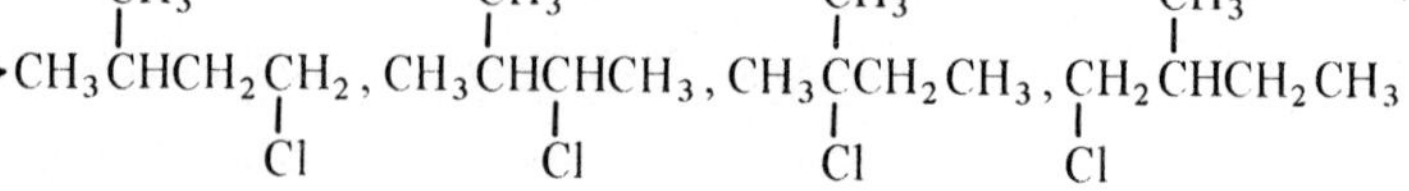

(d) neopentane $\longrightarrow CH_3\underset{\underset{\underset{Cl}{|}}{CH_2}}{\overset{\overset{CH_3}{|}}{C}}CH_2-Cl$, $CH_3\underset{\underset{CH_3}{|}}{\overset{\overset{CH_3}{|}}{C}}CH{\overset{\diagup Cl}{\diagdown Cl}}$

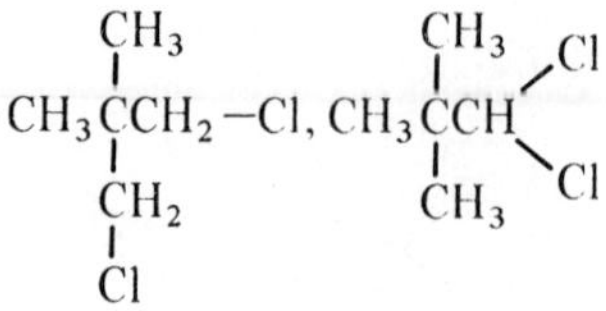

3.13

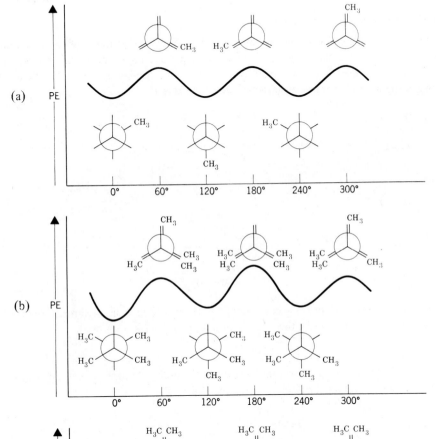

The methyl groups are larger than the hydrogen atom. The resulting mutual repulsions among the methyl groups cause a larger than tetrahedral bond angle.

3.14

(a) PE

(b) PE

(c) PE

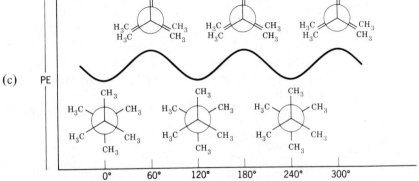

3.15

(a) Hexane. Branched chain hydrocarbons have lower boiling points than their normal isomers.

(b) Hexane. Boiling point increases with molecular weight.

(c) Pentane. (See (a) above).

(d) Chloroethane, because it has a higher molecular weight, and is more polar.

3.16

(a) The *trans*-isomer is more stable.

(b) Since they both yield the same combustion products and in the same molar amounts, the one that has the larger heat of combustion has the higher potential energy, and is therefore less stable (The *cis*-isomer is less stable because of the crowding that exists between the methyl groups on the same side of the ring.)

3.17

(a)

(1) (e, e)   (2) (a, a)

(b)

(3) (a, e)   (4) (e, a)

(c) (1) is more stable than (2) because in (1), both substituents are equatorial. (3) is more stable than (4) because in (3), the larger group ($CH(CH_3)_2$) is equatorial.

3.18

(a) The *trans*-isomer is more stable because both methyl groups can be equatorial. In *cis*-1, 2-dimethylcyclohexane, one methyl must be axial.

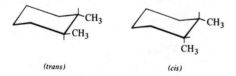

(trans)   *(cis)*

(b) The *cis*-isomer is more stable because both methyl groups are equatorial. In the *trans*-isomer, one methyl must be axial.

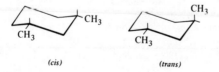

(cis)   *(trans)*

(c) The *trans*-isomer is more stable for the same reason as in (a).

(trans)   *(cis)*

3.19
In *cis*-1, 3-*di-tert*-butylcyclohexane, the two substituents are both equatorial (see problem 3.18 b above), whereas in the *trans*-isomer, one of the *tert*-butyl groups must be axial. The instability of a chair conformation with such a large group in an axial position forces the molecule into a less strained twist conformation:

trans (chair
conformation)

3.20

3.21
(a) Wurtz reaction:

$$2\ CH_3CH_2CH_2-Cl\ +\ 2Na \longrightarrow CH_3CH_2CH_2CH_2CH_2CH_3\ +\ 2\ NaCl$$

Corey-House Synthesis:

$$CH_3CH_2CH_2-Cl\ +\ 2Li \longrightarrow CH_3CH_2CH_2Li\ +\ LiCl$$

$$\downarrow CuI$$

$$(CH_3CH_2CH_2)_2CuLi$$

$$CH_3CH_2CH_2CH_2CH_2CH_3 \xleftarrow{CH_3CH_2CH_2Cl}$$

(b) $CH_3\underset{\underset{Br}{|}}{C}HCH_2CH_2CH_2CH_3 + Zn \xrightarrow{H^+} CH_3CH_2CH_2CH_2CH_2CH_3 + ZnBr_2$

(c) $CH_3CH_2CH_2CH_2CH=CH_2 + H_2 \xrightarrow{Ni} CH_3CH_2CH_2CH_2CH_2CH_3$

3.22

(a)

CH₃
CH₃

(b) From Table 3.5 we find that this is *cis*-1,2-dimethylcyclohexane.

(c) Since catalytic hydrogenation produces the *cis* isomer both hydrogens must have added from the same side of the double bond. (As we will see in Sect. 5.7, this type of addition is called a *syn* addition.)

$\xrightarrow{Pt}$

CH₃ H₃C

H   H

CH₃ 1 1 CH₃
   H H

The *cis*–isomer is produced when both hydrogens add from the same side.

cis–1,2–dimethylcyclohexane

3.23

(a) From Table 3.5 we find that this is *trans*-1,2-dichlorocyclohexane.

(b) Since the product is the *trans*-isomer we can conclude that the chlorine atoms have added from opposite sides of the double bond.

Cl

$\longrightarrow$

Cl

Cl

Cl

The *trans*–isomer is produced when the chlorine atoms add from opposite sides of the double bond.

trans–1,2–dichlorocyclohexane

3.24

(a)

H
O   O
O

(b) Having the hydroxyl axial allows the formation of a hydrogen bond to either ring oxygen as shown above.

3.25

(a) We can better understand the conformational rigidity of *trans*-decalin if we consider one ring (*A*, below) to be a *trans*-1,2-disubstituted cyclohexane where the 1,2-substituents are the two ends of a four-carbon chain, that is, $-CH_2CH_2CH_2CH_2-$.

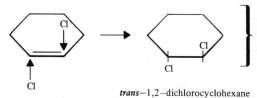

A B $\equiv$ A CH₂
                              CH₂   CH₂
                                  CH₂

Here we consider ring **B** *to be a four-carbon chain. It has no difficulty linking the 1- and 2- positions of ring* **A** *when its ends are diequatorial.*

However if ring **A** of *trans*-decalin were to be flipped into another chair conformation, the carbons of the other ring, **B**, would have to assume a 1,2-diaxial orientation. This is an impossible arrangement for the four carbons of the other ring to assume.

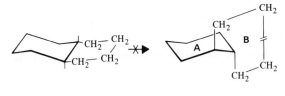

*Here we have flipped ring **A** into another chair conformation. This is an impossible arrangement for the four-carbon chain of ring **B**, however, because it cannot link the 1- and 2- positions when its ends are diaxial.*

(b) No, *cis*-decalin is conformationally "mobile" and can assume two conformations with each ring in a chair formation. In either conformation of *cis*-decalin the four carbons of one ring link the 1- and 2- positions of the other. They can do this easily because one end of the four-carbon chain is always axial and the other equatorial.

3.26

cis-1,2-Dibromocyclohexane must exist in two equivalent conformations with one bromine equatorial and the other axial:

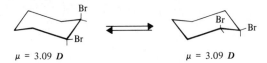

$\mu = 3.09\ D$ $\mu = 3.09\ D$

cis-3-Bromo-*trans*-4-bromo-1-*tert*-butylcyclohexane has a similar dipole moment ($\mu = 3.28\ D$), and since the presence of the large *tert*-butyl group ensures that the bromines are diequatorial, one can conclude that an equatorial-axial arrangement of the bromines and a diequatorial arrangement have roughly the same dipole moments.

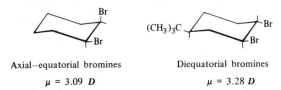

Axial–equatorial bromines   Diequatorial bromines
$\mu = 3.09\ D$ $\mu = 3.28\ D$

Thus were *trans*-1,2-dibromocyclohexane to exist primarily in a diequatorial conformation we would expect it to have a similar dipole moment (i.e., $\sim 3.09\ D$). The fact that the dipole moment of *trans*-1,2-dibromo cyclohexane is much lower (2.11 D) suggests that the diaxial conformation (with $\mu \simeq 0$) is present in appreciable concentration.

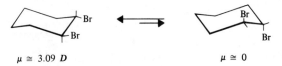

$\mu \simeq 3.09\ D$ $\mu \simeq 0$

(b) Because of bromine's electronegativity the bromine atoms will be partially negatively charged. In the diequatorial conformation, the bromines are closer together and therefore they repel each other. In the diaxial conformation the bromines are farther apart.

4 CHEMICAL REACTIVITY: REACTIONS OF ALKANES AND CYCLOALKANES

4.1

(a) CH_3-H + $F-F$ $\longrightarrow$ CH_3-F + $H-F$
$(D=104)$ $(D=38)$ $\quad\quad$ $(D=108)$ $(D=136)$

+ 142 kcal/mole is required for bond cleavage	− 244 kcal/mole is evolved in bond formation	$\Delta H = +142 - 244$ $= -102$ kcal/mole (exothermic)

(b) CH_3-H + $Cl-Cl$ $\longrightarrow$ CH_3-Cl + $H-Cl$
$(D=104)$ $(D=58)$ $\quad\quad$ $(D=83.5)$ $(D=103)$

+ 162 kcal/mole is required for bond cleavage	− 186.5 kcal/mole is evolved in bond formation	$\Delta H = +162 - 186.5$ $= -24.5$ kcal/mole (exothermic)

(c) CH_3-H + $Br-Br$ $\longrightarrow$ CH_3-Br + $H-Br$
$(D=104)$ $(D=46)$ $\quad\quad$ $(D=70)$ $(D=87.5)$

+ 150 kcal/mole	− 157.5 kcal/mole	= −7.5 kcal/mole = ΔH (exothermic)

(d) CH_3-H + $I-I$ $\longrightarrow$ CH_3-I + $H-I$
$(D\text{-}104)$ $(D\text{-}36)$ $\quad\quad$ $(D=56)$ $(D=71)$

+ 140 kcal/mole	− 127 kcal/mole	= + 13 kcal/mole = ΔH (endothermic)

4.2

(a)

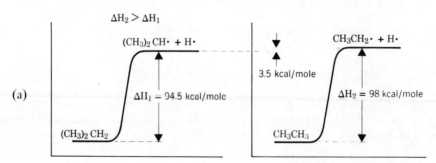

$\Delta H_2 > \Delta H_1$

$(CH_3)_2 CH\cdot + H\cdot$

$(CH_3)_2 CH_2$

$\Delta H_1 = 94.5$ kcal/mole

$CH_3CH_2\cdot + H\cdot$

CH_3CH_3

3.5 kcal/mole

$\Delta H_2 = 98$ kcal/mole

28

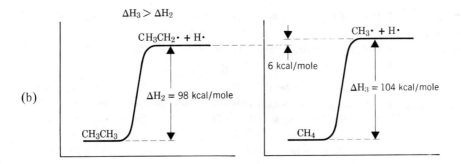

(b)

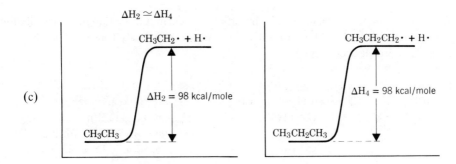

(c)

(d) The radicals produced are both primary radicals, and they are otherwise structurally similar, therefore they are of essentially equal stability.

4.3

Bond dissociation energies of the following C—Cl bonds are:

$CH_3-Cl \longrightarrow CH_3\cdot + Cl\cdot$ $\Delta H = 83.5$ kcal/mole

$CH_3CH_2-Cl \longrightarrow CH_3CH_2\cdot + Cl\cdot$ $\Delta H = 81.5$ kcal/mole

$(CH_3)_2CH-Cl \longrightarrow (CH_3)_2CH\cdot + Cl\cdot$ $\Delta H = 81$ kcal/mole

$(CH_3)_3C-Cl \longrightarrow (CH_3)_3C\cdot + Cl\cdot$ $\Delta H = 78.5$ kcal/mole

Since in each case the same kind of compound (an alkyl chloride) is decomposed into the same kinds of products (an alkyl free radical and a chlorine atom), it follows that the energy required (ΔH) is a measure of the instability of the radical relative to the alkyl halide. In other words, the less stable the free radical, the more energy will be required to break the bond between it and the chlorine atom. Bond dissociation energies for these alkyl chlorides are, respectively, 83.5, 81.5, 81, 78.5. They are in the same order as the stabilities of the free radicals produced: $CH_3\cdot < CH_3CH_2\cdot < (CH_2)_2CH\cdot < (CH_3)_3C\cdot$

4.4

Chain-initiating step	$Br-Br \longrightarrow 2 Br\cdot$ $(D=46)$	$\Delta H = +46$ kcal/mole

Chain-propagating steps

$$Br\cdot + CH_3-H \longrightarrow CH_3\cdot + HBr \qquad \Delta H = +16.5 \text{ kcal/mole}$$
$$(D=104) \qquad\qquad (D=87.5)$$

$$CH_3\cdot + Br-Br \longrightarrow CH_3-Br + Br\cdot \qquad \Delta H = -24 \text{ kcal/mole}$$
$$(D=46) \qquad (D=70)$$

Chain-terminating steps

$$CH_3\cdot + Br\cdot \longrightarrow CH_3-Br \qquad \Delta H = -70 \text{ kcal/mole}$$
$$(D=70)$$

$$CH_3\cdot + CH_3\cdot \longrightarrow CH_3-CH_3 \qquad \Delta H = -88 \text{ kcal/mole}$$
$$(D=88)$$

$$Br\cdot + Br\cdot \longrightarrow Br-Br \qquad \Delta H = -46 \text{ kcal/mole}$$
$$(D=46)$$

4.5

It would be incorrect to include chain-initiation and chain-termination in the calculation of the overall value of ΔH because those steps occur only rarely (once for hundreds or thousands of propagation steps.)

4.6

(a) $\quad CH_3\cdot + H-Cl \longrightarrow CH_3-H + Cl\cdot \qquad \Delta H = -1 \text{ kcal/mole}$
$\qquad\quad (D=103) \qquad (D=104) \qquad\qquad\quad E_{act} = +2.8 \text{ kcal/mole}$

(See text, p 136; E_{act} for the reverse reaction is 3.8 kcal/mole)

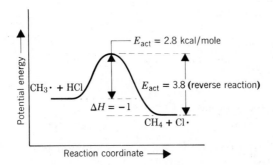

(b) $\quad CH_3\cdot + H-Br \longrightarrow CH_3-H + Br\cdot \qquad \Delta H = -16.5 \text{ kcal/mole}$
$\qquad\quad (D=87.5) \qquad (D=104) \qquad\qquad\qquad E_{act} = +2.1 \text{ kcal/mole}$

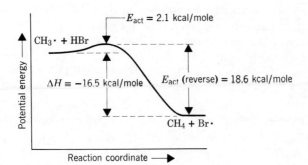

(c) $CH_3—CH_3 \longrightarrow 2\ CH_3 \cdot$ $\Delta H = +88\ kcal/mole$
       $(D=88)$ $E_{act} = +88\ kcal/mole$

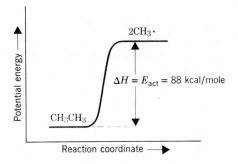

$\Delta H = E_{act}$ for any reaction in which bonds are broken but no bonds are formed.

(d) $Br—Br \longrightarrow 2\ Br \cdot$ $\Delta H = +46\ kcal/mole$
       $(D=46)$ $E_{act} = 46\ kcal/mole$

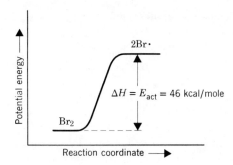

(e) $2\ Cl \cdot \longrightarrow Cl—Cl$ $\Delta H = -58\ kcal/mole$
                        $(D=58)$ $E_{act} = 0\ kcal/mole$

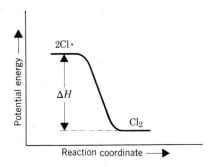

4.7

If all 10 hydrogen atoms of isobutane were equally reactive, the relative amounts of reaction at primary hydrogens and at tertiary hydrogens would be 9/1, i.e., the ratio of isobutyl chloride to *tert*-butyl chloride would be 9/1. Since the ratio is instead 62.5/37.5

(1.67), the tertiary hydrogen atom must be more reactive than the primary hydrogen atoms.

4.8

Laboratory preparation of alkyl halides by direct chlorination can be accomplished in good yield when all hydrogens in the alkane are equivalent. This is true of neopentane, and cyclopentane. (In these cases, the preparation would be practical only for monochlorination, where an excess of hydrocarbon would be employed, or for complete chlorination where an excess of chlorine would be used.)

4.9

(a) $Cl\cdot + CH_3CH_2-H \longrightarrow CH_3CH_2\cdot + H-Cl$
          $(D=98)$ $(D=103)$

   $\Delta H = 98 - 103 = -5$ kcal/mole (exothermic)

(b) $Cl\cdot + (CH_3)_2CH-H \longrightarrow (CH_3)_2CH\cdot + H-Cl$
          $(D=94.5)$ $(D=103)$

   $\Delta H = 94.5 - 103 = -8.5$ kcal/mole (exothermic)

(c) $Cl\cdot + CH_3CH_2CH_2-H \longrightarrow CH_3CH_2CH_2\cdot + H-Cl$
          $(D=98)$ $(D=103)$

   $\Delta H = 98 - 103 = -5$ kcal/mole (exothermic)

4.10

The hydrogen abstraction steps in alkane fluorinations are always highly exothermic. Thus the transition states are even more reactant-like in structure and in energy than they are in alkane chlorinations. The type of C–H bond being broken (1°, 2°, or 3°) has practically no effect on the relative rates of the reactions.

4.11

(a) Homolysis is cleavage of a covalent bond in such a way that the electrons of the ruptured bond are divided equally between the atoms involved:

$$:\overset{..}{\underset{..}{Cl}}-\overset{..}{\underset{..}{Cl}}: \longrightarrow :\overset{..}{\underset{..}{Cl}}\cdot + \cdot\overset{..}{\underset{..}{Cl}}:$$

(b) Heterolysis is cleavage of a covalent bond in such a way that both electrons of the ruptured bond remain with one atom. Ions are formed:

$$H-\overset{..}{\underset{..}{Cl}}: \longrightarrow H^+ + :\overset{..}{\underset{..}{Cl}}:^-$$

(c) Bond dissociation energy (D) is the energy required to dissociate a covalent bond homolytically:

$$H-H \longrightarrow 2\,H\cdot \quad D = 104 \text{ kcal/mole}$$

(d) A free radical is an atom or group that has an unpaired electron.

$$: \overset{..}{\underset{..}{Br}} \cdot \quad \text{or} \quad CH_3 \cdot$$

(e) A carbocation is an ion that has a trivalent carbon atom that bears a positive charge:

$$CH_3 - \overset{\overset{CH_3}{|}}{\underset{\underset{CH_3}{}}{C}} \overset{+}{}$$

(f) A carbanion is an ion that has a trivalent carbon atom that bears an unshared electron pair and a negative charge.

$$H - \overset{\overset{H}{|}}{\underset{\underset{H}{|}}{C}} : ^-$$

4.12

$$CH_3\overset{\overset{CH_3}{|}}{\underset{\bullet}{C}}CH_2CH_3 \; > \; CH_3\overset{\overset{CH_3}{|}}{\underset{\bullet}{CH}}CHCH_3 \; > \; \cdot CH_2\overset{\overset{CH_3}{|}}{CH}CH_2CH_3 \; \cong \; CH_3\overset{\overset{CH_3}{|}}{CH}CH_2CH_2 \cdot$$

4.13

(a) $\quad CH_3\overset{\overset{Br}{|}}{\underset{\underset{CH_3}{|}}{C}}CH_2CH_3$, because the tertiary hydrogen atom is much more reactive than either the primary or secondary hydrogen atoms.

(b) $CH_3\overset{\overset{CH_3}{|}}{CH}CH_2CH_3 + Cl \cdot \longrightarrow \cdot CH_2\overset{\overset{CH_3}{|}}{CH}CH_2CH_3$

$$CH_3 - \overset{\overset{CH_3}{|}}{\underset{\bullet}{C}} - CH_2CH_3$$

$$CH_3 - \overset{\overset{CH_3}{|}}{CH} - \underset{\bullet}{C}HCH_3$$

$$CH_3 - \overset{\overset{CH_3}{|}}{CH} - CH_2CH_2 \cdot$$

$\overset{Cl_2}{\longrightarrow}$

$$Cl - CH_2\overset{\overset{CH_3}{|}}{CH}CH_2CH_3 + Cl \cdot$$
$$+$$
$$CH_3\overset{\overset{CH_3}{|}}{\underset{\underset{Cl}{|}}{C}} - CH_2CH_3 + Cl \cdot$$
$$+$$
$$CH_3\overset{\overset{CH_3}{|}}{CH} - \overset{}{\underset{\underset{Cl}{|}}{C}}HCH_3 + Cl \cdot$$
$$+$$
$$CH_3\overset{\overset{CH_3}{|}}{CH}CH_2CH_2 - Cl + Cl \cdot$$

(c) Chlorine is more reactive than bromine, and is therefore less selective. See pages 146-149.

4.14

(a) $Cl_2 \longrightarrow 2\ Cl\cdot$

(b) $2\ Cl\cdot \longrightarrow Cl_2$

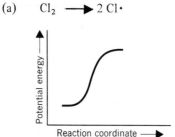 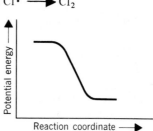

(c) $H\cdot + Cl_2 \longrightarrow HCl + Cl\cdot$

(d) $I\cdot + CH_4 \longrightarrow HI + CH_3\cdot$

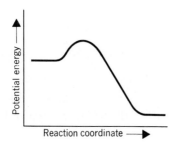 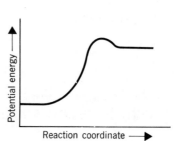

4.15

(a) $Cl_2 \xrightarrow[\text{light}]{\text{heat or}} 2\ Cl\cdot$ Chain-initiating step

(b)
$$\left.\begin{array}{l} \Delta H = -5 \\ \text{kcal/mole} \end{array}\right\}\quad \begin{array}{l} Cl\cdot + CH_3CH_2-H \longrightarrow H-Cl + CH_3CH_2\cdot \\ \qquad\qquad (D=98) \qquad\quad (D=103) \end{array}$$

$$\left.\begin{array}{l} \Delta H = -23.5 \\ \text{kcal/mole} \end{array}\right\}\quad \begin{array}{l} CH_3CH_2\cdot + Cl_2 \longrightarrow CH_3CH_2-Cl + Cl\cdot \\ \qquad\qquad (D=58) \qquad\qquad (D=81.5) \end{array}$$

Chain-propagating steps

$$\left.\begin{array}{l} CH_3CH_2\cdot + Cl\cdot \longrightarrow CH_3CH_2Cl \\ 2CH_3CH_2\cdot \longrightarrow CH_3CH_2CH_2CH_3 \\ 2Cl\cdot \longrightarrow Cl_2 \end{array}\right\}\ \text{Chain-terminating steps}$$

(c) The bond dissociation energy of the CH_3CH_2–H bond (98 kcal/mole) is smaller than that of the CH_3–H bond (104 kcal/mole), therefore ethane reacts with $Cl\cdot$ faster than methane does.

4.16

(a) (2) $Cl\cdot + CH_3-CH_3 \longrightarrow CH_3-Cl + CH_3\cdot$ $\Delta H = +4.5$ kcal/mole
           $(D=88)$ $(D=83.5)$ $E_{act} > +4.5$ kcal/mole

(3) $CH_3\cdot + Cl_2 \longrightarrow CH_3-Cl + Cl\cdot$ $\Delta H = -25.5$ kcal/mole
       $(D=58)$ $(D=83.5)$ E_{act} is small

In this reaction, step 2 is endothermic ($\Delta H = +4.5$ kcal/mole) and thus E_{act} must be greater than $+4.5$ kcal/mole. Although we do not know the exact E_{act} of the reaction that yields ethyl chloride (problem 4.15), we can assume that it is less than 3.8 kcal/mole (E_{act} for the corresponding step in the chlorination of methane). Therefore we conclude that the reaction here with an E_{act} greater than + 4.5 kcal/mole, will not compete with the reaction of problem 4.15.

(b) (1) F–F $\longrightarrow$ 2 F· $\qquad\qquad\qquad\qquad$ ΔH = + 38 kcal/mole
$\quad\quad$ (D=38)

$\quad$ (2) F· + CH$_3$–CH$_3$ $\longrightarrow$ CH$_3$–F + CH$_3$· $\qquad$ ΔH = − 20 kcal/mole
$\quad\quad\quad$ (D=88) $\qquad\qquad$ (D=108) $\qquad\qquad$ E_{act} > 0

$\quad$ (3) CH$_3$· + F–F $\longrightarrow$ CH$_3$–F + F· $\qquad$ ΔH = − 70 kcal/mole
$\quad\quad\quad$ (D=38) $\qquad$ (D=108) $\qquad\qquad$ E_{act} > 0

Since the propagation steps are both highly exothermic it is possible for each E_{act} to be quite small, and therefore for the reaction to take place.

4.17

(a) CH$_3$–H + F–F $\longrightarrow$ CH$_3$· + H–F + F· $\quad$ ΔH = + 6 kcal/mole
$\quad\quad$ (D=104) $\;$ (D=38) $\qquad\qquad\qquad$ (D=136) $\qquad$ E_{act} > 6 kcal/mole

$\quad\;$ CH$_3$· + F· $\longrightarrow$ CH$_3$–F $\qquad\qquad\qquad$ ΔH = −108 kcal/mole
$\quad\quad\quad\quad\quad\quad\quad$ (D=108) $\qquad\qquad\qquad$ E_{act} = 0

If E_{act} for the first step is not much greater than 6 kcal/mole, this mechanism is likely.

(b) CH$_3$–H + Cl–Cl $\longrightarrow$ CH$_3$· + H–Cl + Cl· $\quad$ ΔH = + 59 kcal/mole
$\quad\quad$ (D=104) $\;$ (D=58) $\qquad\qquad\qquad$ (D=103) $\qquad$ E_{act} ≥ 59 kcal/mole

$\quad\;$ CH$_3$· + Cl· $\longrightarrow$ CH$_3$–Cl $\qquad\qquad\qquad$ ΔH = −83.5 kcal/mole
$\quad\quad\quad\quad\quad\quad$ (D=83.5) $\qquad\qquad\qquad$ E_{act} = 0

This mechanism is highly unlikely because the E_{act} for the first step must be ≥ 59 kcal/mole.

4.18

4.19

(b) CH$_3$CH$_2$CH$_2^+$ + I$^-$ $\longrightarrow$ CH$_3$CH$_2$CH$_2$–I

4.20

(a) $\quad$ CH$_3$–H $\quad$ D = 104, $\quad$ CH$_3$CH$_2$–H $\quad$ D = 98 kcal/mole. (Recall that here, E_{act} = D.)

CH_3CH_2-H bond rupture requires less energy, therefore spontaneous homolysis (cracking) occurs at a lower temperature.

(b) CH_3-CH_3 $D = 88$ kcal/mole $= E_{act}$

C—C bond rupture requires less energy than C—H bond rupture, therefore C—C bond rupture occurs at a lower temperature than CH_3CH_2-H bond rupture.

(c) $CH_3CH_2-CH_2CH_3$ $D = 82$ kcal/mole $= E_{act}$

$CH_3CH_2CH_2-CH_3$ $D = 85$ kcal/mole $= E_{act}$

Here again the bond with the lower bond dissociation energy will undergo spontaneous homolysis (cracking) at the lower temperature.

4.21
(1) $CH_3CH_2CH_3 \longrightarrow CH_3CH_2 \cdot + CH_3 \cdot$ $D = 85$ kcal/mole
(2) $CH_3CH_2CH_3 \longrightarrow CH_3CH_2CH_2 \cdot + H \cdot$ $D = 98$ kcal/mole
(3) $CH_3CH_2CH_3 \longrightarrow CH_3\overset{\cdot}{C}HCH_3 + H \cdot$ $D = 94.5$ kcal/mole

(a) Since E_{act} is equal to D, we can assume that (1) is the most likely chain-initiating step.

(b) $CH_3 \cdot + CH_3CH_2CH_3 \longrightarrow CH_3-H + \cdot CH_2CH_2CH_3$ $\Delta H = -6$ kcal/mole
$\qquad\qquad$ (D=98) $\qquad\qquad$ (D=104)

Since ΔH is negative, E_{act} need not be large.

(c) $CH_3 \cdot + CH_3CH_2CH_3 \longrightarrow CH_4 + CH_3\overset{\cdot}{C}HCH_3$ $\Delta H = -9.5$ kcal/mole
$\qquad\qquad$ (D=94.5) $\qquad\qquad$ (D=104)

On the basis of energy requirements, this is a likely alternative to step 1. On the basis of the probability factor, it is less likely because there are only two secondary hydrogen atoms compared with six primary hydrogen atoms.

4.22
(a) The valence shell of a carbocation contains only six electrons; the carbocation needs an additional electron pair to achieve a stable octet.

(b) $CH_3-\overset{\overset{\displaystyle H}{|}}{\underset{\underset{\displaystyle H}{|}}{C^+}} \overset{\curvearrowleft}{} :\ddot{C}l:^- \longrightarrow CH_3-\overset{\overset{\displaystyle H}{|}}{\underset{\underset{\displaystyle H}{|}}{C}}-\ddot{C}l:$

(c) $CH_3-\overset{\overset{\displaystyle H}{|}}{\underset{\underset{\displaystyle H}{|}}{C^+}} \overset{\curvearrowleft}{} :\ddot{O}-\overset{\overset{\displaystyle O}{\|}}{\underset{\underset{\displaystyle O}{\|}}{S}}-OH \longrightarrow CH_3-\overset{\overset{\displaystyle H}{|}}{\underset{\underset{\displaystyle H}{|}}{C}}-O-\overset{\overset{\displaystyle O}{\|}}{\underset{\underset{\displaystyle O}{\|}}{S}}-OH$

(d) $CH_3-\overset{\overset{\displaystyle H}{|}}{\underset{\underset{\displaystyle H}{|}}{C^+}} \overset{\curvearrowleft}{} :\ddot{O}-H \longrightarrow CH_3-\overset{\overset{\displaystyle H}{|}}{\underset{\underset{\displaystyle H}{|}}{C}}-\overset{+}{\underset{\underset{\displaystyle H}{|}}{\ddot{O}}}-H \underset{}{\overset{-H^+}{\rightleftarrows}} CH_3-\overset{\overset{\displaystyle H}{|}}{\underset{\underset{\displaystyle H}{|}}{C}}-\ddot{O}-H$

(e) Loss of a proton from a carbon adjacent to the carbon bearing the positive charge.

(f) $H-\underset{\underset{H}{|}}{\overset{\overset{H}{|}}{C}}\underset{\underset{H}{|}}{\overset{\overset{H}{|}}{C^+}} \xrightarrow{-H^+} \underset{H}{\overset{H}{}}C=C\overset{H}{\underset{H}{}}$

4.23

(a) Oxygen-oxygen single bonds are especially weak, that is,

$$HO-OH \quad D = 51 \, kcal/mole$$
$$CH_3CH_2O-OCH_3 \quad D = 44 \, kcal/mole$$

This means that a peroxide will dissociate into free radicals at a relatively low temperature.

$$RO-OR \xrightarrow{100\text{-}200^\circ} 2RO\cdot$$

Oxygen-hydrogen single bonds, on the other hand, are very strong. (For HO—H, D = 119 kcal/mole.) This means that reactions like the following will be highly exothermic.

$$RO\cdot + R-H \longrightarrow RO-H + R\cdot$$

(b) (1) $(CH_3)_3CO-OC(CH_3)_3 \xrightarrow{heat} 2(CH_3)_3CO\cdot$ } chain-initiating

(2) $(CH_3)_3CO\cdot + R-H \longrightarrow (CH_3)_3COH + R\cdot$ } steps

(3) $R\cdot + Cl-Cl \longrightarrow R-Cl + Cl\cdot$ } chain-propagating

(4) $Cl\cdot + R-H \longrightarrow H-Cl + R\cdot$ } steps

4.24

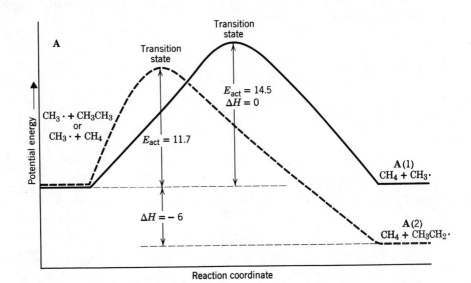

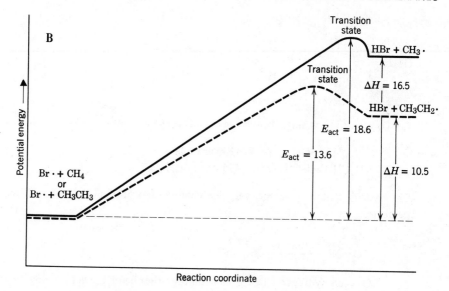

(b) Reaction **A** (2) since it is most exothermic.

(c) Reaction **B** (1) since it is most endothermic.

(d) Since $\Delta H = 0$, bond breaking should be approximately 50% complete.

(e) The reactions of set **B**.

(f) The difference in ΔH simply reflects the difference in the C—H bond strengths of methane and ethane.

(g) Because the reactions of set **B** are highly endothermic the transition states show a strong resemblance to products in structure and *in energy*, and the products differ in energy by 6 kcal/mole. (In this instance, since the difference in E_{act} is five-sixths of the difference in ΔH, we can estimate that bond breaking is about five-sixths complete when the transition states are reached.)

5 ALKENES: STRUCTURE AND SYNTHESES

5.1
Absorption of a photon of the correct frequency can excite a π electron into an anti-bonding orbital. Such an orbital has a nodal plane between the carbon atoms, thus rotation about the C—C bond can occur (see pp. 382-384).

5.2
(a) $CH_2=\underset{\underset{CH_3}{|}}{C}CH_3$ + H_2 $\longrightarrow$ $CH_3\underset{\underset{CH_3}{|}}{C}HCH_3$ (isobutane)

(b) C_4H_8 + $6O_2$ $\longrightarrow$ $4CO_2$ + $4H_2O$

(c) Yes, because the same molar amounts of the same combustion products are formed from all of the C_4H_8 isomers.

(d) $\underset{\underset{\displaystyle CH_2-CH_2}{|\qquad\quad|}}{CH_2-CH_2}$ (cyclobutane) and $\overset{\displaystyle CH_2}{\underset{\displaystyle CH_2 \overline{\qquad} CHCH_3}{\diagup\quad\diagdown}}$ (methylcyclopropane)

(e) Yes. (See answer to (c)).

5.3
(a) No.

(b) Protonation followed by loss of water gives a secondary carbocation which can lose a proton two different ways:

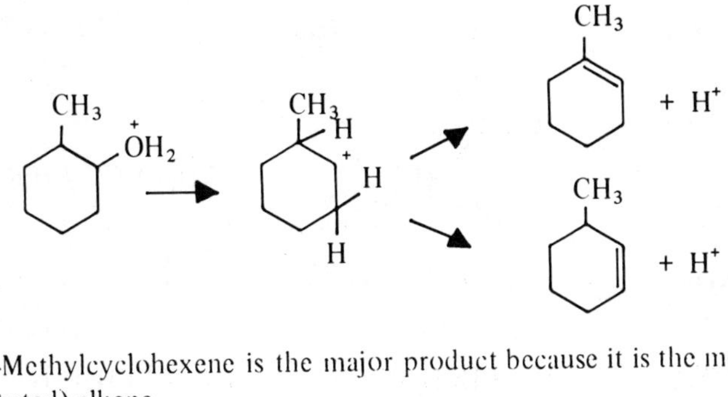

(c) 1-Methylcyclohexene is the major product because it is the more stable (more highly substituted) alkene.

5.4
(a) 2-Methyl-2-butene

(b) *Cis* - 4-octene

(c) 1-Bromo-2-methylpropene

(d) 4-Methylcyclohexene

5.5

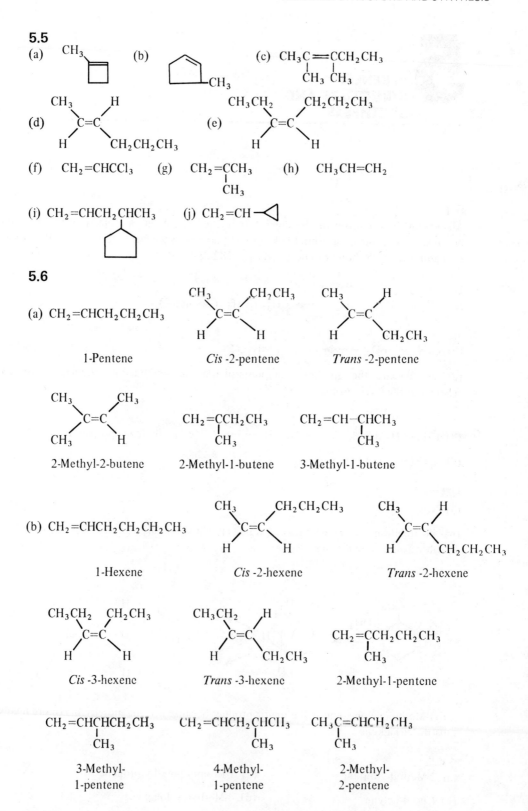

(a) CH$_3$

(b)

CH$_3$

(c) CH$_3$C=CCH$_2$CH$_3$
          | |
       CH$_3$ CH$_3$

(d)
  CH$_3$ H
     C=C
  H   CH$_2$CH$_2$CH$_3$

(e)
  CH$_3$CH$_2$ CH$_2$CH$_2$CH$_3$
       C=C
    H   H

(f)   CH$_2$=CHCCl$_3$ (g) CH$_2$=CCH$_3$ (h)   CH$_3$CH=CH$_2$
                        |
                       CH$_3$

(i) CH$_2$=CHCH$_2$CHCH$_3$

(j) CH$_2$=CH—

5.6

(a) CH$_2$=CHCH$_2$CH$_2$CH$_3$

1-Pentene

  CH$_3$ CH$_2$CH$_3$
    C=C
  H   H

Cis-2-pentene

  CH$_3$ H
    C=C
  H   CH$_2$CH$_3$

Trans-2-pentene

  CH$_3$ CH$_3$
    C=C
  CH$_3$ H

2-Methyl-2-butene

CH$_2$=CCH$_2$CH$_3$
    |
   CH$_3$

2-Methyl-1-butene

CH$_2$=CH—CHCH$_3$
         |
       CH$_3$

3-Methyl-1-butene

(b) CH$_2$=CHCH$_2$CH$_2$CH$_2$CH$_3$

1-Hexene

  CH$_3$ CH$_2$CH$_2$CH$_3$
    C=C
  H   H

Cis-2-hexene

  CH$_3$ H
    C=C
  H   CH$_2$CH$_2$CH$_3$

Trans-2-hexene

  CH$_3$CH$_2$ CH$_2$CH$_3$
      C=C
   H   H

Cis-3-hexene

  CH$_3$CH$_2$ H
      C=C
   H   CH$_2$CH$_3$

Trans-3-hexene

CH$_2$=CCH$_2$CH$_2$CH$_3$
    |
   CH$_3$

2-Methyl-1-pentene

CH$_2$=CHCHCH$_2$CH$_3$
       |
     CH$_3$

3-Methyl-
1-pentene

CH$_2$=CHCH$_2$CHCH$_3$
          |
       CH$_3$

4-Methyl-
1-pentene

CH$_3$C=CHCH$_2$CH$_3$
   |
  CH$_3$

2-Methyl-
2-pentene

$$\begin{array}{c} CH_3 \\ \diagdown \\ C=C \\ \diagup \diagdown \\ H CH_3 \end{array} \begin{array}{c} CH_2CH_3 \end{array}$$

Cis-3-methyl-
2-pentene*

$$\begin{array}{c} CH_3 \\ \diagdown \\ C=C \\ \diagup \diagdown \\ H CH_2CH_3 \end{array} \begin{array}{c} CH_3 \end{array}$$

Trans-3-methyl-
2-pentene*

$$\begin{array}{c} CH_3 \\ CHCH_3 \\ CH_3 \diagdown \\ C=C \\ \diagup \diagdown \\ H H \end{array}$$

Cis-4-methyl-
2-pentene

$$CH_2{=}C\begin{array}{c} CH_2CH_3 \\ \diagdown \\ CH_2CH_3 \end{array}$$

2-Ethyl-
1-butene

$$\begin{array}{c} CH_3 \\ \diagdown \\ C=C \\ \diagup \diagdown \\ H CHCH_3 \\ CH_3 \end{array} \begin{array}{c} H \end{array}$$

Trans-4-methyl-
2-pentene

$$CH_2{=}\overset{CH_3}{\underset{CH_3}{\overset{|}{\underset{|}{C}}}}CHCH_3$$

2, 3-dimethyl-
1-Butene

$$CH_2{=}CH\overset{CH_3}{\underset{CH_3}{\overset{|}{\underset{|}{C}}}}CH_3$$

3, 3-dimethyl-
1-Butene

$$\begin{array}{c} CH_3 CH_3 \\ \diagdown \diagup \\ C=C \\ \diagup \diagdown \\ CH_3 CH_3 \end{array}$$

2, 3-Dimethyl-
2-butene

(c)

C_5H_{10} :

C_6H_{12} :

5.7

(a) $CH_3\overset{OH}{\underset{CH_3}{\overset{|}{\underset{|}{C}}}}CH_3$ or $CH_3\overset{}{\underset{CH_3}{\overset{|}{CH}}}CH_2OH$

(c)

(b) $CH_3\overset{OH}{\underset{CH_3}{\overset{|}{\underset{|}{C}}}}{-}CHCH_3$ or $CH_3\overset{CH_3}{\underset{CH_3}{\overset{|}{\underset{|}{C}}}}{-}\overset{}{\underset{OH}{\overset{}{CH}}}{-}CH_3$

(d)

* The *cis-trans* designation here is ambiguous. See pp. 274 - 276.

5.8

$$CH_2\!=\!CHCH_2CH_2CH_2CH_3 \; + \; Br_2 \quad \xrightarrow[\substack{\text{(dark)}\\ \text{R.T.}}]{\text{CCl}_4} \quad CH_2CHCH_2CH_2CH_2CH_3 \;\text{(colorless)}$$

with Br, Br below the product.

Cyclohexane does not react with Br_2 in the dark at room temperature, thus the red-brown color of the bromine will persist in the solution.

5.9

(a) No   (b) No

(c) Yes

(d) No   (e) No   (f) No   (g) Yes

(h) Yes

(i) No

(j) No   (k) No   (l) Yes

5.10

(a) 2,3-dimethyl-2-butene > 2-methyl-2-pentene > *trans*-3-hexene > *cis*-2-hexene > 1-hexene.

(b) The only alkenes whose relative stabilities could be measured by comparative heats of hydrogenation are those that yield the same hydrogenation product; i.e., *trans*-3-hexene, 1-hexene, *cis*-2-hexene all yield hexane on hydrogenation.

5.11

Although *trans* molecules are usually more stable than their *cis* isomers, in the case of cyclooctene, the *trans* isomer is probably more strained than the *cis* isomer because the ring is too small to allow a strain-free *trans* configuration. Therefore we would expect the *trans* isomer to have the higher heat of hydrogenation.

5.12

(a) *Cis-trans* isomerization caused by rupture of the π-bond.

(b) Equilibrium should favor the *trans* isomer because it is more stable than the *cis* isomer.

5.13

(a) An sp^2 hybridized carbon atom has more s-character than one that is sp^3 hybridized, thus electrons in these sp^2 orbitals are closer to the nucleus than those in sp^3 orbitals. A bond between an sp^2 hybridized carbon and an sp^3 hybridized carbon is polarized toward the sp^2 hybridized carbon.

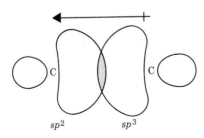

(b) The greater acidity of ethene can be explained by the greater stability of its conjugate base, $CH_2{=}CH^-$. The greater stability of $CH_2{=}CH^-$ compared with $CH_3CH_2^-$ can be explained by the greater s-character of the sp^2 orbital that holds the unbonded electron pair in $CH_2{=}CH^-$. In $CH_3CH_2^-$, the electrons are in an sp^3 orbital. Being farther from the nucleus, the sp^3 electrons have a higher potential energy.

5.14

(a) $CH_3CH{=}\overset{\underset{|}{CH_3}}{C}CH_3$ (major) $+$ $CH_2{=}CH\overset{\underset{|}{CH_3}}{C}HCH_3$

(b) $CH_3CH_2\overset{\underset{|}{CH_3}}{C}{=}CH_2$ (c) $CH_3CH{=}CHCH_2CH_3$ (*trans*-predominates)

(d)

$$\underset{H}{\overset{CH_3}{\diagdown}}C{=}C\underset{CH_2CH_3}{\overset{H}{\diagup}}$$ (major) $+$ $$\underset{H}{\overset{CH_3}{\diagdown}}C{=}C\underset{H}{\overset{CH_2CH_3}{\diagup}}$$ $+$ $CH_2{=}CHCH_2CH_2CH_3$

(e) 〈 〉${=}CH_2$ (f) 〈 〉${-}CH_3$ (major product) $+$ 〈 〉${=}CH_2$

5.15

$$\underset{\underset{CH_3}{|}}{\overset{\overset{OH}{|}}{CH_3CCH_2CH_3}} > \underset{\underset{CH_3}{|}}{\overset{\overset{OH}{|}}{CH_3CHCHCH_3}} > \underset{\underset{CH_3}{|}}{CH_3CHCH_2CH_2OH}$$

The order of reactivity is dictated by the order of stability of the intermediate carbocations: tertiary > secondary > primary

5.16

(a) *Cis*-1, 2-dimethylcyclopentane

(b) *Cis*-1, 2-dimethylcyclohexane:

(c) *Cis*-1, 2-dideuteriocyclohexane:

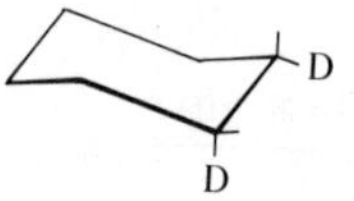

5.17

(a) (1) $CH_3-\underset{\underset{CH_3}{|}}{\overset{\overset{CH_3}{|}}{C}}-CH_2-OH + H_3O^+ \rightleftharpoons CH_3-\underset{\underset{CH_3}{|}}{\overset{\overset{CH_3}{|}}{C}}-CH_2-OH_2^+ + H_2O$

(2) $CH_3-\underset{\underset{CH_3}{|}}{\overset{\overset{CH_3}{|}}{C}}-CH_2\overset{\frown}{-}OH_2^+ \longrightarrow CH_3-\underset{\underset{CH_3}{|}}{\overset{\overset{CH_3}{|}}{C}}-CH_2^+ + H_2O$

(3) $CH_3-\underset{\underset{CH_3}{|}}{\overset{\overset{CH_3}{|}}{C}}-CH_2^+ \longrightarrow CH_3-\overset{+}{\underset{\underset{CH_3}{|}}{C}}-CH_2-CH_3$

(4) $CH_3-\overset{+}{\underset{\underset{CH_3}{|}}{C}}\overset{H}{\overset{\frown}{-}}CH-CH_3 +:\overset{H}{\underset{|}{O}}-H \longrightarrow \underset{CH_3}{\overset{CH_3}{}}C=CHCH_3$ (more substituted alkene)

$+ H_3O^+$

(4a) $\overset{H}{\underset{\underset{CH_3}{|}}{C}}H_2-\overset{+}{C}-CH_2-CH_3 + :\overset{H}{\underset{|}{O}}-H \longrightarrow CH_2=C\overset{CH_2CH_3}{\underset{CH_3}{}}$ (less substituted alkene)

$+ H_3O^+$

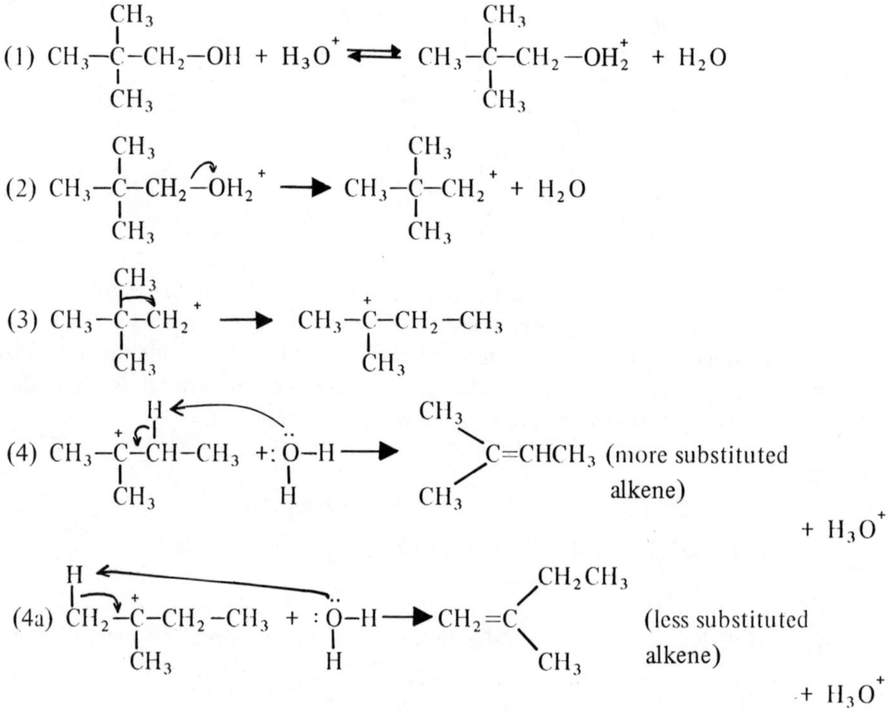

(Steps 2 and 3 may occur at the same time.)

(b)

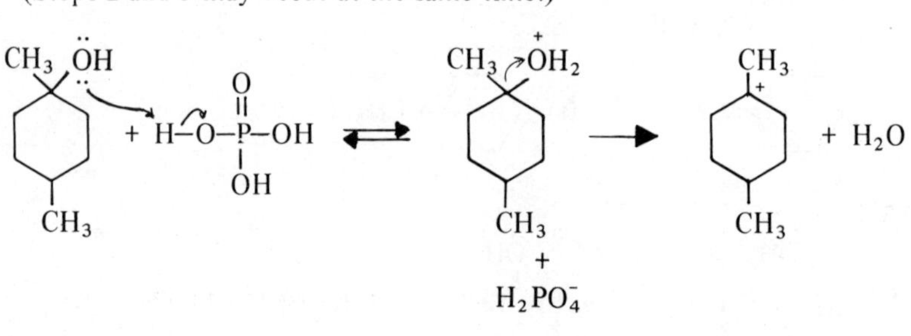

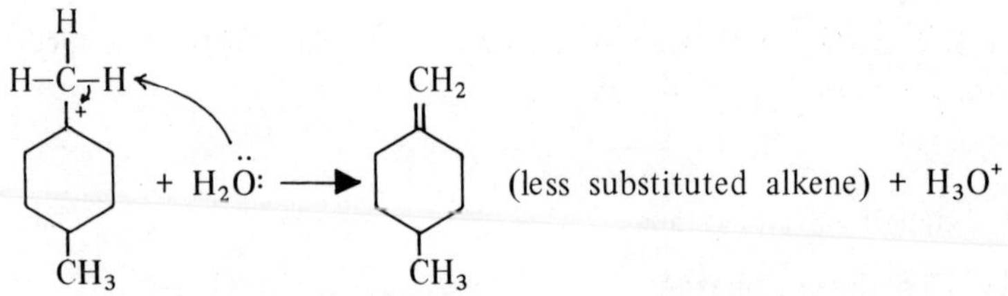

(less substituted alkene) $+ H_3O^+$

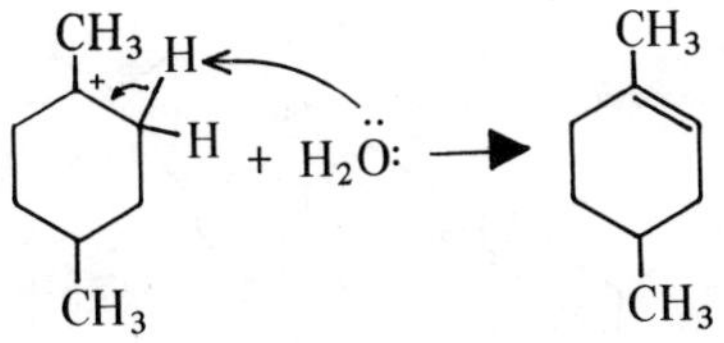

(more substituted alkene) + H_3O^+

(c)

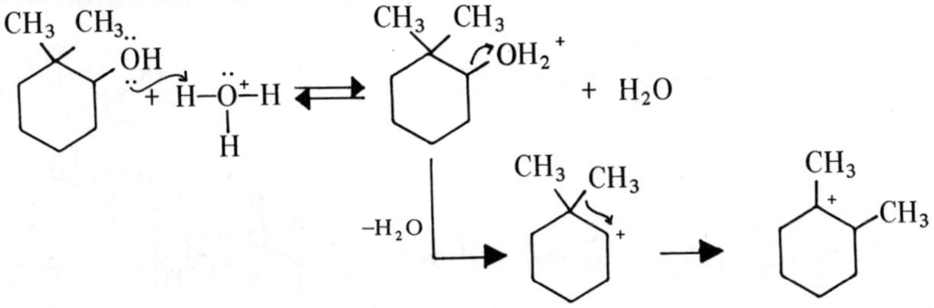

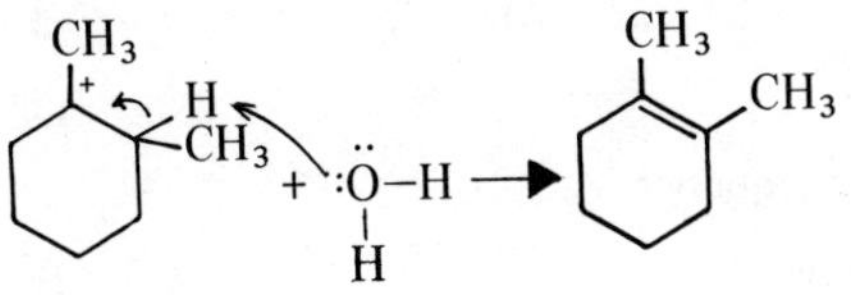

+ H_2O

$-H_2O$

+ H_3O^+ (most substituted alkene)

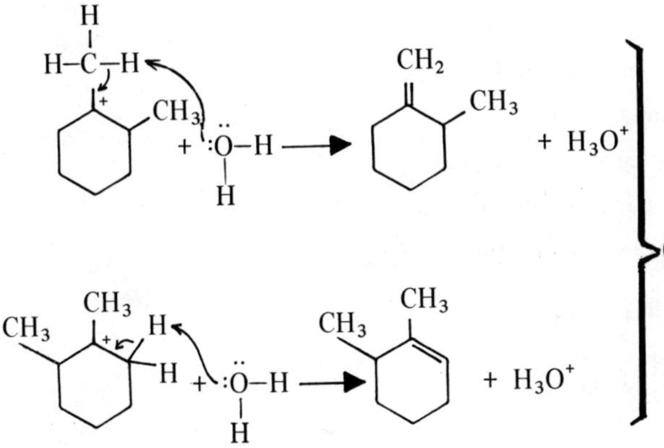

+ H_3O^+

(less substituted alkenes)

+ H_3O^+

5.18

Cholesterol

$\xrightarrow{\text{Br}_2 \atop \text{CHCl}_3}$

(crude)

$\xrightarrow{\text{crystalization}}$

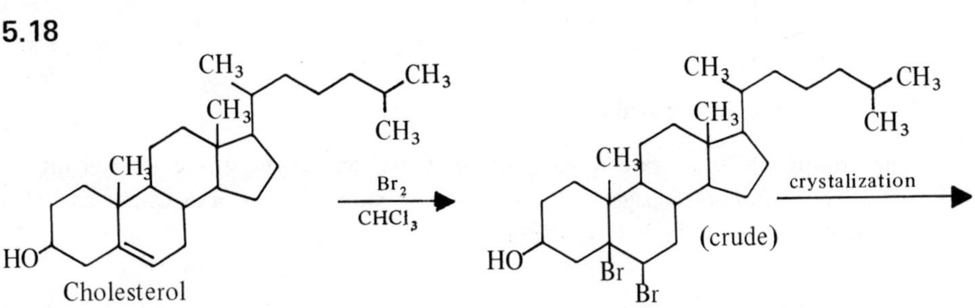

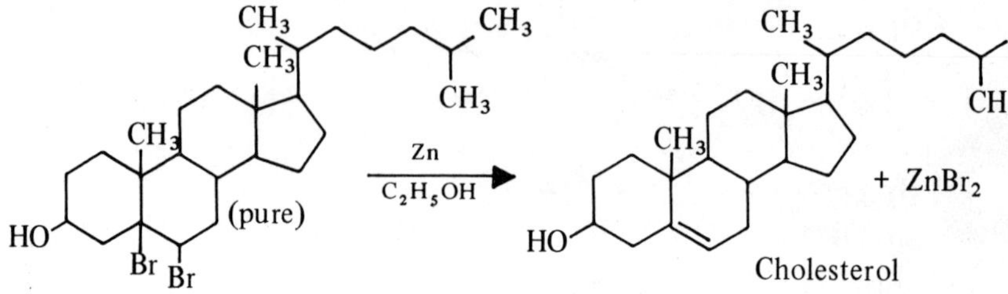

Cholesterol

5.19

3β-Friedelanol → (see below)

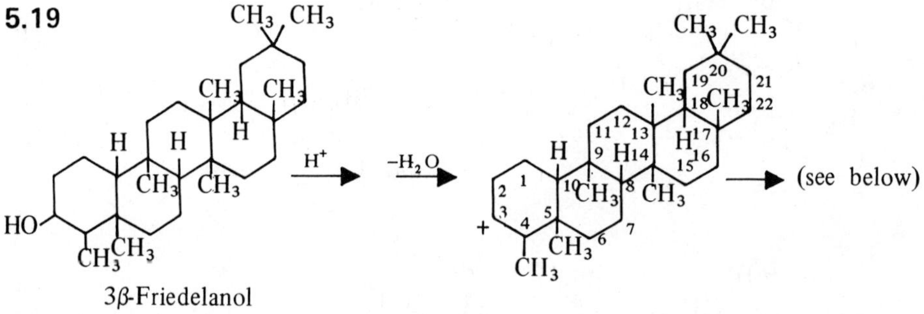

The migrations occur in the following sequence:

(1) H:⁻ from C_4 to C_3 leaves (+) at C_4.

(2) CH_3:⁻ from C_5 to C_4 leaves (+) at C_5.

(3) H:⁻ from C_{10} to C_5 leaves (+) at C_{10}.

(4) CH_3:⁻ from C_9 to C_{10} leaves (+) at C_9.

(5) H:⁻ from C_8 to C_9 leaves (+) at C_8.

(6) CH_3:⁻ from C_{14} to C_8 leaves (+) at C_{14}.

(7) CH_3:⁻ from C_{13} to C_{14} leaves (+) at C_{13}.

(8) Loss of H⁺ from C_{18} leaves double bond at C_{13}–C_{18}:

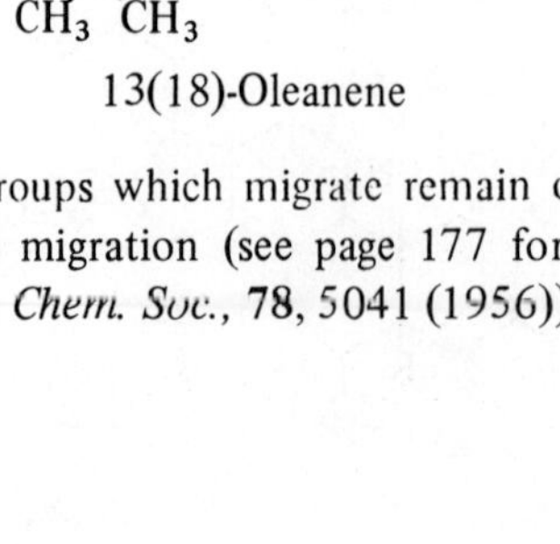

13(18)-Oleanene

The groups which migrate remain on the same face of the molecule after migration as before migration (see page 177 for transition state and see also Corey and Ursprung, *J. Am. Chem. Soc.*, **78**, 5041 (1956)).

6

REACTIONS OF ALKENES: ADDITION REACTIONS OF THE CARBON - CARBON DOUBLE BOND

6.1

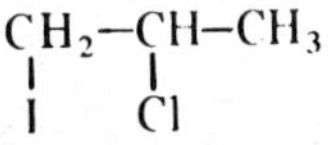

$$CH_2-CH-CH_3$$

with substituents I (on CH_2) and Cl (on CH)

2-chloro-1-iodopropane

6.2

(a) 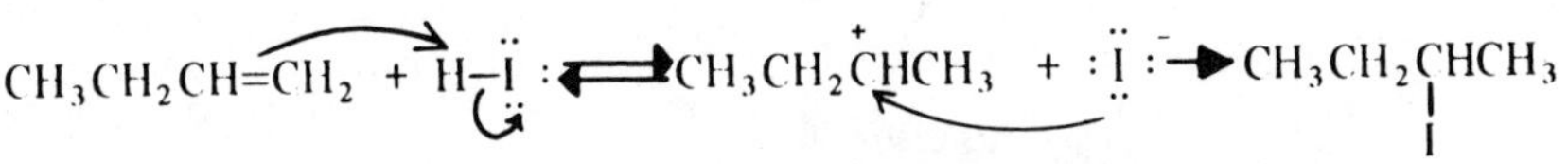 $CH_3CH_2CH{=}CH_2 + H{-}\ddot{I}: \rightleftarrows CH_3CH_2\overset{+}{C}HCH_3 + :\ddot{I}:^- \longrightarrow CH_3CH_2CHCH_3$ (with I substituent)

(b)

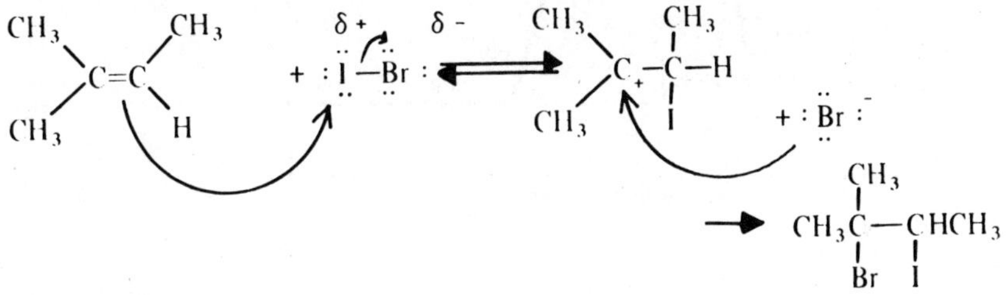

(c)

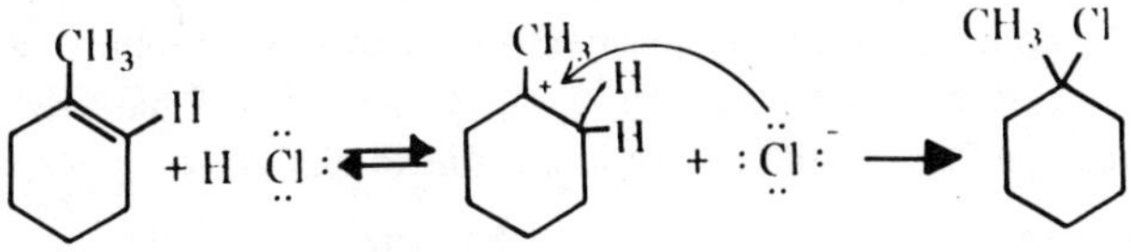

6.3

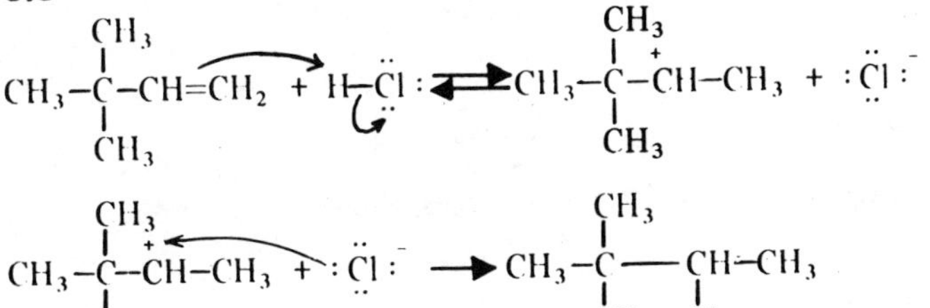

or

45

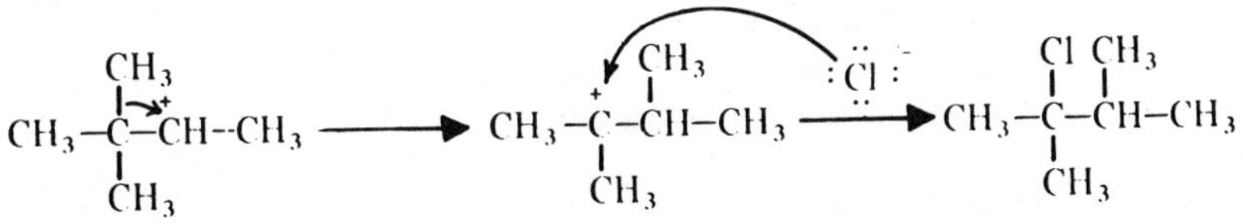

5.20

(a) We are given (on p. 163) the following heats of hydrogenation:

$$cis\text{-2-Butene} + H_2 \longrightarrow butane \qquad \Delta H = -28.6 \text{ kcal/mole}$$

$$trans\text{-2-Butene} + H_2 \longrightarrow butane \qquad \Delta H = -27.6 \text{ kcal/mole}$$

thus for

$$cis\text{-2-Butene} \longrightarrow trans\text{-2-butene} \qquad \Delta H = -1.0 \text{ kcal/mole}$$

(b) Converting *cis*-2-butene into *trans*-2-butene involves breaking the π bond. Therefore we would expect the energy of activation to be at least as large as the π-bond strength, that is, at least 60 kcal/mole.

(c)

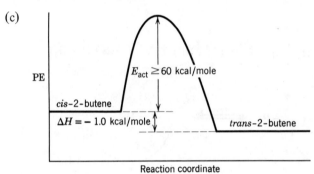

Reaction coordinate

5.21

In answering questions (a) to (h) we determine the index of hydrogen deficiency by comparing the formula of the compound under consideration with the formula of an alkane with the same number of carbons. In parts (a) to (d) the compounds all have the formula C_6H_{12}, so we make our comparisons with the formula for hexane, C_6H_{14}.

(a)-(d) $CH_2{=}CHCH_2CH_2CH_2CH_3$ $\qquad$ $CH_3CH{=}CHCH_2CH_2CH_3$
$\qquad$ 1-Hexene $\qquad\qquad\qquad\qquad$ 2-Hexene
$\qquad$ (C_6H_{12}) $\qquad\qquad\qquad\qquad\quad$ (C_6H_{12})

$\qquad$ Cyclohexane $\qquad\qquad\qquad$ Methylcyclopentane
$\qquad$ (C_6H_{12}) $\qquad\qquad\qquad\qquad$ (C_6H_{12})

All of these compounds have the molecular formula C_6H_{12}, thus the index of hydrogen deficiency for each is equal to 1.

$$\begin{array}{ll} C_6H_{14} & = \text{ formula of alkane (hexane)} \\ \underline{C_6H_{12}} & = \text{ formula of compound} \\ H_2 & = \text{ difference } = 1 \text{ pair of hydrogens} \end{array}$$

Index of hydrogen deficiency = 1

(e) and (f) 1-Methylcyclopentene and 1,3-hexadiene both have the formula C_6H_{10}. Thus we find that for each compound the index of hydrogen deficiency is equal to 2.

$$CH_2=CHCH=CHCH_2CH_3$$

1-Methylcyclopentene   1,3-Hexadiene
    (C_6H_{10}) (C_6H_{10})

C_6H_{14} = formula of alkane (hexane)
$\underline{C_6H_{10}}$ = formula of compound
  H_4 = difference = 2 pairs of hydrogens

Index of hydrogen deficiency = 2

(g) with 1,3,5-hexatriene the index of hydrogen deficiency equals 3.

$$CH_2=CHCH=CHCH=CH_2$$
1,3,5-Hexatriene
  (C_6H_8)

C_6H_{14} = formula of alkane (hexane)
$\underline{C_6H_8}$ = formula of compound
  H_6 = difference = 3 pairs of hydrogens

Index of hydrogen deficiency = 3

(h) 13(18)-Oleanene has the formula $C_{30}H_{50}$ so we make our comparison with the formula for a 30-carbon alkane, $C_{30}H_{62}$.

$C_{30}H_{62}$ = formula of alkane (triacontane)
$\underline{C_{30}H_{50}}$ = formula of compound
  H_{12} = difference = 6 pairs of hydrogens

Index of hydrogen deficiency = 6

(i) The index of hydrogen deficiency equals the number of double bonds plus the number of rings in each molecule.

(j) and (k) The index of hydrogen deficiency tells us nothing about ring size or double bond location.

(l) For $C_{10}H_{16}$ the index of hydrogen deficiency equals 3:

$C_{10}H_{22}$ = formula of alkane (decane)
$\underline{C_{10}H_{16}}$ = formula of compound
  H_6 = difference = 3 pairs of hydrogens

Index of hydrogen deficiency = 3

Thus molecules of the compound could have three double bonds and no rings, two double bonds and one ring, one double bond and two rings, or they could have three rings.

5.22

(a) $C_{15}H_{32}$ = formula of alkane
   $C_{15}H_{24}$ = formula of zingiberene
   $\overline{H_8}$ = difference = 4 pairs of hydrogens

Index of hydrogen deficiency = 4

(b) Since one mole of zingiberene absorbs three moles of hydrogen, one molecule of zingiberene must contain three double bonds. (We are assuming here that molecules of zingiberene do not contain any triple bonds.)

(c) If a molecule of zingiberene has three double bonds and an index of hydrogen deficiency equal to 4, it must have one ring. (The structural formula for zingiberene can be found in Sect. 11.4.)

5.23

(a) Caryophyllene has the same molecular formula as zingiberene (problem 5.22), thus it, too, has an index of hydrogen deficiency equal to 4. That one mole of caryophyllene absorbs two moles of hydrogen on catalytic hydrogenation indicates the presence of two double bonds per molecule.

(b) Caryophyllene molecules must also have two rings. (See Sect. 11.4 for the structure of caryophyllene.)

5.24

(a) $C_{30}H_{62}$ = formula of alkane
   $C_{30}H_{50}$ = formula of squalene
   $\overline{H_{12}}$ = difference = 6 pairs of hydrogen

Index of hydrogen deficiency = 6

(b) Molecules of squalene contain six double bonds.

(c) Squalene molecules contain no rings. (See Sect. 11.4 for the structural formula of squalene.)

6.4

(a) $CH_3-CH=CH_2$ + $H-\overset{..+}{\underset{|}{O}}-H$ $\rightleftharpoons$ $CH_3-\overset{+}{C}H-CH_3$ + H_2O

$CH_3-\overset{+}{C}H-CH_2$ + $:\overset{..}{\underset{|}{O}}-H$ $\rightleftharpoons$ $CH_3-\overset{\overset{\displaystyle H}{\underset{|}{:\underset{|}{O}-H}}}{CH}-CH_3$

$CH_3-\overset{\overset{\displaystyle H}{\underset{|}{:\overset{+}{\underset{|}{O}}-H}}}{CH}-CH_3$ + $:\overset{..}{\underset{|}{O}}-H$ $\rightleftharpoons$ $CH_3-\overset{\overset{\displaystyle OH}{|}}{CH}-CH_3$ + H_3O^+

(b) The product is isopropyl alcohol because the more stable isopropyl carbocation is produced in the first step. The formation of *n*-propyl alcohol would require the production of the less stable *n*-propyl carbocation.

6.5

$CH_3CH_2CH_2CH_2CH=CH_2$ + Hg^+OAc $\rightarrow$ $CH_3CH_2CH_2CH_2-\overset{+}{C}H-CH_2-HgOAc$

$CH_3CH_2CH_2CH_2-\overset{+}{\underset{\underset{\displaystyle HgOAc}{|}}{C}H}-CH_2$ + $:\overset{\overset{\displaystyle H}{|}}{\underset{..}{O}}-CH_2CH_3$ $\rightarrow$ $CH_3CH_2CH_2CH_2-\overset{\overset{\overset{\displaystyle H}{|}}{:\overset{+}{\underset{|}{O}}-CH_2CH_3}}{C}H-CH_2-HgOAc$

$\downarrow -H^+$

$CH_3CH_2CH_2CH_2-\overset{\overset{\displaystyle O-CH_2CH_3}{|}}{C}H-CH_3$ $\overset{NaBH_4}{\underset{OH^-}{\longleftarrow}}$ $CH_3CH_2CH_2CH_2-\overset{\overset{\displaystyle OCH_2CH_3}{|}}{C}H-CH_2-HgOAc$

6.6

(a) $CH_3-\overset{\overset{\displaystyle }{\underset{\underset{\displaystyle CH_3}{|}}{C}}}=CH-CH_3$ + $Hg(OAc)_2$ + H_2O $\overset{THF}{\longrightarrow}$ $CH_3-\overset{\overset{\displaystyle OH}{|}}{\underset{\underset{\displaystyle CH_3}{|}}{C}}---\overset{\overset{\displaystyle CH_3}{|}}{C}H-HgOAc$

$\overset{NaBH_4}{\underset{OH^-}{\big\downarrow}}$

$CH_3-\overset{\overset{\displaystyle OH}{|}}{\underset{\underset{\displaystyle CH_3}{|}}{C}}-CH_2-CH_3$

(b) + $Hg(OAc)_2$ + H_2O $\overset{THF}{\longrightarrow}$ $\overset{NaBH_4}{\underset{OH^-}{\longrightarrow}}$

(c) + $Hg(OAc)_2$ + CH_3OH $\overset{THF}{\longrightarrow}$ $\overset{NaBH_4}{\underset{OH^-}{\longrightarrow}}$

(d) $CH_3-\underset{\underset{CH_3}{|}}{\overset{\overset{CH_3}{|}}{C}}-CH=CH_2$ + $Hg(OAc)_2$ + $CH_3OH \xrightarrow{THF}$ $CH_3-\underset{\underset{CH_3}{|}}{\overset{\overset{CH_3}{|}}{C}}-\underset{\underset{OCH_3}{|}}{CH}-\overset{\overset{HgOAc}{|}}{CH_2}$

$\xrightarrow[OH^-]{NaBH_4}$ $CH_3-\underset{\underset{CH_3}{|}}{\overset{\overset{CH_3}{|}}{C}}-\underset{\underset{OCH_3}{|}}{CH}-CH_3$

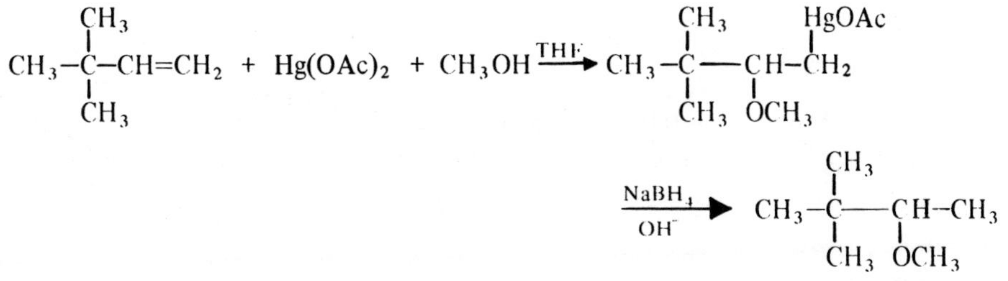

6.7

(a) $3CH_3CH_2CH=CH_2$ + $\frac{1}{2}(BH_3)_2 \longrightarrow (CH_3CH_2CH_2CH_2)_3B \xrightarrow[OH^-]{H_2O_2}$

$3CH_3CH_2CH_2CH_2OH$ + H_3BO_3

(b) $3CH_3-\overset{\overset{CH_3}{|}}{C}=CH-CH_3$ + $\frac{1}{2}(BH_3)_2 \longrightarrow (CH_3-\overset{\overset{CH_3}{|}}{CH}-\overset{\overset{CH_3}{|}}{CH})_3-B \xrightarrow[OH^-]{H_2O_2}$

$3CH_3-\underset{\underset{}{}}{\overset{\overset{CH_3}{|}}{CH}}-\overset{\overset{OH}{|}}{CH}-CH_3$ + H_3BO_3

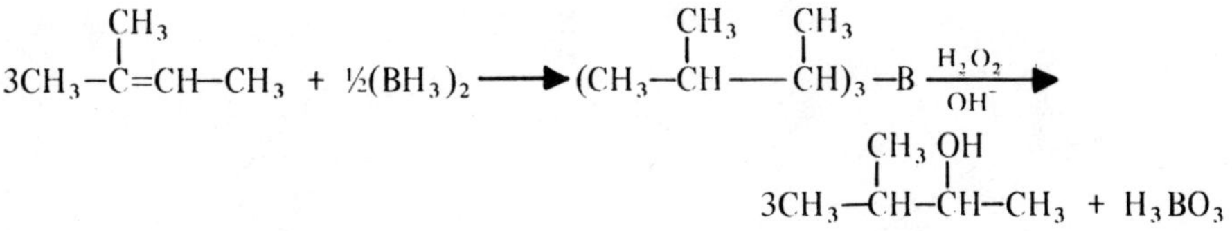

(c)

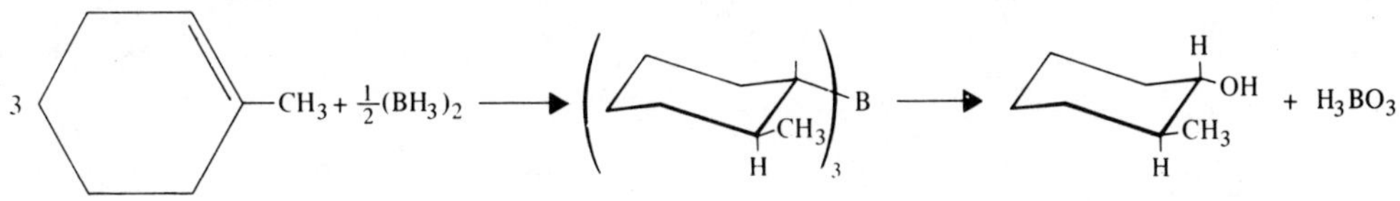

6.8

(a) $3CH_3-\overset{\overset{CH_3}{|}}{C}=CH_2$ + $\frac{1}{2}(BH_3)_2 \longrightarrow (CH_3 \overset{\overset{CH_3}{|}}{CH} CH_2)_3-B$

$\xrightarrow[heat]{CH_3COOD}$ $3CH_3 \overset{\overset{CH_3}{|}}{CH} CH_2D$ + $(CH_3COO)_3B$

(b)

3 $-CH_2D$ + $(CH_3COO)_3B$

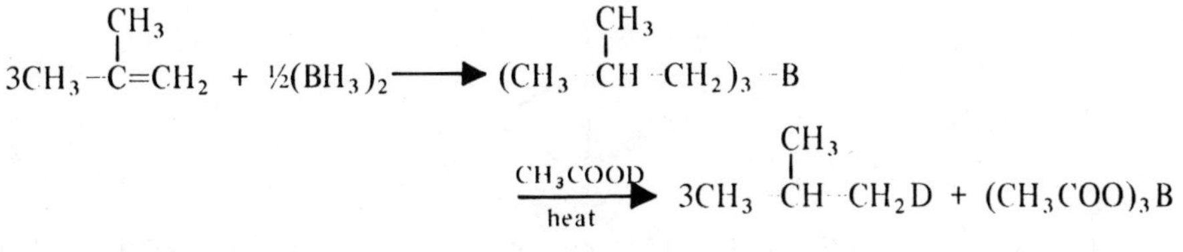

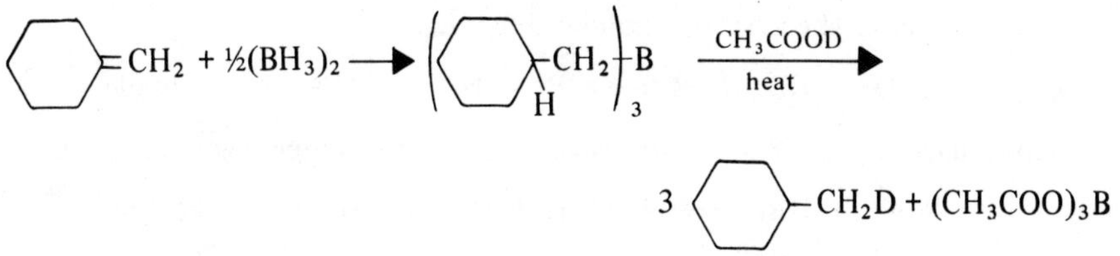

(c)

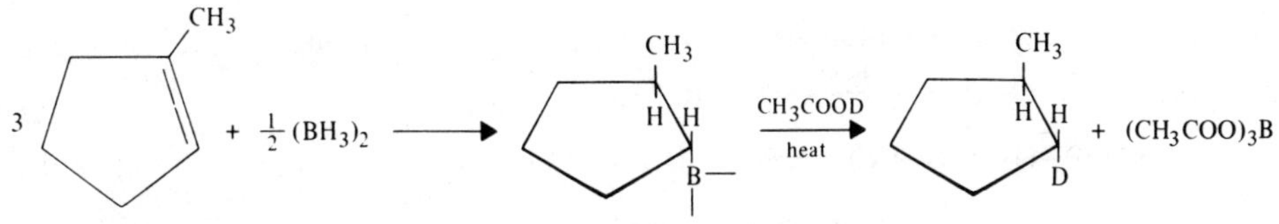

(d)

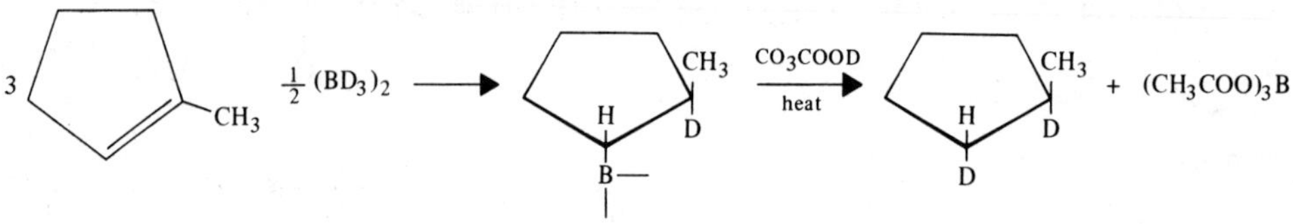

6.9

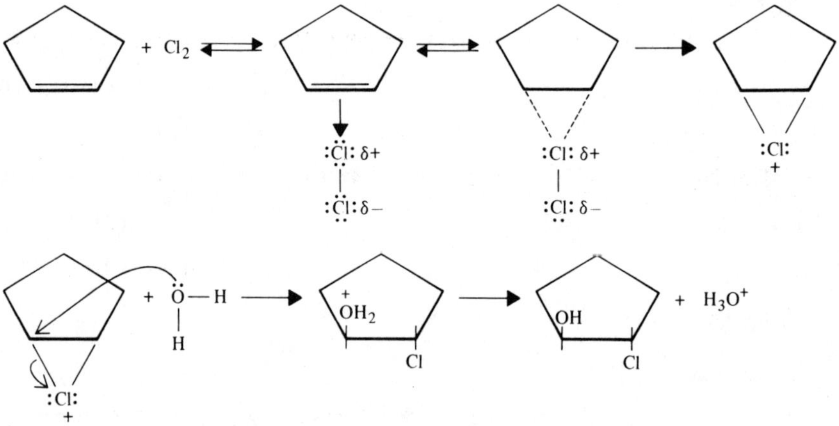

6.10

(a) $CH_3CH_2CH=CHCH_3$ (c) $CH_3CH_2\overset{\displaystyle |}{\underset{\displaystyle CH_3}{C}}HCH=CH_2$

(b) $CH_3\overset{\displaystyle CH_3}{\overset{\displaystyle |}{C}}=\overset{\displaystyle CH_3}{\overset{\displaystyle |}{C}}CH_3$ (d)

6.11

The initiation reaction (step 1) has no competitor and all of the subsequent steps must occur rapidly. The reaction disallowed for step 2,

$$R-\ddot{\underset{..}{O}}\cdot + HBr \nrightarrow R-\ddot{\underset{..}{O}}-Br + H\cdot \qquad \Delta H \cong 39 \text{ kcal/mole}$$

must have an $E_{act} > 39$ kcal/mole. Thus it cannot compete with step (2),

$$R-\ddot{\underset{..}{O}}\cdot + HBr \longrightarrow R-\ddot{\underset{..}{O}}-H + Br\cdot \qquad \Delta H \cong -13 \text{ kcal/mole}$$

which has a much lower E_{act}.

6.12

Chain-initiating steps

(a) (1) $R-\ddot{\underset{..}{O}}-\ddot{\underset{..}{O}}-R \xrightarrow{\text{heat}} 2 R-\ddot{\underset{..}{O}}\cdot$

(2) $R-\ddot{\underset{..}{O}}\cdot + H-CCl_3 \longrightarrow R-\ddot{\underset{..}{O}}H + \cdot CCl_3$

Chain-propagating steps

(3) $CH_3CH_2CH_2CH=CH_2 + \cdot CCl_3 \longrightarrow CH_3CH_2CH_2\overset{\displaystyle \cdot}{C}H-CH_2CCl_3$

(4) $CH_3CH_2CH_2\overset{\curvearrowright}{\underset{\cdot}{C}}H\overset{\curvearrowleft}{CH_2CCl_3}$ + $\overset{\curvearrowleft}{H-CCl_3}\longrightarrow CH_3CH_2CH_2CH_2CH_2CCl_3$

$+ \cdot CCl_3$

then (3), (4), (3), (4), etc.

(b) (1) $R-\overset{..}{\underset{..}{O}}-\overset{..}{\underset{..}{O}}-R \xrightarrow{\text{heat}} 2\ R-\overset{..}{\underset{..}{O}}\cdot$

(2) $R-\overset{..}{\underset{..}{O}}\cdot + CH_3CH_2-\overset{..}{\underset{..}{S}}-H \longrightarrow R-\overset{..}{O}H + CH_3CH_2-\overset{..}{\underset{..}{S}}\cdot$

Chain-propagating steps

(3) $\underset{\underset{CH_3}{|}}{CH_3C}=CH_2$ + $\cdot\overset{..}{\underset{..}{S}}CH_2CH_3\longrightarrow \underset{\underset{CH_3}{|}}{CH_3\underset{\cdot}{C}}-CH_2-\overset{..}{\underset{..}{S}}-CH_2CH_3$

(4) $\underset{\underset{CH_3}{|}}{CH_3\underset{\cdot}{C}}CH_2SCH_2CH_3 + HSCH_2CH_3\longrightarrow \underset{\underset{CH_3}{|}}{CH_3\overset{}{C}}HCH_2SCH_2CH_3 + \cdot\overset{..}{\underset{..}{S}}CH_2CH_3$

then (3), (4), (3), (4), etc.

(c) (1) $R-\overset{..}{\underset{..}{O}}-\overset{..}{\underset{..}{O}}-R \xrightarrow{\text{heat}} 2\ R-\overset{..}{\underset{..}{O}}\cdot$

(2) $R-\overset{..}{\underset{..}{O}}\cdot + Cl-CCl_3\longrightarrow R-\overset{..}{\underset{..}{O}}-Cl + \cdot CCl_3$

Chain-propagating steps

(3) $\underset{\underset{CH_3}{|}}{CH_3CH_2C}=CH_2$ + $\cdot CCl_3\longrightarrow \underset{\underset{CH_3}{|}}{CH_3CH_2\underset{\cdot}{C}}-CH_2CCl_3$

(4) $\underset{\underset{CH_3}{|}}{CH_3CH_2\underset{\cdot}{C}}CH_2CCl_3$ + $CCl_4\longrightarrow \underset{\underset{Cl}{|}}{\overset{\overset{CH_3}{|}}{CH_3CH_2C}}CH_2CCl_3$ + $\cdot CCl_3$

then (3), (4), (3), (4), etc.

6.13

(a) $\underset{\underset{Br}{|}}{CH_3CH_2\overset{}{C}H}CH_2Br$

(b)

cis

(c) Same as (b)

(d)

trans

(e) $CH_3CH_2-O-\overset{\overset{O}{\|}}{\underset{\underset{O}{\|}}{S}}-OH$

(f) CH_3CH_2OH

(g) $(CH_3\underset{\underset{CH_3}{|}}{\overset{}{C}H}CH_2)_3B$

(h) $CH_3\underset{\underset{CH_3}{|}}{\overset{}{C}H}CH_2OH$

(i) (major product)

(j)

(k) $CH_2=CHCH_2CH_2CH_3$

(l) $CH_3\underset{\underset{Br}{|}}{\overset{\overset{CH_3}{|}}{C}}CH(CH_3)_2$

(m) $CH_3\overset{\overset{\displaystyle Cl}{|}}{C}HCH_2CH_2CH_2CH_3$

(n) $BrCH_2CH_2CH_2CH_2CH_2CH_3$

(o) $CH_3CH_2CH\overset{\displaystyle O}{\underset{\displaystyle O\!-\!O}{<}}CHCH_2CH_3$

(p) $2\ CH_3CH_2CHO$

(q) $2CH_3CH_2\overset{\overset{\displaystyle O}{||}}{C}-O^-$

(r)

(s) $CH_3\overset{\overset{\displaystyle O}{||}}{C}CH_2CH_2CH_2CHO$

(t) $CH_3\overset{\overset{\displaystyle O}{||}}{C}CH_2CH_2CH_2\overset{\overset{\displaystyle O}{||}}{C}-OH$

(u)

(v)

(w) $CH_3\overset{\overset{\displaystyle CH_3}{|}}{\underset{\underset{\displaystyle I}{|}}{C}}CH_2CH_2CH_3$

(x)

6.14

(a) $CH_3CH=CH_2 + H_3O^+ \ \rightleftharpoons\ CH_3\overset{+}{C}HCH_3 + H_2O$

$\overset{\displaystyle CH_3}{\underset{\displaystyle CH_3}{>}}\overset{+}{C}H + CH_2=CHCH_3 \longrightarrow CH_3\overset{\overset{\displaystyle CH_3}{|}}{C}HCH_2\overset{+}{C}HCH_3$

$CH_3\overset{\overset{\displaystyle CH_3}{|}}{C}HCH\overset{+}{-}CH-CH_2 + :\ddot{O}-H \longrightarrow$

$\begin{array}{l} a\ \blacktriangleright\ CH_3\overset{\overset{\displaystyle CH_3}{|}}{C}HCH=CHCH_3 \\[4pt] b\ \blacktriangleright\ CH_3\overset{\overset{\displaystyle CH_3}{|}}{C}HCH_2CH=CH_2 \end{array}$

(b) $CH_3\overset{\overset{\displaystyle CH_3}{|}}{C}HCH=CHCH_3$ (more substituted alkene)

6.15

$CH_3CH=CHCH_3 + HCl \ \rightleftharpoons\ CH_3CH_2\overset{+}{C}HCH_3 + :\overset{..}{\underset{..}{Cl}}:^-$

$CH_3CH_2\overset{+}{C}HCH_3 + :\overset{..}{O}-CH_2CH_3 \longrightarrow CH_3CH_2\overset{\overset{\displaystyle \overset{|}{+}:\overset{..}{O}-CH_2CH_3}{|}}{C}HCH_3 \xrightarrow{-H^+} CH_3CH_2\overset{\overset{\displaystyle OCH_2CH_3}{|}}{C}HCH_3$

6.16

(a)

(* D may also be axial)

(b) 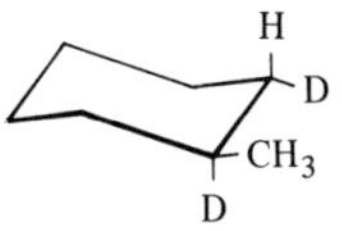 (c)

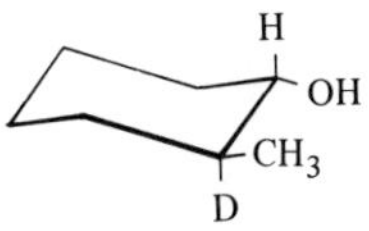

6.17

$$\left(2CH_3\overset{O}{\overset{\|}{C}}CH_3 \right), 4\left(O=CHCH_2CH_2\overset{CH_3}{\underset{|}{C}}=O \right), \quad O=CHCH_2CH_2CH=O$$

6.18

$$CH_3\underset{\underset{CH_3}{|}}{C}=CH_2 > CH_3CH=CH_2 > CH_2=CH_2$$

The order is the same as the order of stability of the carbocations formed by protonation of the alkenes.

$$CH_3\underset{\underset{CH_3}{|}}{\overset{+}{C}}-CH_3 > CH_3\overset{+}{C}H-CH_3 > \overset{+}{C}H_2-CH_3$$

6.19

(a) $CH_3CH=CHCH_3 + \overset{+}{H} \rightleftharpoons CH_3\overset{+}{C}HCH_2CH_3$

(*cis* or *trans*)

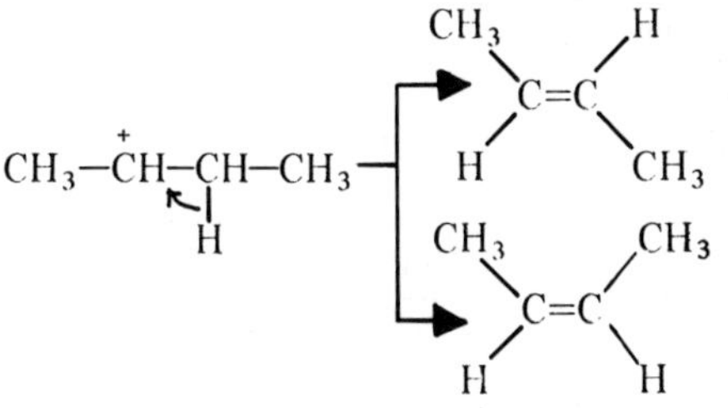

$$CH_3-CH_2-\overset{+}{C}H\overset{H}{\overset{\curvearrowleft|}{C}}H_2 \longrightarrow CH_3CH_2CH=CH_2 + \overset{+}{H}$$

The most stable (most substituted) alkene is formed in greatest amount; i.e., 2-butenes > 1-butene, and *trans*-2-butene > *cis*-2-butene.

(b) 1-Butene, on protonation, gives the same intermediate carbocation: $CH_3CH_2\overset{+}{C}HCH_3$.

(c) The carbocation, $CH_3CH_2\overset{+}{C}HCH_3$, cannot easily rearrange to the branched chain compound because to do so would require the formation of an intermediate primary carbocation, $\overset{+}{C}H_2\underset{\underset{CH_3}{|}}{C}HCH_3$.

6.20

$$CH_3-\underset{\underset{OH}{|}}{C}H-\underset{\underset{CH_3}{|}}{\overset{\overset{CH_3}{|}}{C}}-CH_3 \;+\; H-Cl \;\rightleftharpoons\; CH_3-\underset{\underset{{}^{+}OH_2}{|}}{C}H-\underset{\underset{CH_3}{|}}{\overset{\overset{CH_3}{|}}{C}}-CH_3 \;+\; :\overset{..}{\underset{..}{Cl}}:^-$$

$$\longrightarrow\; CH_3-\overset{+}{C}H-\underset{\underset{CH_3}{|}}{\overset{\overset{CH_3}{|}}{C}}-CH_3 \;+\; H_2O$$

$$CH_3-\overset{+}{C}H\overset{\curvearrowleft}{-}\underset{\underset{CH_3}{|}}{\overset{\overset{CH_3}{|}}{C}}-CH_3 \;\longrightarrow\; CH_3-\underset{\underset{CH_3}{|}}{C}H-\overset{\overset{CH_3}{|}}{\overset{+}{C}}-CH_3 \;\xrightarrow{\;:\overset{..}{Cl}:^-\;}\; CH_3-\underset{\underset{CH_3}{|}}{C}H-\underset{\underset{CH_3}{|}}{\overset{\overset{Cl}{|}}{C}}-CH_3$$

6.21

6.22

(a) $CH_3CH_2CH_3 \;+\; Br_2 \xrightarrow[\text{light}]{CCl_4}$ $\left. \begin{array}{l} CH_3CH_2CH_2Br \\ + \\ CH_3\underset{\underset{Br}{|}}{C}HCH_3 \end{array} \right\} \xrightarrow[CH_3CH_2OH]{KOH} CH_3CH=CH_2$

(excess)

(b) $CH_3CH=CH_2$ (above) $+$ HBr $\xrightarrow[\text{inhibitor}]{\text{free-radical}}$ $CH_3\underset{\underset{Br}{|}}{C}HCH_3$

(c) $CH_3CH=CH_2$ (from (a)) $+$ HBr $\xrightarrow{\text{Peroxides}}$ $CH_3CH_2CH_2Br$

(d) $CH_3\underset{\underset{}{\overset{\overset{CH_3}{|}}{}}}{C}HCH_3 \;+\; Br_2 \xrightarrow[\text{light}]{\text{heat}} CH_3\underset{\underset{Br}{|}}{\overset{\overset{CH_3}{|}}{C}}CH_3 \xrightarrow[CH_3CH_2OH]{KOH} CH_3\overset{\overset{CH_3}{|}}{C}=CH_2$

(excess)

(e) $CH_3\overset{\overset{CH_3}{|}}{C}=CH_2$ (from (d)) $+$ H_2O $\xrightarrow{H_3O^+}$ $CH_3\underset{\underset{OH}{|}}{\overset{\overset{CH_3}{|}}{C}}CH_3$

(f) $CH_3CH_2CH_2CH_2Cl \;+\; KOH \xrightarrow{CH_3CH_2OH} CH_3CH_2CH=CH_2 \xrightarrow[\substack{CCl_4 \\ \text{dark}}]{Cl_2} CH_3CH_2\underset{\underset{Cl}{|}}{C}HCH_2Cl$

(g) $CH_3CH_2Br \;+\; KOH \xrightarrow{CH_3CH_2OH} CH_2=CH_2 \xrightarrow{Br_2 \;+\; H_2O} \underset{\underset{OH\;\;Br}{|\;\;\;\;|}}{CH_2CH_2}$

(h) $CH_3\overset{\overset{CH_3}{|}}{C}=CH_2$ (from (d)) $+$ HBr $\xrightarrow{\text{Peroxides}}$ $CH_3\overset{\overset{CH_3}{|}}{C}HCH_2Br \xrightarrow{Li} CH_3\overset{\overset{CH_3}{|}}{C}HCH_2Li$

$CH_3\overset{\overset{CH_3}{|}}{C}HCH_2CH_2\overset{\overset{CH_3}{|}}{C}HCH_3 \xleftarrow{CH_3\overset{\overset{CH_3}{|}}{C}HCH_2Br} (CH_3\overset{\overset{CH_3}{|}}{C}HCH_2)_2CuLi \xleftarrow{CuI}$

(or use Wurtz reaction as in (k).)

(i) [cyclopentane] $+ Br_2$ $\xrightarrow[\text{light}]{CCl_4}$ [cyclopentane with Br] $\xrightarrow[CH_3CH_2OH]{KOH}$ [cyclopentene]

(excess)

[cyclopentene] $+ Cl_2 + H_2O \longrightarrow$ [cyclopentane with Cl and OH (trans)]

(j) $CH_3CH_2CH_2CH_2Br$ $\xrightarrow[CH_3CH_2OH]{KOH}$ $CH_3CH_2CH=CH_2$ $\xrightarrow[\substack{\text{free radical} \\ \text{inhibitor}}]{HBr}$ $CH_3CH_2\underset{\underset{Br}{|}}{C}HCH_3$

(k) $CH_3CH_2CH_2CH_2Cl$ $\xrightarrow[CH_3CH_2OH]{KOH}$ $CH_3CH_2CH=CH_2$ $\xrightarrow{HCl}$ $CH_3CH_2\underset{\underset{Cl}{|}}{C}HCH_3$

$2CH_3CH_2\underset{\underset{CH_3}{|}}{C}H-Cl + Na \xrightarrow{\text{(Wurtz)}} CH_3CH_2\underset{\underset{CH_3}{|}}{C}H-\underset{\underset{CH_3}{|}}{C}HCH_2CH_3$

(or use $(R)_2CuLi$ as in (h).)

(l) $CH_3\underset{\underset{CH_3}{|}}{\overset{\overset{CH_3}{|}}{C}}-OH \xrightarrow[\text{heat}]{H_2SO_4} CH_3\overset{\overset{CH_3}{|}}{C}=CH_2 \xrightarrow{(BH_3)_2} (CH_3\overset{\overset{CH_3}{|}}{C}HCH_2)_3B$

$(CH_3\overset{\overset{CH_3}{|}}{C}HCH_2)_3B \xrightarrow[OH^-]{H_2O_2} CH_3\overset{\overset{CH_3}{|}}{C}HCH_2OH$

6.23

(a) $CH_3CH_2CH_2CH=CH_2$ $+$ Br_2 $\longrightarrow$ $CH_3CH_2CH_2\underset{\underset{Br}{|}}{C}HCH_2Br$

  MW=70.12   MW=159.8

159.8 g Br_2 will react with 70.12 g pentene. Therefore $\sim$ 16 g Br_2 will react with 7.0 g pentene.

(b) Since bromine and an alkene react in equimolar proportions:

$$\frac{3.20 \text{ g}}{160 \text{ g/mole}} = 0.02 \text{ mole } Br_2 = 0.02 \text{ mole alkene}$$

2.24 g alkene = (0.02 mole) (Mol. Wt.)

$$\therefore \text{Mol. Wt.} = \frac{2.24 \text{ g}}{0.02 \text{ mole}} = 112 \text{ g/mole alkene}$$

6.24

Rewriting the starting compound, we can better see the required reaction:

6.25

The intermediate I is competitively attacked by Cl^-, Br^- and H_2O.

6.26

The rate of addition of a proton to an alkene is faster the more stable the carbocation that is produced. The order of carbocation stability is:

$$\overset{+}{C}H_2-CH_3 < CH_3\overset{+}{C}HCH_3 < CH_3\overset{+}{C}HCH_2CH_3 < CH_3\underset{\underset{CH_3}{|}}{\overset{+}{C}}CH_3$$

Proton addition is the rate-limiting step in the hydration reaction.

6.27

6.28

6.29

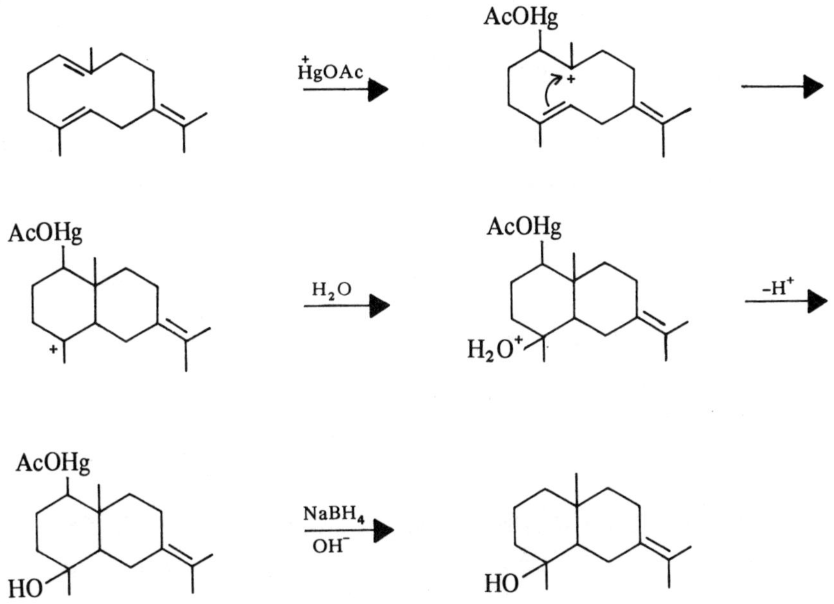

6.30

(a) $H-SH \xrightarrow{h\nu} H\cdot + \cdot SH$ **Chain-initiating step**

$R-CH=CH_2 + \cdot SH \longrightarrow R\overset{\cdot}{C}H-CH_2SH$

$R-\overset{\cdot}{C}HCH_2SH + H-SH \longrightarrow RCH_2CH_2SH$ $\left.\phantom{\begin{matrix}1\\1\end{matrix}}\right\}$ **Chain-propagating steps**

(b) $R\overset{\cdot}{C}HCH_2SH + RCH_2CH_2SH \longrightarrow RCH_2CH_2SH + RCH_2CH_2S\cdot$

$RCH=CH_2 + \cdot SCH_2CH_2R \longrightarrow R\overset{\cdot}{C}HCH_2SCH_2CH_2R$

$R\overset{\cdot}{C}HCH_2SCH_2CH_2R + RCH_2CH_2SH \longrightarrow (RCH_2CH_2)_2S + RCH_2CH_2S\cdot$

6.31

One dimer (the major product) gives the following products on ozonolysis.

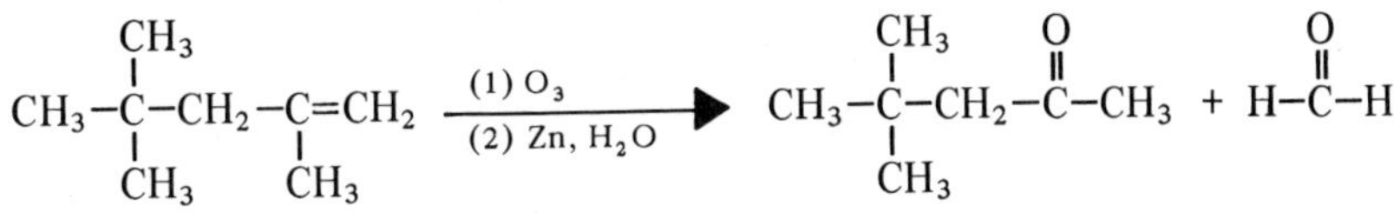

The other dimer gives different products.

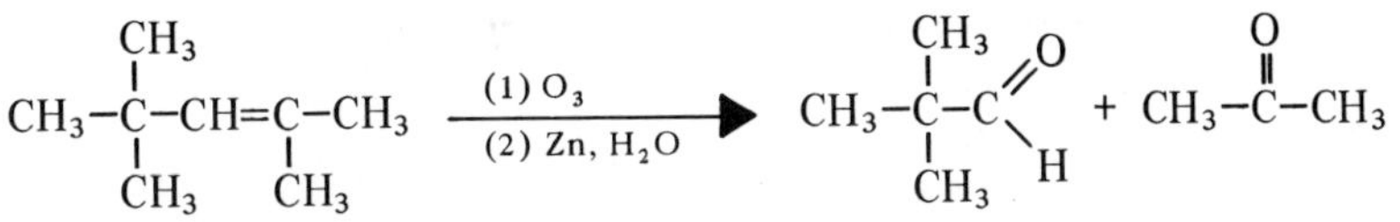

By isolating and identifying the products of each reaction, Whitmore and his students were able to deduce the structures of the diisobutylenes.

6.32

The isomers are propene tetramers formed by an acid-catalyzed reaction:

$$CH_3CH=CH_2 + H_3PO_4 \rightleftharpoons CH_3\overset{+}{C}HCH_3 + H_2PO_4^-$$

$$CH_3\overset{+}{C}H\underset{CH_3}{} + CH_2=CHCH_3 \longrightarrow CH_3\underset{CH_3}{C}HCH_2\overset{+}{C}H\underset{CH_3}{}$$

$$CH_3\underset{CH_3}{C}HCH_2\overset{+}{C}H\underset{CH_3}{} + CH_2=CHCH_3 \longrightarrow CH_3\underset{CH_3}{C}HCH_2\underset{CH_3}{C}HCH_2\overset{+}{C}H\underset{CH_3}{}$$

$$CH_3\underset{CH_3}{C}HCH_2\underset{CH_3}{C}HCH_2\overset{+}{C}H\underset{CH_3}{} + CH_2=CHCH_3 \longrightarrow CH_3\underset{CH_3}{C}HCH_2\underset{CH_3}{C}HCH_2\underset{CH_3}{C}HCH_2\overset{+}{C}HCH_3$$

$$CH_3\underset{CH_3}{C}HCH_2\underset{CH_3}{C}HCH_2\underset{CH_3}{C}HCH_2\overset{+}{C}HCH_3 \xrightarrow{-H^+} CH_3\underset{CH_3}{C}HCH_2\underset{CH_3}{C}HCH_2\underset{CH_3}{C}HCH=CHCH_3$$

$$+$$

$$CH_3\underset{CH_3}{C}HCH_2\underset{CH_3}{C}HCH_2\underset{CH_3}{C}HCH_2CH=CH_2$$

6.33

The halogen does not go to "the less hydrogenated carbon" as predicted by the original version of Markovnikov's rule, but the reaction does proceed through the *formation of the more stable carbocation* as required by the modern form of the rule. Of the two paths below, the reaction follows path (1) because the carbocation formed in the path (1) is more stable. This is true even though the carbocation is a 1° cation and the carbocation in path (2) is a 2° cation. The positive charge of the carbocation formed in (2) is located on a carbon that is directly attached to a highly electron-withdrawing group, the CF₃-group. (The group is highly electron withdrawing because of the combined electronegativities of the three fluorines.) The electron-withdrawing CF₃- group intensifies the positive charge of CF₃ĊHCH₃ and makes it highly unstable.

(1) $CF_3CH=CH_2 \xrightarrow{H^+} CF_3CH_2\overset{+}{C}H_2 \xrightarrow{Cl^-} CF_3CH_2CH_2Cl$
more stable even though 1°

(2) $CF_3CH=CH_2 \xrightarrow{H^+} CF_3\overset{+}{C}HCH_3 \xrightarrow{Cl^-} CF_3\underset{Cl}{C}HCH_3$
less stable even though 2° *(not formed)*

We examine this reaction again in Sect. 13.9.

7

STEREOCHEMISTRY

7.1
Chiral   (a) screw, (e) foot, (f) ear, (g) shoe, (h) spiral staircase
Achiral   (b) plain spoon, (c) fork, (d) cup

7.2
(b) Yes.   (c) No.   (d) No.

7.3
(b) No.   (c) No.

7.4
(a) 1-Chloropropane, (c) 2-methyl-1-chloropropane, (d) 2-methyl-2-chloropropane, (f) 1-chloropentane, and (h) 3-chloropentane are all achiral.

(b)

$$H-\overset{\text{Cl}}{\underset{\text{I}}{\text{C}}}-Br \qquad Br-\overset{\text{Cl}}{\underset{\text{I}}{\text{C}}}-H$$

    I   II

(e)

$$H-\overset{\text{CH}_3}{\underset{\text{CH}_2\text{CH}_3}{\text{C}}}-Br \qquad Br-\overset{\text{CH}_3}{\underset{\text{CH}_2\text{CH}_3}{\text{C}}}-H$$

    I   II

(g)

$$H-\overset{\text{CH}_3}{\underset{\text{CH}_2\text{CH}_2\text{CH}_3}{\text{C}}}-Cl \qquad Cl-\overset{\text{CH}_3}{\underset{\text{CH}_2\text{CH}_2\text{CH}_3}{\text{C}}}-H$$

    I   II

7.5

(a) 1. $H-\overset{\text{X}}{\underset{\text{H}}{\text{C}}}-H$ 2. $X-\overset{\text{H}}{\underset{\text{H}}{\text{C}}}-X$ $X-\overset{\text{H}}{\underset{\text{H}}{\text{C}}}-Y$ 3. $Z-\overset{\text{H}}{\underset{\text{Y}}{\text{C}}}-X$ $X-\overset{\text{H}}{\underset{\text{Y}}{\text{C}}}-Z$

(b) 1. One

2. Two: $\overset{H---X}{\underset{H---X}{\text{C}}}$ and $\overset{X---H}{\underset{H---X}{\text{C}}}$, or $\overset{H---X}{\underset{H---Y}{\text{C}}}$ and $\overset{H---X}{\underset{Y---H}{\text{C}}}$

3. Three: (three structures shown)

(c) **1.** One

2. Three: (three structures) or (three structures)

3. Six: (six structures)

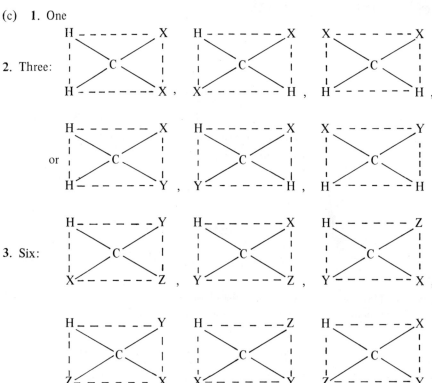

(d) **1.** One

2. Two or three: (structures) or (structures)

3. Six: (six structures labeled 1, 2, 3, 4, 5, 6)

enantiomeric pairs

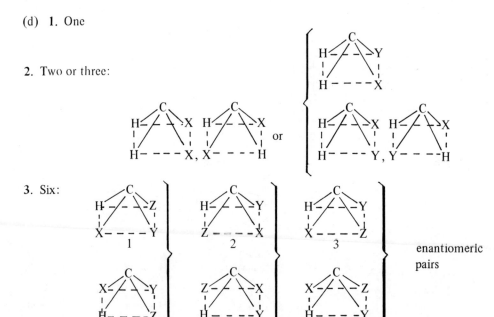

7.6

(b) plain spoon, (c) fork, (d) cup all possess a plane of symmetry.

7.7

(a)

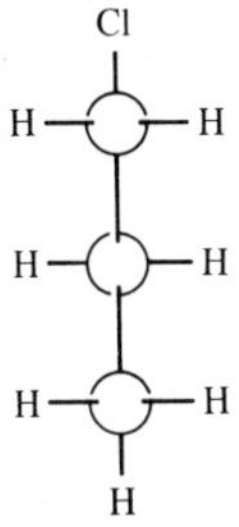

The plane of symmetry is perpendicular to page and passes through Cl and 3C's

(c)

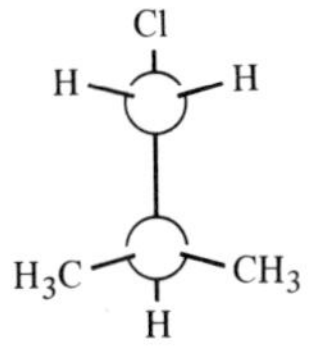

The plane of symmetry is perpendicular to page and passes through Cl and 2C's

(d)

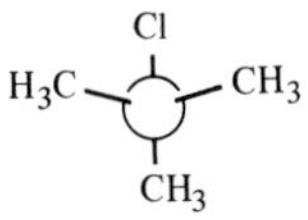

A vertical plane perpendicular to page passes through Cl, tertiary C, and CH$_3$ at bottom

(f)

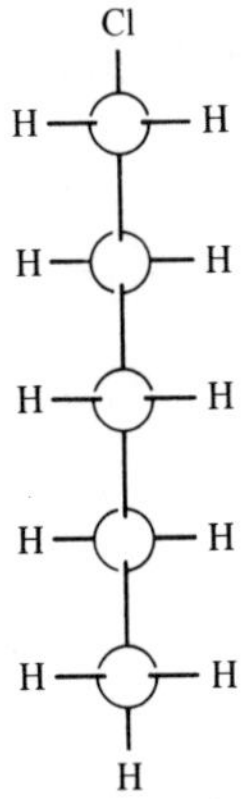

A plane perpendicular to page passes through Cl and 5C's

(h)

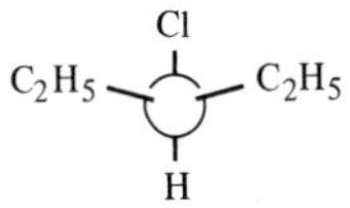

A plane perpendicular to page passes through Cl, C, and H

7.8

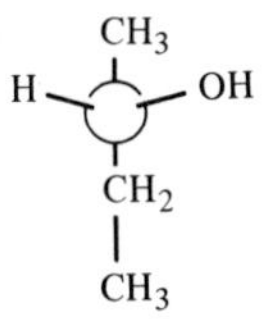

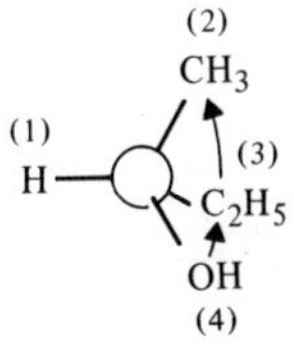

From priority 4 to 3 to 2, the direction is counterclockwise, therefore II is *S*−2−butanol

7.9

(b)

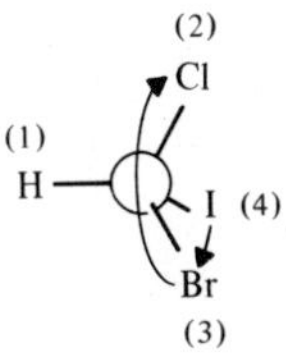

I = *R*

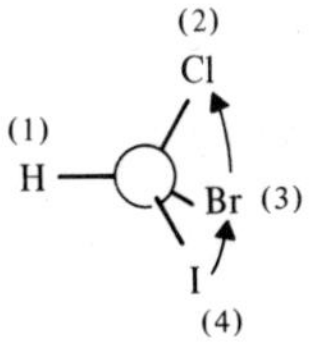

II = *S*

(e)

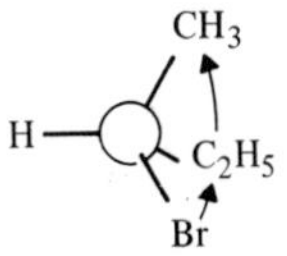

I = *S*

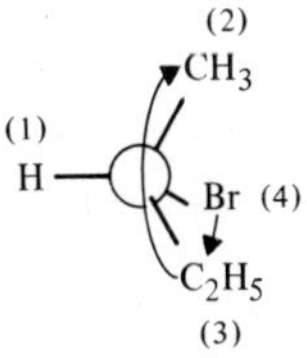

II = *R*

(g)

$$I = S \qquad\qquad II = R$$

7.10

(a) R　　(b) R　　(c) R

7.11

The optical purity is 50% (see previous paragraph in text). That means that the sample contains 50% of the S-enantiomer and 50% of the racemic mixture. The racemic mixture is 50% S- and 50% R. Therefore the total percentage of S-enantiomer in the sample is 75%, the percentage of R-enantiomer is 25%.

7.12

(a) ($\pm$) $CH_3CH_2\underset{\underset{OH}{|}}{C}HCH_3$　　　　(b) Same as (a)

　　(Racemic modification)

(c) ($\pm$) $CH_3CH_2CH_2\underset{\underset{CH_3}{|}}{C}HCH_2CH_3$　　(d) Same as (a)

　　　(Racemic modification)　　　(e) Same as (a)

7.13

(a) Diastereomers　　(b) Diastereomers　　(c) Diastereomers

(d)

	1	2	3	4
1		enantiomers	diastereomers	diastereomers
2	enantiomers		diastereomers	diastereomers
3	diastereomers	diastereomers		enantiomers
4	diastereomers	diastereomers	enantiomers	

(e) Yes.　　(f) No.

7.14

(a) **5** alone would be optically active.

(b) **6** alone would be optically active.

(c) **7** would not be optically active because it is a meso compound.

(d) An equimolar mixture of **5** and **6** would not be optically active because it is a racemic modification.

7.15

(a)

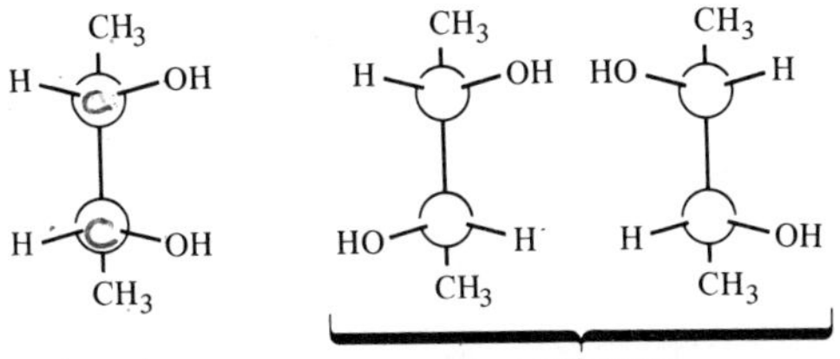

(meso)

enantiomers

(b)

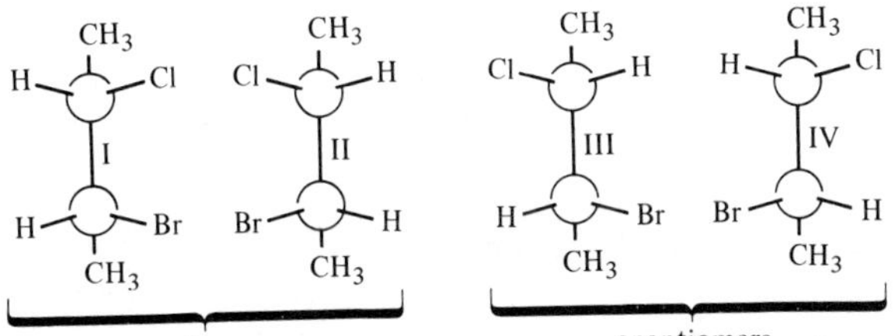

enantiomers   enantiomers

Diastereomers are I and III, I and IV, II and III, and II and IV.

(c)

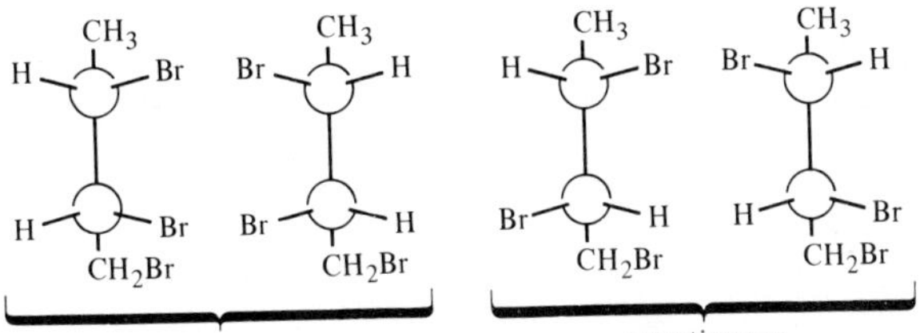

enantiomers   enantiomers

(d)

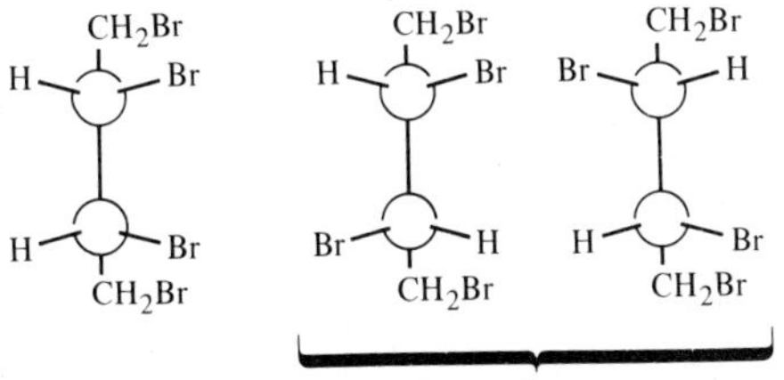

meso   enantiomers

(e)

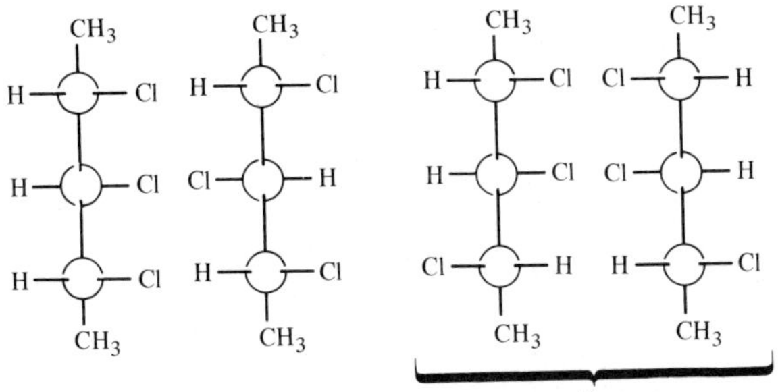

meso   meso   enantiomers

7.16
(a) No   (b) Yes   (c) No   (d) No   (e) Diastereomers   (f) Diastereomers

7.17

(a) *Trans*-1,2-dibromocyclopentanes   (b) Racemic modification

(c) *Cis*-1,2-dibromocyclopentane (meso)

7.18

2   2 R, 3 S -2, 3-dibromopentane
3   2 S, 3 S -2, 3-dibromopentane
4   2 R, 3 R -2, 3-dibromopentane
5   2 S, 3 S -2, 3-dibromobutane
6   2 R, 3 R -2, 3-dibromobutane
7   2 R, 3 S -2, 3-dibromobutane; same as (2 S , 3 R)

7.19

(a) 2, 3-Dibromobutane (racemic modification):

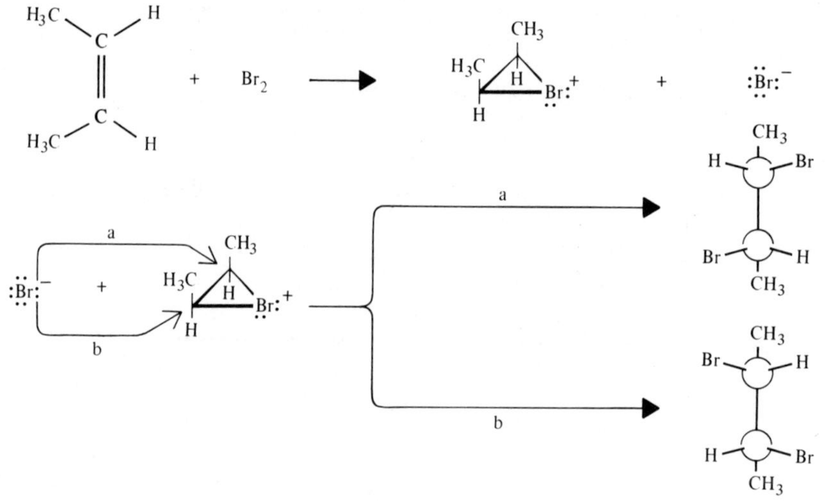

(b) *Meso* -2, 3-dibromobutane:

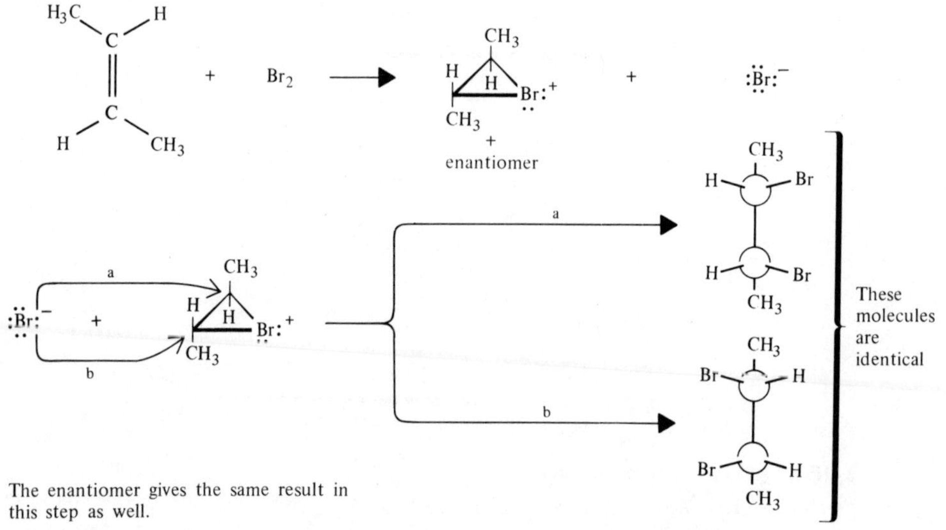

These
molecules
are
identical

The enantiomer gives the same result in
this step as well.

7.20

(a)

CH₃
H—OH
H—OH
CH₃

(meso)

(b)

CH₃
H—OH
HO—H
CH₃

+

CH₃
HO—H
H—OH
CH₃

(racemic modification)

7.21

(a)(b)

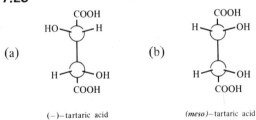

O=C—OH
H—OH
CH₂OH

(R)–(–)–glyceric acid

(a)(c)

O=C—OH
H—OH
CH₂Br

(S)–(–)–3–bromo–
2–hydroxypropanoic
acid

(d)

O=C—OH
H—OH
CH₃

(R)–(–)–lactic acid

7.22

O=C—OCH₃
HO—H
CH₃

(S)–(–)–methyl lactate

7.23

(a)

COOH
HO—H
H—OH
COOH

(–)–tartaric acid

(b)

COOH
H—OH
H—OH
COOH

(meso)–tartaric acid

(c) No, *meso*-tartaric acid would *not* be optically active.

7.24

They are diastereomers.

7.25

(a) (Z) - 1-Bromo-1-chloro-1-butene

(b) (Z) - 2-Bromo-1-chloro-1-iodopropene

(c) (E) - 3-Ethyl-4-methyl-2-pentene

(d) (E) - 1-Chloro-1-fluoro-2-methyl-1-butene

7.26

(a) Isomers are different compounds that have the same molecular formula. C_2H_6O: CH_3CH_2OH and CH_3OCH_3

(b) Structural isomers are isomers that differ because their atoms are joined in a different order. C_4H_{10}: $CH_3CH_2CH_2CH_3$ and CH_3CHCH_3
$\quad\quad\quad\quad\quad\quad\quad\quad\quad\quad\quad\quad\quad\quad\quad\quad\;\;|$
$\quad\quad\quad\quad\quad\quad\quad\quad\quad\quad\quad\quad\quad\quad\quad\quad CH_3$

(c) Stereoisomers are isomers that differ only in the arrangement of their atoms in space: *cis*- and *trans* -2-butene.

(d) Diastereomers are stereoisomers that are not mirror reflections of each other: *cis*- and *trans* -2-butene, or (2 S, 3 S)- and (2 S, 3 R)- 2, 3-dibromobutane.

(e) Enantiomers are stereoisomers that are **non-superposable** mirror reflections of each other: (2 S, 3 S)- and (2 R, 3 R)- 2, 3-dibromobutane.

(f) A meso compound is made up of achiral molecules that contain chiral centers: (2 S, 3 R)- 2, 3-dibromobutane.

(g) A racemic modification is an equimolar mixture of a pair of enantiomers.

(h) A plane of symmetry is an imaginary plane that bisects a molecule in such a way that the two halves of the molecule are mirror reflections of each other. (See Fig. 7.7.)

(i) A chiral center is any tetrahedral atom that has four different groups attached to it.

(j) A chiral molecule is one that is not superposable on its mirror reflection.

(k) An achiral molecule is superposable on its mirror reflection.

(l) Optical activity is the rotation of the plane of polarization of plane polarized light by a substance placed in the light path.

(m) A dextrorotatory substance is one that rotates the plane of polarization of plane polarized light in a clockwise direction.

(n) A reaction occurs with retention of configuration when all the groups around the chiral atom retain the same relative configuration after the reaction that they had before the reaction.

7.27

(a) Enantiomers　　(b) Same　　(c) Enantiomers　　(d) Diastereomers　　(e) Same

(f) Structural isomers　　(g) Same　　(h) Diastereomers　　(i) Same

(j) Enantiomers　　(k) Same　　(l) Enantiomers　　(m) Same　　(n) Structural isomers　　(o) Same　　(p) Diastereomers　　(q) Enantiomers

7.28

(a)

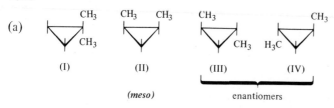

(I)　　　　(II)　　　　(III)　　　　(IV)

(meso)　　　　　　enantiomers

(b) III and IV   (c) II   (d) Three: I, II, and a mixture of III and IV.   (e) None, since the only chiral molecules are **III** and **IV**, and they would be obtained in the same amounts as a racemic modification.

7.29

(a)

```
    CH3
H ──┼── OH
    |
H ──┼── OH      and enantiomer,
    CH2CH3
```

(b)

```
    CH3
 H ──┼── OH
     |
HO ──┼── H      and enantiomer,
    CH2CH3
```

(c)

```
    CH3
 H ──┼── OH
     |
HO ──┼── H      and enantiomer,
    CH2CH3
```

(d)

```
    CH3
H ──┼── OH
    |
H ──┼── OH      and enantiomer,
    CH2CH3
```

(e)

```
    CH3
H ──┼── Br
    |
H ──┼── Br      and enantiomer,
    CH2CH3
```

(f)

```
    CH3
 H ──┼── Br
     |
Br ──┼── H      and enantiomer,
    CH2CH3
```

7.30

(a) 2S, 3R - (the enantiomer is 2R,3S)   (b) 2 S, 3 S - (the enantiomer is 2 R, 3 R-)

(c) Same as (b)   (d) Same as (a)   (e) 2 S, 3 R- (the enantiomer is 2 R, 3 S-)

(f) 2 S, 3 S- (the enantiomer is 2 R, 3 R-)

7.31

(b) and (c) *must* occur with retention of configuration because no bonds to the chiral carbon are broken in either reaction.

7.32

(a)

```
            O                          O
            ‖                          ‖
    CH3   ╱O─C              CH3      ╱O─C
 H ─┼─ CH2  H ─┼─ OH     H ─┼─ CH2  HO ─┼─ H
    |          |            |            |
   CH2CH3     CH3          CH2CH3       CH3
   (S)        (R)    and   (S)          (S)
```

(b) They are diasteromers.

(c) The boiling points of these esters *will* be different. If there is a large enough difference in boiling points, then separation by fractional distillation will be possible.

(d) Yes.

(e) After separation of the diastereomeric esters by fractional distillation, they could each be hydrolyzed to yield the separate enantiomeric acids.

7.33

(a) 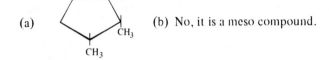   (b) Yes.

7.34

(a)   (b) No, it is a meso compound.

7.35

(a)

(b)

The BH₃ group may attack the double bond from either side of the ring

$$\downarrow H_2O_2,\ OH^-$$

7.36

(a) Four

(b)

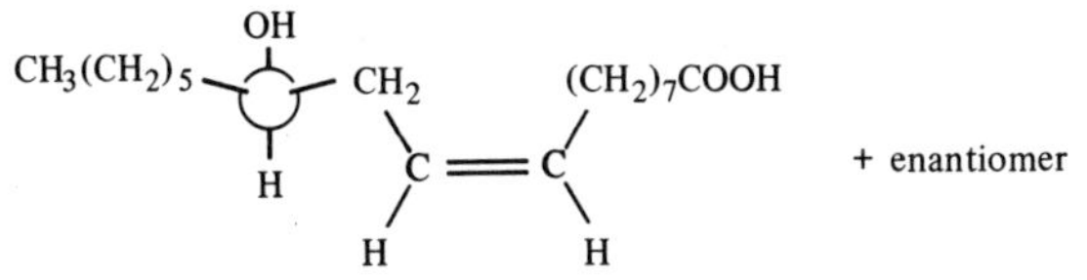

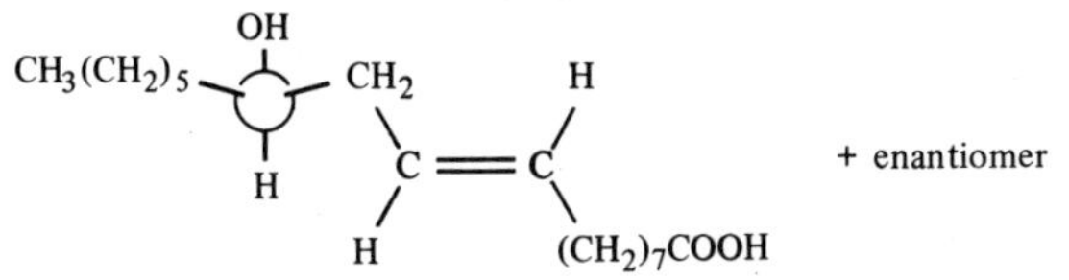

$CH_3(CH_2)_5$ — [structure with OH, H, CH_2, C=C, $(CH_2)_7COOH$, H, H] + enantiomer

$CH_3(CH_2)_5$ — [structure with OH, H, CH_2, C=C, H, H, $(CH_2)_7COOH$] + enantiomer

7.37

Hydroxylations by $KMnO_4$ are *syn* hydroxylations (cf p. 215). Thus, maleic acid must be the *cis*-dicarboxylic acid:

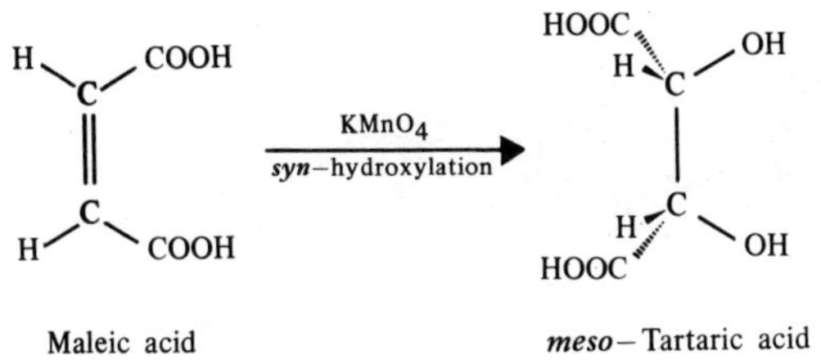

[structure: Maleic acid H, COOH / H, COOH with C=C]

$\xrightarrow[\text{syn—hydroxylation}]{KMnO_4}$

[structure: HOOC, H, OH / H, HOOC, OH]

Maleic acid   *meso*—Tartaric acid

Fumaric acid must be the *trans*-dicarboxylic acid:

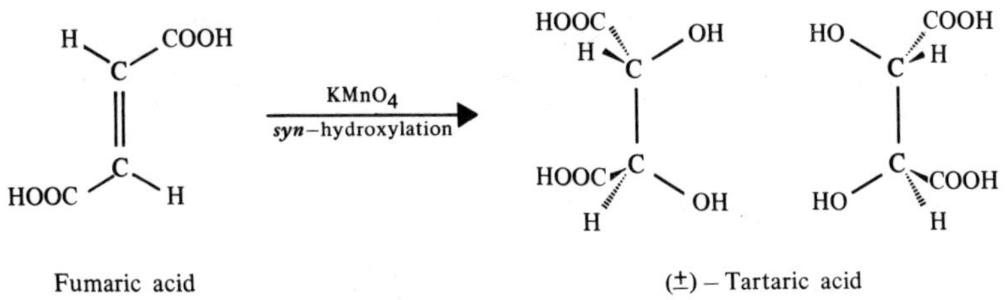

[structure: Fumaric acid H, COOH / HOOC, H with C=C]

$\xrightarrow[\text{syn—hydroxylation}]{KMnO_4}$

[structures: HOOC, H, OH / HOOC, H, OH   HO, COOH, H / HO, COOH, H]

Fumaric acid   (±) – Tartaric acid

7.38

(a) The addition of bromine is an *anti* addition. Thus fumaric acid yields a *meso* compound.

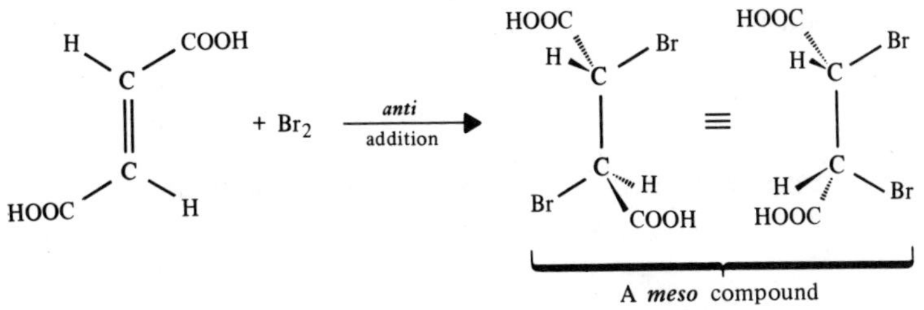

[structure: Fumaric acid H, COOH / HOOC, H with C=C] $+ Br_2$ $\xrightarrow[\text{addition}]{anti}$ [structures: HOOC, H, Br / Br, H, COOH ≡ HOOC, H, Br / H, HOOC, Br]

A *meso* compound

(b) Maleic acid adds bromine to yield a racemic modification.

7.39

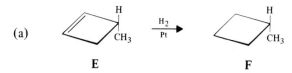

7.40

(a)

E

F

Optically active
(the enantiomeric form
is an equally valid
answer)

Optically inactive and
nonresolvable

(b) CH_3CH_2‑C=C=C‑ with CH_3 and H $\xrightarrow[\text{Pt}]{H_2}$ $CH_3CH_2CH_2CH_2CH_2CH_3$

G

H

Optically active
(the enantiomeric form is
an equally valid answer)

Optically inactive and
nonresolvable

7.41

That **I** and **J** rotate plane-polarized light in the same direction tells us that **I** and **J** are not enantiomers of each other. Thus, the following are possible structures for **I**, **J** and **K**. (The enantiomers of **I**, **J** and **K** would form another set of structures, and other answers are possible as well.)

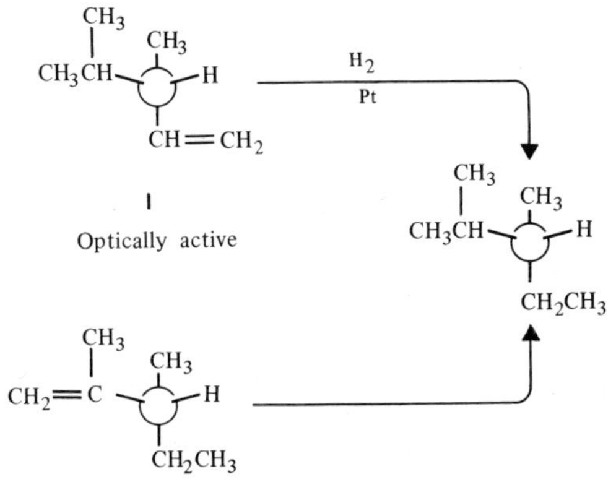

I
Optically active

K
Optically active

J
Optically active

7.42

Possible structures are:

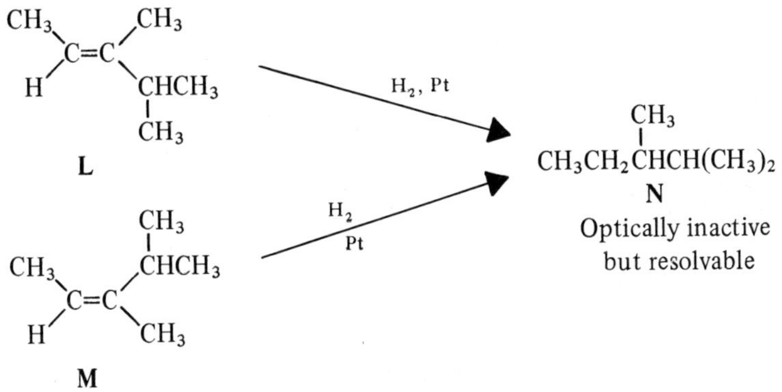

L

M

N
Optically inactive
but resolvable

(Other answers are possible as well.)

7.43

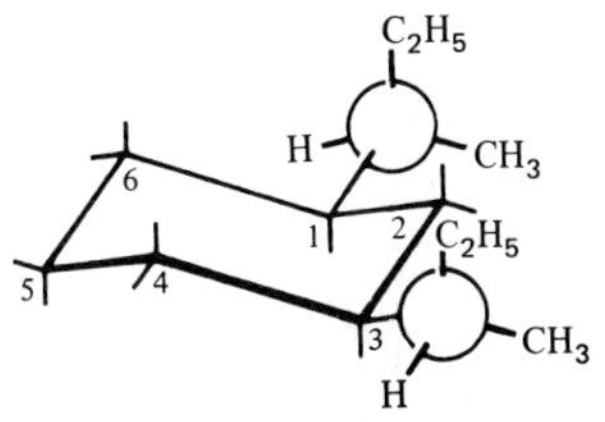

The (1R, 3S)-*cis*-di-*sec*-butylcyclohexane shown above is optically inactive because it is a *meso*-compound. Since the *sec*-butyl groups are of opposite chirality, it has a plane of symmetry perpendicular to the ring and passing through atoms 2 and 5.

7.44

The reactions proceed through the formation of bromonium ions identical to those formed in the bromination of *trans*- and *cis*-2-butene (see problem 7.19).

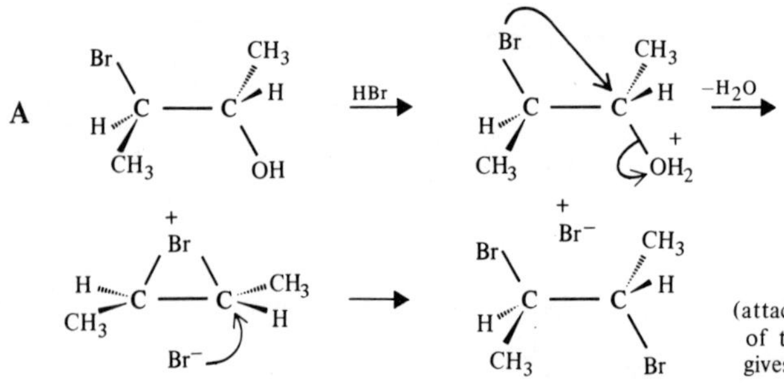

meso–2,3–dibromobutane

(attack at the other carbon of the bromonium ion gives the same product)

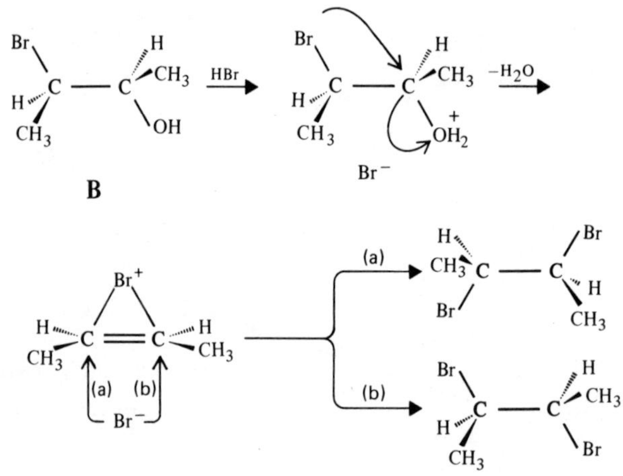

(±)–2,3–dibromobutane

8
SPECIAL TOPICS I

8.1

(a) $X \longrightarrow Y$, $K_{eq} = \dfrac{[Y]}{[X]} = 10$

Initial $[X] = 1.0$
Equilibrium $[Y] = a$
Equilibrium $[X] = 1.0 - a$

then $K_{eq} = 10 = \dfrac{a}{1.0 - a}$

$10 - 10a = a$

$-11a = -10$

$a = \dfrac{10}{11} = 0.91$ mole/liter

At equilibrium, $[Y] = 0.91$ mole/liter,
$\qquad\qquad\quad [X] = 0.09$ mole/liter,
and 91% of X is converted to product, Y.

(b) If $K_{eq} = 1$, $1 = \dfrac{a}{1.0 - a}$

$1 - a = a$

$-2a = -1$

$a = 0.5$

$\therefore$ At equilibrium, $[Y] = 0.5$ mole/liter,
$\qquad\qquad\qquad [X] = 0.5$ mole/liter,
and 50% of X is converted to product, Y.

(c) If $K_{eq} = 10^{-3}$, $10^{-3} = \dfrac{a}{1 - a}$

$10^{-3} - 10^{-3}\,a = a$

$-1.001a = -10^{-3}$

$a = \dfrac{10^{-3}}{1.001} \cong 10^{-3}$

At equilibrium, $[Y] = 10^{-3}$ mole/liter,
$\qquad\qquad\qquad [X] = 0.999$ mole/liter,
and 0.1% of X is converted to product, Y.

8.2

(a) $\Delta G = \Delta H - T\Delta S$

$\Delta G = -41700$ cal/mole $- 300$deg(-26.6 cal/deg mole)

$\Delta G = -41,700 + 7980 = -33,720$ cal/mole

or $\Delta G = -33.72$ kcal/mole

(b) Yes, because a negative value of ΔG tells us that the products are favored at equilibrium.

(c) No, a negative entropy tells us that the products are more ordered, and therefore less favored than the reactants.

(d) There are fewer degrees of freedom in the product molecule, ethene, than in the separate and independent molecules, ethyne and hydrogen.

8.3

(a) Yes, as shown by the negative value of ΔG.

(b) Yes, as shown by the positive value of ΔS.

(c) The extra degrees of freedom associated with rotation about the C—C single bond of ethane are not possible in ethene.

8.4

The reaction will proceed through the most stable free radical that can be produced. Head-to-head polymerization, as shown, will lead to a primary radical,

$$R-CH_2-\underset{\underset{CH_3}{|}}{CH}\cdot + \underset{\underset{CH_3}{|}}{CH}=CH_2 \longrightarrow R-CH_2-\underset{\underset{CH_3}{|}}{CH}-\underset{\underset{CH_3}{|}}{CH}-CH_2\cdot,$$

which is less stable than the secondary radical that is produced by head-to-tail polymerization:

$$R-CH_2-\underset{\underset{CH_3}{|}}{CH}\cdot + \underset{\underset{CH_3}{|}}{CH_2}=CH \longrightarrow R-CH_2-\underset{\underset{CH_3}{|}}{CH}-CH_2-\underset{\underset{CH_3}{|}}{CH}\cdot$$

8.5

(a) nCH$_2$=$\underset{\underset{F}{|}}{CH}$ $\xrightarrow[\text{peroxide}]{\text{organic}}$ $\left(\!\!CH_2-\underset{\underset{F}{|}}{CH}\!\!\right)_n$

(b) nCF$_2$=$\underset{\underset{Cl}{|}}{CF}$ $\xrightarrow[\text{peroxide}]{\text{organic}}$ $\left(\!\!CF_2-\underset{\underset{Cl}{|}}{CF}\!\!\right)_n$

(c) nCF$_2$=$\underset{\underset{CF_3}{|}}{CF}$ + mCH$_2$=CF$_2$ $\xrightarrow[\text{peroxide}]{\text{organic}}$ $\left(\!\!CF_2-\underset{\underset{CF_3}{|}}{CF}\!\!\right)_n\!\!\left(\!\!CH_2-CF_2\!\!\right)_m$

Note that the units are randomly ordered, and not necessarily joined to their own kind as shown.

8.6

Polymerization will occur to produce the most stable carbocation possible. The scheme shown in this problem involves formation of the primary carbocations,

$$
\underset{\underset{CH_3}{|}}{\overset{\overset{CH_3}{|}}{CH}}-CH_2\overset{+}{} \; , \quad \underset{\underset{CH_3}{|}}{\overset{\overset{CH_3}{|}}{CH}}-CH_2-\underset{\underset{CH_3}{|}}{\overset{\overset{CH_3}{|}}{C}}-CH_2\overset{+}{} \; , \text{etc.}
$$

instead of the tertiary carbocations,

$$
CH_3-\underset{\underset{CH_3}{|}}{\overset{\overset{CH_3}{|}}{C}}-CH_2-\underset{\underset{CH_3}{|}}{\overset{\overset{CH_3}{|}}{C}}\!{\scriptstyle +}
$$

8.7

(a) By proton transfer from water to the strongly basic carbanion,

$$
\underset{\underset{CN}{|}}{R-CH_2-CH}\!:^- \; + \; H-\underset{\underset{H}{|}}{\ddot{O}}:\; \longrightarrow \; R-CH_2-\underset{\underset{CN}{|}}{CH_2} \; + \; :\ddot{O}H^-
$$

(b) $\underset{\underset{R}{|}}{(CH_2CH)}\!\!\!\!\overset{}{\underset{n}{}}\!\!-CH_2\underset{\underset{R}{|}}{CH}\!:^- \; + \; (m+1)CH_2\!\!\!\overset{\diagdown}{\underset{O}{}}\!\!\!\!\diagup\!CH_2 \; \longrightarrow$

$$
\underset{\underset{R}{|}}{(CH_2CH)}\!\!\overset{}{\underset{n+1}{}}\!\!(CH_2-CH_2-O)\!\!\overset{}{\underset{m}{}}\!\!CH_2-CH_2-\ddot{O}\!:^-
$$

$$
\overset{H_2O}{\longrightarrow} \underset{\underset{R}{|}}{(CH_2CH)}\!\!\overset{}{\underset{n+1}{}}\!\!(CH_2-CH_2-O)\!\!\overset{}{\underset{m}{}}\!\!CH_2-CH_2-OH
$$

In this polymer, each chain consists of a long uninterrupted segment of the first repeating unit, $\underset{\underset{R}{|}}{(CH_2CH)}\!\!\overset{}{\underset{n+1}{}}$, followed by a long uninterrupted segment of the second repeating unit, $(CH_2-CH_2-O)\!\!\overset{}{\underset{m}{}}$.

8.8

(a)

(b)

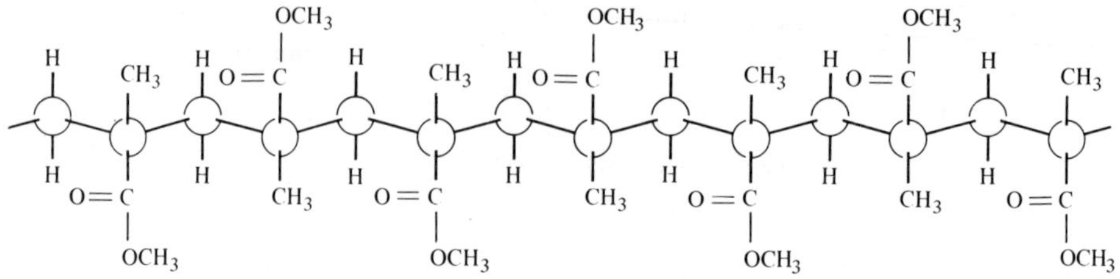

(c)

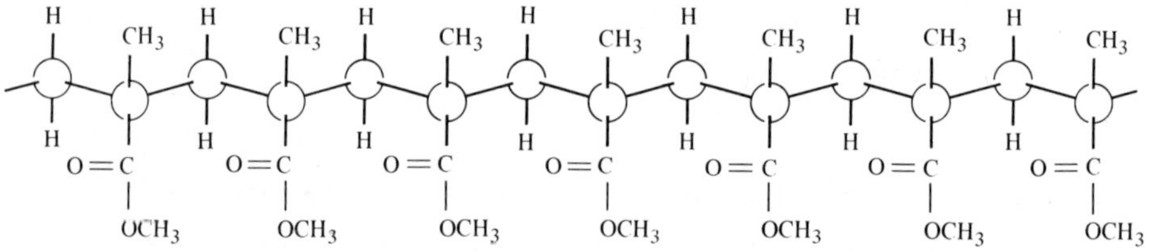

8.9
The singlet methylene reacts with the double bond at the same rate from either side as shown:

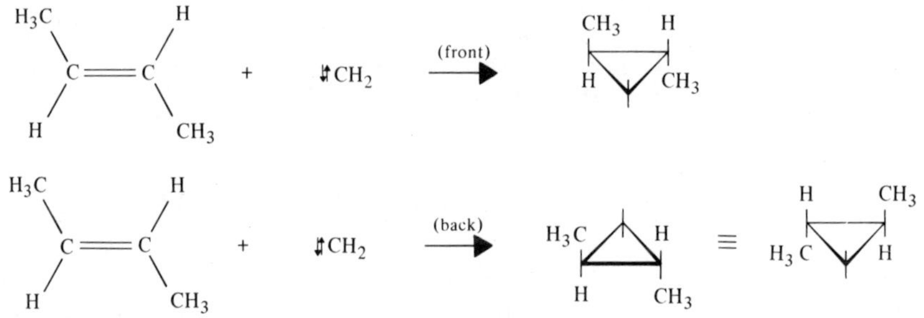

8.10

(a)

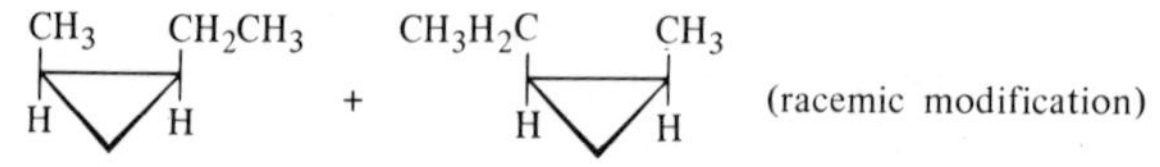

(b)

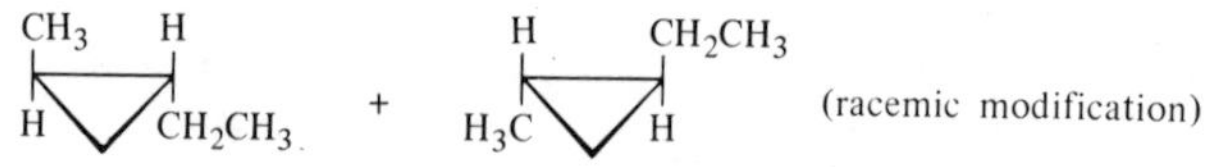

(c)

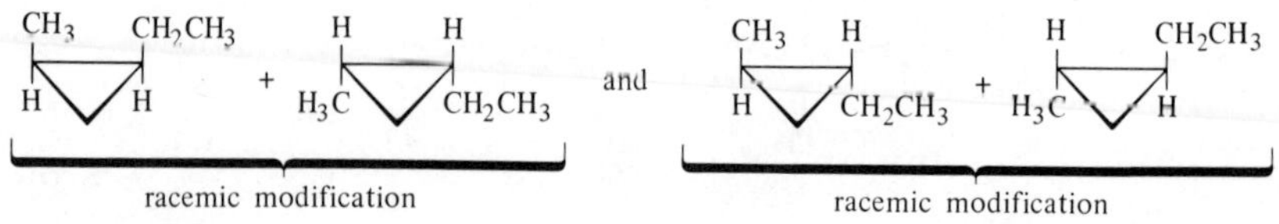

(d) Same mixture as in (c).

8.11
Recombination of radicals should yield ethane, butane, hexane, isobutane, 2-methylpentane, 2,3-dimethylbutane.

8.12

(a)

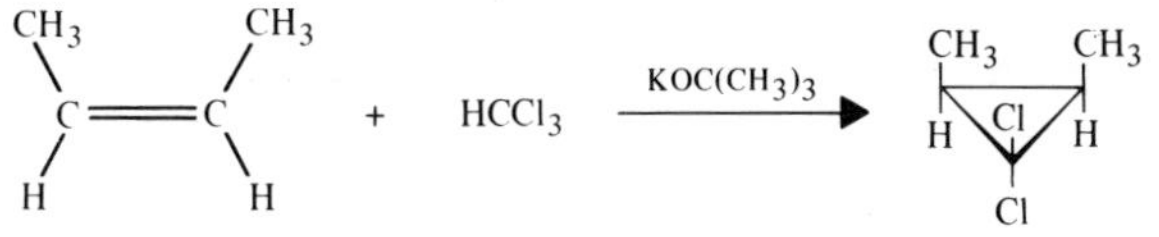

(b)

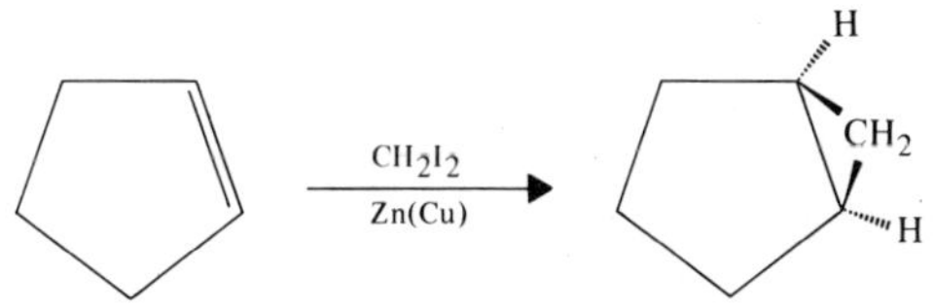

(c)

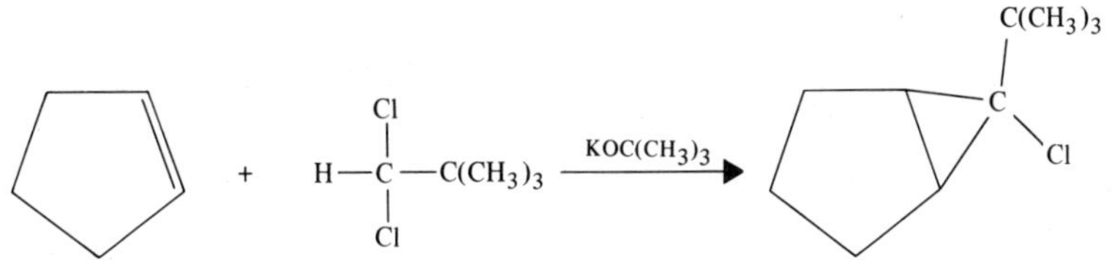

8.13
The secondary carbocation that is formed initially in the polymerization reaction rearranges to a tertiary carbocation (via a hydride shift) before each new monomer unit is added to the growing chain.

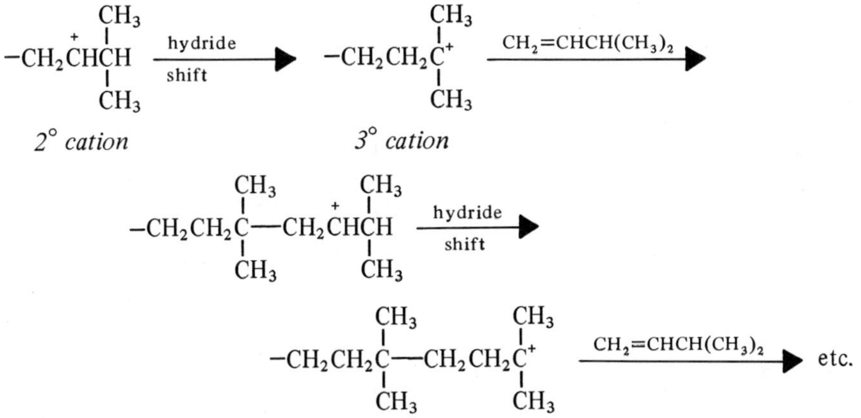

9 ALKYNES

9.1

(a) C_4H_6 : $CH_3CH_2C{\equiv}CH$, $CH_3C{\equiv}CCH_3$

1-Butyne 2-Butyne

(b) C_5H_8 : $CH_3CH_2CH_2C{\equiv}CH$ $CH_3CH_2C{\equiv}CCH_3$

1-Pentyne 2-Pentyne

$$\begin{array}{c} CH_3 \\ | \\ CH_3CHC{\equiv}CH \end{array}$$

3-Methyl-1-butyne

(c) $CH_3CH_2CH_2CH_2C{\equiv}CH$ $CH_3CH_2CH_2C{\equiv}CCH_3$

1-Hexyne 2-Hexyne

$CH_3CH_2C{\equiv}CCH_2CH_3$

3-Hexyne

$$\begin{array}{l} CH_3CHCH_2C{\equiv}CH \\ \quad\;\; | \\ \quad\; CH_3 \end{array}$$

4-Methyl-1-pentyne

$$\begin{array}{l} CH_3C{\equiv}CCHCH_3 \\ \qquad\quad\;\; | \\ \qquad\quad CH_3 \end{array}$$

4-Methyl-2-pentyne

$$\begin{array}{l} HC{\equiv}CCHCH_2CH_3 \\ \qquad\;\; | \\ \qquad CH_3 \end{array}$$

3-Methyl-1-pentyne

$$\begin{array}{c} CH_3 \\ | \\ HC{\equiv}CCCH_3 \\ | \\ CH_3 \end{array}$$

3,3-Dimethyl-
1-Butyne

9.2

(a) $HC{\equiv}CH + :\ddot{N}H_2^- \rightleftarrows HC{\equiv}C:^- + :NH_3$

stronger acid	stronger base	weaker base	weaker acid

(No appreciable amount of reactants are present at equilibrium.)

(b) $CH_2=CH_2$ + $:\overset{..}{N}H_2^-$ $\longleftrightarrow$ $CH_2=\overset{..}{C}H^-$ + $:NH_3$ $\left(\begin{array}{l}\text{No appreciable amount of pro-}\\\text{ducts are present at equilibrium.}\end{array}\right)$

    weaker   weaker   stronger   stronger

    acid   base   base   acid

(c) CH_3CH_3 + $:\overset{..}{N}H_2^-$ $\longleftrightarrow$ $CH_3\overset{..}{C}H_2^-$ + $:NH_3$ $\left(\begin{array}{l}\text{No appreciable amount of pro-}\\\text{ducts are present at equilibrium.}\end{array}\right)$

    weaker   weaker   stronger   stronger

    acid   base   base   acid

(d) $HC\equiv C:^-$ + $CH_3CH_2\overset{..}{O}H$ $\longleftarrow$ $HC\equiv CH$ + $CH_3CH_2\overset{..}{\underset{..}{O}}:^-$ $\left(\begin{array}{l}\text{No appreciable amount of}\\\text{reactants are present at equi-}\\\text{librium.}\end{array}\right)$

    stronger   stronger   weaker   weaker

    base   acid   acid   base

(e) $HC\equiv C:^-$ + $H\overset{|}{\underset{H}{\overset{..}{O}}}:$ $\longleftarrow$ $HC\equiv CH$ + $:\overset{..}{O}H^-$ $\left(\begin{array}{l}\text{No appreciable amount of}\\\text{reactants are present at equi-}\\\text{librium.}\end{array}\right)$

    stronger   stronger   weaker   weaker

    base   acid   acid   base

9.3

$\overset{\overset{\displaystyle CH_3}{|}}{CH_3CH}\overset{\overset{\displaystyle CH_3}{|}}{—CH}—OH$ from the Sia_2BH.

9.4

(a) $CH_3CH_2C\equiv CH \xrightarrow[\substack{HgSO_4\\H_2SO_4}]{H_2O} CH_3CH_2\overset{\overset{\displaystyle O}{\|}}{C}CH_3$

(b) $\square\!\!-C\equiv CH$ + $Sia_2BH \xrightarrow{0°} \square\!\!-CH=CH-BSia_2 \xrightarrow[OH^-]{H_2O_2} \square\!\!-CH_2\overset{\overset{\displaystyle O}{\|}}{C}H$

(c) $3CH_3C\equiv CCH_3$ + $\frac{1}{2}(BD_3)_2 \xrightarrow{0°} \left(\underset{\displaystyle D}{\overset{\displaystyle CH_3}{C}}=\underset{\displaystyle B}{\overset{\displaystyle CH_3}{C}}\right)_3 \xrightarrow[0°]{CH_3COOD} \underset{\displaystyle D}{\overset{\displaystyle CH_3}{C}}=\underset{\displaystyle D}{\overset{\displaystyle CH_3}{C}}$

or $CH_3C\equiv C-CH_3$ + $D_2 \xrightarrow{Ni_2B(P-2)} \underset{\displaystyle D}{\overset{\displaystyle CH_3}{C}}=\underset{\displaystyle D}{\overset{\displaystyle CH_3}{C}}$

(d) $CH_3CH_2C\equiv CH$ + $Sia_2BH \longrightarrow \underset{\displaystyle H}{\overset{\displaystyle CH_3CH_2}{C}}=\underset{\displaystyle BSia_2}{\overset{\displaystyle H}{C}} \xrightarrow{CH_3COOD} \underset{\displaystyle H}{\overset{\displaystyle CH_3CH_2}{C}}=\underset{\displaystyle D}{\overset{\displaystyle H}{C}}$

9.5

$$CH_3-\underset{\underset{CH_3}{|}}{\overset{\overset{CH_3}{|}}{C}}-C\equiv CH + NaNH_2 \longrightarrow CH_3-\underset{\underset{CH_3}{|}}{\overset{\overset{CH_3}{|}}{C}}-C\equiv C\overset{-}{:} Na^+ + NH_3$$

$$\xrightarrow[\text{CH}_3\text{CH}_2\text{Br}]{} \quad CH_3-\underset{\underset{CH_3}{|}}{\overset{\overset{CH_3}{|}}{C}}-C\equiv C-CH_2-CH_3$$

A reaction between $CH_3CH_2C\equiv C\overset{-}{:}\overset{+}{Na}$ and $CH_3-\underset{\underset{CH_3}{|}}{\overset{\overset{CH_3}{|}}{C}}-Br$ would result in elimination to pro-

duce $CH_2=\underset{\underset{CH_3}{|}}{C}-CH_3 + CH_3CH_2C\equiv CH.$

9.6

(a)

$$\text{Cyclohexyl}-C\equiv CH \xrightarrow[\substack{HgSO_4 \\ H_2SO_4}]{H_2O} \text{Cyclohexyl}-\overset{\overset{O}{\|}}{C}-CH_3$$

(b)

$$\text{Cyclohexyl}-\overset{\overset{O}{\|}}{C}-CH_3 \xrightarrow[O^\circ]{PCl_5} \text{Cyclohexyl}-\underset{\underset{Cl}{|}}{\overset{\overset{Cl}{|}}{C}}-CH_3 \xrightarrow[\substack{\text{mineral} \\ \text{oil, heat}}]{NaNH_2} \text{Cyclohexyl}-C\equiv CH$$

$$\text{Cyclohexyl}-C\equiv CH + Sia_2BH \longrightarrow \text{Cyclohexyl}-CH=CH-BSia_2 \xrightarrow[OH^-]{H_2O_2} \text{Cyclohexyl}-CH_2\overset{\overset{O}{\|}}{C}H$$

9.7

(a) $CH_3\underset{\underset{CH_3}{|}}{C}HC\equiv CCH_2CH_3$

2-Methyl-3-hexyne

(b)

Cyclooctyne

(c) $HC\equiv CCH_2CH_2CH_2CH_2CH_3$

1-Heptyne

9.8

(a) 3-Methyl-1-butyne

(b) 2,2-Dimethyl-3-hexyne

(c) 3-Nonyne

(d) 3-Hexyne

(e) 2,2,5,5-Tetramethyl-3-hexyne

(f) 2,5-Dimethyl-3-hexyne

(g) 2-Hexyne

(h) 4-Methyl-2-hexyne

(i) 2,7-Dimethyl-4-octyne

(j) 1-Octyne

73
ALKYNES

9.9

a, j

9.10

(a) d, g   (b) None

9.11

(a) $3\,C + CaO \xrightarrow{2500°} CaC_2 + CO$
  (coke) (lime)

  $CaC_2 + 2H_2O \xrightarrow[\text{temperature}]{\text{room}} HC\equiv CH + Ca(OH)_2$

(b) $HC\equiv CH + H_2 \xrightarrow{Ni_2B} CH_2=CH_2$

(c) $HC\equiv CH + NaNH_2 \xrightarrow[NH_3]{\text{liquid}} HC\equiv C:^-Na^+ + NH_3$

  $CH_4 + Br_2 \xrightarrow[\text{heat}]{\text{light}} CH_3Br + HBr$
  (excess)

  $HC\equiv C:^-Na^+ + CH_3Br \longrightarrow HC\equiv C-CH_3 + Na^+Br^-$

(d) $CH_3C\equiv CH + H_2 \xrightarrow{Ni_2B} CH_3CH=CH_2$

(e) $CH_3C\equiv CH + H_2O \xrightarrow[H_2SO_4]{HgSO_4} CH_3\overset{\overset{\text{O}}{\|}}{C}CH_3$

(f) $CH_3C\equiv CH + NaNH_2 \xrightarrow[NH_3]{\text{liquid}} CH_3C\equiv C:^-Na^+ + NH_3$

  $\xrightarrow{CH_3Br} CH_3C\equiv CCH_3 + Na^+Br^-$

(g) $CH_2=CH_2 + HBr \longrightarrow CH_3CH_2Br \xrightarrow{HC\equiv \bar{C}:Na^+} HC\equiv CCH_2CH_3 + Na^+Br^-$

(h) $CH_3C\equiv CCH_3 + H_2O \xrightarrow[H_2SO_4]{HgSO_4} CH_3\overset{\overset{\text{O}}{\|}}{C}CH_2CH_3$

(i) See (g) above.

(j) $CH_3C\equiv C:^-Na^+ + CH_3CH_2Br \longrightarrow CH_3C\equiv CCH_2CH_3 + Na^+Br^-$

(k) $CH_2=CH_2 + H_2O \xrightarrow{H_2SO_4} CH_3CH_2OH$

(l) $CH_2=CH_2 + Br_2 \xrightarrow[\text{dark}]{CCl_4} CH_2BrCH_2Br$

(m) $CH_3C\equiv CH + 2HCl \longrightarrow CH_3-\overset{\overset{\text{Cl}}{|}}{\underset{\underset{\text{Cl}}{|}}{C}}-CH_3$

(n) $CH_3C{\equiv}CCH_3 + H_2 \xrightarrow{Ni_2B}$
$$\underset{H}{\overset{CH_3}{>}}C{=}C\underset{H}{\overset{CH_3}{<}}$$

(o) $CH_3C{\equiv}CCH_3 \xrightarrow[-78°]{Li\ +\ C_2H_5NH_2}$
$$\underset{H}{\overset{CH_3}{>}}C{=}C\underset{CH_3}{\overset{H}{<}}$$

(p) $CH_3CH_2C{\equiv}CH + H_2 \xrightarrow{Ni_2B} CH_3CH_2CH{=}CH_2$

(q) $CH_3CH_2CH{=}CH_2 + HBr \xrightarrow[inhibitor]{peroxide} CH_3CH_2\overset{Br}{\underset{|}{C}}HCH_3$

(r) $CH_3CH_2CH{=}CH_2 + HBr \xrightarrow{peroxide} CH_3CH_2CH_2CH_2Br$

(s) $CH_3CH_2CH{=}CH_2 + (BH_3)_2 \longrightarrow (CH_3CH_2CH_2CH_2)_3B \xrightarrow[OH^-]{H_2O_2}$
$$CH_3CH_2CH_2CH_2OH$$

(t) $CH_3CH_2CH{=}CH_2 + Hg(OAc)_2 \longrightarrow CH_3CH_2\underset{HgOAc}{\underset{|}{C}}HCH_3 \xrightarrow[OH^-]{NaBH_4} CH_3CH_2\underset{OH}{\underset{|}{C}}HCH_3$

or $CH_3CH_2CH{=}CH_2 + H_2O \xrightarrow{H_2SO_4} CH_3CH_2\underset{OH}{\underset{|}{C}}H{-}CH_3$

(u)
$$\underset{H}{\overset{H_3C}{>}}C{=}C\underset{CH_3}{\overset{H}{<}} + Br_2 \xrightarrow[dark]{CCl_4}$$

(v)
$$\underset{H}{\overset{H_3C}{>}}C{=}C\underset{H}{\overset{CH_3}{<}} + Br_2 \xrightarrow[dark]{CCl_4}$$ + enantiomer

(w) $CH_3C{\equiv}CCH_3 + HCl \xrightarrow[CH_3COOH]{Cl^-,\ 25°}$
$$\underset{Cl}{\overset{CH_3}{>}}C{=}C\underset{CH_3}{\overset{H}{<}}$$

(x) $CH_3CH{=}CH_2 + HBr \xrightarrow{peroxide} CH_3CH_2CH_2Br$

(y) $CH_3CH_2C{\equiv}CH + NaNH_2 \xrightarrow[NH_3]{liquid} CH_3CH_2C{\equiv}C:^-\overset{+}{Na} + NH_3$
$$\xrightarrow{CH_3CH_2Br} CH_3CH_2C{\equiv}CCH_2CH_3$$

(z) $CH_3CH_2C\equiv CH + HBr \xrightarrow{peroxide} CH_3CH_2CH=CHBr$

9.12

(a)
$$\begin{array}{ccc} CH_3CH_2CH_2 & & Br \\ & C=C & \\ Br & & H \end{array}$$

(b)
$$\begin{array}{cc} CH_3CH_2CH_2 & \\ & C=CH_2 \\ Cl & \end{array}$$

(c) $CH_3CH_2CH_2\overset{\overset{\displaystyle Cl}{|}}{\underset{\underset{\displaystyle Cl}{|}}{C}}CH_3$

(d) $CH_3CH_2CH_2CH=CHBr$

(e) $CH_3CH_2CH_2\overset{\overset{\displaystyle O}{\|}}{C}CH_3$

(f) $CH_3CH_2CH_2CH=CH_2$

(g) $CH_3CH_2CH_2CH=CH_2$

(h) $CH_3CH_2CH_2CH_2\overset{\overset{\displaystyle O}{\|}}{C}H$

(i) $CH_3CH_2CH_2C\equiv C\!:\overset{-}{}\,\overset{+}{Na}$

(j) $CH_3CH_2CH_2C\equiv CCH_3$

(k) $CH_3CH_2CH_2C\equiv C-C\equiv CCH_2CH_2CH_3$

(l) $CH_3CH_2CH_2C\equiv CAg$

(m) $CH_3CH_2CH_2C\equiv CCu$

(n) $CH_3CH_2CH_2COOH$
$\qquad\qquad + CO_2$

9.13

(a)
$$\begin{array}{ccc} CH_3CH_2 & & H \\ & C=C & \\ Cl & & CH_2CH_3 \end{array}$$

(b) $CH_3CH_2\overset{\overset{\displaystyle Cl}{|}}{\underset{\underset{\displaystyle Cl}{|}}{C}}CH_2CH_2CH_3$

(c)
$$\begin{array}{ccc} CH_3CH_2 & & Br \\ & C=C & \\ Br & & CH_2CH_3 \end{array}$$

(d) $CH_3CH_2\overset{\overset{\displaystyle Br}{|}}{\underset{\underset{\displaystyle Br}{|}}{C}}-\overset{\overset{\displaystyle Br}{|}}{\underset{\underset{\displaystyle Br}{|}}{C}}CH_2CH_3$

(e)
$$\begin{array}{ccc} CH_3CH_2 & & CH_2CH_3 \\ & C=C & \\ H & & H \end{array}$$

(f) Same as (e)

(g)
$$\begin{array}{ccc} CH_3CH_2 & & H \\ & C=C & \\ H & & CH_2CH_3 \end{array}$$

(h) $CH_3CH_2\overset{\overset{\displaystyle O}{\|}}{C}CH_2CH_2CH_3$

(i) No reaction

(j) No reaction

(k) Same as (e)

(l) Same as (h)

(m) No reaction

(n) $CH_3CH_2CH_2CH_2CH_2CH_3$

(o) $2CH_3CH_2COOH$

(p) $2CH_3CH_2COOH$

(q) No reaction

9.14

(a) $CH_3CH_2CH_2CH=CH_2 + Br_2 \longrightarrow CH_3CH_2CH_2\overset{\underset{\underset{\displaystyle Br}{|}}{}}{C}HCH_2Br$

$\xrightarrow[\text{liq } NH_3]{2NaNH_2} CH_3CH_2CH_2C\equiv CH$

(b) $CH_3CH_2CH_2CH_2CH_2Cl \xrightarrow[CH_3CH_2OH]{KOH} CH_3CH_2CH_2CH=CH_2$

then proceed as in (a) above.

(c) $CH_3CH_2CH_2CH=CHCl \xrightarrow[NH_3]{NaNH_2} CH_3CH_2CH_2C\equiv CH$

(d) $CH_3CH_2CH_2CH_2CHCl_2 \xrightarrow[NH_3]{2NaNH_2} CH_3CH_2CH_2C\equiv CH$

(e) $HC\equiv CH \xrightarrow[liq.\ NH_3]{NaNH_2} HC\equiv C:^-Na^+ \xrightarrow{CH_3CH_2CH_2Br} HC\equiv CCH_2CH_2CH_3$

9.15

(a) Propyne is soluble in cold, concentrated H_2SO_4; propane is not. Other tests are Br_2/CCl_4 and $KMnO_4^-/H_2O$.

(b) $Ag(NH_3)_2^+ OH^-$ gives a precipitate with propyne, not with propene.

(c) Dilute $KMnO_4$ oxidizes 1-bromopropene and not 2-bromopropane.

(d) $Ag(NH_3)_2^+ OH^-$ gives a precipitate with 1-butyne, not with 2-bromo-2-butene.

(e) Sodium fusion followed by acidification with dilute HNO_3 and addition of $AgNO_3$ gives a AgBr precipitate with 2-bromo-2-butene, not with 2-butyne.

(f) Br_2/CCl_4 is decolorized by 2-butyne, not by *n*-butyl alcohol.

(g) $AgNO_3/C_2H_5OH$ gives a AgBr precipitate wth 2-bromobutane, not with 2-butyne.

(h) Br_2/CCl_4 is decolorized by $CH_3C\equiv CCH_2OH$, not by $CH_3CH_2CH_2CH_2OH$.

(i) Br_2/CCl_4 is decolorized by $CH_3CH=CHCH_2OH$, not by $CH_3CH_2CH_2CH_2OH$.

(In many cases above other tests are possible.)

9.16

(a) A = $CH_3CH_2CH_2C\equiv CH$, B = $CH_3CH_2C\equiv CCH_3$, C =

(b) Yes, B may also be $CH_3CH=CH-CH=CH_2$ or $CH_2=CH-CH_2-CH=CH_2$.

C may also be CH_3 , CH_3 , or CH_2 or CH_2CH_3 , etc.

(c) B = $CH_3CH_2C\equiv CCH_3$

(d)

9.17

(a) $CH_3\overset{\underset{\displaystyle |}{CH_3}}{C}HC\equiv CH$ + HCl (1 mole) $\longrightarrow$ $CH_3\overset{\underset{\displaystyle |}{CH_3}}{C}HC=CH_2$ (with Cl below)

(b) $CH_3\overset{\underset{\displaystyle |}{CH_3}}{C}HC\equiv CH \xrightarrow[quinoline]{H_2/Pd,BaSO_4} CH_3\overset{\underset{\displaystyle |}{CH_3}}{C}HCH=CH_2 \xrightarrow[peroxides]{HBr} CH_3\overset{\underset{\displaystyle |}{CH_3}}{C}HCH_2CH_2Br$

(c) Product of (a) $\xrightarrow{H_2/Ni}$ $CH_3\overset{\overset{\displaystyle CH_3}{|}}{\underset{\underset{\displaystyle Cl}{|}}{CH}}CHCH_3$

(d) Product of (a) $\xrightarrow[\text{dark}]{Cl_2/CCl_4}$ $CH_3\overset{\overset{\displaystyle CH_3}{|}}{CH}-\overset{\overset{\displaystyle Cl}{|}}{\underset{\underset{\displaystyle Cl}{|}}{C}}-CH_2Cl$

(e) Product of (a) $\xrightarrow[\substack{\text{peroxide}\\\text{inhibitor}}]{HBr}$ $CH_3\overset{\overset{\displaystyle CH_3}{|}}{CH}-\overset{\overset{\displaystyle Cl}{|}}{\underset{\underset{\displaystyle Br}{|}}{C}}-CH_3$

(f) $CH_3\overset{\overset{\displaystyle CH_3}{|}}{CH}C\equiv CH \xrightarrow[\text{then}\ \ H^+]{KMnO_4/OH^-,}$ $CH_3\overset{\overset{\displaystyle CH_3}{|}}{CH}COOH + CO_2$

9.18

(a) $CH_3CH_2CH_2C\equiv CH + D_2 \xrightarrow{Ni_2B}$

$$\underset{D}{\overset{CH_3CH_2CH_2}{\diagdown}}C=C\underset{D}{\overset{H}{\diagup}}$$

(b) $CH_3CH_2CH_2C\equiv CH + Sia_2BH \longrightarrow$

$$\underset{H}{\overset{CH_3CH_2CH_2}{\diagdown}}C=C\underset{BSia_2}{\overset{H}{\diagup}}$$

$\xrightarrow{CH_3COOD}$ $\underset{H}{\overset{CH_3CH_2CH_2}{\diagdown}}C=C\underset{D}{\overset{H}{\diagup}}$

(c) $CH_3CH_2CH_2C\equiv CH + DCl \longrightarrow$

$$\underset{Cl}{\overset{CH_3CH_2CH_2}{\diagdown}}C=C\underset{H}{\overset{D}{\diagup}}$$

(d) $CH_3CH_2CH_2C\equiv CH + Sia_2BD \longrightarrow$ $\underset{D}{\overset{CH_3CH_2CH_2}{\diagdown}}C=C\underset{BSia_2}{\overset{H}{\diagup}}$

$\xrightarrow[OH^-]{H_2O_2}$ $CH_3CH_2CH_2\overset{\overset{\displaystyle O}{\|}}{\underset{\underset{\displaystyle D}{|}}{CH}}CH$

9.19
The syntheses involve combinations of chlorination, dehydrochlorination, and hydrogenation as follows:

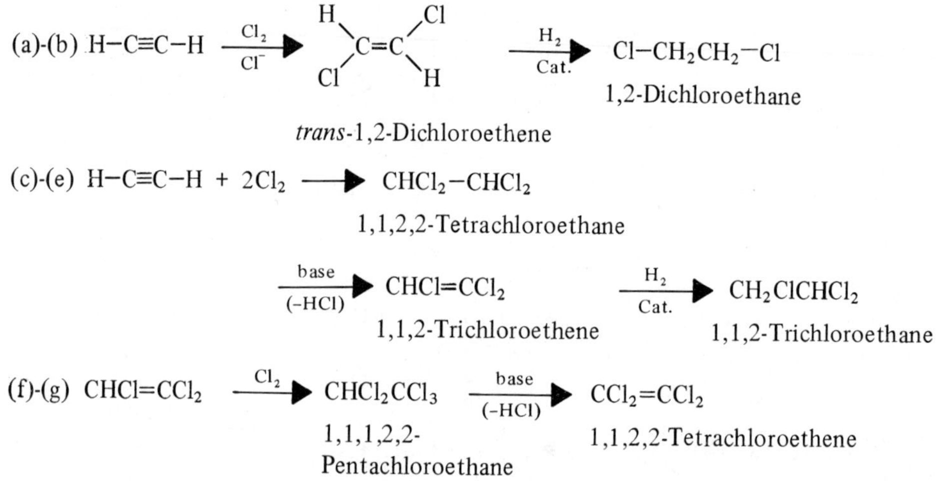

(a)-(b) $H-C\equiv C-H$ $\xrightarrow[Cl^-]{Cl_2}$

trans-1,2-Dichloroethene

$\xrightarrow{H_2}{Cat.}$ $Cl-CH_2CH_2-Cl$

1,2-Dichloroethane

(c)-(e) $H-C\equiv C-H + 2Cl_2 \longrightarrow CHCl_2-CHCl_2$

1,1,2,2-Tetrachloroethane

$\xrightarrow[(-HCl)]{base}$ $CHCl=CCl_2$

1,1,2-Trichloroethene

$\xrightarrow[Cat.]{H_2}$ $CH_2ClCHCl_2$

1,1,2-Trichloroethane

(f)-(g) $CHCl=CCl_2$ $\xrightarrow{Cl_2}$ $CHCl_2CCl_3$

1,1,1,2,2-Pentachloroethane

$\xrightarrow[(-HCl)]{base}$ $CCl_2=CCl_2$

1,1,2,2-Tetrachloroethene

9.20

(a) $(BH_3)_2 \rightleftharpoons 2BH_3$

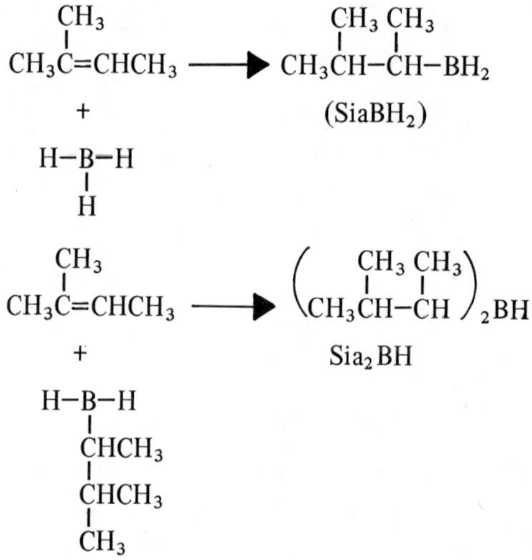

(SiaBH$_2$)

$\left(\begin{array}{c} CH_3\ CH_3 \\ | \quad | \\ CH_3CH-CH \end{array}\right)_2 BH$

Sia$_2$BH

(b) Steric hindrance. It is difficult to place three bulky groups around the boron atom.

9.21

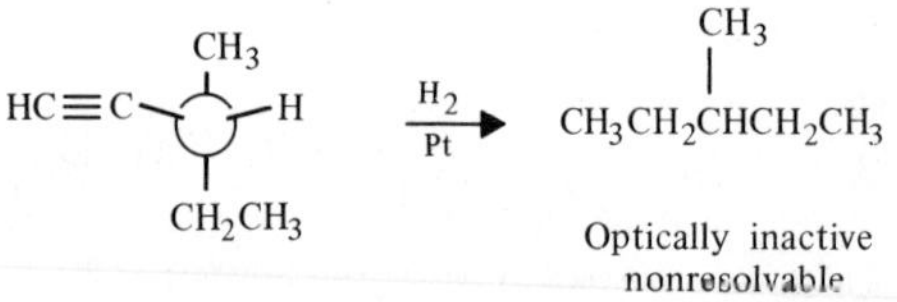

D

Optically active
(the other enantiomer
is an equally valid
answer)

$\xrightarrow{H_2}{Pt}$

$CH_3CH_2CHCH_2CH_3$ with CH$_3$ branch

Optically inactive
nonresolvable

9.22

Ordinary alkenes *are* more reactive toward electrophilic reagents. But, the alkenes obtained from the addition of an electrophilic reagent to an alkyne have at least one electronegative atom (Cl, Br, etc.) attached to a carbon of the double bond.

$$-C{\equiv}C- \xrightarrow{\text{HX}} \underset{H}{\overset{}{>}}C{=}C\overset{X}{<}$$

or

$$-C{\equiv}C- \xrightarrow{X_2} \underset{X}{\overset{}{>}}C{=}C\overset{X}{<}$$

These alkenes are less reactive than alkynes toward electrophilic addition because the electronegative group makes the double bond "electron poor."

9.23

(a) That hydrogenation of erythrogenic acid produces the unbranched product, $CH_3(CH_2)_{16}COOH$, tells us that the carbon chain of erythrogenic acid is unbranched. That five moles of hydrogen are absorbed tells us that the carbon chain of erythrogenic acid contains combinations of two triple bonds and one double bond, one triple bond and three double bonds, or five double bonds.

(b) $CH_2{=}CHCH_2C{\equiv}C{-}C{\equiv}C{-}(CH_2)_{10}COOH$

or

$CH_3CH{=}CHC{\equiv}C{-}C{\equiv}C(CH_2)_{10}COOH$

or

$CH_2{=}CHCH{=}CHCH{=}CHC{\equiv}C(CH_2)_9COOH$

or

$CH_3C{\equiv}CCH_2CH{=}CHCH_2CH{=}CHCH_2CH{=}CH(CH_2)_5COOH$

etc.

(c) Either of the compounds shown below.

$CH_2{=}CH(CH_2)_4C{\equiv}C{-}C{\equiv}C(CH_2)_7COOH$

or

$CH_2{=}CHC{\equiv}C(CH_2)_4C{\equiv}C(CH_2)_7COOH$

(d) Here we have an oxidative coupling of two different alkynes.

$$CH_2{=}CH(CH_2)_4C{\equiv}CH + HC{\equiv}C(CH_2)_7COOH \xrightarrow[O_2]{\text{CuCl, NH}_3}$$

$CH_2{=}CH(CH_2)_4C{\equiv}C{-}C{\equiv}C(CH_2)_4CH{=}CH_2$ + $CH_2{=}CH(CH_2)_4C{\equiv}C{-}C{\equiv}C(CH_2)_7COOH$ +

Nonacidic product　　　　　　　　　　　　　　　　Erythrogenic acid

$HOOC(CH_2)_7C{\equiv}C{-}C{\equiv}C(CH_2)_7COOH$

Dicarboxylic acid

10

CONJUGATED UNSATURATED SYSTEMS VISIBLE - ULTRAVIOLET SPECTROSCOPY

10.1

(a) $^{14}CH_2=CHCH_2X$ and $CH_2-CH^{14}CH_2X$

(b) The reaction proceeds through the resonance stabilized free radical,

$$^{14}\overset{\bullet}{CH_2}=CH-CH_2 \longleftrightarrow \quad ^{14}CH_2-CH=\overset{\bullet}{CH_2} \text{ or } ^{14}\overset{\delta\bullet}{CH_2}=CH=\overset{\delta\bullet}{CH_2}$$

Thus attack on X_2 can occur by the carbon at either end of the chain since they are equivalent.

(c) 50:50

10.2

(a)
$$\underset{D}{\overset{4}{CH_3}-\overset{3}{\underset{+}{CH}}\diagup\overset{2}{CH}\diagdown\overset{1}{CH_2}} \longleftrightarrow \underset{E}{\overset{4}{CH_3}-\overset{3}{\underset{+}{CH}}\diagup\overset{2}{CH}\diagdown\overset{1}{CH_2}} \text{ or } \underset{F}{\overset{4}{CH_3}-\overset{3}{\underset{\delta+}{CH}}\diagup\overset{2}{CH}\diagdown\overset{1}{\underset{\delta+}{CH_2}}}$$

(b) We know that the allyl cation is almost as stable as a tertiary carbocation. Here we find not only the resonance stabilization of an allyl cation but the additional stabilization that arises from contributor **D** in which the plus charge is on a secondary carbon.

(c) $CH_3-\overset{\overset{\textstyle Cl}{|}}{CH}-CH=CH_2$ and $CH_3-CH=CH-CH_2-Cl$

10.3

(a) *Cis*-1,3-pentadiene, *trans,trans*-2,4-hexadiene, *cis,trans*-2,4-hexadiene, and 1,3-cyclo-hexadiene are conjugated dienes.

(b) 1,4-Cyclohexadiene is an isolated diene.

(c) 1-Penten-4-yne is an isolated enyne.

10.4

(a) $CH_3CH_2\overset{\overset{\textstyle}{|}}{\underset{\overset{|}{Cl}}{C}}HCH=CHCH_3$ and $CH_3CH_2CH=CH\overset{\overset{\textstyle}{|}}{\underset{\overset{|}{Cl}}{C}}HCH_3$

(b) The most stable cation is a hybrid of equivalent forms: $CH_3\overset{}{\underset{+}{C}}HCH=CHCH_3 \longleftrightarrow$

$CH_3CH=CHCHCH_3$. Thus 1,4 and 1,2- addition yield the same product,

$$\overset{+}{C}H_3CHCH=CHCH_3$$
$$|$$
$$Cl$$

10.5

(a) Addition of the proton gives the resonance hybrid of

$$CH_3-\overset{+}{C}H-CH=CH_2 \longleftrightarrow CH_3-CH=CH-\overset{+}{C}H_2$$

$$\text{I} \qquad\qquad\qquad \text{II}$$

The inductive effect of the methyl group in I stabilizes the positive charge on the adjacent carbon. Such stabilization of the positive charge does not occur in II. Because I contributes more heavily to the resonance hybrid than does II, C-2 bears a greater positive charge and reacts faster with the bromide ion.

(b) In the 1,4-addition product, the double bond is more highly substituted than in the 1,2-addition product, hence it is the more stable alkene.

10.6

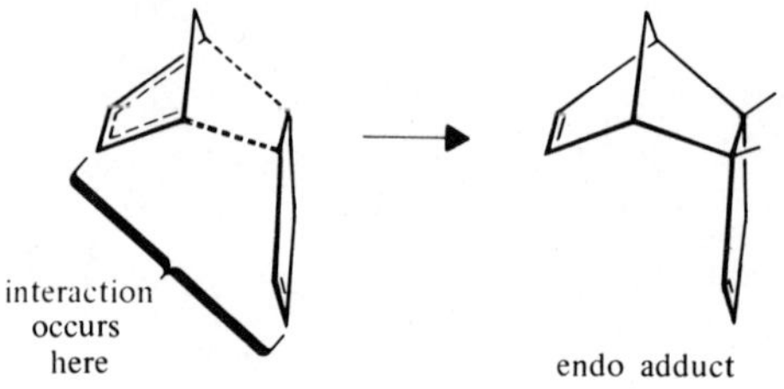

interaction
occurs
here

endo adduct

10.7

(a)

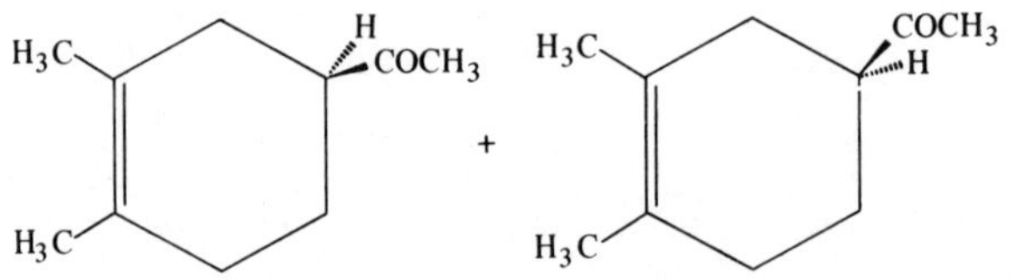

+

(b)

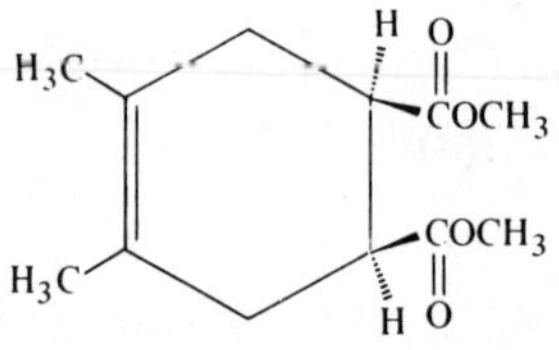

(c)

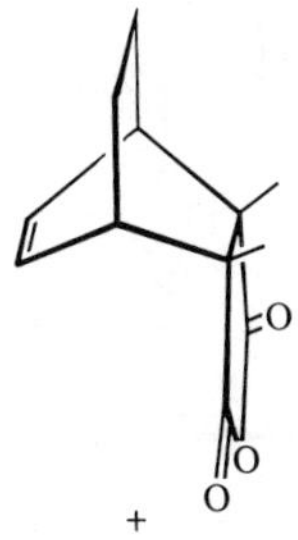

(major product)

+

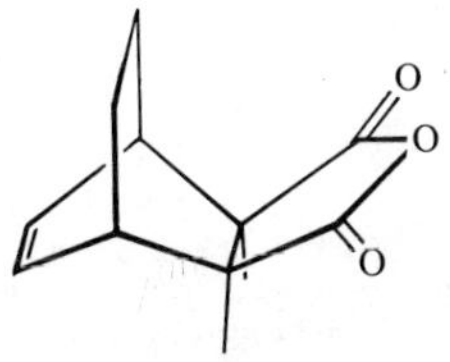

(minor product)

10.8

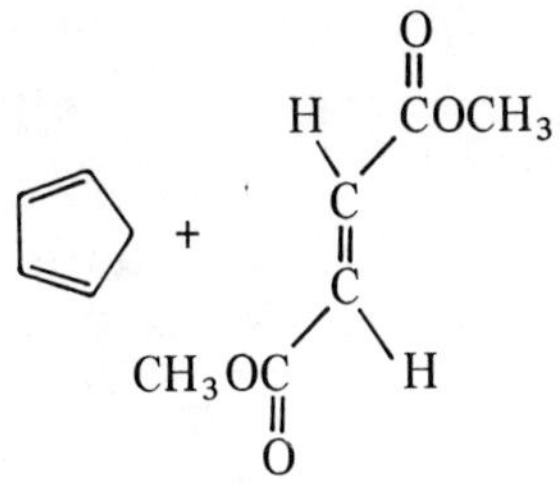

10.9

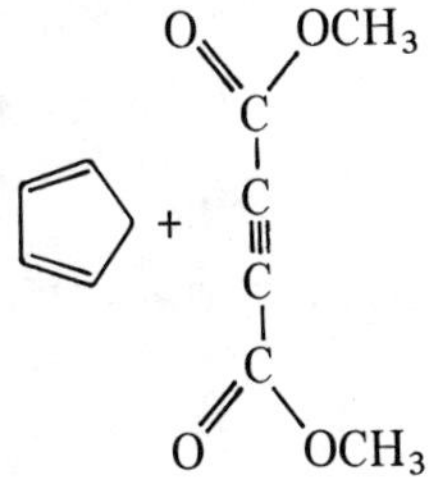

10.10

(a) $BrCH_2CH_2CH_2CH_2Br \xrightarrow[\text{(CH}_3)_3\text{COH}]{\text{(CH}_3)_3\text{COK}} CH_2{=}CH{-}CH{=}CH_2$

(b) $HOCH_2CH_2CH_2CH_2OH \xrightarrow[\text{heat}]{\text{conc.}\,H_2SO_4} CH_2{=}CH{-}CH{=}CH_2$

(c) $CH_2{=}CH{-}CH_2CH_2{-}OH \xrightarrow[\text{heat}]{\text{conc. }H_2SO_4} CH_2{=}CH{-}CH{=}CH_2$

(d) $CH_2{=}CH{-}CH_2CH_2{-}Cl \xrightarrow[\text{(CH}_3)_3\text{COH}]{\text{(CH}_3)_3\text{COK}} CH_2{=}CH{-}CH{=}CH_2$

(e) $CH_2{=}CH{-}\underset{\underset{Cl}{|}}{CH}{-}CH_3 \xrightarrow[\text{(CH}_3)_3\text{COH}]{\text{(CH}_3)_3\text{COK}} CH_2{=}CH{-}CH{=}CH_2$

(f) $CH_2=CH-\underset{\underset{OH}{|}}{CH}-CH_3 \xrightarrow[\text{heat}]{\text{conc. } H_2SO_4} CH_2=CH-CH=CH_2$

(g) $HC\equiv C-CH=CH_2 + H_2 \xrightarrow[\text{quinoline}]{\text{Pd. } BaSO_4} CH_2=CH-CH=CH_2$

10.11

$CH_2=\underset{\underset{CH_3}{|}}{C}-\underset{\underset{CH_3}{|}}{C}=CH_2$

10.12

(a) $Cl-CH_2\underset{\underset{Cl}{|}}{CH}CH=CH_2 + Cl-CH_2-CH=CH-CH_2-Cl$

(b) $\underset{\underset{Cl}{|}}{CH_2}-\underset{\underset{Cl}{|}}{CH}-\underset{\underset{Cl}{|}}{CH}-\underset{\underset{Cl}{|}}{CH_2}$

(c) $\underset{\underset{Br}{|}}{CH_2}-\underset{\underset{Br}{|}}{CH}-\underset{\underset{Br}{|}}{CH}-\underset{\underset{Br}{|}}{CH_2}$

(d) $CH_3-CH_2-CH_2-CH_3$

(e) No reaction

(f) $Cl-CH_2-\underset{\underset{OH}{|}}{CH}-CH=CH_2 + Cl-CH_2-CH=CH-CH_2-OH$

(g) $4CO_2$ (Note: $KMnO_4$ oxidizes $HOOC-COOH$ to $2CO_2$)

(h) $CH_3-\underset{\underset{OH}{|}}{CH}-CH=CH_2 + CH_3-CH=CH-CH_2OH$

10.13

(a) $CH_2=CH-CH_2-CH_3$ +

(NBS)

$\xrightarrow{CCl_4} CH_2=CH-\underset{\overset{|}{Br}}{CH}-CH_3$

$\xrightarrow[\text{(CH}_3)_3COH]{\text{(CH}_3)_3COK} CH_2=CH-CH=CH_2$

(b) $CH_2=CH-CH_2CH_2CH_3$ + NBS $\xrightarrow{CCl_4} CH_2=CH-\underset{\overset{|}{Br}}{CH}CH_2CH_3$

$\xrightarrow[\text{(CH}_3)_3COH]{\text{(CH}_3)_3COK} CH_2=CH-CH=CH-CH_3$

(c) $CH_3CH_2CH_2CH_2OH \xrightarrow[\text{heat}]{\text{conc. } H_2SO_4} CH_3CH_2CH=CH_2 \xrightarrow{\text{as in (a)}} CH_2=CH-CH=CH_2$

$\underset{\overset{|}{Br}}{CH_2}-CH=CH-\underset{\overset{|}{Br}}{CH_2} \xleftarrow[\text{heat}]{Br_2}$

(d) $CH_3-CH=CH-CH_3$ + NBS $\xrightarrow{CCl_4}$ $CH_3-CH=CH-CH_2-Br$

(e) + Br_2 $\xrightarrow[\text{heat}]{\text{light}}$ $\xrightarrow[(CH_3)_3COH]{(CH_3)_3COK}$ $\xrightarrow[CCl_4]{NBS}$

(excess)

(f) $\xrightarrow[(CH_3)_3COH]{(CH_3)_3COK}$ $\left(\text{same as} \quad \bigpentagon \right)$

10.14

$$R-\overset{..}{\underset{..}{O}}-\overset{..}{\underset{..}{O}}-R \xrightarrow[\text{or heat}]{\text{light}} 2R-\overset{..}{\underset{..}{O}}\cdot$$

$$R-\overset{..}{\underset{..}{O}}\cdot + H-\overset{..}{\underset{..}{Br}}: \longrightarrow R-\overset{..}{\underset{..}{O}}-H + \cdot\overset{..}{\underset{..}{Br}}:$$

$$CH_2=CH-CH=CH_2 + \cdot\overset{..}{\underset{..}{Br}}: \longrightarrow \left[CH_2=CH-\overset{\bullet}{C}H-\underset{\underset{Br}{|}}{C}H_2 \longleftrightarrow \overset{\bullet}{C}H_2-CH=CH-\underset{\underset{Br}{|}}{C}H_2 \right]$$

$$\xrightarrow{HBr} CH_2=CH-\underset{\underset{H}{|}}{C}H-\underset{\underset{Br}{|}}{C}H_2 + CH_2-CH=CH-\underset{\underset{Br}{|}}{C}H_2 + \cdot\overset{..}{\underset{..}{Br}}:$$

(*cis* and *trans*)

10.15

(a) $Ag(NH_3)_2OH$ gives a precipitate with 1-butyne only.

(b) 1,3-Butadiene decolorizes bromine solution; *n*-butane does not.

(c) CrO_3/H_2SO_4 oxidizes the alcohol. The solution changes from orange to green. No color change with 1,3-butadiene.

(d) $AgNO_3$ in C_2H_5OH gives a AgBr precipitate with $CH_2=CHCH_2CH_2Br$. No reaction with 1,3-butadiene.

(e) $AgNO_3$ in C_2H_5OH gives a AgBr precipitate with $BrCH_2CH=CHCH_2Br$ (it is an allylic bromide), but not with $CH_3\underset{\underset{Br}{|}}{C}H=\underset{\underset{Br}{|}}{C}HCH_3$ (a vinyl bromide).

10.16

(a) Because a highly resonance-stabilized free radical is formed:

$$CH_2=CH-\overset{\bullet}{C}H-CH=CH_2 \longleftrightarrow \overset{\bullet}{C}H_2=CH-CH=CH-\overset{\bullet}{C}H_2 \longleftrightarrow \overset{\bullet}{C}H_2-CH=CH-CH=CH_2$$

(b) Because the carbanion is more stable:

$$CH_2=CH-\overset{..}{\underset{..}{C}}H-CH=CH_2 \longleftrightarrow CH_2=CH-CH=CH-\overset{..}{\underset{..}{C}}H_2 \longleftrightarrow \overset{..}{\underset{..}{C}}H_2-CH=CH-CH=CH_2$$

i.e., we can write more resonance structures of nearly equal energies.

10.17

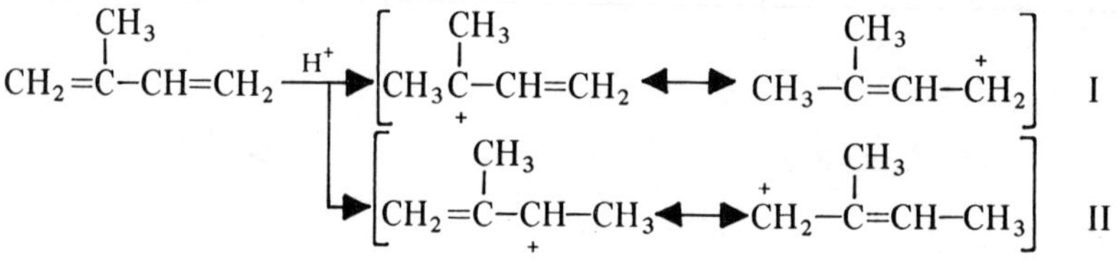

$$CH_2{=}\overset{\overset{\displaystyle CH_3}{|}}{C}{-}CH{=}CH_2 \xrightarrow{H^+}$$

$$\left[CH_3\overset{\overset{\displaystyle CH_3}{|}}{\underset{+}{C}}{-}CH{=}CH_2 \longleftrightarrow CH_3{-}\overset{\overset{\displaystyle CH_3}{|}}{C}{=}CH{-}\overset{+}{C}H_2 \right] \quad I$$

$$\left[CH_2{=}\overset{\overset{\displaystyle CH_3}{|}}{\underset{+}{C}}{-}CH{-}CH_3 \longleftrightarrow \overset{+}{C}H_2{-}\overset{\overset{\displaystyle CH_3}{|}}{C}{=}CH{-}CH_3 \right] \quad II$$

The resonance hybrid, I, has the positive charge, in part, on the tertiary carbon; in II, the positive charge is on primary and secondary carbons only. Therefore hybrid I is more stable, and will be the intermediate carbocation. 1,4-addition to I gives

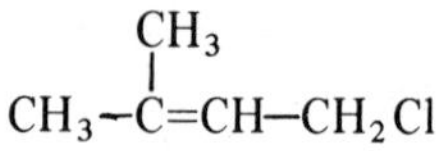

$$CH_3{-}\overset{\overset{\displaystyle CH_3}{|}}{C}{=}CH{-}CH_2Cl$$

10.18

(a)

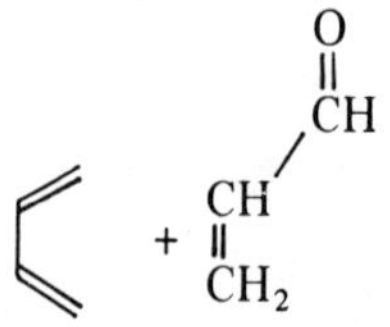

(d)

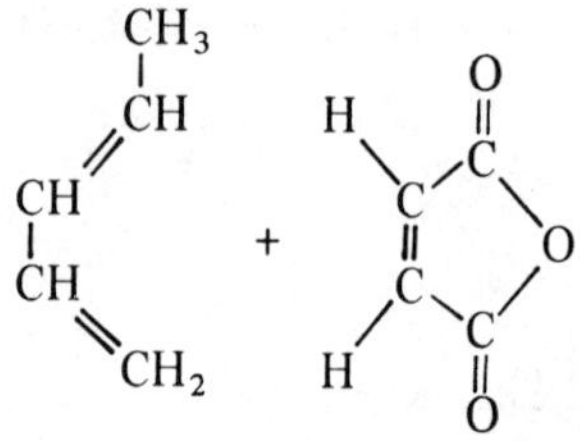

(b)

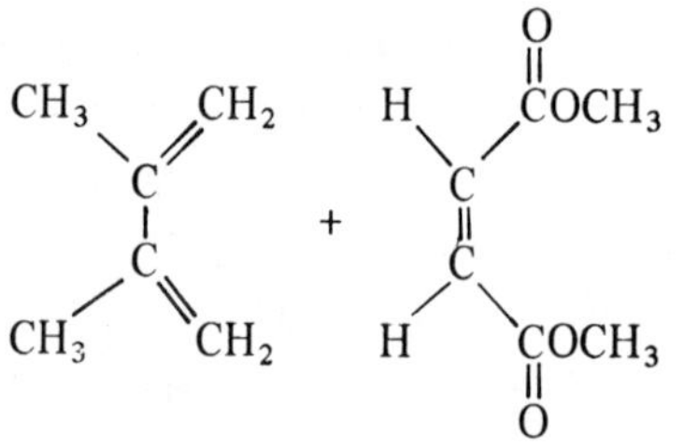

(e)

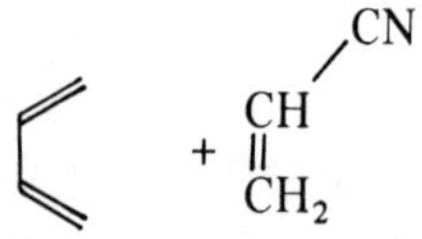

(c)

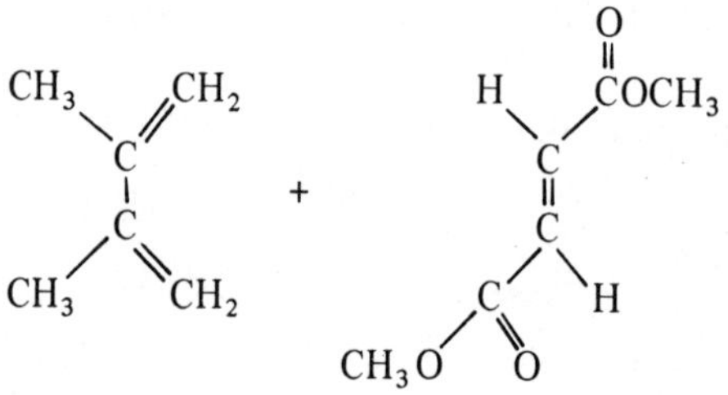

10.19

Neither compound can assume the s-*cis*-conformation. 1,3-Butadiyne is linear, and

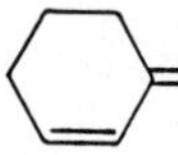

$=CH_2$ is forced into the s-*trans* conformation by the requirements of the ring.

10.20

(a) [structure: cyclohexadiene ring with two C(=O)–OCH$_3$ groups]   (b) [structure: cyclohexadiene ring with two CF$_3$ groups]

10.21
The formula, C_6H_8, tells us that **A** and **B** have six hydrogens less than an alkane. This unsaturation may be due to three double bonds, one triple bond and one double bond, or combinations of two double bonds and a ring, or one triple bond and a ring. Since both **A** and **B** react with two moles of H_2 to yield cyclohexane, they are either cyclohexyne or cyclohexadienes. The absorption maximum of 256 nm for **A** tells us that it is conjugated. **B**, with no absorption maximum beyond 200, possesses isolated double bonds. We can rule out cyclohexyne because of ring strain caused by the requirement of linearity of the $-C\equiv C-$ system. Therefore **A** is 1,3-cyclohexadiene; **B** is 1,4-cyclohexadiene.

10.22
In the dimer, the remaining double bonds are isolated:

The rate of dimerization can be followed by observing the rate of disappearance of the absorption maximum at 239 nm that is due to the conjugated double bonds in cyclopentadiene (see Table 10.3). The product does not absorb at 239 nm.

10.23
All three compounds have an unbranched five-carbon chain. The formula, C_5H_6, suggests that they have one double bond and one triple bond. **D**, **E**, and **F** must differ, therefore, in the way the multiple bonds are distributed in the chain. **E** and **F** have a terminal $-C\equiv CH$ (reaction with $Ag(NH_3)_2{}^+OH^-$). The absorption maximum near 230 nm for **D** and **E** suggests that in these compounds, the multiple bonds are conjugated. The structures are:

$$CH_3-C\equiv C-CH=CH_2 \qquad HC\equiv C-CH=CH-CH_3 \qquad HC\equiv C-CH_2-CH=CH_2$$
$$\textbf{D} \qquad\qquad\qquad \textbf{E} \qquad\qquad\qquad \textbf{F}$$

10.24
The C_2-C_3 bond has partial double bond character.

10.25
The *endo* adduct is less stable than the *exo*, but is produced at a faster rate at 25°. At 90°, equilibrium is established, and the more stable *exo* adduct predominates.

10.26

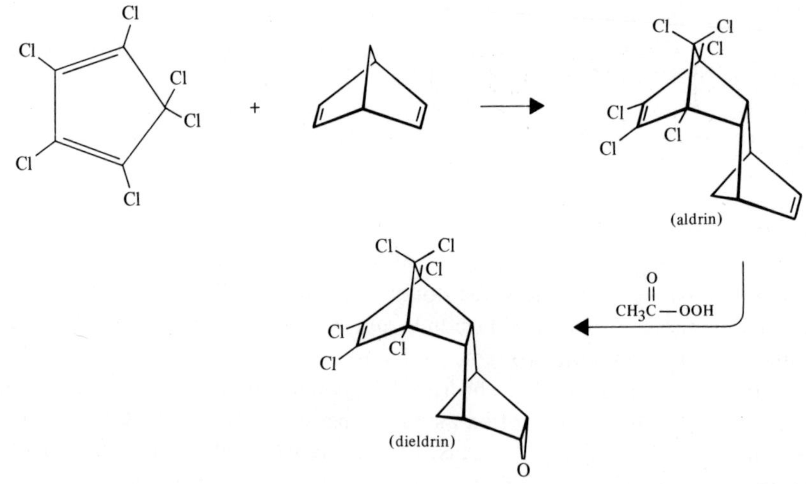
Cl
+
(aldrin)
$CH_3C(=O)—OOH$
(dieldrin)
O

10.27

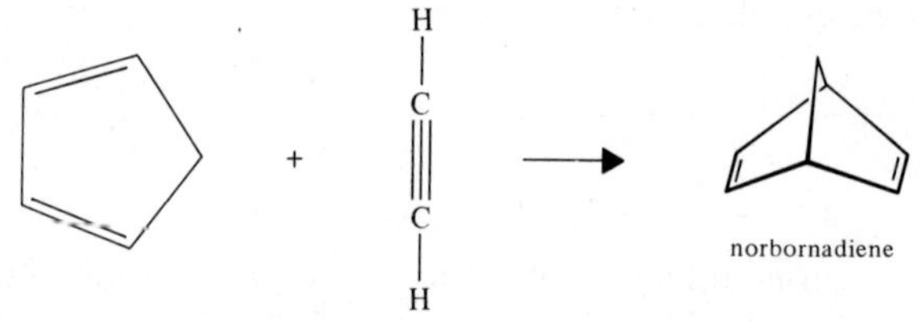
+
H
C
C
H
norbornadiene

10.28

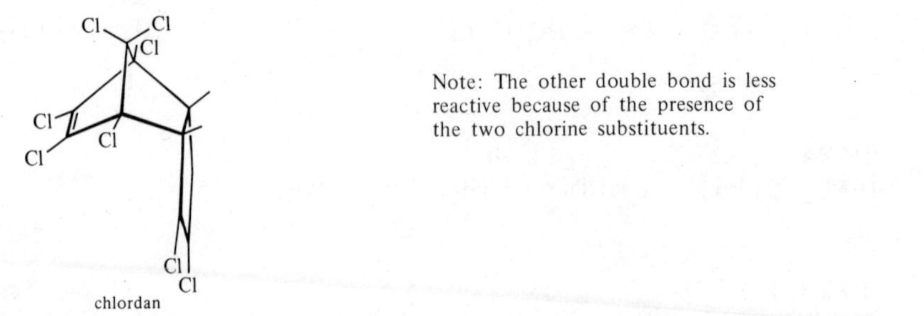
Cl
Cl_2
(dark)
chlordan

Note: The other double bond is less reactive because of the presence of the two chlorine substituents.

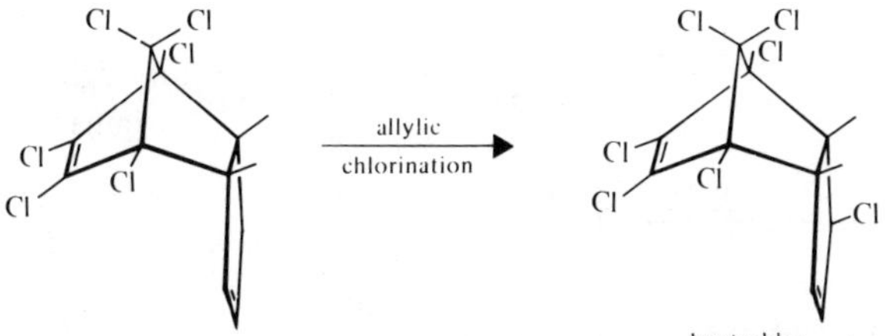

heptachlor

10.29

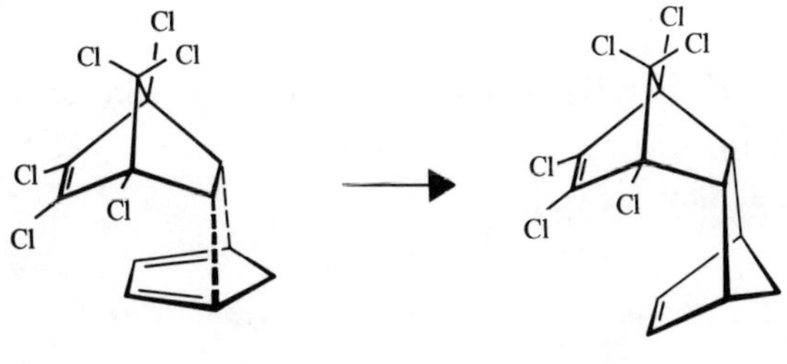

Endrin

10.30

Protonation of the alcohol leads to an allylic cation that can react with a chloride ion at either carbon-1 or carbon-3.

$$CH_3CH{=}CHCH_2OH \xrightarrow{\ H^+\ } CH_3CH{=}CHCH_2{-}\overset{\overset{\displaystyle H}{|}}{O}{}^+{-}H \xrightarrow{\ -H_2O\ }$$

$$CH_3CH{=}CHCH_2{}^+ \longleftrightarrow CH_3\overset{+}{C}HCH{=}CH_2 \xrightarrow{\ Cl^-\ }$$

$$CH_3CH{=}CHCH_2Cl \ + \ CH_3\underset{\underset{\displaystyle Cl}{|}}{C}HCH{=}CH_2$$

10.31

(1) $CH_2{=}CH{-}CH{=}CH_2 \ + \ Cl_2 \longrightarrow ClCH_2{-}\overset{+}{C}H{-}CH{=}CH_2$

$$\Updownarrow$$

$$ClCH_2{-}CH{=}CH{-}\overset{+}{C}H_2$$

$$ClCH_2{-}\overset{\delta+}{C}H{\cdots}CH{\cdots}\overset{\delta+}{C}H_2$$

(2) $ClCH_2{-}\overset{\delta+}{C}H{\cdots}CH{\cdots}\overset{\delta+}{C}H_2 \xrightarrow[(-H^+)]{\ CH_3OH\ } ClCH_2{-}\underset{\underset{\displaystyle OCH_3}{|}}{C}H{-}CH{=}CH_2$

$$+ \ ClCH_2{-}CH{=}CH{-}CH_2OCH_3$$

10.32

A six-membered ring cannot accommodate a triple bond because of the strain that would be introduced.

10.33

The products are $CH_3CH_2CHCH=CH_2$ and $CH_3CH_2CH=CHCH_2Br$. They are formed
$\hspace{6.5cm}|$
$\hspace{6.3cm}Br$
from an allylic radical in the following way:

$$Br_2 \longrightarrow 2\ Br\cdot$$
(from
NBS)

$$Br\cdot\ +\ CH_3CH_2CH_2CH=CH_2 \longrightarrow CH_3CH_2\overset{\cdot}{C}HCH=CH_2$$

$$\updownarrow \qquad\qquad + HBr$$

$$CH_3CH_2CH=CH\overset{\cdot}{C}H_2$$

$$CH_3CH_2\overset{\delta\cdot}{CH}\!=\!=\!=\!CH\!=\!=\!=\!\overset{\delta\cdot}{CH_2}\ +\ Br_2 \longrightarrow CH_3CHCHCH=CH_2$$
$$\hspace{7cm}|$$
$$\hspace{7cm}Br$$

$$+\ CH_3CH_2CH=CHCH_2Br\ +\ Br\cdot$$

10.34

The first crystalline solid is the Diels-Alder adduct below, mp 125°,

On melting, this adduct undergoes a reverse Diels-Alder reaction yielding furan (which vaporizes) and maleic anhydride, mp 56°.

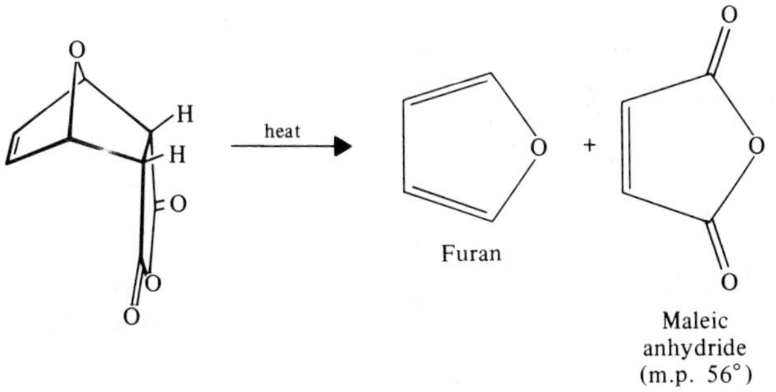

Furan

Maleic
anhydride
(m.p. 56°)

10.35

(a) *trans*-1,3-Pentadiene assumes the required *s-cis* conformation readily. With *cis*-1,3-pentadiene the concentration of the *s-cis* conformation is much lower because of steric hindrance presented by the internal methyl group.

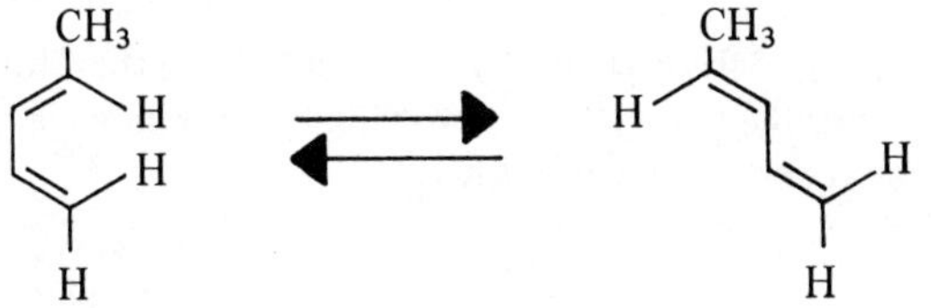

trans-1,3-Pentadiene

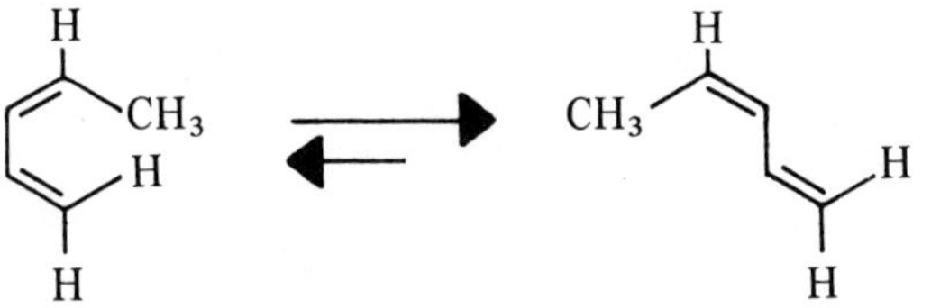

cis-1,3-pentadiene

(b) The large *tert*-butyl group at the 2-position of 1,3-butadiene favors the formation of the *s-cis* conformation. (It also increases reactivity by releasing electrons.) On the other hand, a *tert*-butyl group at the 1-position of 1,3-butadiene prevents the formation of significant concentrations of the *s-cis* conformation and thus no Diels-Alder reactions can take place.

2-*Tert*-butyl-1,3-butadiene

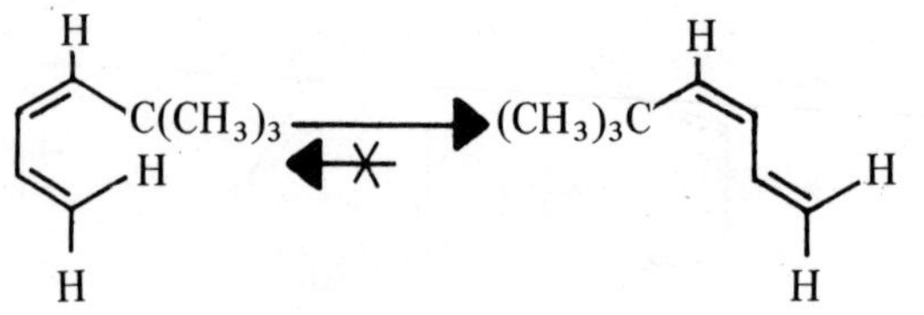

Cis-5,5-dimethyl-1,3-hexadiene
(*Cis*-1-*tert*-butyl-1,3-butadiene)

10.36

The following reaction takes place:

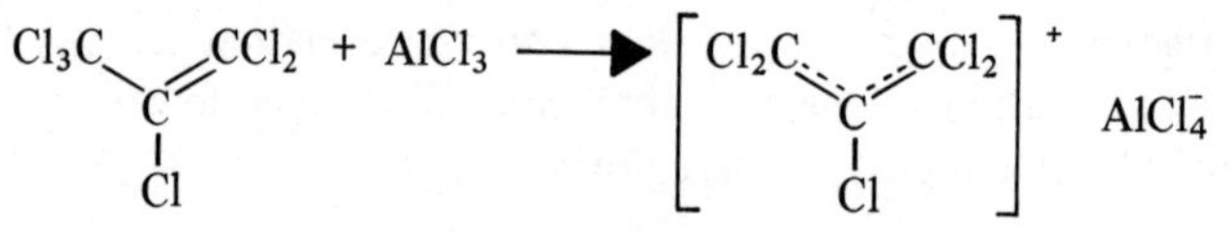

The salt forms because the organic cation is an allylic cation and is, therefore, unusually stable. It is also stabilized by resonance structures involving nonbonding electron pairs of the chlorine atoms such as the second structure below.

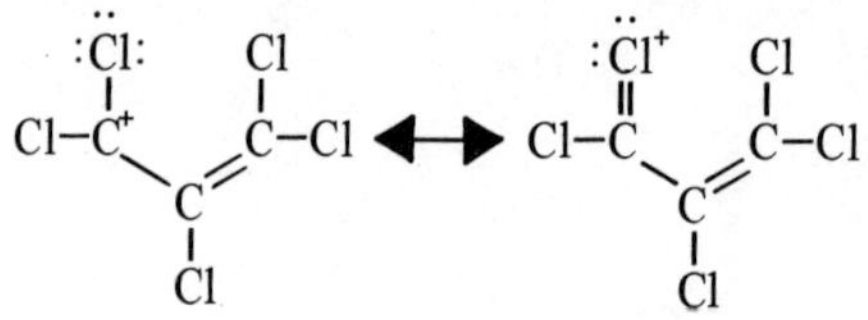

11

SPECIAL
TOPICS II

11.1

Conrotatory motion of the type shown would lead to increasingly unfavorable interaction of the methyl groups as the transition state is approached. Thus this path is not followed to any appreciable extent.

11.2

According to the Woodward-Hoffmann rule for electrocyclic reactions of $4n$ π electron systems (p. 390), the photoehemical cyclization of *cis,trans*-2,4-hexadiene should proceed with *disrotatory motion*. Thus it should yield *trans*-3,4-dimethylcyclobutene:

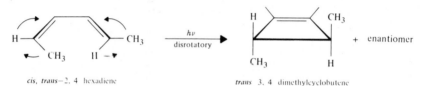

cis, trans—2, 4 hexadiene   *trans* 3, 4 dimethylcyclobutene

+ enantiomer

11.3

(a)

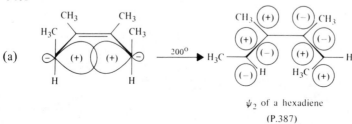

ψ_2 of a hexadiene

(P.387)

(b) This is a thermal electrocyclic reaction of a $4n$ π electron system; it should, *and does*, proceed with conrotatory motion.

11.4

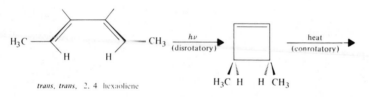

trans, trans, 2, 4 hexaoliene

cis 3, 4 dimethylcyclobutene

(structure at top)

cis, trans−2, 4−hexadiene

Here we find that two consecutive electrocyclic reactions (the first photochemical, the second thermal), provide a stereospecific synthesis of *cis,trans*-2,4-hexadiene from *trans,trans*-2,4-hexadiene.

11.5

(a) This is a photochemical electrocyclic reaction of an eight π electron system a $4n$ π system where $n = 2$. It should, therefore, proceed with disrotatory motion.

cis−7, 8−dimethyl−1, 3, 5−cyclooctatriene

(b) This is a thermal electrocyclic reaction of the eight π electron system. It should proceed with conrotatory motion.

cis−7, 8−dimethyl−1, 3, 5−cyclooctatriene

11.6

(a) This is conrotatory motion and since this is a $4n$ π electron system (where $n = 1$) it should occur under the influence of heat.

(structure with heat (conrotatory) arrow)

(b) This is conrotatory motion and since this is also a $4n$ π electron system (where $n = 2$) it should occur under the influence of heat.

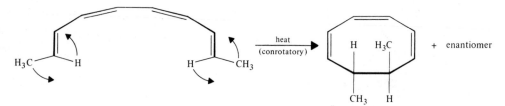

(c) This is disrotatory motion. This, too is a $4n$ π electron system (where $n = 1$), thus it should occur under the influence of light.

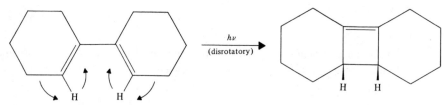

11.7

(a) This is a $4n + 2$ π electron system (where $n = 1$); a thermal reaction should take place with disrotatory motion:

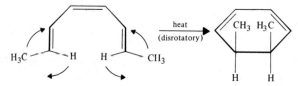

(b) This is also a $4n + 2$ π electron system; a photochemical reaction should take place with conrotatory motion.

11.8

Here we need a conrotatory ring-opening of *trans*-5,6-dimethyl-1,3-cyclohexadiene (to produce *trans,cis,trans*-2,4,6-octatriene), then we need a disrotatory cyclization to produce *cis*-5,6-dimethyl-1,3-cyclohexadiene.

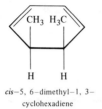

cis−5, 6−dimethyl−1, 3−
cyclohexadiene

Since both reactions involve $4n + 2$ π electron systems we apply light to accomplish the first step and heat to accomplish the second. It would also be possible to use heat to produce *trans,cis,cis*-2,4,6-octatriene then use light to produce the desired product.

11.9

The first electrocyclic reaction is a thermal, conrotatory ring opening of a $4n$ π electron system. The second electrocyclic reaction is a thermal, disrotatory ring closure of a $4n + 2$ π electron system.

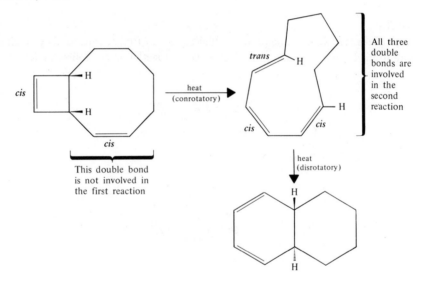

11.10

(a) This reaction involves two π electrons, thus it is a $4n + 2$ π system where $n = 0$.

(b) An allylic cation is formed.

(c) The cyclopropyl anion is a $4n$ π system (where $n = 1$), thus a thermal reaction should take place with conrotatory motion.

11.11

(a) This is a $4n + 2$ π electron system undergoing disrotatory motion. Heat is required.

(b) This is a $4n + 2$ π electron system undergoing conrotatory motion. Light is required.

(c) This is a $4n + 2$ π electron system undergoing conrotatory motion. Light is required.

(d) This is a $4n + 2$ π electron system undergoing disrotatory motion. Heat is required.

11.12

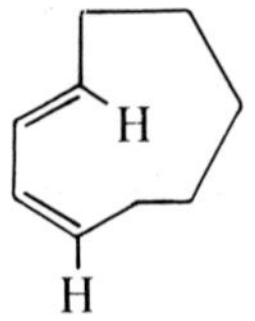

1

Cis,trans-cyclonona-1,3-diene

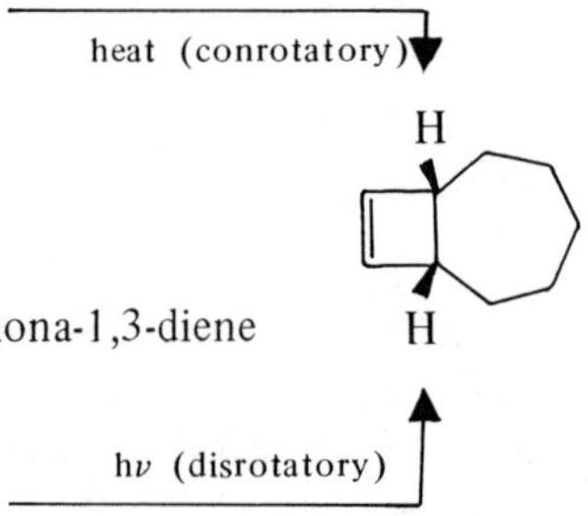
heat (conrotatory)
hν (disrotatory)

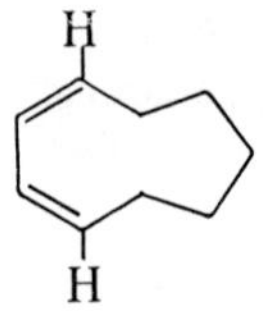

2

Cis,cis-cyclonona-1,3-diene

11.13

(a) There are two possible products that can result from a concerted cycloaddition. They are formed when *cis*-2-butene molecules come together in the following ways:

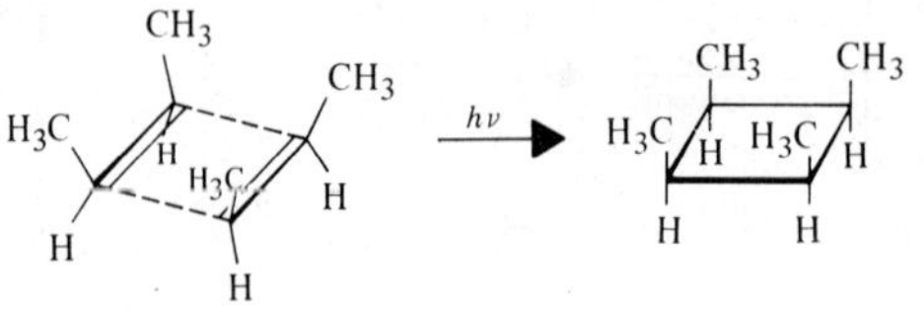
hν

and

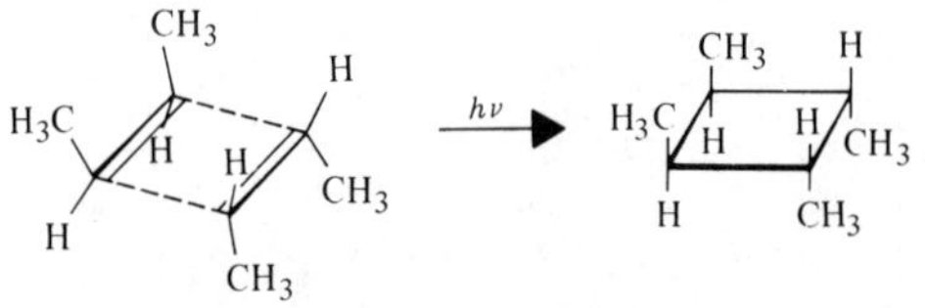
hν

(b) There are two possible products that can be obtained from *trans*-2-butene as well.

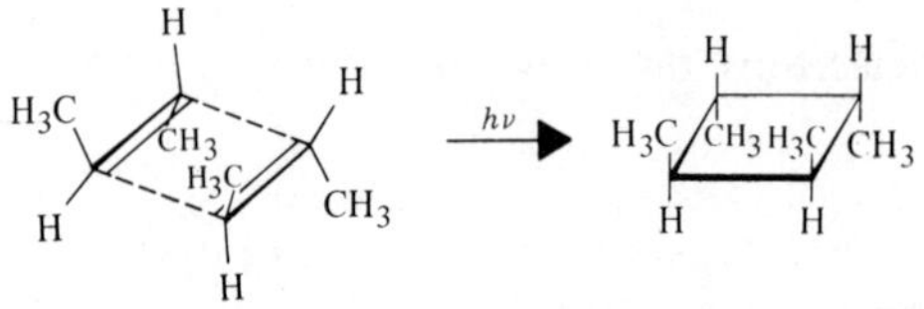
hν

and

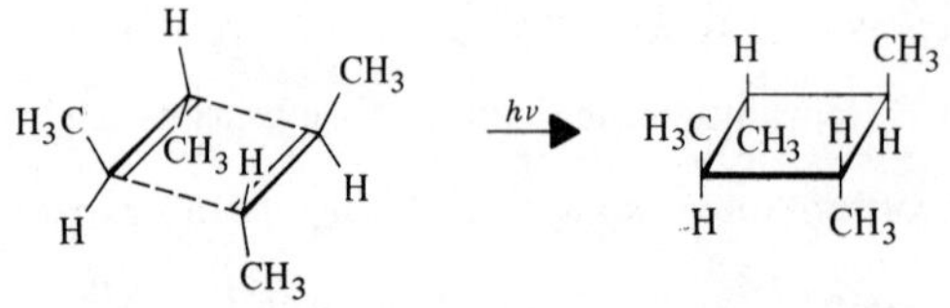
hν

11.14

This is an intramolecular [2 + 2] cycloaddition.

11.15

(a) No. A thermal [2 + 2] concerted cycloaddition is symmetry forbidden and therefore, is likely to have a very high activation energy.

(b) The formation of a relatively stable intermediate in which both unpaired electrons are involved in allylic systems explains the preference for head-to-head cyclization.

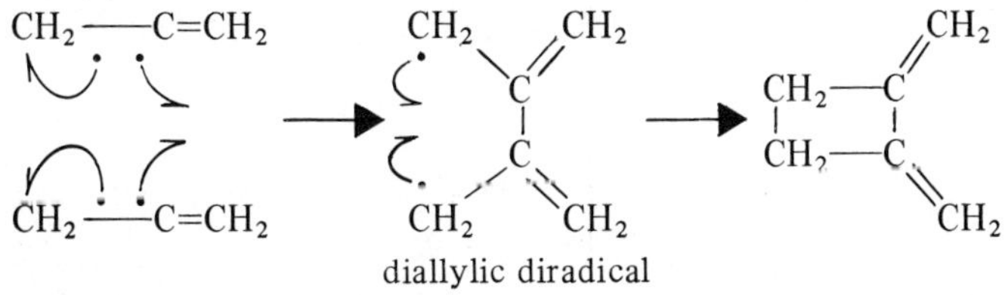

diallylic diradical

Cyclization in a head-to-tail manner would require the formation of a less stable diradical intermediate – only one electron would be involved in an allylic system:

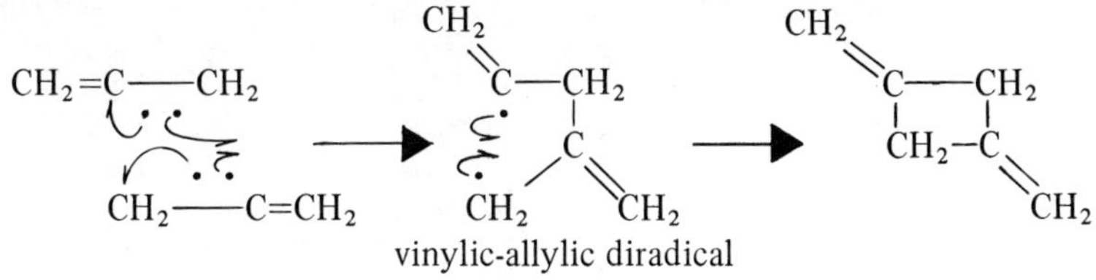

vinylic-allylic diradical

11.16

(a)

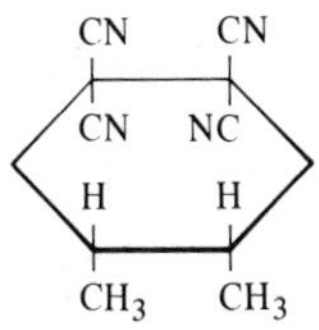

(b)

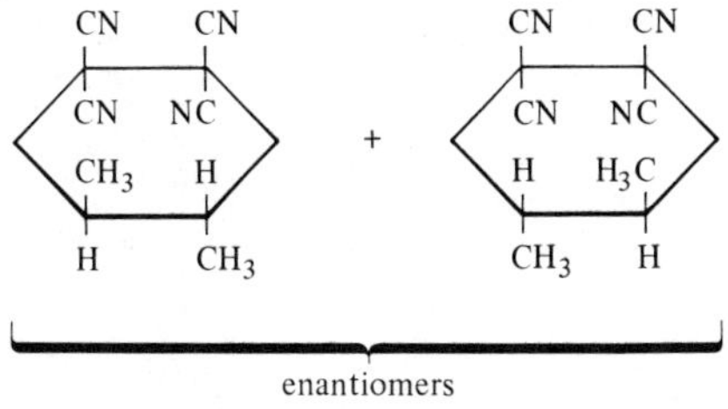

enantiomers

11.17

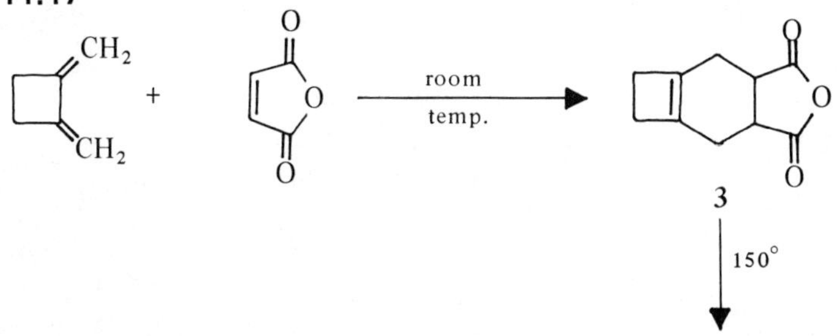

room temp.

3

150°

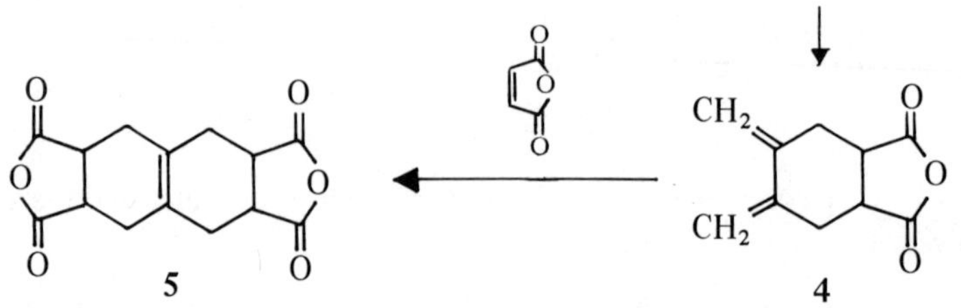

5   4

11.18

Compound **7** (below) results from a conrotatory ring opening; it then reacts as the diene component of a Diels-Alder reaction.

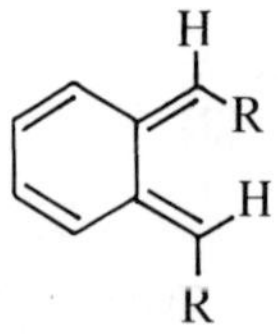

7

11.19

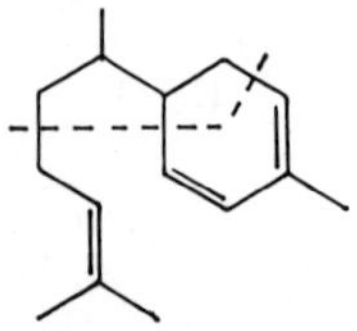

Zingiberene
(a sesquiterpene)

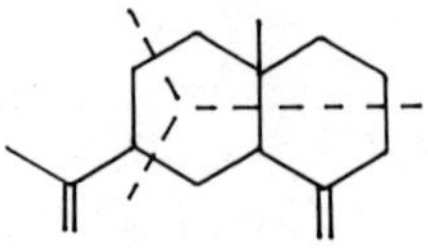

β-Selinene
(a sesquiterpene)

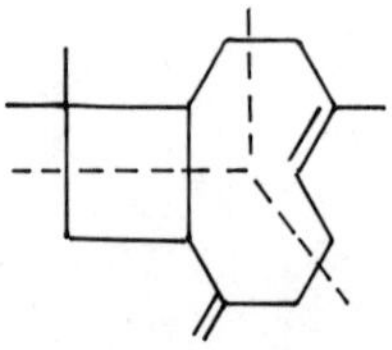

caryophyllene
(a sesquiterpene)

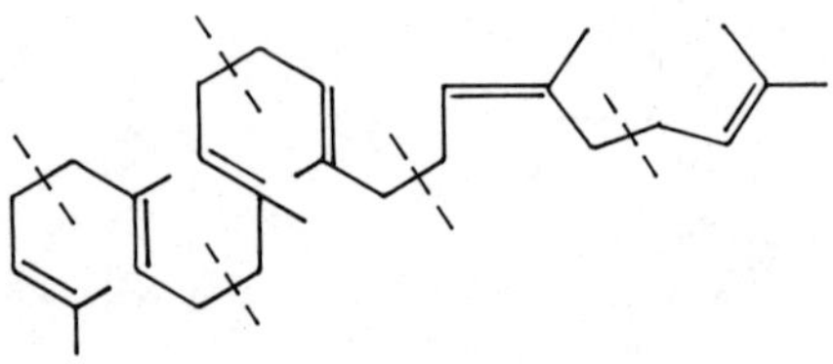

squalene
(a triterpene)

11.20

(a)

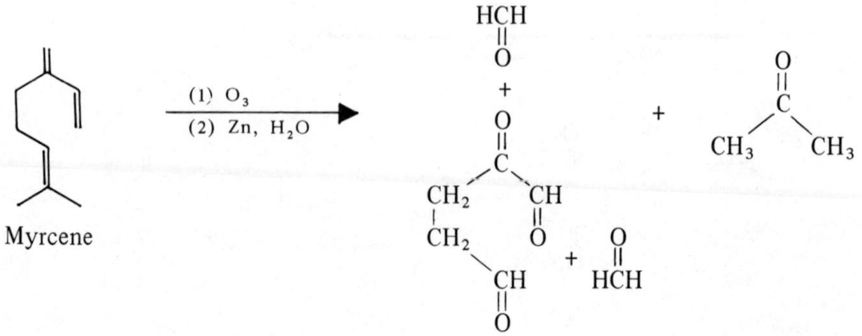

$\xrightarrow{\text{(1) O}_3 \quad \text{(2) Zn, H}_2\text{O}}$

Myrcene

(b)

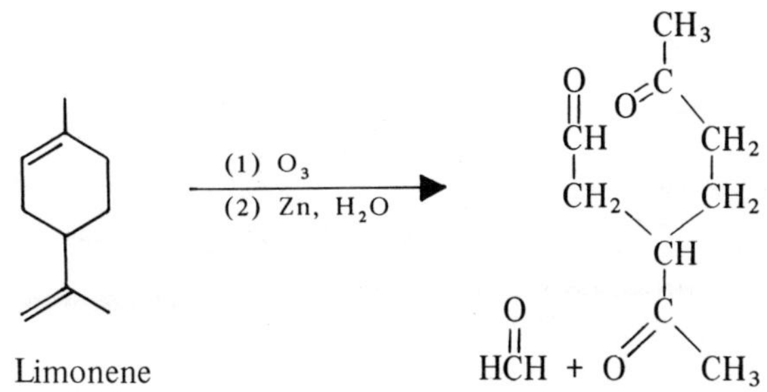

Limonene
$\xrightarrow[\text{(2) Zn, H}_2\text{O}]{\text{(1) O}_3}$

(c) α-Farnesene $\xrightarrow[\text{(2) Zn, H}_2\text{O}]{\text{(1) O}_3}$ $CH_3\overset{O}{\overset{||}{C}}CH_3$ + $H\overset{O}{\overset{||}{C}}CH_2CH_2\overset{O}{\overset{||}{C}}CH_3$
(See p. 404)

+ $H\overset{O}{\overset{||}{C}}CH_2\overset{O}{\overset{||}{C}}H$ + $H\overset{O}{\overset{||}{C}}-\overset{O}{\overset{||}{C}}CH_3$ + $H\overset{O}{\overset{||}{C}}H$

(d) Geraniol $\xrightarrow[\text{(2) Zn, H}_2\text{O}]{\text{(1) O}_3}$ $CH_3\overset{O}{\overset{||}{C}}CH_3$ + $H\overset{O}{\overset{||}{C}}CH_2CH_2\overset{O}{\overset{||}{C}}CH_3$
(See p. 405)

+ $H\overset{O}{\overset{||}{C}}CH_2OH$

(e) Squalene $\xrightarrow[\text{(2) Zn, H}_2\text{O}]{\text{(1) O}_3}$ $2CH_3\overset{O}{\overset{||}{C}}CH_3$ + $H\overset{O}{\overset{||}{C}}CH_2CH_2\overset{O}{\overset{||}{C}}H$
(See p. 405)

+ $4CH_3\overset{O}{\overset{||}{C}}CH_2CH_2\overset{O}{\overset{||}{C}}H$

11.21

(a)

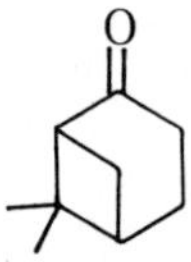

+ CO_2

(c)

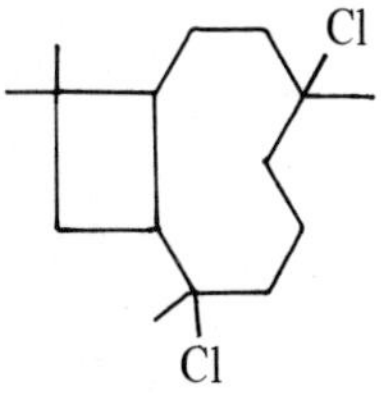

(b)

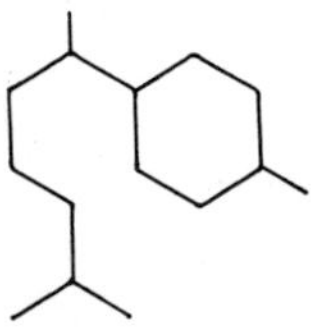

(d)

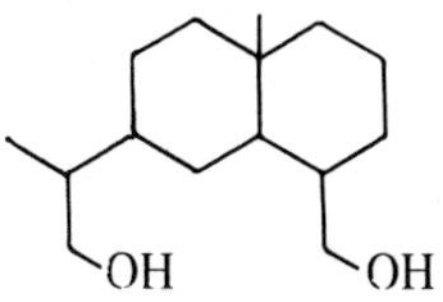

11.22

Br_2 in CCl_4 or $KMnO_4$ in H_2O. Either reagent would give a positive result with geraniol and a negative result with menthol.

11.23

(a)

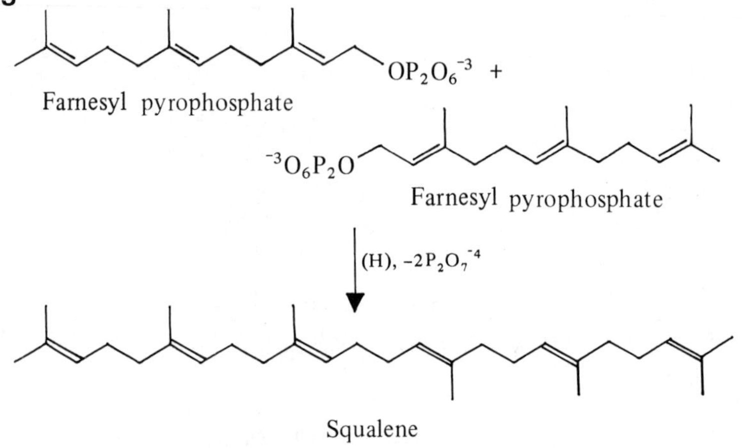

Farnesyl pyrophosphate

$^{-3}O_6P_2O$

Farnesyl pyrophosphate

$(H), -2P_2O_7^{-4}$

Squalene

(b)

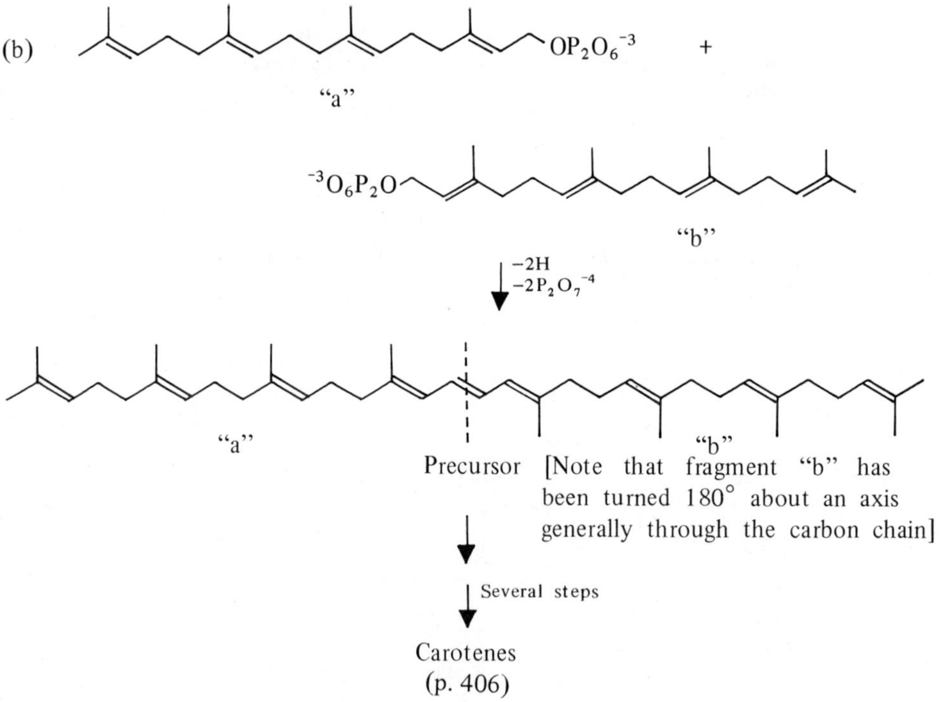

"a"

$^{-3}O_6P_2O$

"b"

$-2H$
$-2P_2O_7^{-4}$

"a"

Precursor [Note that fragment "b" has
been turned 180° about an axis
generally through the carbon chain]

"b"

Several steps

Carotenes
(p. 406)

11.24

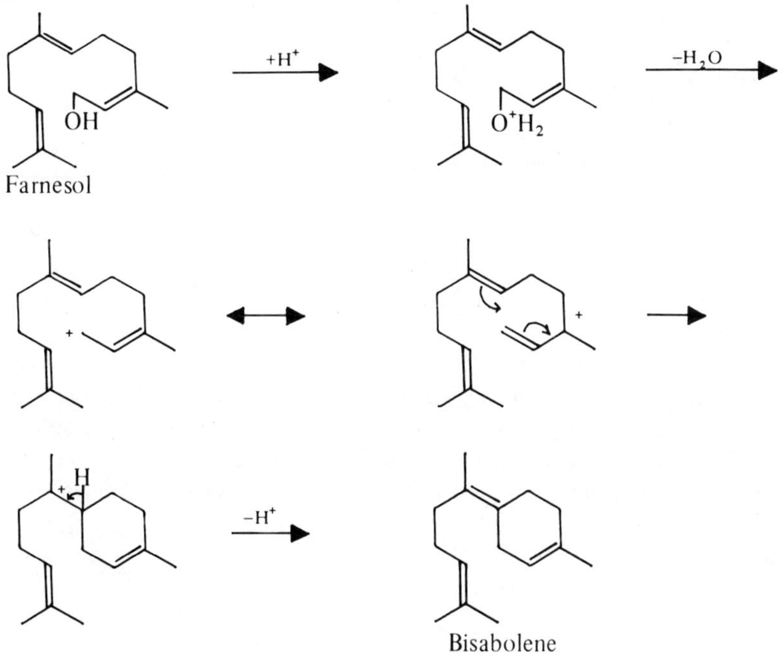
+H$^+$
OH
Farnesol
O$^+$H$_2$
−H$_2$O
+
+
H
+
−H$^+$
Bisabolene

11.25

A reverse Diels-Alder reaction takes place.

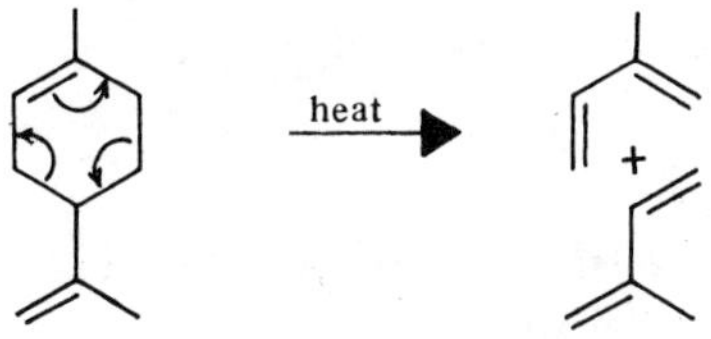
heat
+

11.26

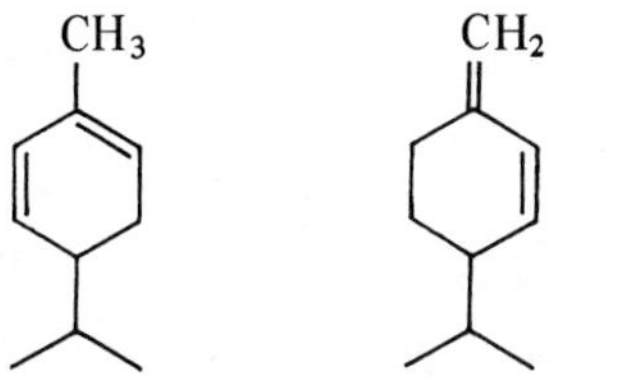
CH$_3$
CH$_2$

α-Phellandrene β-Phellandrene

Note: on permanganate oxidation, the $=CH_2$ group of β-phellandrene is converted to CO_2 and thus is not detected in the reaction.

11.27

The Diels-Alder reaction requires that the diene units assume an *s-cis* conformation (p. 364). Vitamin A can do this easily; however, for Neovitamin A steric hindrance is considerable and thus the concentration of the *s-cis* conformation is very small.

$C_{14}H_{21}$

Vitamin A

$C_{14}H_{21}$

Neovitamin A

11.28

(a)

$\xrightarrow[\text{disrotatory}]{h\nu}$

(b) Although the product is highly strained, a concerted thermal reversal of the electrocyclic reaction would require conrotatory ring opening and the production of an impossibly strained cyclopentadiene ring with one *trans* double bond. The ring opening, therefore, is probably nonconcerted and has a relatively high activation energy.

11.29

A is

B and **C** are

and

11.30

A is the product of a disrotatory thermal electrocyclic reaction involving a 6π electron segment of cyclooctatetraene. **B** is the Diels-Alder adduct.

H

H

A

B

***11.31**

The first step is a nonconcerted dimerization of allene (cf. problem 11.15):

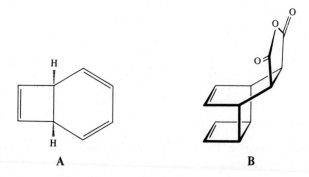

$$2\ CH_2{=}C{=}CH_2 \xrightarrow{\text{heat}}$$

CH_2

CH_2

This is followed by a Diels-Alder reaction:

Then a 4π electron electrocyclic ring-opening occurs:

Then Diels-Alder reactions take place again.

12

AROMATIC COMPOUNDS I: THE PHENOMENON OF AROMATICITY

12.1

(a) d   (b) None

12.2

$C=C$ = 1.34 Å $C-C$ = 1.47 Å (See page 354 of text for sp^2-sp^2 carbon-carbon single bond length.)

12.3

(a)

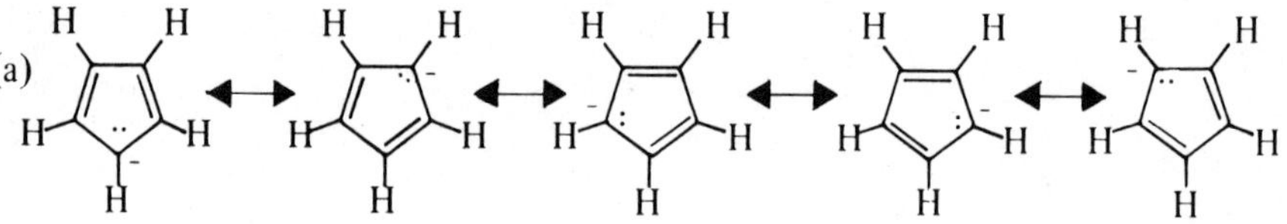

(b) Yes, all of the five resonance structures are equivalent, and all five hydrogen atoms are equivalent.

12.4

(a)

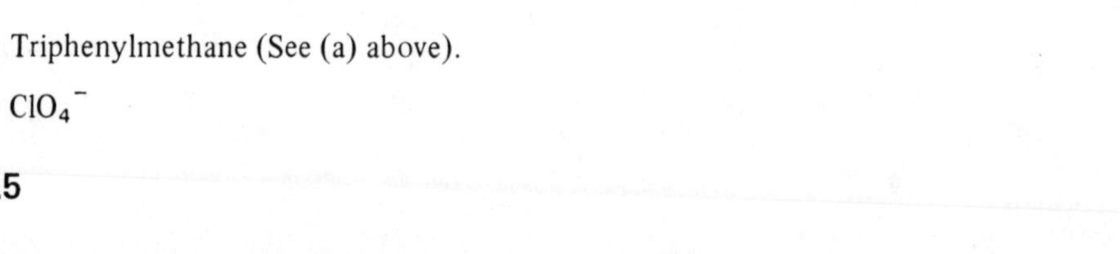

(b) Triphenylmethane (See (a) above).

(c) ClO_4^-

12.5

Tropylium bromide is ionic and has the structure, + Br⁻. The ring is aromatic.

12.6

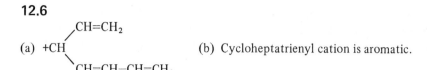

(a) $+CH$ with $CH=CH_2$ and $CH=CH-CH=CH_2$

(b) Cycloheptatrienyl cation is aromatic.

12.7

(a) $+CH$ with $CH=CH_2$ and $CH=CH_2$

$\xrightarrow[\text{energy increases}]{\pi \text{ electron}}$ $+$ (cyclopentadienyl) $+ H_2$

The π electron energy of cyclopentadienyl cation is higher than that of the open chain counterpart.

(b) $+CH_2$ with $CH=CH_2$ and CH_2

$\xrightarrow[\text{energy decreases}]{\pi \text{ electron}}$ $\overset{+}{\triangle}$ $+ H_2$

The π electron energy of cyclopropenyl cation is lower than that of the open chain counterpart.

(c) $4n+2 = 2$ when $n = 0$. Hückel's rule predicts that cyclopropenyl cation is aromatic.

(d) The π electron energy of cyclopropenyl anion is higher than that of the open chain counterpart.

12.8

(a)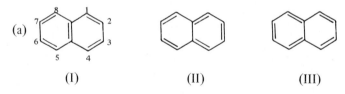

(I)   (II)   (III)

(b) Two of the structures (I and III) have a double bond between the C_1-C_2 carbons, whereas only structure II has a double bond between the C_2-C_3 carbons. Assuming that the three structures contribute nearly equally, the C_1-C_2 bond should be more like a double bond and therefore should be shorter than the C_2-C_3 bond.

12.9

C_6H_5 and C_6H_5

$C=C$

$C+$

$:\ddot{O}:^-$

III

III is a more important contributor to the resonance hybrid of I than a corresponding ionic structure of II is to the hybrid of II. III is an important contributor to the hybrid of diphenylcyclopropenone because it resembles the aromatic cyclopropenyl cation (Cf problem 12.7); i.e., the ring in structure III has 2 π electrons and is a $4n + 2$ system where $n = 0$.

12.10

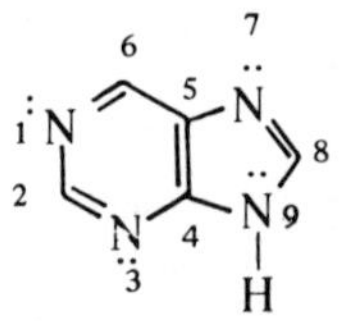

1,3,7 are pyridine type nitrogens; 9 is a pyrrole type nitrogen

12.11

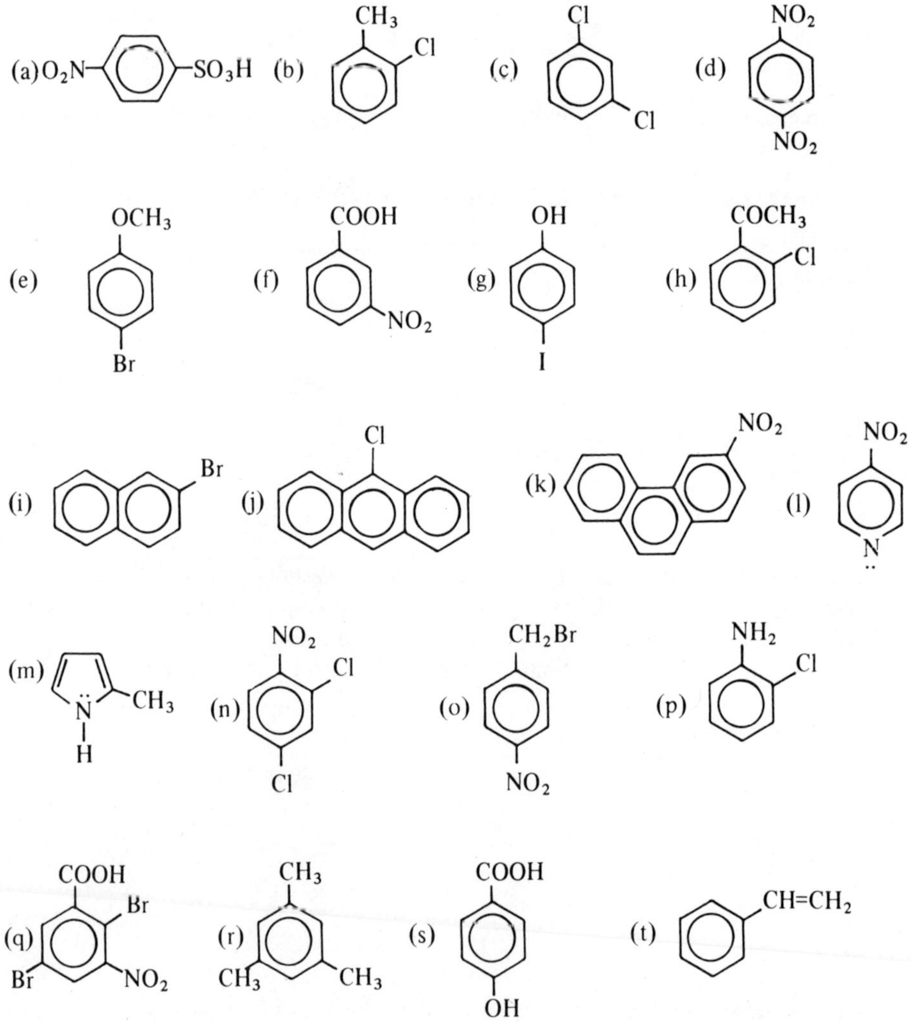

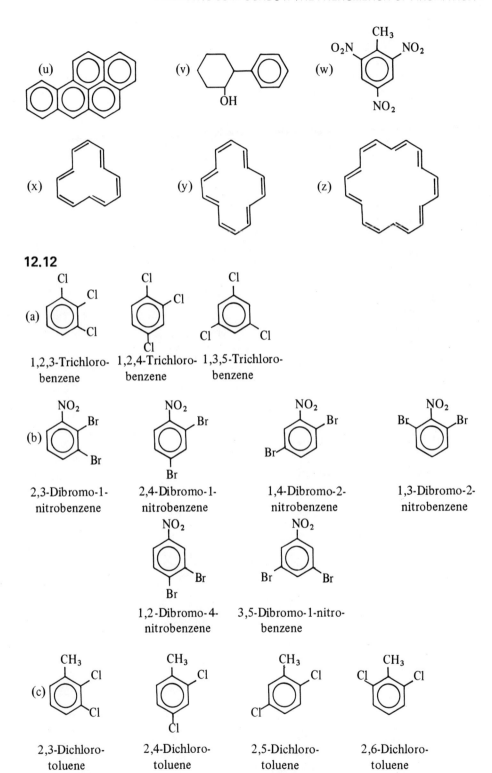

(u)

(v)

(w)

(x)

(y)

(z)

12.12

(a)

1,2,3-Trichloro-
benzene   1,2,4-Trichloro-
benzene   1,3,5-Trichloro-
benzene

(b)

2,3-Dibromo-1-
nitrobenzene   2,4-Dibromo-1-
nitrobenzene   1,4-Dibromo-2-
nitrobenzene   1,3-Dibromo-2-
nitrobenzene

1,2-Dibromo-4-
nitrobenzene   3,5-Dibromo-1-nitro-
benzene

(c)

2,3-Dichloro-
toluene   2,4-Dichloro-
toluene   2,5-Dichloro-
toluene   2,6-Dichloro-
toluene

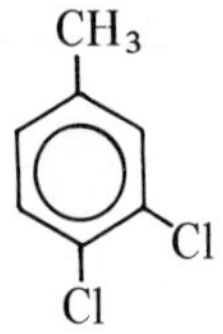

3,4-Dichloro-
toluene

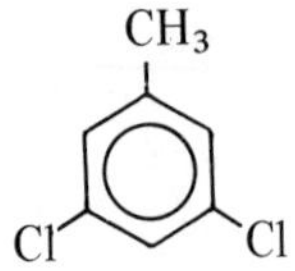

3,5-Dichloro-
toluene

(d) 1-Chloronaphthalene

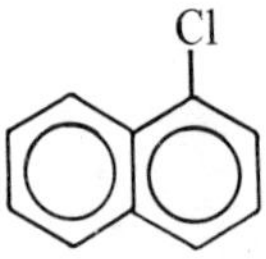

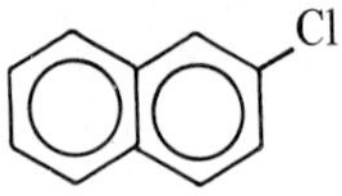

2-Chloronaphthalene

(e) 2-Nitro-
pyridine

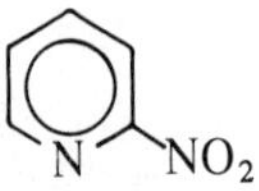

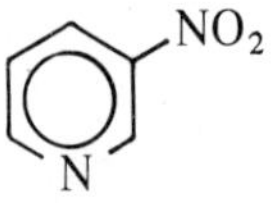

3-Nitro-
pyridine

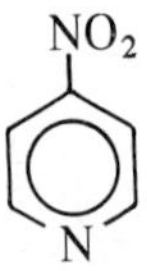

4-Nitro-
pyridine

(f) 2-Methylfuran

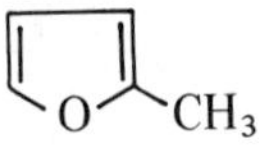

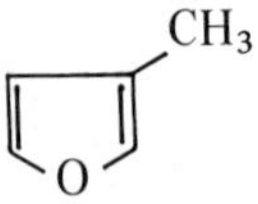

3-Methylfuran

(g) 1-Chloro-
2,3-dinitrobenzene

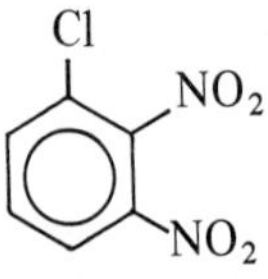

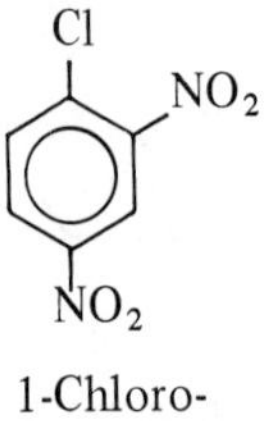

1-Chloro-
2,4-dinitrobenzene

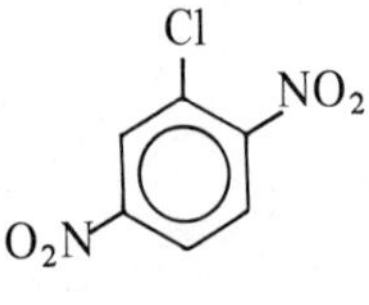

2-Chloro-
1,4-dinitrobenzene

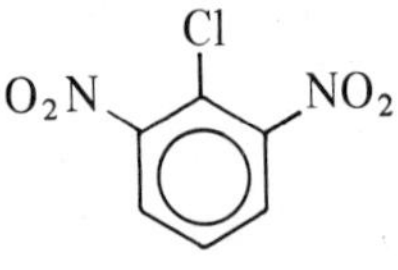

2-Chloro-
1,3-dinitrobenzene

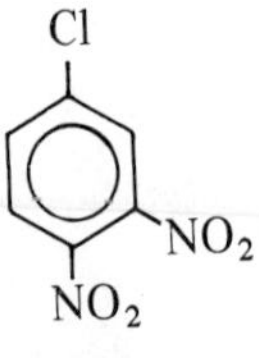

4-Chloro-
1,2-dinitrobenzene

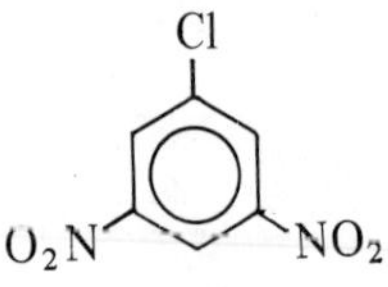

1-Chloro-
3,5-dinitrobenzene

(h)

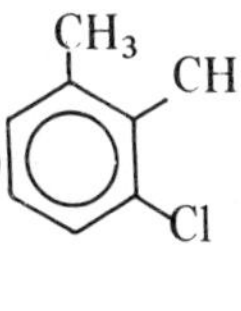

1-Chloro-
2,3-dimethylbenzene

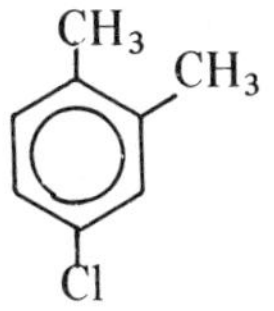

4-Chloro-
1,2-dimethylbenzene

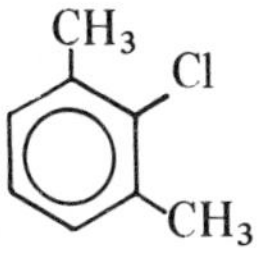

2-Chloro-
1,3-dimethylbenzene

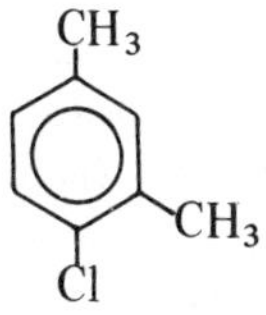

1-Chloro-
2,4-dimethylbenzene

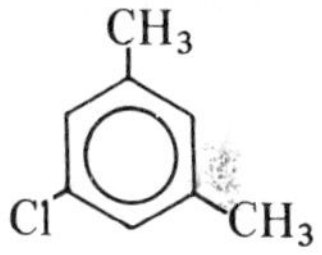

1-Chloro-
3,5-dimethylbenzene

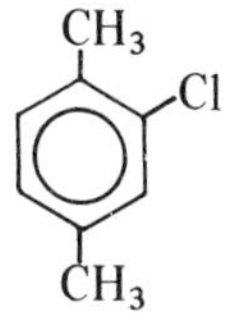

2-Chloro-
1,4-dimethylbenzene

(i)

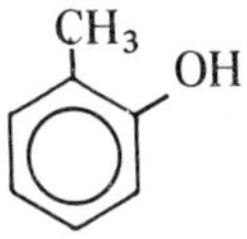

o-Cresol

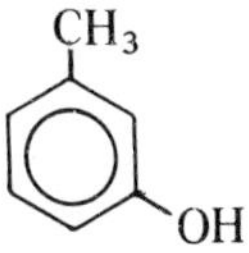

m-Cresol

p-Cresol

12.13

(a)

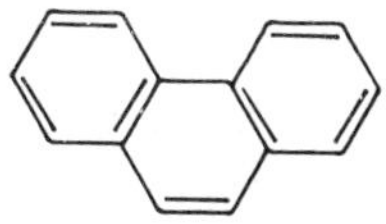

I

II

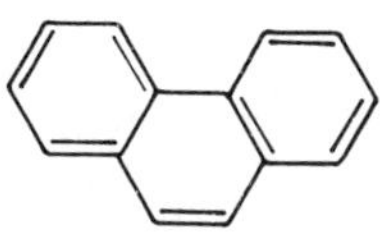

III

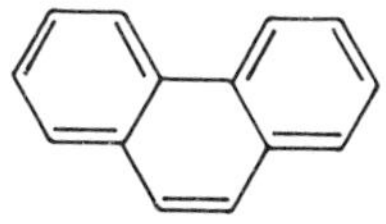

IV

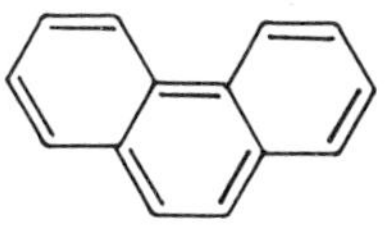

V

(b) The 9,10 bond should be close to that of a double bond, 1.33Å, since in four of the five contributors it is a double bond.

(c) Almost that of an actual double bond.

(d) Bromine adds to the 9,10 double bond because of its large double bond character and because addition disrupts only one of three aromatic rings.

12.14

(a)  (mp + 87°) (b) 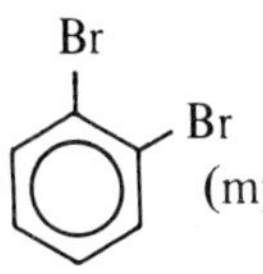 (mp + 6°) (c) 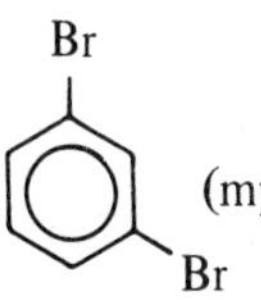 (mp − 7°)

12.15
By mononitrating each xylene and separating the mononitration products. The *ortho*-xylene will yield two mononitroxylenes, *meta*-xylene will yield three, and *para*-xylene will yield only one.

12.16

Three. 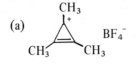 $\longrightarrow$ 2 mononitro products

$\longrightarrow$ 3 mononitro products

$\longrightarrow$ 1 mononitro product

12.17

(a)

CH_3

BF_4^-

CH_3 CH_3

(b) The trimethylcyclopropenyl cation is aromatic.

12.18

C_6H_5 C_6H_5

Br^-

OH

or

C_6H_5 C_6H_5

$(+)$

Br^-

OH

12.19

(a) 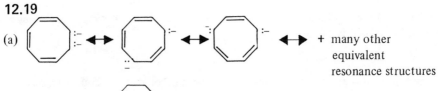 $\longleftrightarrow$ + many other equivalent resonance structures

(It may be drawn as ⬡ .)

(b) It has $4n + 2 = 10$ π electrons ($n = 2$), and is therefore aromatic; i.e., it obeys Huckel's rule.

12.20

(a) Cyclononatetraenyl anion: ⬡ Li^+ .

(b)

The product anion has 10 π-electrons and is therefore aromatic.

12.21

(a) $+ H_2 \longrightarrow$ $\Delta H = (-49.8) - (-55.4) = +5.6 \text{ kcal/mole}$

(b) 2 $\longrightarrow$ $+$
$\Delta H = -5.6 + (-55.4 - (-28.6))$
$= -5.6 - 26.8 = -32.4 \text{ kcal/mole}$

(c) Reaction (a) is endothermic, and the competing reaction (reaction (b)) is exothermic. Therefore we must conclude that reaction (a) is unlikely under conditions (i.e., the presence of a catalyst) that will allow equilibrium to be established.

12.22

(a)
mp 104°   mp 63°
mp 142°

(b) mp 104°:   mp 142°:

mp 63°:

12.23

(a) Would not be aromatic. It is a monocyclic system of 12 π electrons and thus does not obey Huckel's rule.

(b) Would not be aromatic; it is not a conjugated system.

(c) Would not be aromatic; it is an 8 π electron monocyclic system and thus does not obey Huckel's rule.

(d) Would not be aromatic; it is a 16 π electron monocyclic system and thus does not obey Huckels rule.

(e) Would be aromatic because of resonance structures (below) that consist of a cycloheptatrienyl cation and cyclopentadienyl anion.

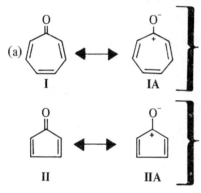

(f)   Would be aromatic; it is a planar monocyclic system of 14 π electrons. (We count only two electrons of the triple bond because only two are in p-orbitals that overlap with those of the double bonds on either side.)

(g) Would be aromatic; it is a planar monocyclic system of 10 π electrons.

(h) Would be aromatic; it is a planar monocyclic system of 10 π electrons. (The bridging $-CH_2-$ groups allows the ring system to be planar.)

12.24

Resonance contributors that involve the carbonyl group of **I** resemble the *aromatic* cycloheptatrienyl cation and thus stabilize **I**. Similar contributors to the hybrid of **II** resemble the *antiaromatic* cyclopentadienyl cation (see Prob. 12.7) and thus destabilize **II**.

(a)

Contributors like this are exceptionally stable because they resemble an aromatic compound. They therefore make large stabilizing contributions to the hybrid.

Contributors like this are exceptionally unstable because they resemble an antiaromatic compound. Any contribution they make to the hybrid is destabilizing.

(b)

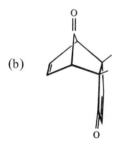

12.25

C is [14]-annulene. The steps in its synthesis are the following:

B

C

13

AROMATIC COMPOUNDS II: REACTIONS OF AROMATIC COMPOUNDS WITH ELECTROPHILES

13.1

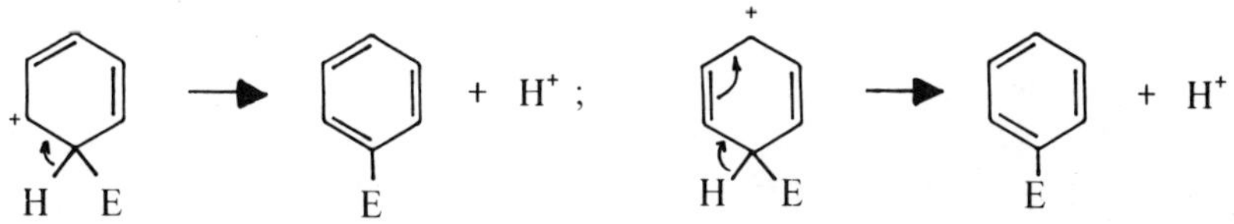

13.2

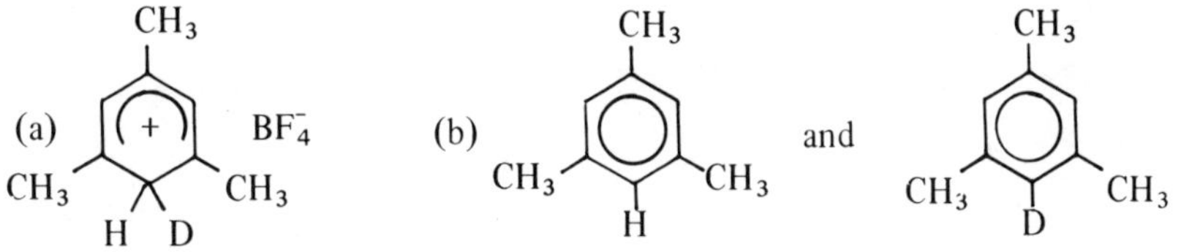

(a)   (b)   and

13.3

Oxidation of the Fe by X_2 generates the ferric salt:

$$2Fe + 3X_2 \longrightarrow 2FeX_3$$

13.4

$$H-\ddot{O}-NO_2 + H-\ddot{O}-NO_2 \longrightarrow H-\overset{+}{\underset{H}{\ddot{O}}}-NO_2 + NO_3^-$$

$$H-\overset{+}{\underset{H}{\ddot{O}}}-NO_2 + HONO_2 \longrightarrow NO_2^+ + H_3O^+ + NO_3^-$$

13.5

(a)

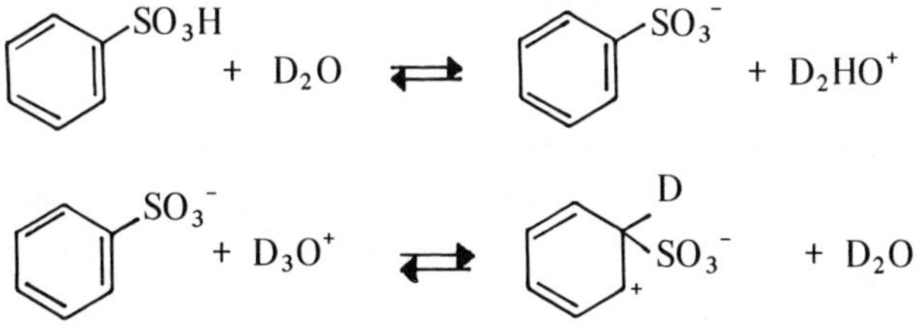

$$\text{C}_6\text{H}_5\text{SO}_3\text{H} + D_2O \rightleftharpoons \text{C}_6\text{H}_5\text{SO}_3^- + D_2HO^+$$

$$\text{C}_6\text{H}_5\text{SO}_3^- + D_3O^+ \rightleftharpoons \text{(arenium)} SO_3^- + D_2O$$

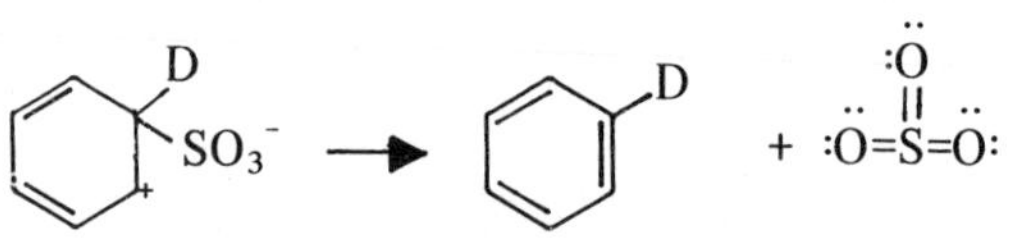

(b) $:\overset{..}{Br}-\overset{..}{Br}-Fe Br_3$ is the electrophile.

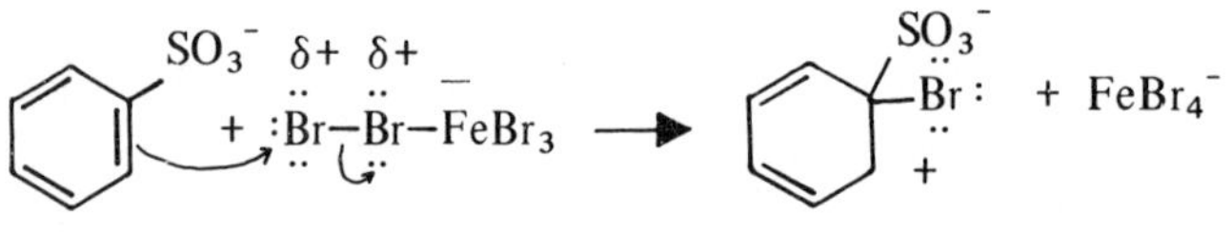

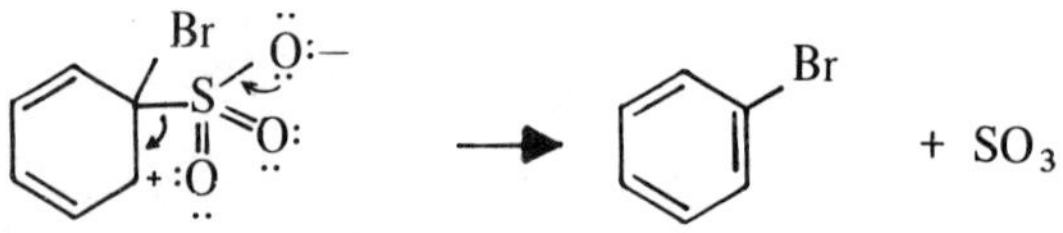

13.6

(a)

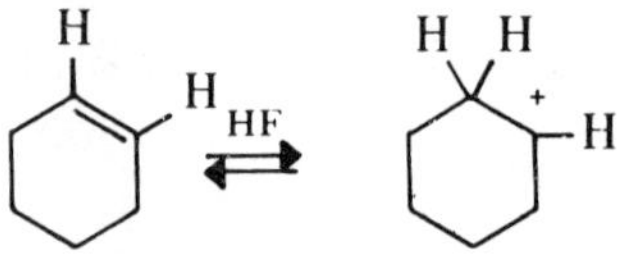

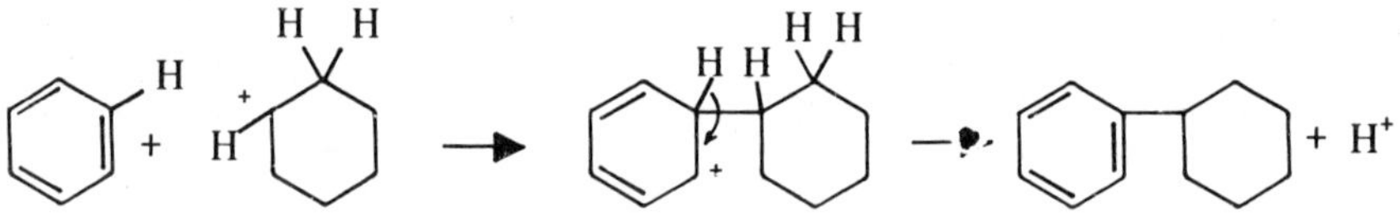

(b)

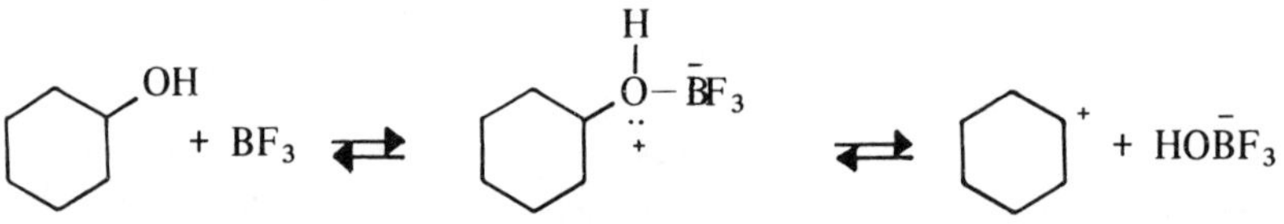

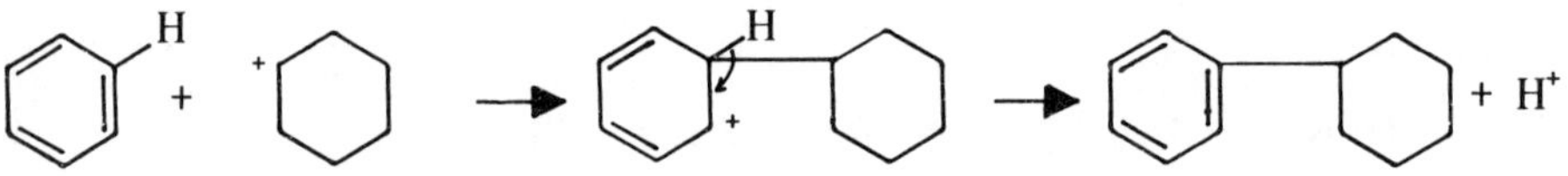

13.7

(a) $CH_3 CH_2 CH_2 CH_2 -Br + AlCl_3 \rightleftharpoons CH_3 CH_2 CH_2 CH_2{}^+ + \overset{-}{A}lCl_3 Br$

$CH_3 CH_2 CH_2 CH_2{}^+ \xrightarrow[\text{shift}]{\text{Hydride}} CH_3 CH_2 \overset{+}{C}HCH_3$ (secondary carbocation is more stable than primary)

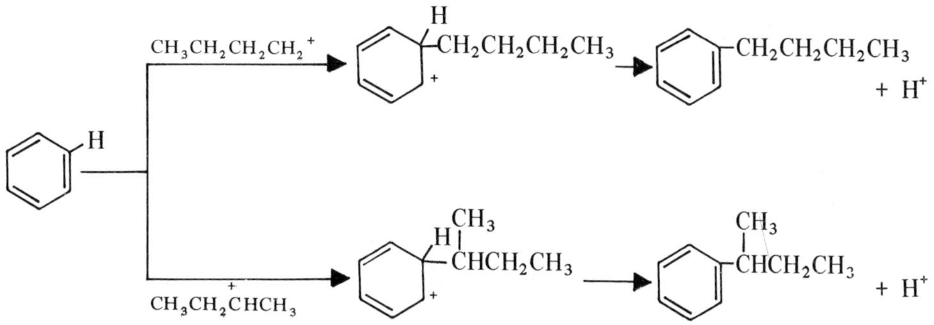

(b) $CH_3CH_2CH_2-OH + BF_3 \rightleftharpoons CH_3CH_2CH_2^+ + HOBF_3^-$.

The *n*-propyl cation can rearrange to an isopropyl cation:

$$CH_3CH_2\overset{+}{C}H_2 \xrightarrow[\text{Shift}]{\text{Hydride}} CH_3\overset{+}{C}HCH_3$$

Both cations can then attack the benzene ring.

13.8

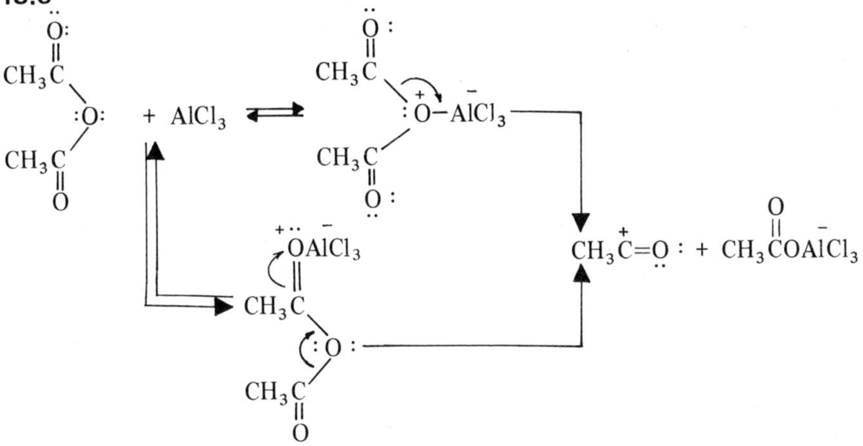

13.9

(a)

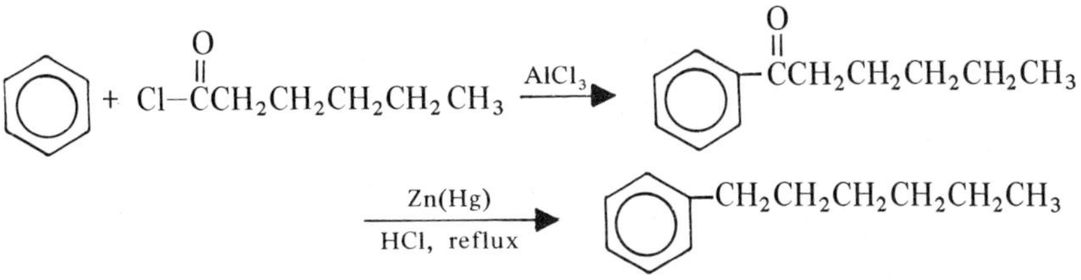

(b)

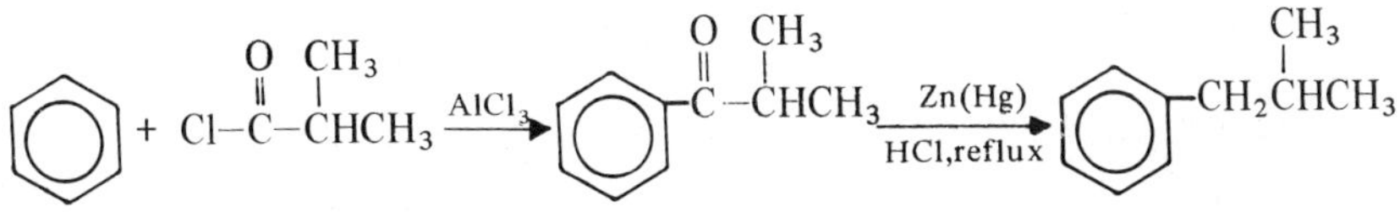

13.10
40% ortho, 40% meta, 20% para.

13.11
(a) At the lower temperature the proportions are determined by the relative reaction rates. At the higher temperature the proportions are determined by the relative stabilities; i.e., the more stable isomer predominates even if it is produced at the slower rate because all steps are reversible.

(b) p-Toluenesulfonic acid.

13.12

13.13

(f)

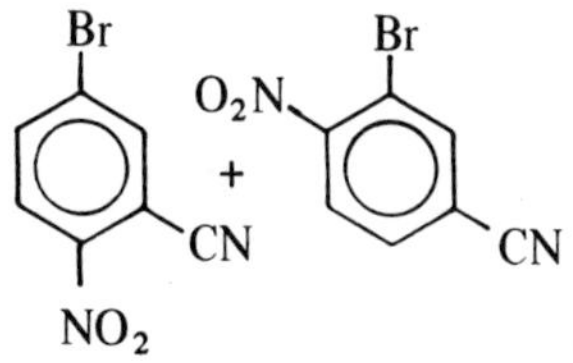

(g)

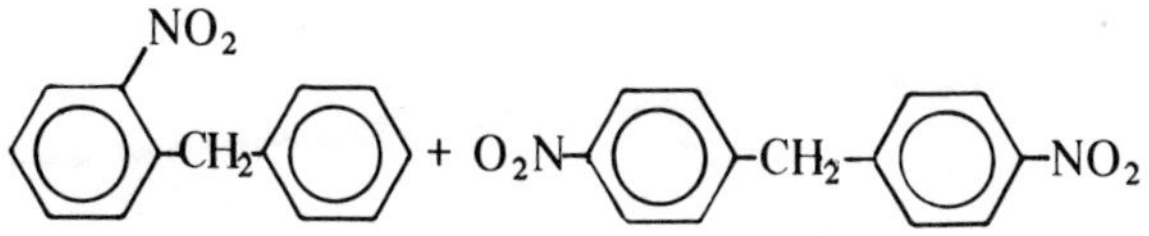

13.14

(a) $Cl-CH_2-CH_2-\overset{+}{N}(CH_3)_3\overset{-}{Cl}$. This is anti-Markovnikov addition because the intermediate carbocation formed by Markovnikov addition would have positive charges on adjacent atoms, i.e.,

$$CH_3-\overset{+}{C}H-\overset{+}{N}(CH_3)_3\overset{-}{Cl}$$

The anti-Markovnikov carbocation is $\overset{+}{C}H_2-CH_2-\overset{+}{N}(CH_3)_3\overset{-}{Cl}$. Although this is a primary carbocation, it is more stable because of the greater separation of charges.

(b) Slower because it requires formation of an intermediate primary cation with two positive charges.

(c) The positive charge of the trimethylammonium ion makes it a powerful electron-withdrawing group.

(d)

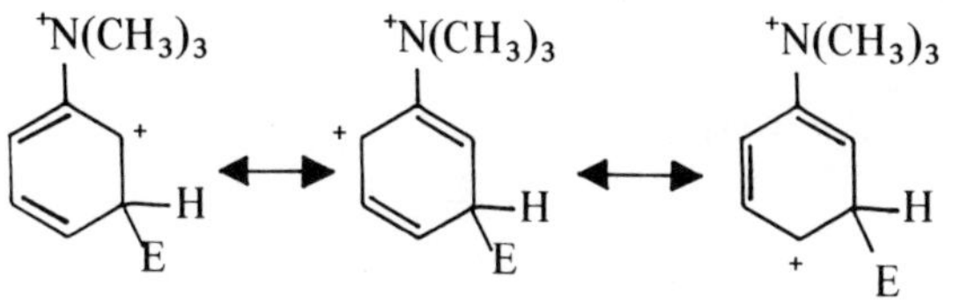

None of these structures is especially unstable, whereas in *ortho* or *para* attack at least one structure would be.

13.15

(a) The electronic influence of the trimethylammonium ion on the ring is an inductive effect that causes *meta* orientation and it is diminished by every methylene group that separates it from the ring.

(b) The increasing number of chlorines makes the methyl group increasingly electron-withdrawing, and hence increasingly *meta*-directing.

13.16

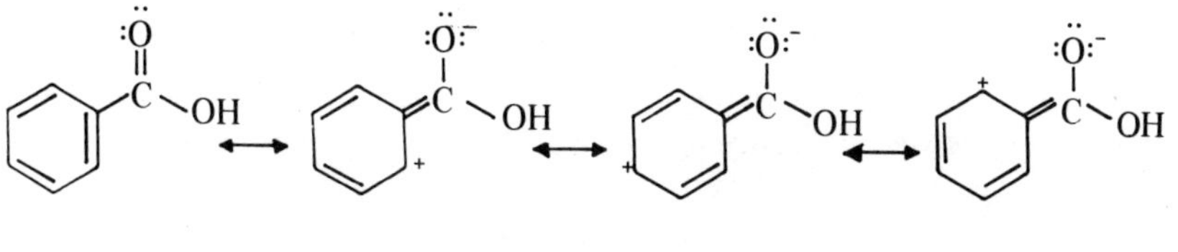

13.17

(a) ortho:

meta:

para:

(b) The electron-releasing ability of the —OH group through resonance increases the electron density of the ring, and it stabilizes the positive charge of the intermediate carbo-cation.

(c) An exceptionally stable structure (above) contributes to the intermediate carbocation only when attack is *ortho* or *para*.

(d,e) More reactive because the negatively charged —O⁻ group of the phenoxide ion is an even more powerful electron-releasing group than the —OH group of phenol.

13.18

(a)

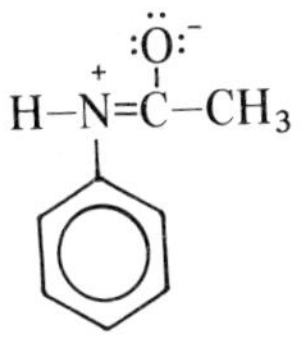

$$\underset{}{H-\overset{+}{N}=\overset{\overset{\displaystyle :\ddot{O}:^-}{|}}{C}-CH_3}$$

(b) The $-\overset{\overset{\displaystyle O}{\|}}{C}-CH_3$ group competes with the ring for the electron pair on N, therefore stabilization of the intermediate carbocation is less effective than in aniline.

(c)

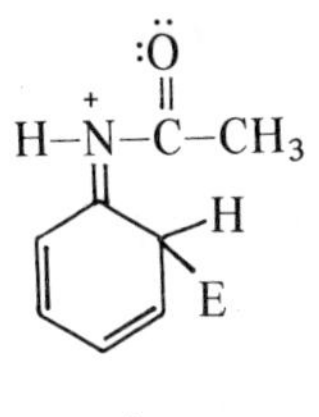

$$H-\overset{+}{N}-\overset{\overset{\displaystyle :\ddot{O}}{\|}}{C}-CH_3$$

A

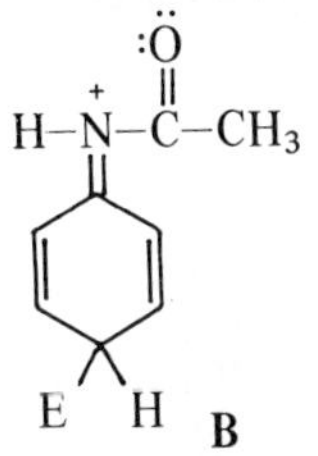

$$H-\overset{+}{N}-\overset{\overset{\displaystyle :\ddot{O}}{\|}}{C}-CH_3$$

E   H   **B**

Yes, resonance accounts for an electron release from nitrogen to the ring, and exceptionally stable structures (**A** and **B** above) contribute to the sigma complexes formed when attack takes place at an *ortho* or *para* carbon.

(d) Phenyl acetate should be less reactive than phenol because the $-COCH_3$ group competes with the ring for electrons on oxygen as shown in the following structure.

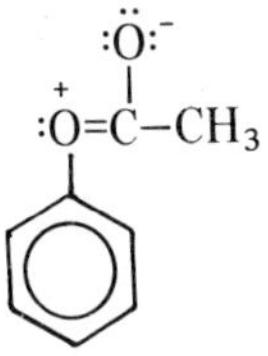

$$:O=\overset{\overset{\displaystyle :\ddot{O}:^-}{|}}{\overset{+}{C}}-CH_3$$

Notice that this structure also places a positive charge on the oxygen attached to the ring.

(e) Ortho-para

(f) More reactive

13.19

In each case the orientation results from the formation of the more stable intermediate carbocation.

With 3,3,3-tri fluoropropene the possibilities are:

$$CF_3-CH=CH_2 + H^+$$

$$\rightarrow CF_3-\overset{+}{CH}-CH_3$$
I
(less stable)

$$\rightarrow CF_3-CH_2-\overset{+}{CH_2} \xrightarrow{Cl^-} CF_3-CH_2-CH_2Cl$$
II
(more stable)

Although II is a primary carbocation and I is a secondary carbocation, II is more stable because in it the positive charge is separated from the highly electron-withdrawing CF_3 group by an intervening carbon.

With chloroethene the possibilities are:

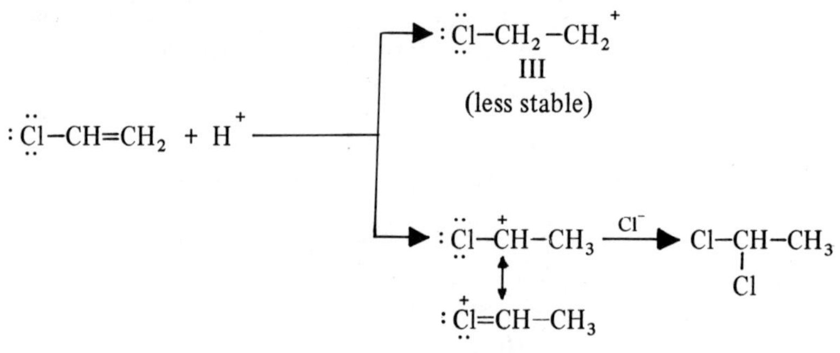

$$: \overset{..}{\underset{..}{Cl}}-CH=CH_2 + H^+$$

$$\longrightarrow : \overset{..}{\underset{..}{Cl}}-CH_2-CH_2^+$$

III
(less stable)

$$\longrightarrow : \overset{..}{\underset{..}{Cl}}-\overset{+}{C}H-CH_3 \xrightarrow{Cl^-} Cl-\underset{\underset{Cl}{|}}{CH}-CH_3$$

$$\updownarrow$$

$$: \overset{..}{\underset{..}{Cl}}=\overset{+}{C}H-CH_3$$

IV
(more stable)

Here carbocation IV is more stable than III because of resonance involving an electron pair of the chlorine atom.

13.20

(a) ortho:

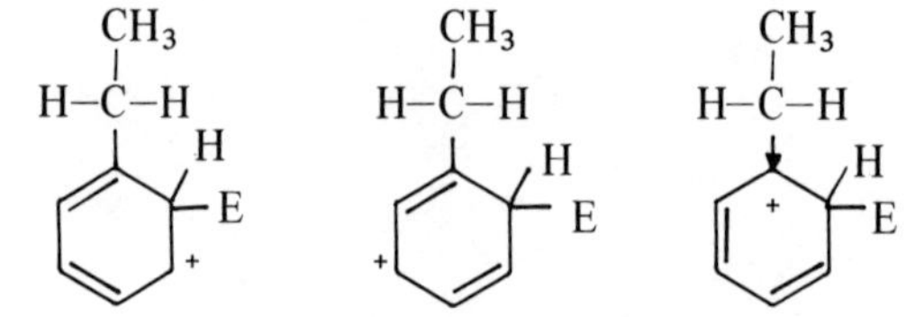

(exceptionally stable)

meta:

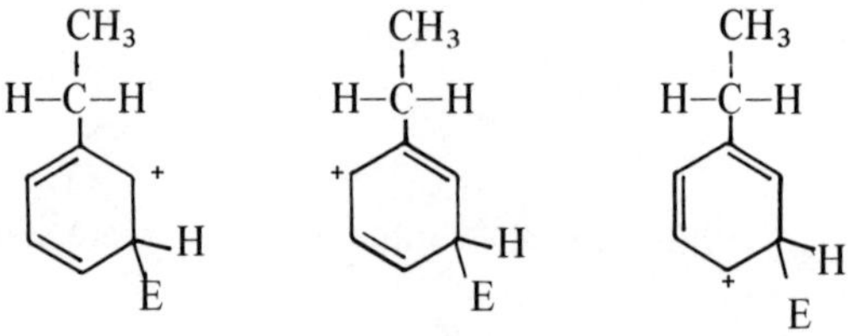

para:

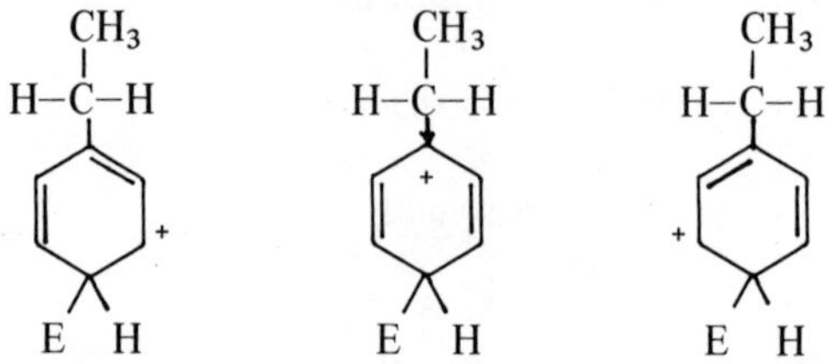

(exceptionally stable)

(b) Yes, substitution at an ortho or para position yields a carbocation that is stabilized by the contribution of an exceptionally stable structure. The exceptionally stable structures (above) have a positive charge on the ring carbon that bears the ethyl group.

(c) Because the ethyl group is electron-releasing it stabilizes the intermediate sigma complexes.

13.21

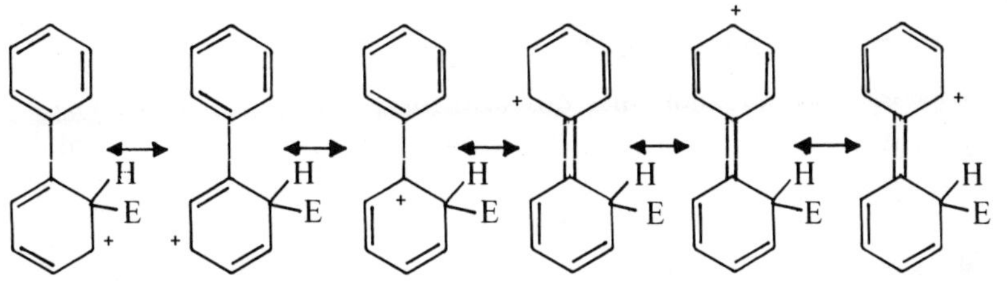

The carbocation above has the positive charge delocalized over both rings and thus it is exceptionally stable. Similar structures can be drawn for the sigma complex formed when substitution takes place at a para position. However, when electrophilic attack takes place at the meta position it produces a carbocation whose positive charge cannot be delocalized over both rings:

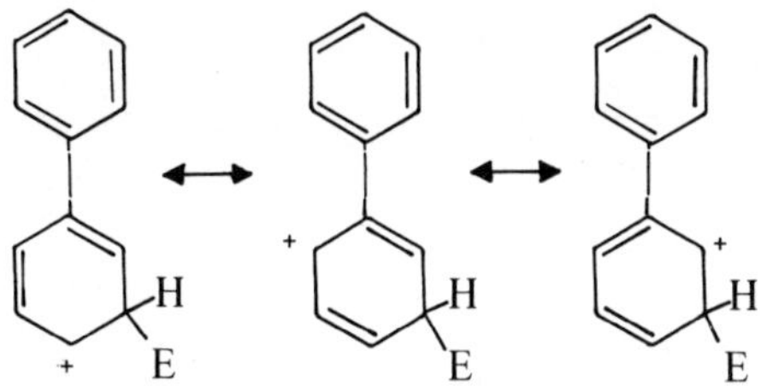

13.22
(a) 1-Chloro-1-phenylethane results from the more stable free radical. The more stable

free radical is $\bigcirc$–ĊHCH$_3$ because the unshared electron is conjugated with the ring:

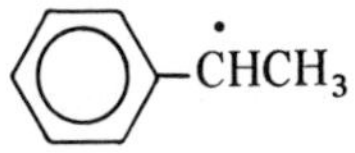

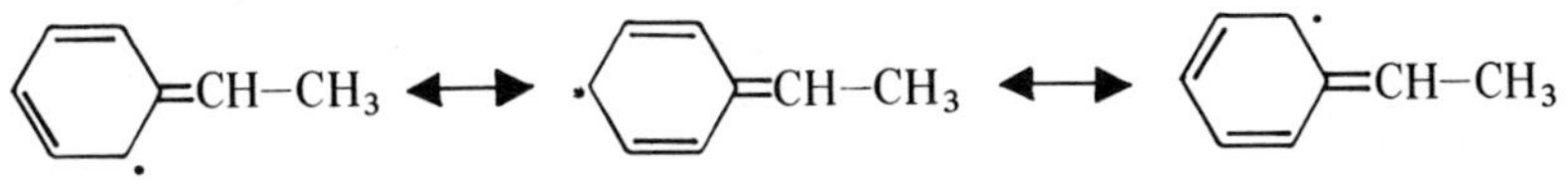

(b) A primary free radical is involved, and the unpaired electron is not conjugated with

the ring: $\bigcirc$–CH$_2$–CH$_2$·

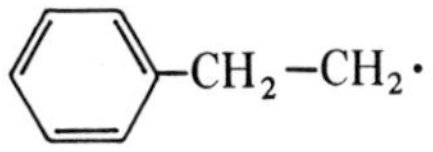

(c) 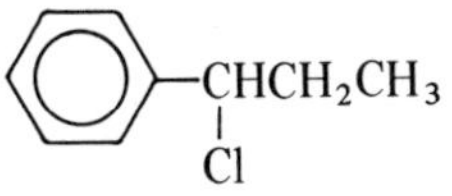 $\bigcirc$–CHCH$_2$CH$_3$ (1-chloro-1-phenylpropane)
        |
        Cl

13.23

(a)

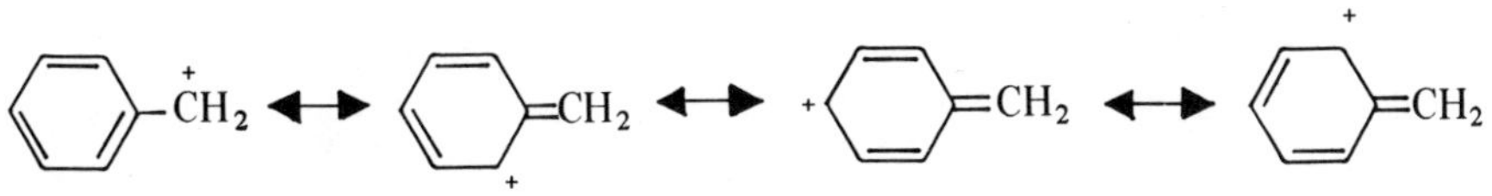

(b) The benzyl cation is stabilized by resonance.

(c) Yes, because the benzyl cation is a hybrid of the structures given in (a).

(d) Over the ortho and para ring carbons and the benzylic carbon.

(e) π_1, π_2, π_3 (see Figure 13.6).

(f) Since π_4 is vacant, molecular orbital theory predicts the same thing as resonance theory—that the positive charge is delocalized over the benzylic carbon and the ortho and para ring carbons.

13.24

(a) Anti addition gives: Syn-addition gives:

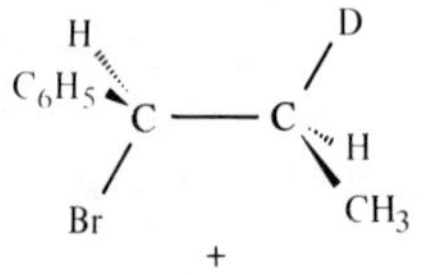

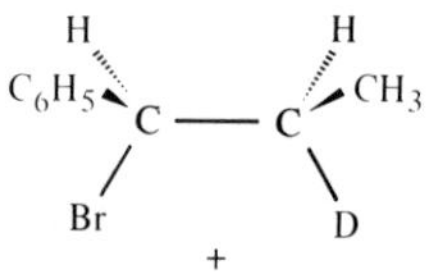

enantiomer   enantiomer

(b) Yes, anti-addition and syn-addition give diastereomeric products.

13.25

(a)   (b)

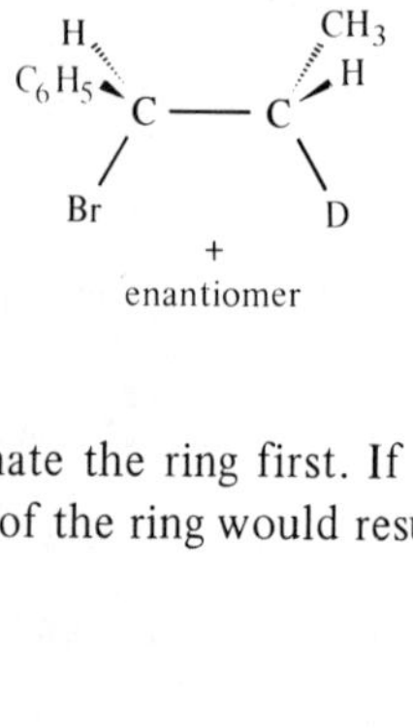

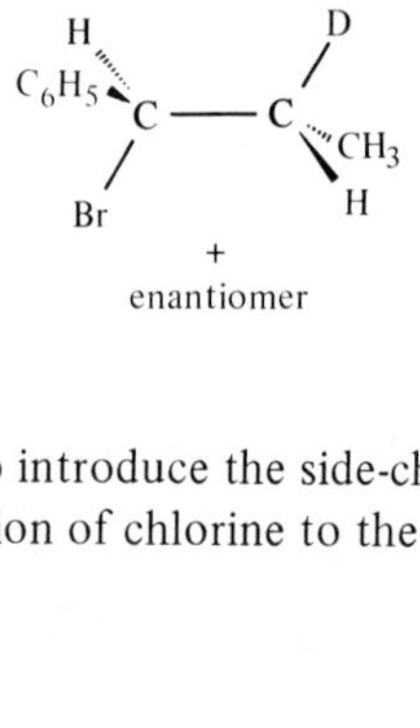

enantiomer   enantiomer

13.26

Chlorinate the ring first. If we were to introduce the side-chain double bond first, chlorination of the ring would result in addition of chlorine to the side-chain double bond.

13.27

(a)

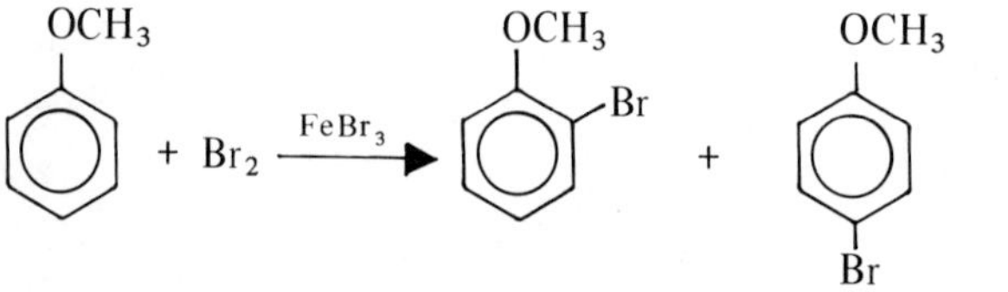

o-Bromoanisole p-Bromoanisole

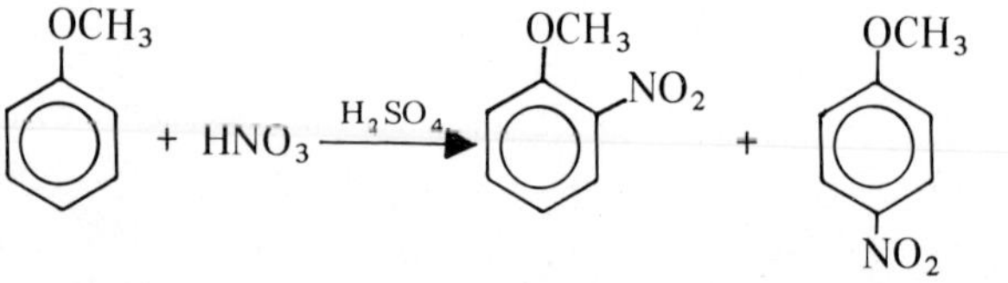

o-nitroanisole p-nitroanisole

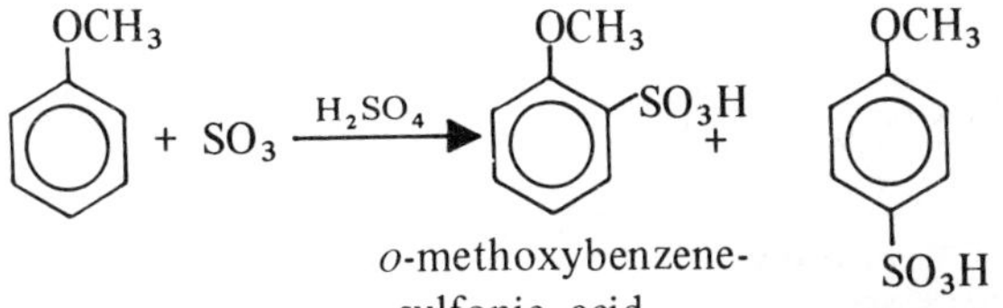

o-methoxybenzene-
sulfonic acid

p-methoxybenzene-
sulfonic acid

Reactions are faster than the corresponding reactions of benzene.

(b)

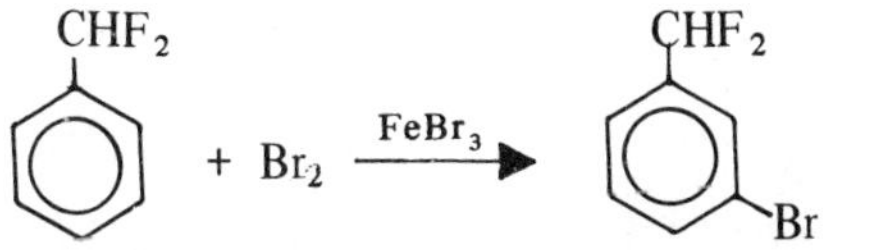

m-Bromobenzal
difluoride

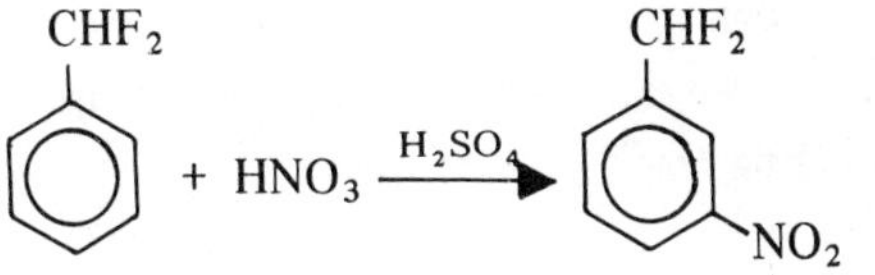

m-Nitrobenzal
difluoride

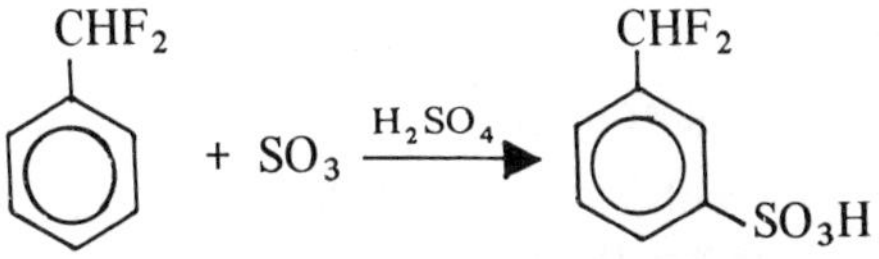

m-Difluoromethyl-
benzenesulfonic acid

Reactions are slower than corresponding reactions of benzene.

(c)

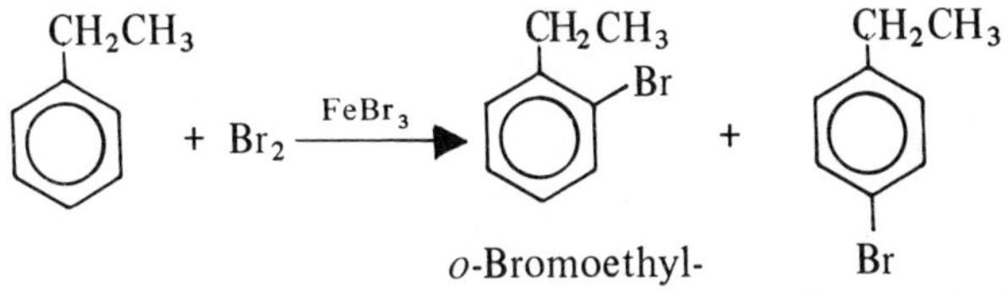

o-Bromoethyl-
benzene

p-Bromoethyl-
benzene

nitration ⟶ o-nitroethylbenzene and p-nitroethylbenzene

sulfonation ⟶ o-ethylbenzenesulfonic acid and p-ethylbenzenesulfonic acid.

Reactions are faster than corresponding reactions of benzene.

(d)

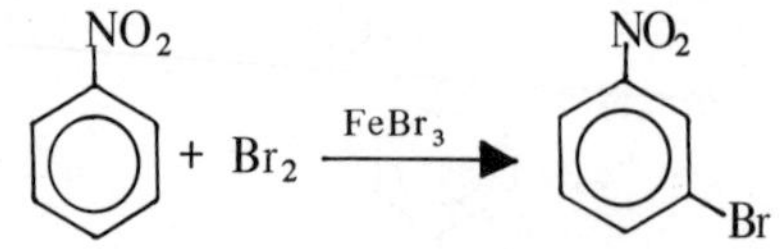

m-Bromonitrobenzene

nitration ———▶ *m*-dinitrobenzene

sulfonation ———▶ *m*-nitrobenzenesulfonic acid

Reactions are slower than corresponding reactions of benzene.

(e)

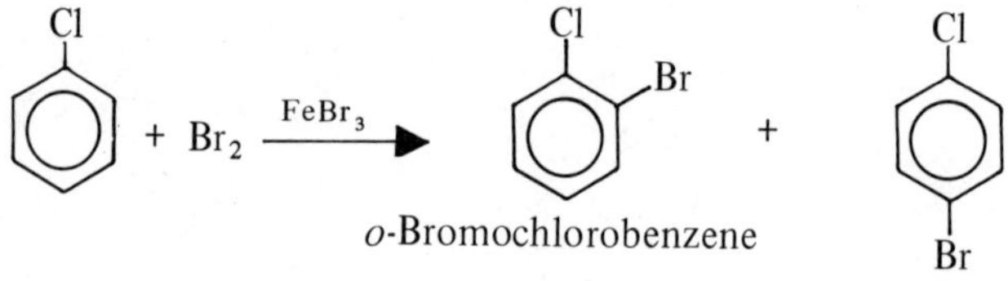

o-Bromochlorobenzene

p-bromochloro-
benzene

nitration———▶*o*-nitrochlorobenzene + *p*-nitrochlorobenzene

sulfonation———▶*o*-chlorobenzenesulfonic acid + *p*-chlorobenzenesulfonic acid

Reactions are slower than corresponding reactions of benzene.

(f)

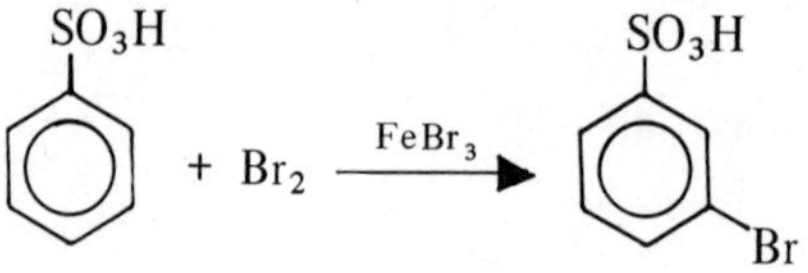

m-Bromobenzenesulfonic acid

nitration ———▶ *m*-nitrobenzene sulfonic acid

sulfonation ———▶ *m*-benzenedisulfonic acid

Reactions are slower than corresponding reactions of benzene.

(g)

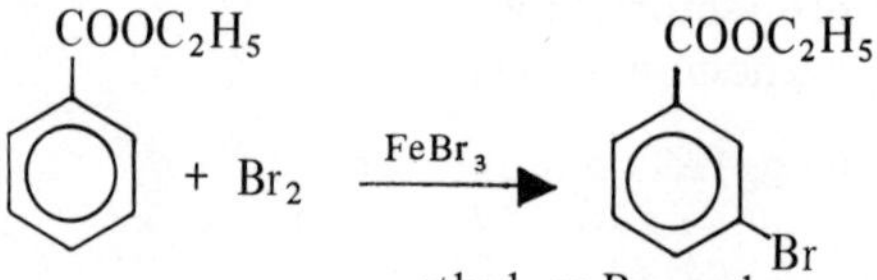

ethyl *m*-Bromobenzoate*

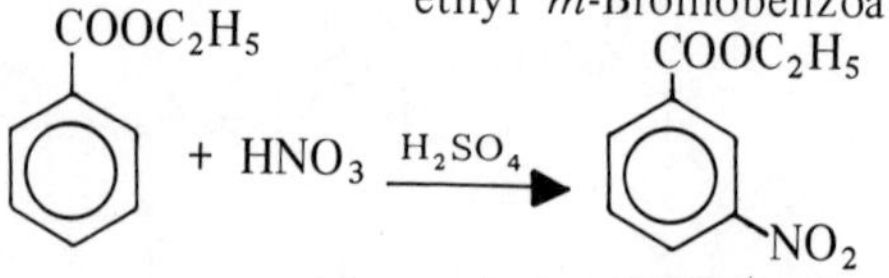

ethyl *m*-Nitrobenzoate*

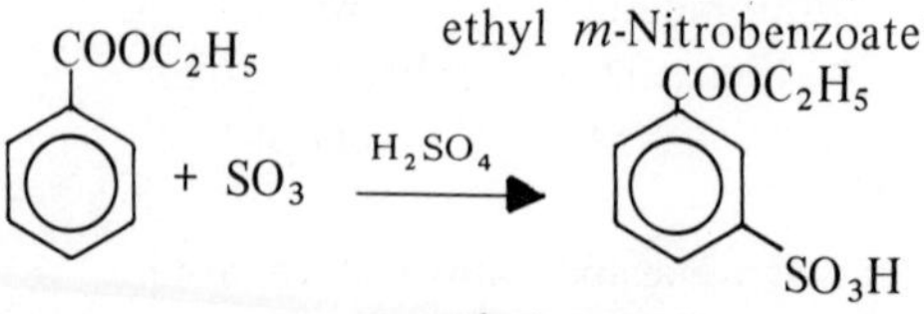

ethyl *m*-Sulfobenzoate*

Reactions are slower than corresponding reactions of benzene.

*Author's apology: Rules for naming the starred compounds in this problem have not been given in the text.

(h)

$$\text{C}_6\text{H}_5-\text{O}-\text{C}_6\text{H}_5 + Br_2 \xrightarrow{FeBr_3}$$

o-bromophenoxybenzene   p-bromophenoxybenzene

nitration ⟶ o-nitrophenoxybenzene + p-nitrophenoxybenzene

sulfonation ⟶ o-phenoxybenzenesulfonic acid + p-phenoxybenzenesulfonic acid

Reactions are faster than corresponding reactions of benzene.

(i)

$$\text{C}_6\text{H}_5-\text{C}_6\text{H}_5 + Br_2 \xrightarrow{FeBr_3}$$

2-Bromobiphenyl*   4-Bromobiphenyl*

nitration ⟶ 2-nitrobiphenyl + 4-nitrobiphenyl

sulfonation ⟶ o-phenylbenzenesulfonic acid + p-phenylbenzenesulfonic acid

Reactions are faster than corresponding reactions of benzene.

(j)

$$CH_3-\underset{\underset{CH_3}{|}}{\overset{\overset{CH_3}{|}}{C}}-\text{C}_6\text{H}_5 + Br_2 \xrightarrow{FeBr_3}$$

o-Bromo-tert-
butylbenzene   p-Bromo-tert-
butylbenzene

nitration ⟶ o-nitro-tert-butylbenzene + p-nitro-tert-butylbenzene

sulfonation ⟶ o-tert-butylbenzenesulfonic acid + p-tert-butylbenzenesulfonic acid

Reactions are faster than corresponding reactions of benzene.

(k)

$$\text{C}_6\text{H}_5\text{F} + Br_2 \xrightarrow{FeBr_3}$$

o-Bromofluorobenzene   p-Bromofluorobenzene

nitration ⟶ o-nitrofluorobenzene + p-nitrofluorobenzene

sulfonation ⟶ o-fluorobenzenesulfonic acid + p-fluorobenzenesulfonic acid

Reactions are slower than corresponding reactions of benzene.

(l)

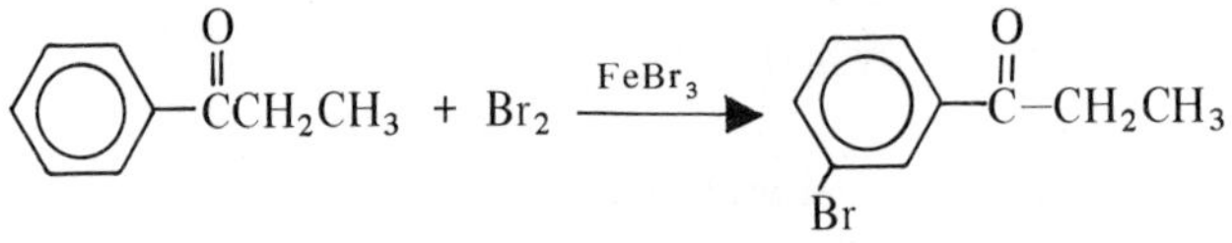

m-Bromopropanoylbenzene

nitration ⟶ *m*-Nitropropanoylbenzene
sulfonation ⟶ *m*-Propanoylbenzenesulfonic acid

Reactions are slower than corresponding reactions of benzene.

(m)

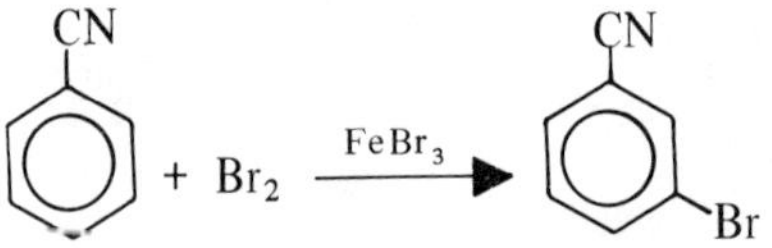

m-Bromobenzonitrile

nitration ⟶ *m*-nitrobenzonitrile
sulfonation ⟶ *m*-cyanobenzenesulfonic acid

Reactions are slower than corresponding reactions of benzene.

(n)

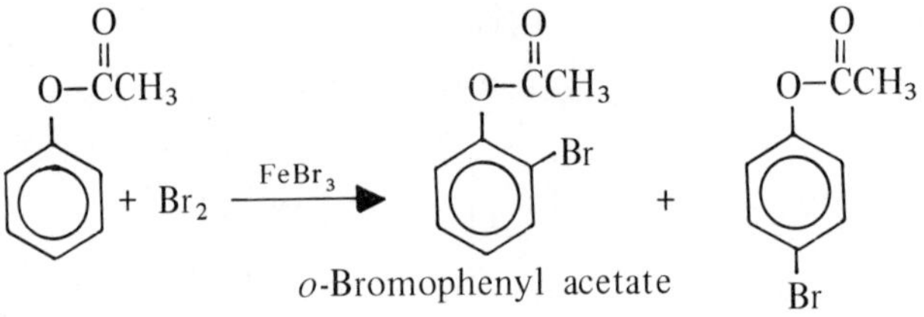

o-Bromophenyl acetate   +   *p*-Bromophenyl acetate

nitration ⟶ *o*-nitrophenyl acetate + *p*-nitrophenyl acetate

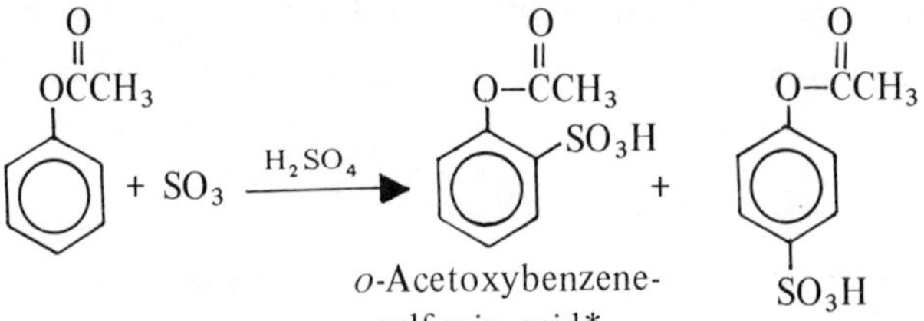

o-Acetoxybenzene-
sulfonic acid*   +   *p*-Acetoxybenzene-
sulfonic acid*

Reactions are faster than corresponding reactions of benzene.

(o)

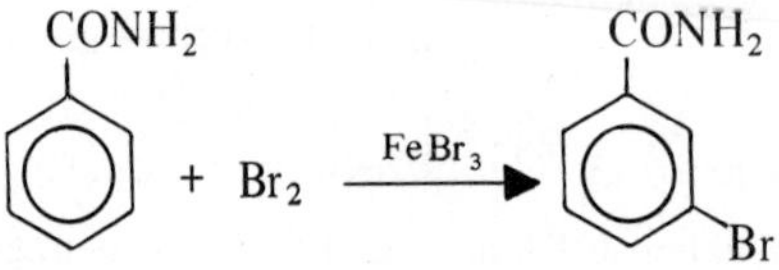

m-Bromobenzamide

nitration ——▶ *m*-nitrobenzamide

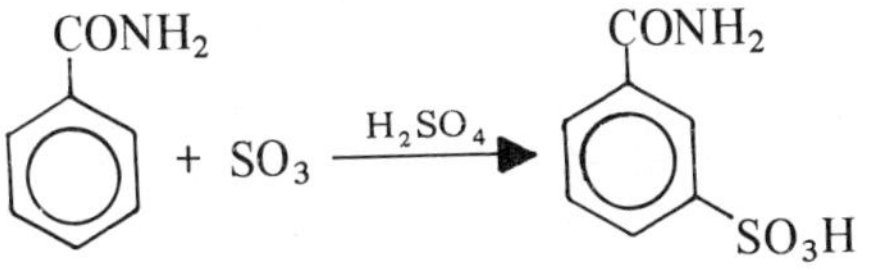

$$\text{(CONH}_2\text{ benzene)} + SO_3 \xrightarrow{H_2SO_4} \text{(CONH}_2\text{, SO}_3\text{H benzene)}$$

m-Carbamoylbenzene-
sulfonic acid*

Reactions are slower than corresponding reactions of benzene.

(p) 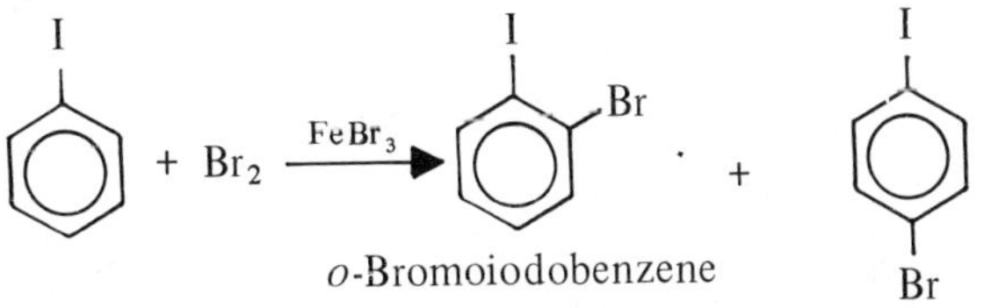 $$\text{(I benzene)} + Br_2 \xrightarrow{FeBr_3} \text{(I, Br benzene)} \quad + \quad \text{(I, Br benzene)}$$

o-Bromoiodobenzene   *p*-Bromoiodobenzene

nitration ——▶ *o*-nitroiodobenzene + *p*-nitroiodobenzene

sulfonation ——▶ *o*-iodobenzenesulfonic acid + *p*-iodobenzenesulfonic acid

Reactions are slower than corresponding reactions of benzene.

13.28

(a) 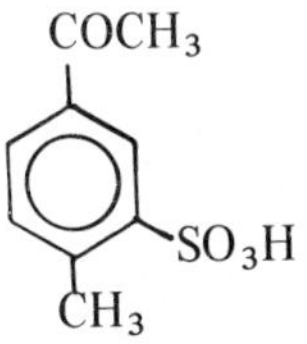 COCH₃ / SO₃H / CH₃ benzene

(b) 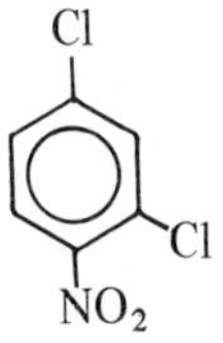 Cl / Cl / NO₂ benzene

(c) 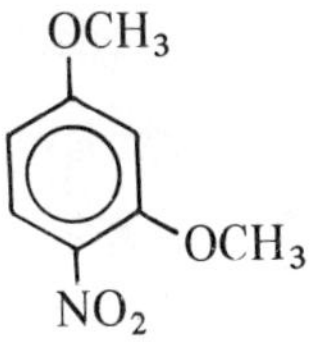 OCH₃ / OCH₃ / NO₂ benzene

(d) 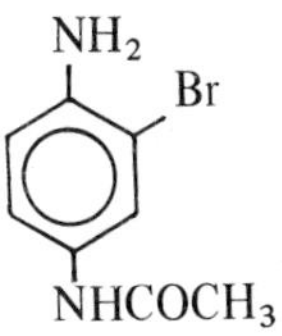 NH₂ / Br / NHCOCH₃ benzene

(e) 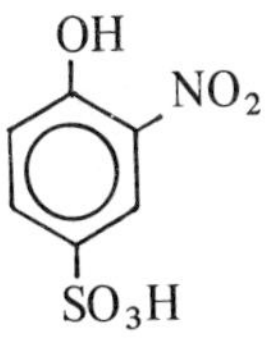 OH / NO₂ / SO₃H benzene

(f) 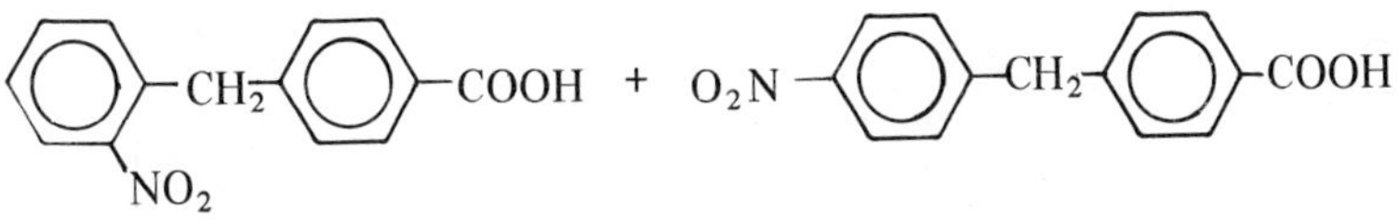 (benzene)-CH₂-(benzene)-COOH with NO₂   +   O₂N-(benzene)-CH₂-(benzene)-COOH

(g) 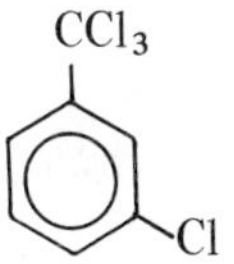 CCl₃ / Cl benzene

13.29

(a)

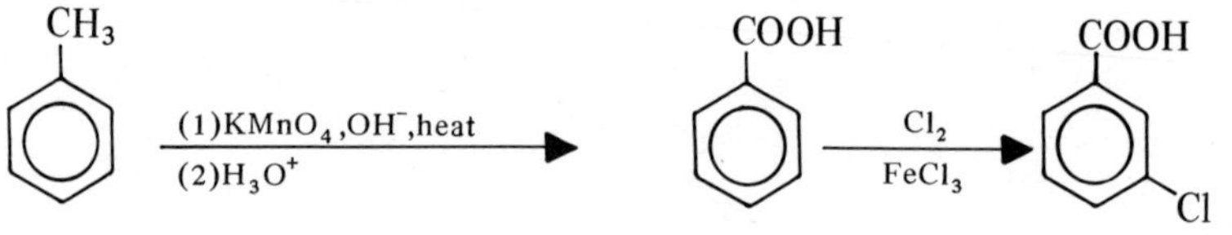

Toluene $\xrightarrow[\text{(2)}H_3O^+]{\text{(1)}KMnO_4,OH^-,\text{heat}}$ benzoic acid $\xrightarrow[FeCl_3]{Cl_2}$ 3-chlorobenzoic acid

(b)

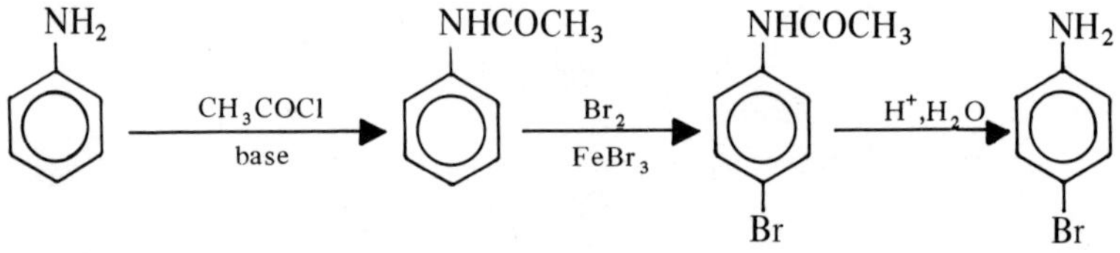

Aniline $\xrightarrow[\text{base}]{CH_3COCl}$ acetanilide $\xrightarrow[FeBr_3]{Br_2}$ 4-bromoacetanilide $\xrightarrow{H^+,H_2O}$ 4-bromoaniline

(c)

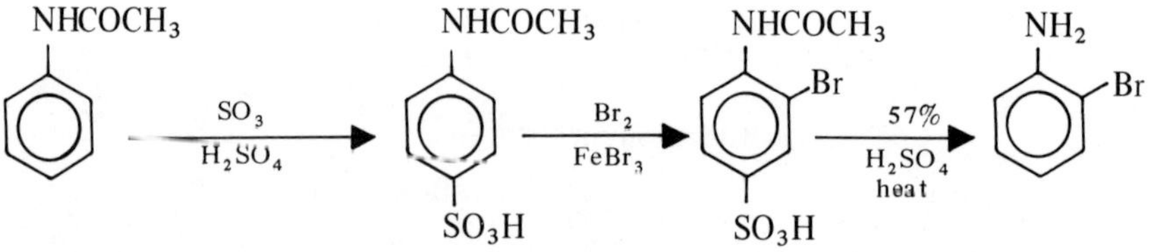

acetanilide $\xrightarrow{SO_3 / H_2SO_4}$ → $\xrightarrow[FeBr_3]{Br_2}$ → $\xrightarrow[\text{heat}]{57\% \; H_2SO_4}$ 2-bromoaniline

(d)

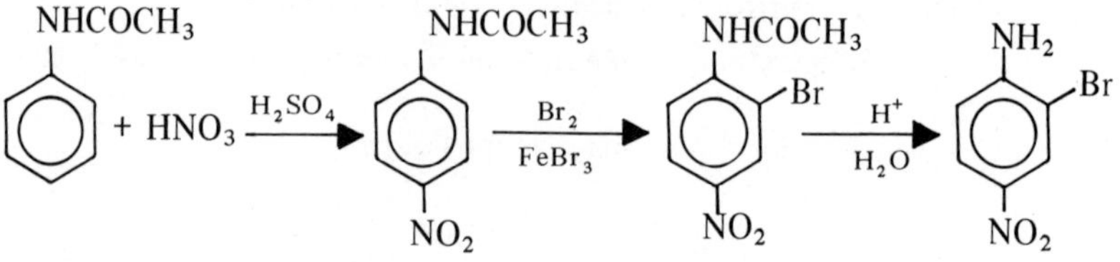

acetanilide $+ HNO_3 \xrightarrow[FeBr_3... H_2SO_4]{}$ ...

(e)

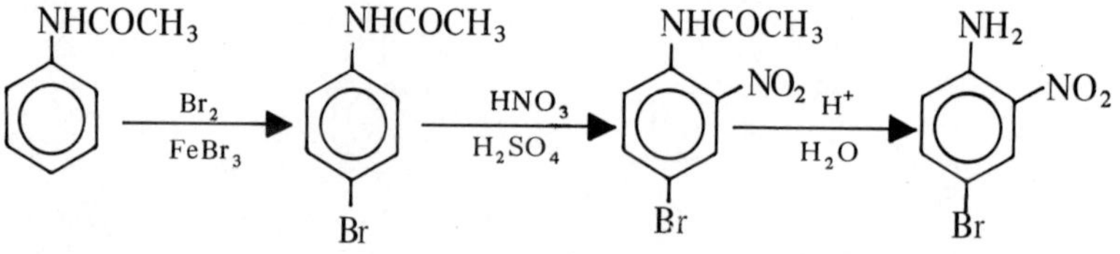

(f)

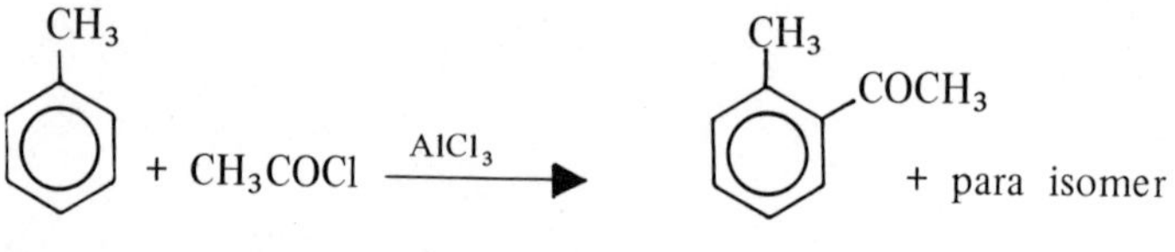

toluene $+ CH_3COCl \xrightarrow{AlCl_3}$ 2-methylacetophenone $+$ para isomer

(g)

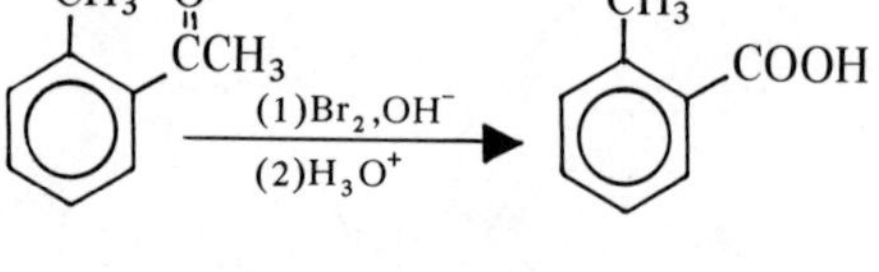

$\xrightarrow[\text{(2)}H_3O^+]{\text{(1)}Br_2,OH^-}$ 2-methylbenzoic acid

(h)

benzene $+ I_2 + AgClO_4 \xrightarrow{25°}$ iodobenzene $\xrightarrow{SO_3 / H_2SO_4}$ + Ortho isomer

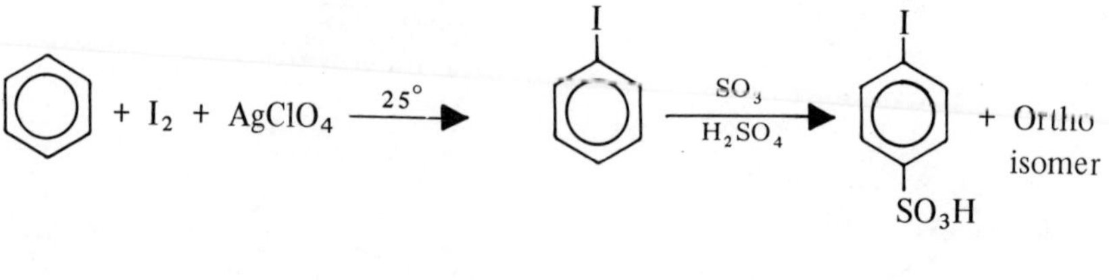

(i) 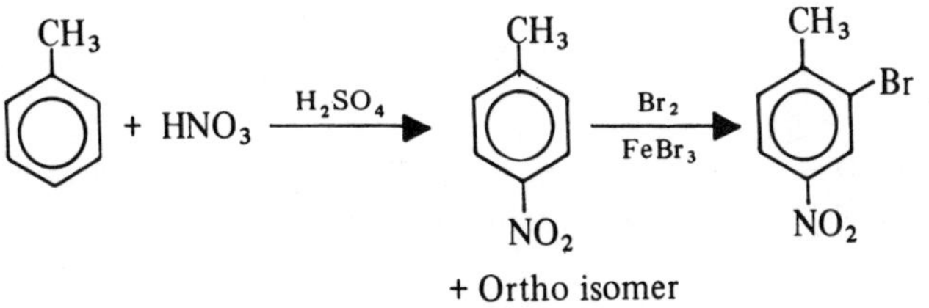
CH3
+ HNO3
H2SO4
CH3
NO2
+ Ortho isomer
Br2
FeBr3
CH3
Br
NO2

(j) 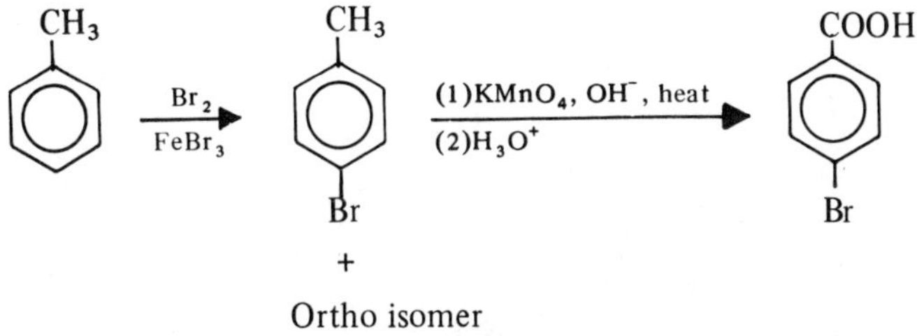
CH3
Br2
FeBr3
CH3
Br
+
Ortho isomer
(1)KMnO4, OH⁻, heat
(2)H3O⁺
COOH
Br

(k) 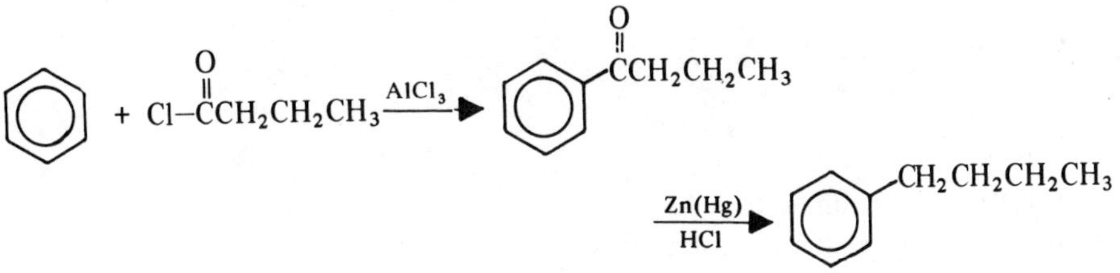
+ Cl–CCH2CH2CH3
O
AlCl3
O
CCH2CH2CH3
Zn(Hg)
HCl
CH2CH2CH2CH3

(l) 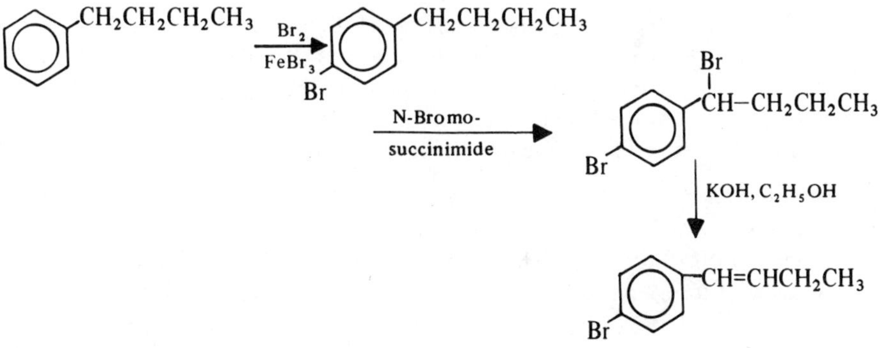
CH2CH2CH2CH3
Br2
FeBr3
CH2CH2CH2CH3
Br
N-Bromo-
succinimide
Br
CH–CH2CH2CH3
Br
KOH, C2H5OH
CH=CHCH2CH3
Br

(m) 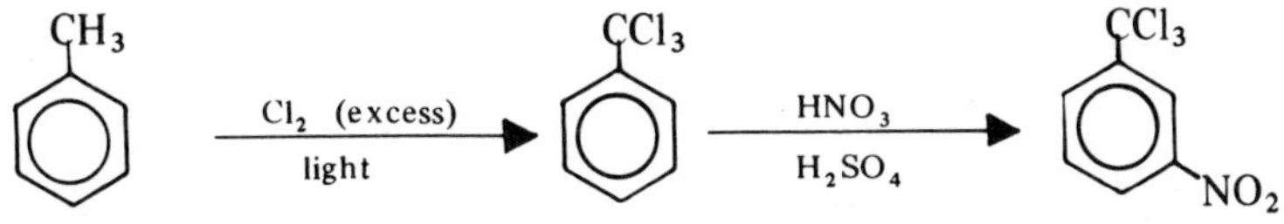
CH3
Cl2 (excess)
light
CCl3
HNO3
H2SO4
CCl3
NO2

(n) 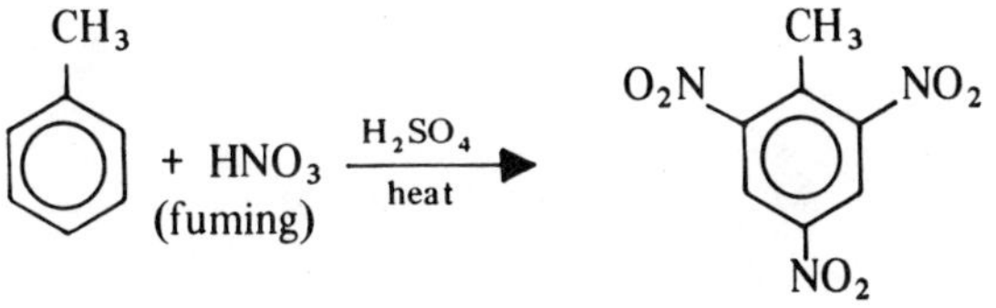
CH3
+ HNO3
(fuming)
H2SO4
heat
CH3
O2N
NO2
NO2

13.30

(a)

A   B

Ring B undergoes electrophilic substitution more readily than ring A.

(b) Resonance structures such as the one below stabilize the intermediate carbocation:

E

13.31

13.32

(a) We observe an isotope effect of this kind only when C–H or C–D bond-breaking occurs in the rate-limiting step. When benzene (or C_6D_6) is nitrated, the slowest step is the formation of the sigma complex. Once the sigma complex is formed it loses a proton (or deuteron) rapidly to form nitrobenzene (or $C_6D_5NO_2$). Nitrations, moreover, are essentially irreversible. This means that once the sigma complex is formed it goes on to form products and does not revert to reactants (See Fig. 13.2, page 450). This means, therefore, that the formation of the sigma complex is a truly rate-limiting step and thus that the exact type of bond being broken (C–H or C–D) in the fast step (loss of H^+ or D^+) will have no effect on the overall rate of reaction.

(b) Sulfonations, on the other hand, *are reversible*. (See Fig. 13.3, page 450). In sulfonations some molecules of the sigma complex lose SO_3 and revert to reactants while others go on to products. When C_6D_6 is sulfonated, more molecules of the sigma complex revert to reactants than with C_6H_6 because the loss of a deuteron occurs more slowly than the loss of a proton. This means that, overall, C_6D_6 will undergo sulfonation more slowly than C_6H_6.

13.33

(a)

(b) No   (c) Lindane is a meso compound.

(d)

13.34

If we consider resonance structures for the ring that undergoes electrophilic attack, two structures are possible for the sigma complex that forms when attack takes place at the 1-position,

whereas only one is possible when attack takes place at the 2-position,

Attack at the 1-position, therefore, takes place faster.

13.35

13.36

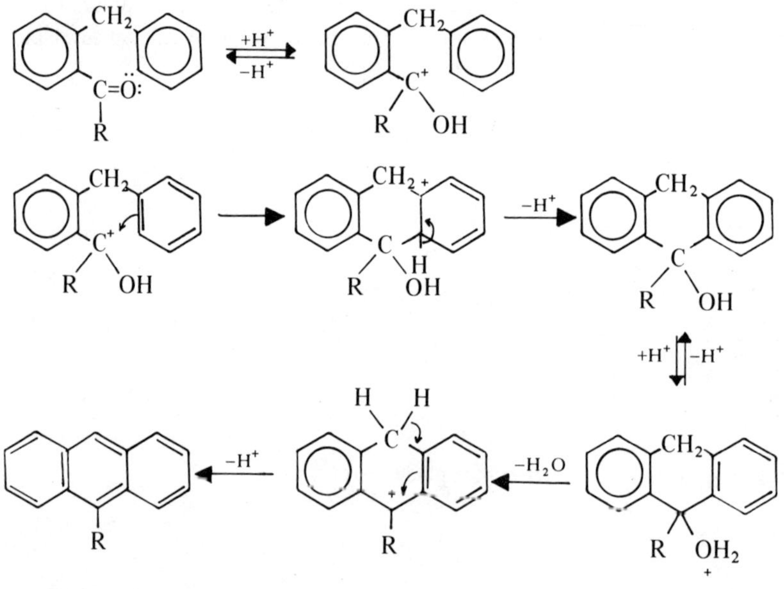

13.37

This problem serves as another illustration of the use of a sulfonic acid group as a blocking groups in a synthetic sequence. Here we are able to bring about nitration between two *meta* substituents.

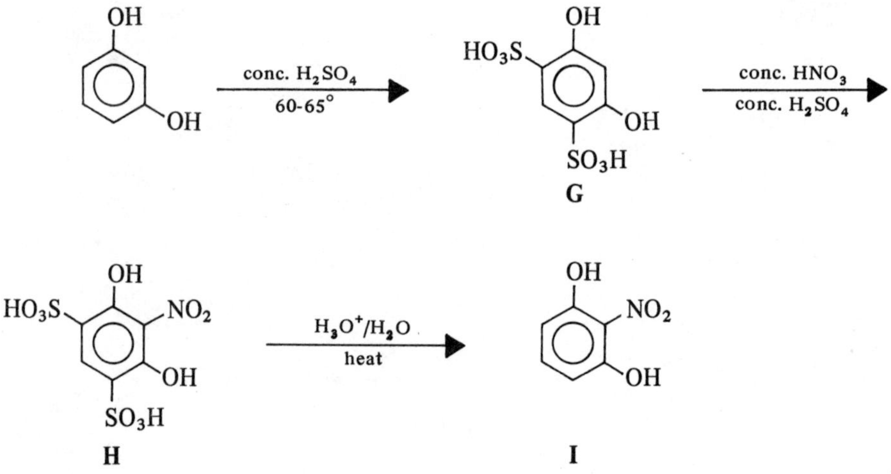

13.38

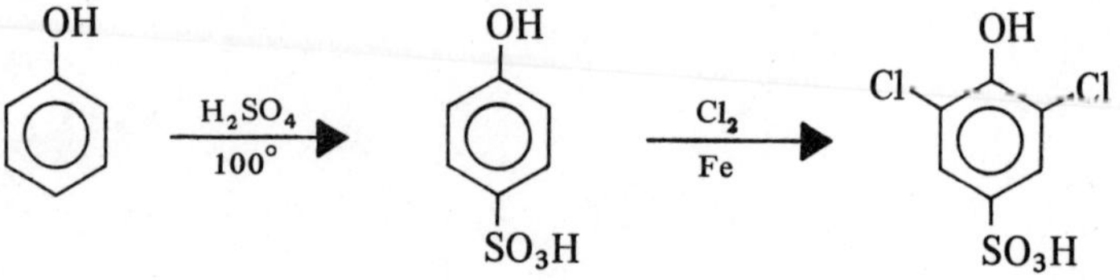

$$\xrightarrow[\text{heat}]{\text{H}_3\text{O}^+}$$

(structure: phenol with OH, two Cl at 2 and 6 positions — 2,6-dichlorophenol)

13.39

$$\text{CH}_3\overset{\text{O}}{\overset{\|}{\text{C}}}\text{Cl} + \text{AlCl}_3 \rightleftharpoons \text{CH}_3\overset{+}{\text{C}}\!=\!\text{O} + \text{AlCl}_4^-$$

$$\text{CH}_3\overset{+}{\text{C}}\!=\!\text{O} + \text{CH}_2\!=\!\text{CHCH}_3 \longrightarrow \text{CH}_3\overset{\text{O}}{\overset{\|}{\text{C}}}\text{CH}_2\overset{+}{\text{C}}\text{HCH}_3$$

$$\text{CH}_3\overset{\text{O}}{\overset{\|}{\text{C}}}\text{CH}_2\overset{+}{\text{C}}\text{HCH}_3 \;\longrightarrow\;$$

$$\longrightarrow \text{CH}_3\overset{\text{O}}{\overset{\|}{\text{C}}}\text{CH}=\text{CHCH}_3 + \text{H}^+$$

A

$$\longrightarrow \text{CH}_3\overset{\text{O}}{\overset{\|}{\text{C}}}\text{CH}_2\text{CH}=\text{CH}_2 + \text{H}^+$$

B

$$\xrightarrow{\text{Cl}^-} \text{CH}_3\overset{\text{O}}{\overset{\|}{\text{C}}}\text{CH}_2\underset{\underset{\text{Cl}}{|}}{\text{C}}\text{HCH}_3$$

C

13.40

(a)
(1) $\text{C}_6\text{H}_5\text{CH}=\text{CH}-\text{CH}=\text{CH}_2 \xrightarrow{\text{H}^+} \text{C}_6\text{H}_5\text{CH}=\text{CH}-\overset{+}{\text{C}}\text{H}-\text{CH}_3$

$$\updownarrow$$

$$\text{C}_6\text{H}_5\overset{+}{\text{C}}\text{H}-\text{CH}=\text{CH}-\text{CH}_3$$

$$\text{C}_6\text{H}_5\overset{\delta+}{\text{C}}\text{H}\text{---}\text{CH}\text{---}\overset{\delta+}{\text{C}}\text{H}-\text{CH}_3$$

(2) $\text{C}_6\text{H}_5\overset{\delta+}{\text{C}}\text{H}\text{---}\text{CH}\text{---}\overset{\delta+}{\text{C}}\text{H}-\text{CH}_3 \xrightarrow{\text{X}^-} \text{C}_6\text{H}_5\text{CH}=\text{CH}-\underset{\underset{\text{X}}{|}}{\text{C}}\text{H}-\text{CH}_3$

(b) 1,2-Addition (or in this instance actually 2,3-addition).

(c) Yes. The carbocation given in (a) is a hybrid of *secondary allylic and benzylic* contributors and is therefore more stable than any other possibility; for example

$$\text{C}_6\text{H}_5\text{CH}=\text{CH}-\text{CH}=\text{CH}_2 \xrightarrow{\text{H}^+} \text{C}_6\text{H}_5\text{CH}_2-\overset{+}{\text{C}}\text{H}-\text{CH}=\text{CH}_2$$

$$\updownarrow$$

$$\text{C}_6\text{H}_5\text{CH}_2-\text{CH}=\text{CH}-\overset{+}{\text{C}}\text{H}_2$$

a hybrid of allylic contributors only

(d) Since the reaction produces only *the more stable isomer*—that is, the one in which the double bond is conjugated with the benzene ring—the reaction must be under equilibrium control (cf. pp. 361):

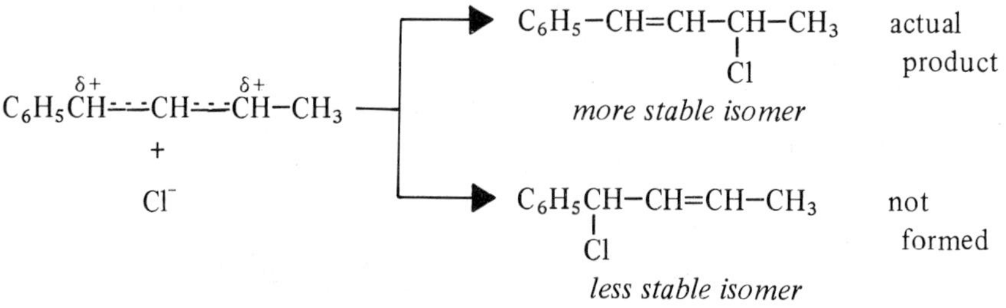

$$C_6H_5\overset{\delta+}{CH}=\overset{\delta+}{CH}=\overset{}{CH}-CH_3$$
$$+$$
$$Cl^-$$

$$\longrightarrow \quad C_6H_5-CH=CH-\underset{\underset{Cl}{|}}{CH}-CH_3 \qquad \text{actual product}$$

more stable isomer

$$\longrightarrow \quad C_6H_5\underset{\underset{Cl}{|}}{CH}-CH=CH-CH_3 \qquad \text{not formed}$$

less stable isomer

13.41

(a) The addition product is the *meso* compound:

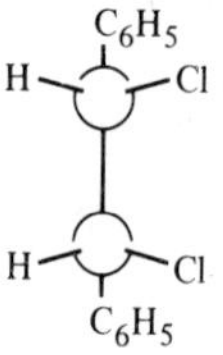

meso−1,2−dichloro−1,2−diphenylethane

(b) The reaction is a *syn* addition:

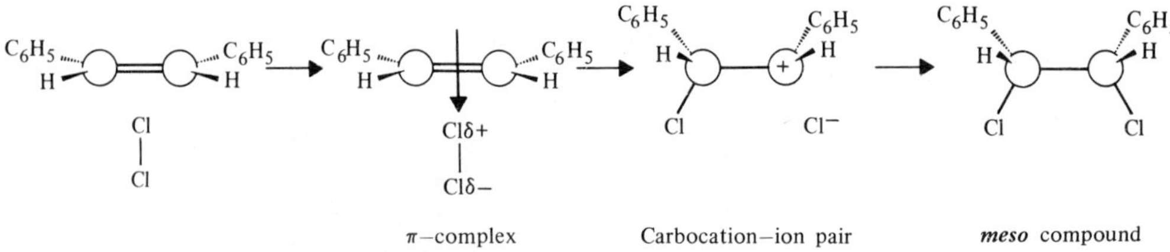

π−complex   Carbocation–ion pair   ***meso* compound**

(c) A similar reaction starting with *trans*-stilbene would yield a racemic modification of 1,2-dichloro-1,2-diphenylethane: *syn* addition at one face would yield one enantiomer; *syn* addition at the other face would yield the other enantiomer. The racemic modification would be optically inactive, but it could be resolved into the separate (and optically active) enantiomers.

13.42

(a)

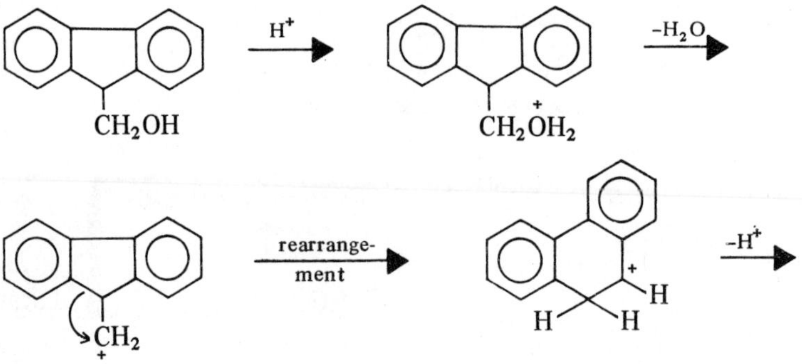

(b) $CH_3-C=CH_2$ $\xrightarrow{\ H^+\ }$ $CH_3-\overset{+}{C}-CH_3$
$\quad\quad\ \ |$ $\quad\quad\quad\quad\quad\quad\ |$
$\quad\quad\ \ C_6H_5$ $\quad\quad\quad\quad\quad\ C_6H_5$

$CH_2=C-CH_3$
$\quad\ \ \ |$
$\quad\ \ \ C_6H_5$
$\xrightarrow{\quad\quad}$

$CH_3\quad CH_3$

$\quad\quad\ \ \overset{+}{\quad}$
$\quad CH_3\ \ C_6H_5$
$\xrightarrow{\quad\quad}$

$CH_3\quad CH_3$

$\quad\quad\quad C_6H_5$
$H\,CH_3$
$\xrightarrow{\ -H^+\ }$

$CH_3\quad CH_3$

$\quad\quad\quad C_6H_5$
$\quad CH_3$

13.43

$CH_3CH\left(CH_2CH\right)CH=CHCH_3$
$\quad\ |\quad\quad\ |$
$\quad CH_3\quad CH_3 \Big/_2$
$\quad\quad\quad\quad + \text{ and}$
$CH_3CH\left(CH_2CH\right)CH_2CH=CH_2$
$\quad\ |\quad\quad\ |$
$\quad CH_3\quad CH_3 \Big/_2$
$\xrightarrow[35\text{-}45°]{AlCl_3}$

$CH_3CH\left(CH_2CH\right)CH_2CH-$
$\quad\ |\quad\quad\ |\quad\quad\quad |$
$\quad CH_3\quad CH_3 \Big/_2\ CH_3$
$\quad\quad\quad\text{and}$

$CH_3CH\left(CH_2CH\right)\ CH-$
$\quad\ |\quad\quad\ |\quad\quad |$
$\quad CH_3\quad CH_3 \Big/_2\ CH_2$
$\quad\quad\quad\quad\quad\quad\quad |$
$\quad\quad\quad\quad\quad\quad\quad CH_3$
$\xrightarrow[heat]{H_2SO_4}$

$CH_3CH\left(CH_2CH\right)CH_2CH-\!\!\bigcirc\!\!-SO_3H$
$\quad\ |\quad\quad\ |\quad\quad\quad |$
$\quad CH_3\quad CH_3 \Big/_2\ CH_3$
$\quad\quad\quad\text{and}$

$CH_3CH\left(CH_2CH\right)\ CH-\!\!\bigcirc\!\!-SO_3H$
$\quad\ |\quad\quad\ |\quad\quad |$
$\quad CH_3\quad CH_3 \Big/_2\ CH_2$
$\quad\quad\quad\quad\quad\quad\quad |$
$\quad\quad\quad\quad\quad\quad\quad CH_3$
$\xrightarrow{\ NaOH\ }$

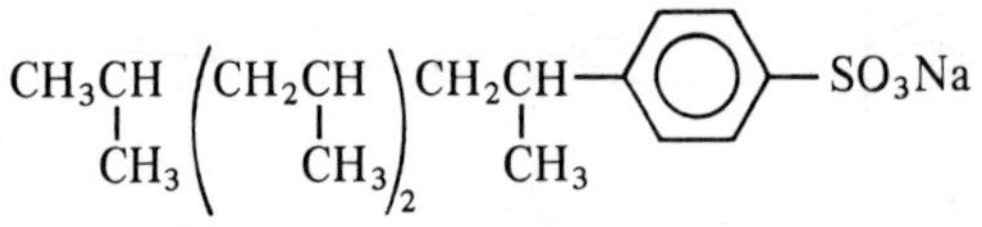

$$CH_3CH \left(\begin{array}{c} CH_2CH \\ | \\ CH_3 \end{array}\right)_2 CH_2CH \text{—}\bigcirc\text{—} SO_3Na$$
$$\quad\quad | \quad\quad\quad\quad\quad | $$
$$\quad CH_3 \quad\quad\quad\quad CH_3$$

and

$$CH_3CH \left(\begin{array}{c} CH_2CH \\ | \\ CH_3 \end{array}\right)_2 CH\text{—}\bigcirc\text{—} SO_3Na$$
$$\quad\quad | \quad\quad\quad\quad\quad | $$
$$\quad CH_3 \quad\quad\quad\quad CH_2$$
$$\quad\quad\quad\quad\quad\quad\quad\quad | $$
$$\quad\quad\quad\quad\quad\quad\quad\quad CH_3$$

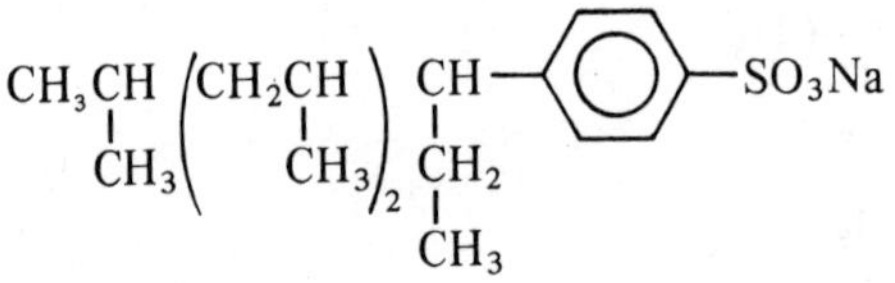

13.44

(a) Large *ortho* substituents prevent the two rings from becoming coplanar and if the correct substitution patterns are present, the molecule as a whole will be chiral. Thus enantiomeric forms are possible even though the molecules do not have a chiral carbon. The compound with 2-NO₂, 6-COOH, 2′-NO₂, 6′-COOH is an example.

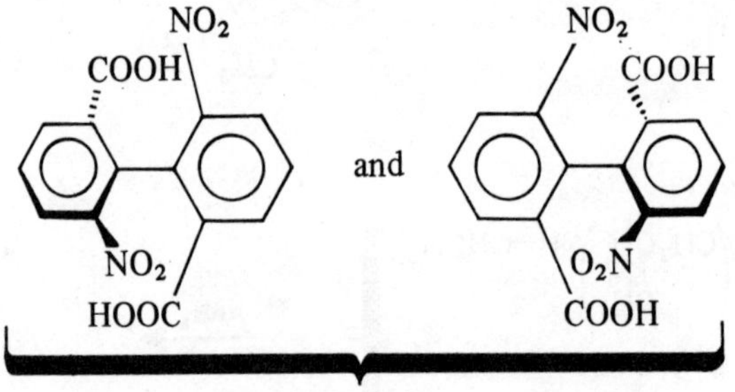

and

These molecules are nonsuperposable mirror reflections and, thus, are enantiomers.

(b) Yes

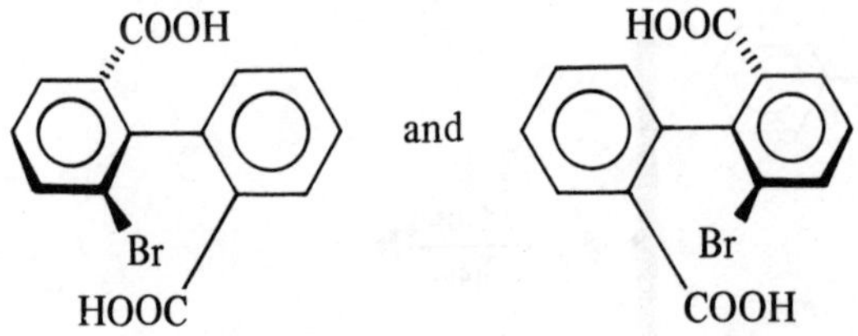

and

(c) This molecule has a plane of symmetry.

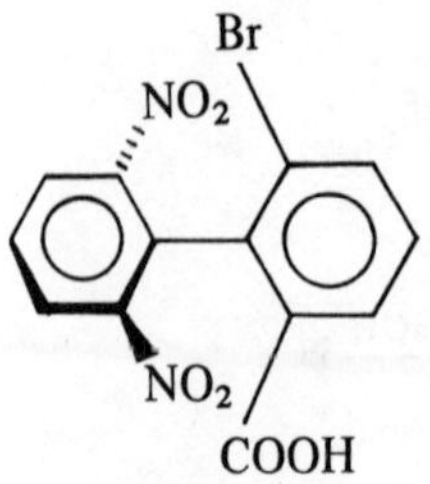

The plane of the page is a plane of symmetry.

14.1
The methyl protons of 15,16-dimethylpyrene are highly shielded by the induced field in the center of the aromatic system where the induced field opposes the applied field. (p 509).

14.2
(a) The six protons (hydrogens) of ethane are equivalent:

$$\overset{a}{CH_3} - \overset{a}{CH_3}$$

Ethane gives a single signal in its pmr spectrum.

(b) Propane has two different sets of equivalent protons:

$$\overset{a}{CH_3} - \overset{b}{CH_2} - \overset{a}{CH_3}$$

Propane gives two signals.

(c) The six protons of dimethyl ether are equivalent:

$$\overset{a}{CH_3} - O - \overset{a}{CH_3}$$

One signal.

(d) Three different sets of equivalent protons:

Three signals.

(e) Two different sets of equivalent protons:

$$\overset{a}{CH_3} - \overset{O}{\underset{\|}{C}} - O - \overset{b}{CH_3}$$

Two signals.

(f) Three different sets of equivalent protons:

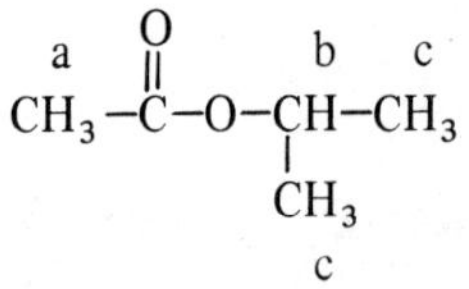

$$\underset{a}{CH_3}-\overset{\overset{O}{\|}}{C}-O-\underset{\underset{\underset{c}{CH_3}}{|}}{\underset{b}{CH}}-\underset{c}{CH_3}$$

Three signals.

14.3

(a)

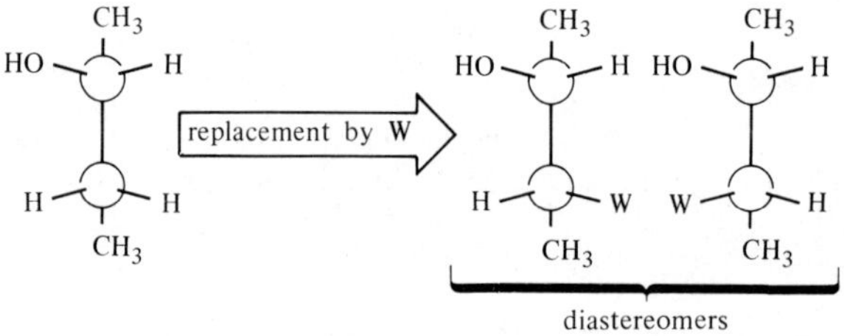

replacement by **W**

diastereomers

(b) Six,

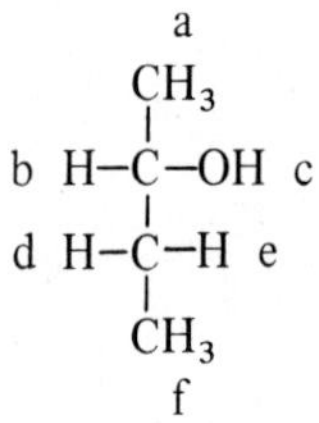

$$\begin{array}{c} \overset{a}{CH_3} \\ | \\ b\ H-\overset{|}{C}-OH\ c \\ | \\ d\ H-\overset{|}{C}-H\ e \\ | \\ \underset{f}{CH_3} \end{array}$$

(c) Six signals.

14.4

(a) Two, $\overset{a}{CH_3}-\overset{b}{CH_2}-\overset{b}{CH_2}-\overset{a}{CH_3}$

(b) Three, $\overset{a}{CH_3}-\overset{b}{CH_2}-O-\overset{c}{H}$

(c) Four,
$$\begin{array}{ccc} \overset{a}{CH_3} & & \overset{c}{H} \\ & C=C & \\ \underset{b}{H} & & \underset{d}{H} \end{array}$$

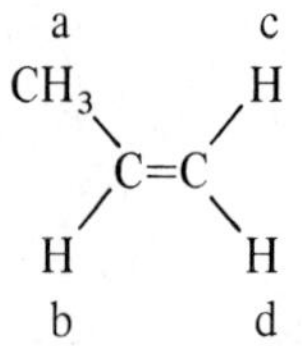

(d) Two,
$$\begin{array}{ccc} \overset{a}{CH_3} & & \overset{b}{H} \\ & C=C & \\ \underset{b}{H} & & \underset{a}{CH_3} \end{array}$$

(e) Four, $\overset{a}{CH_3}-\overset{b}{CH}Br-\underset{\underset{d}{H}}{\overset{\overset{c}{H}}{\underset{|}{\overset{|}{C}}}}-Br$

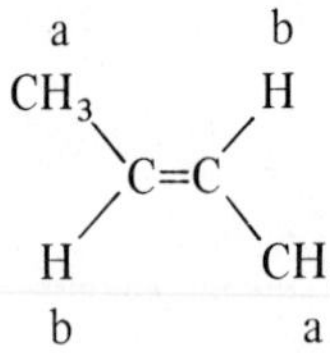

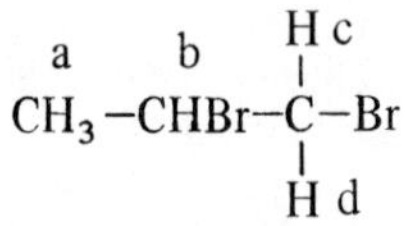

(f) Two,

(g) Three,

(h) Four,

(i) Six,

14.5

The pmr spectrum of $CHBr_2CHCl_2$ consists of two doublets. The doublet from the proton of the $-CHCl_2$ group should occur at lowest magnetic field strength because the greater electronegativity of chlorine reduces the electron density in the vicinity of the $-CHCl_2$ proton, and consequently, reduces its shielding relative to $-CHBr_2$.

14.6

The determining factors here are the number of chlorine atoms attached to the carbons bearing protons and the deshielding that results from chlorine's electronegativity. In 1,1,2-trichloroethane the proton that gives rise to the triplet is on a carbon that bears two chlorines, and the signal from this proton is downfield. In 1,1,2,3,3-pentachloropropane the proton that gives rise to the triplet is on a carbon that bears only one chlorine; the signal from this proton is upfield.

14.7

The signal from the three equivalent protons designated **a** should be split into a doublet by the proton **b**. This doublet, because of the electronegativity of the attached chlorines, should occur downfield.

$$\overset{a}{(Cl_2CH)_3}-\overset{b}{CH}$$

The proton designated **b** should be split into a quartet by the three equivalent protons **a**. The quartet should occur upfield.

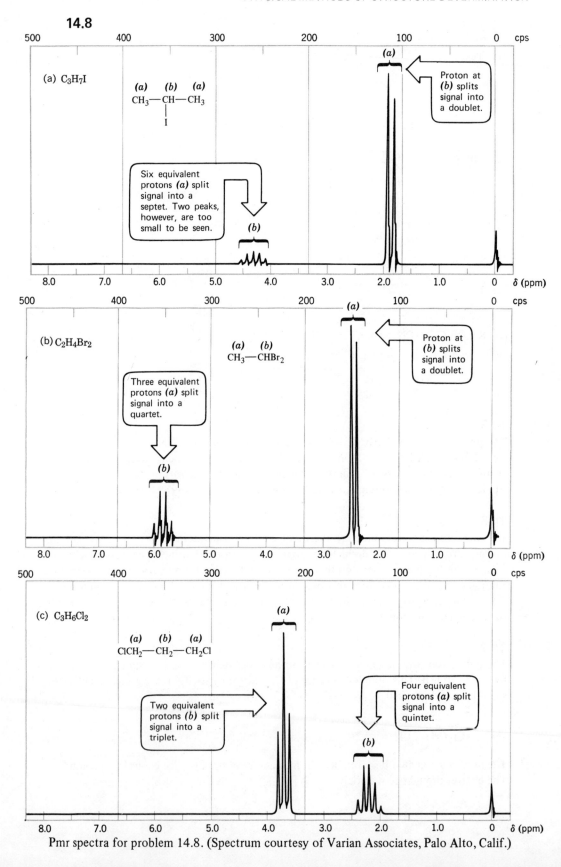

Pmr spectra for problem 14.8. (Spectrum courtesy of Varian Associates, Palo Alto, Calif.)

14.9

(a) $J_{ab} = 2J_{bc}$

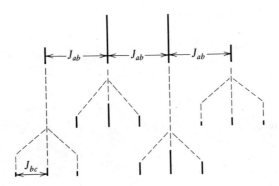

Result: | | | | | | | | | (Nine peaks)

(b) $J_{ab} = J_{bc}$

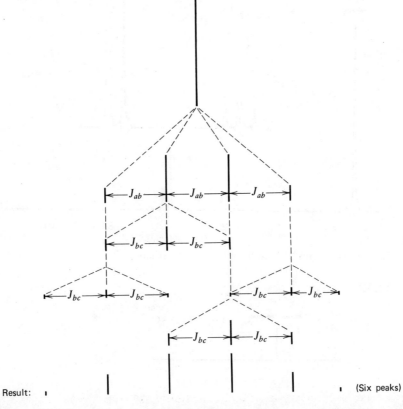

Result: | | | | | | (Six peaks)

14.10

(a) $C_6H_5CH(CH_3)_2$

(b) $C_6H_5\underset{\underset{NH_2}{|}}{C}HCH_3$

(c)

Pmr spectra for problem 14.10 are on p. 133.

14.11

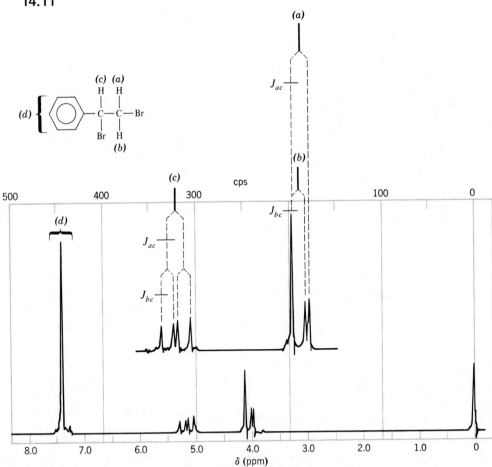

The pmr spectrum of 1,2-dibromo-1-phenylethane.

14.12

(a)

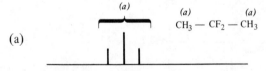

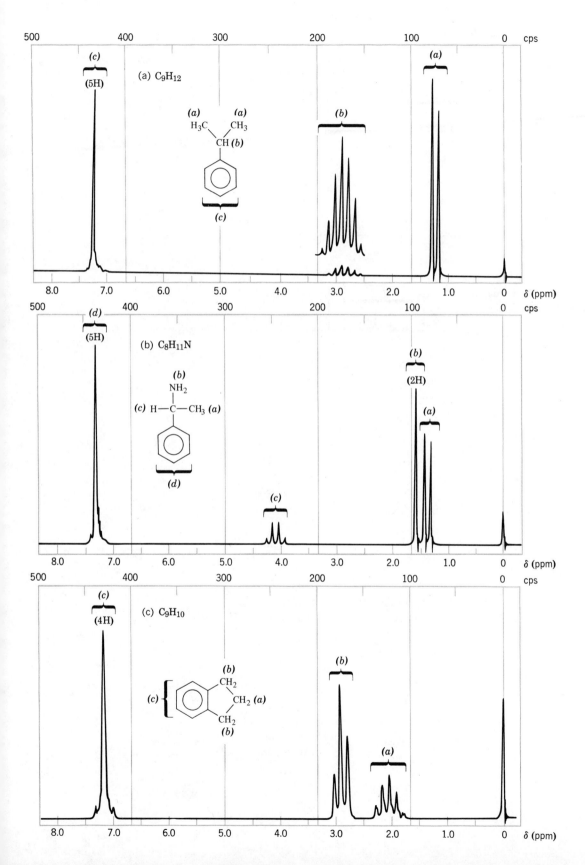

14.12 continued

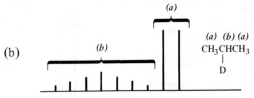

(a) (b) (a)
CH_3CHCH_3
|
D

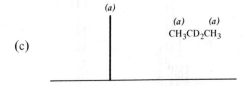

(a)   (a)
$CH_3CD_2CH_3$

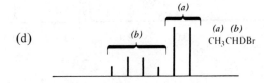

(a) (b)
CH_3CHDBr

14.13
A single unsplit signal.

14.14

a Singlet, $\delta 2.35$ (9H)
b Singlet, $\delta 6.70$ (3H)

a Singlet, $\delta 2.8$ (6H)
b Singlet, $\delta 2.9$ (3H)
c Singlet, $\delta 4.6$ (2H)
d Singlet, $\delta 7.7$ (2H)

14.15

a Doublet, $\delta 1.48$ (6H)
b Multiplet, $\delta 4.45$ (1H)
c Multiplet, $\delta 8.0$ (10H)

14.16

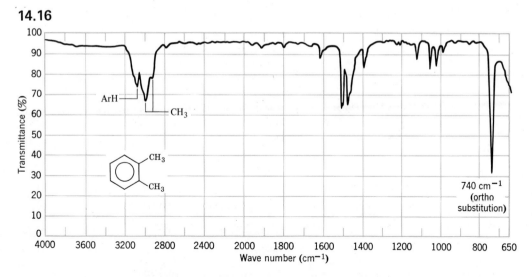

The infrared spectrum of o-xylene.
Similar assignments can be made for m-xylene and p-xylene.

14.17

CH$_3$	CH$_3$	CH$_3$	CH$_2$Br
Br	Br	Br	
A	**B**	**C**	**D**

14.18

(a)   a
       CH$_3$
  a |
(a) CH$_3$–C–OH b   a Singlet, δ1.28 (9H)
       | b Singlet, δ1.35 (1H)
       CH$_3$
       a

    a   b   a
(b) CH$_3$–CH–CH$_3$ a Doublet, δ1.71 (6H)
       | b Septet, δ4.32 (1H)
       Br

    b   O   c   a
        ‖
(c) CH$_3$–C–CH$_2$–CH$_3$ a Triplet, δ1.05 (3H)
                           b Singlet, δ2.13 (3H)
                           c Quartet, δ2.47 (2H)
                           C=O, 1720 cm^{-1}

(d)

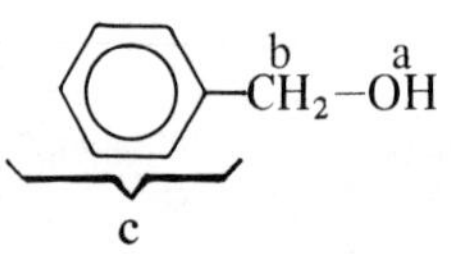

a Singlet, $\delta 2.43$ (1H)
b Singlet, $\delta 4.58$ (2H)
c Multiplet, $\delta 7.28$ (5H)
O–H, 3200-3600 cm^{-1}

(e) CH$_3$–CH–CH$_2$Cl

a Doublet, $\delta 1.04$ (6H)
b Multiplet, $\delta 1.95$ (1H)
c Doublet, $\delta 3.35$ (2H)

(f) C$_6$H$_5$–CH–C–CH$_3$

a Singlet, $\delta 2.20$ (3H)
b Singlet, $\delta 5.08$ (1H)
c Multiplet, $\delta 7.25$ (10H)
C=O, near 1720 cm^{-1}

(g) CH$_3$–CH$_2$–CHCOOH

a Triplet, $\delta 1.08$ (3H)
b Multiplet, $\delta 2.07$ (2H)
c Triplet, $\delta 4.23$ (1H)
d Singlet, $\delta 10.97$ (1H)
O–H, 2500-3000 cm^{-1}

(h)

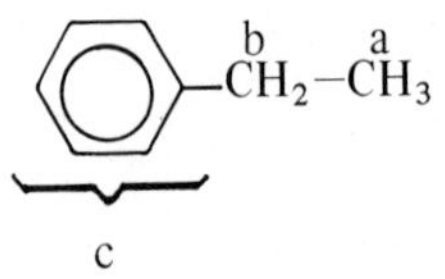

a Triplet, $\delta 1.25$ (3H)
b Quartet, $\delta 2.68$ (2H)
c Multiplet, $\delta 7.23$ (5H)

(i) CH$_3$–CH$_2$–O–CH$_2$–COOH

a Triplet, $\delta 1.27$ (3H)
b Quartet, $\delta 3.66$ (2H)
c Singlet, $\delta 4.13$ (2H)
d Singlet, $\delta 10.95$ (1H)
O–H, 2500-3000 cm^{-1}

(j) CH$_3$–CH–CH$_3$

a Doublet, $\delta 1.55$ (6H)
b Septet, $\delta 4.67$ (1H)

(k) CH$_3$O–CH$_2$CH$_2$–OCH$_3$

a Singlet, $\delta 3.25$ (6H)
b Singlet, $\delta 3.45$ (4H)

(l) CH$_3$–C–CH–CH$_3$

a Doublet, $\delta 1.10$ (6H)
b Singlet, $\delta 2.10$ (3H)
c Septet, $\delta 2.50$ (1H)
C=O, near 1720 cm^{-1}

(m)   a Doublet, $\delta 2.0$ (3H)
       b Quartet, $\delta 5.15$ (1H)
       c Multiplet, $\delta 7.35$ (5H)

14.19

Compound **E** is phenylacetylene, $C_6H_5C{\equiv}CH$. We can make the following assignments in the infrared spectrum:

$\sim$ 3300 cm^{-1} , ${\equiv}C{-}H$

$\sim$ 3030 cm^{-1} , $Ar{-}H$

$\sim$ 2100 cm^{-1} (weak) , $-C{\equiv}C-$

$\sim$ 690 cm^{-1} and $\sim$ 710 cm^{-1} ,

14.20

A pmr signal this far upfield indicates that cyclooctatetraene is a cyclic polyene and is not aromatic.

14.21

Both [14] annulene and dehydro[14] annulene are aromatic as shown by the signals at $\delta 7.78$ (10H) and at $\delta 8.0$ (10H) respectively. [14] Annulene has four "internal" protons (δ -0.61) and dehydro[14] annulene has only two ($\delta 0.0$).

14.22

Compound **F** is p-isopropyltoluene.

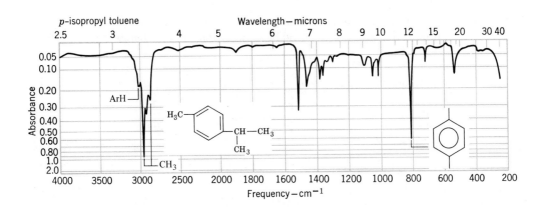

The IR and (on next page) pmr spectra of compound F, problem 14.22. (pmr spectrum adapted from Varian Associates, Palo Alto, Calif. IR spectrum adapted from Sadtler Research Laboratories, Philadelphia, Pa.)

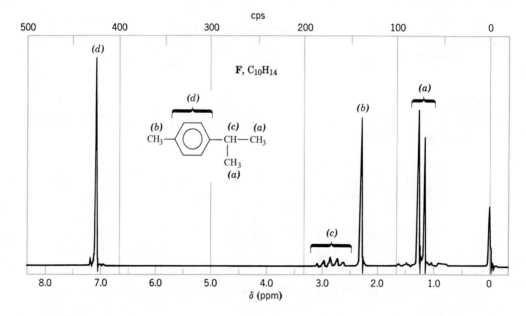

14.23

(a) In SbF_5 the carbocations formed initially apparently undergo a complex series of rearrangements to the more stable *tert*-butyl cation.

(b) All of the cations formed intially rearrange to the more stable *tert*-pentyl cation,

$$CH_3CH_2\overset{\overset{\displaystyle CH_3}{|}}{\underset{\underset{\displaystyle CH_3}{|}}{C}}+$$

The spectrum of the *tert*-pentyl cation should consist of a singlet (6H), a quartet (2H), and a triplet (3H). The triplet should be most upfield and the quartet most downfield.

14.24

(a) Four unsplit signals,

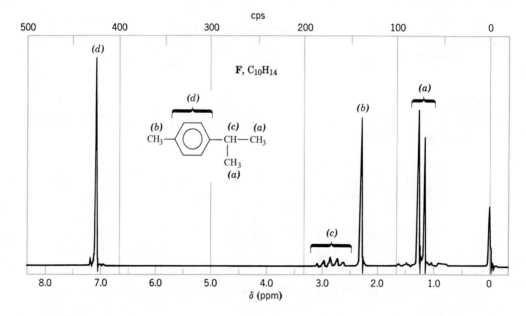

(b) Absorptions arising from: $=C-H$, CH_3 , and $C=O$ groups.

14.25

Compound **G** is 2-bromobutane.

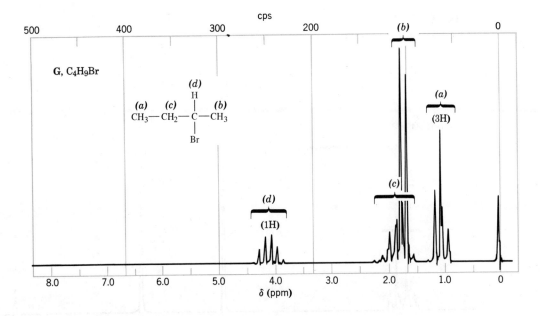

The pmr spectrum of compound **G** (problem 14.25). (Spectrum courtesy of Varian Associates, Palo Alto, Calif.)

Compound **H** is 2,3-dibromopropene.

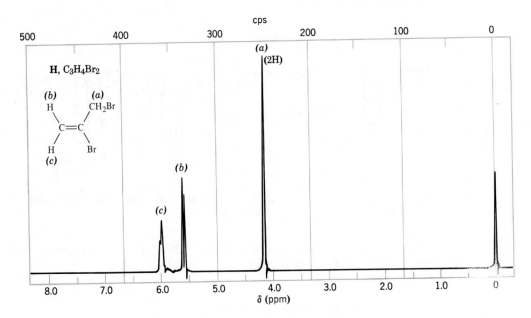

The pmr spectrum of compound **H** (problem 14.25). (Spectrum courtesy of Varian Associates, Palo Alto, Calif.).

14.26

Compound **I** is *p*-methoxytoluene.

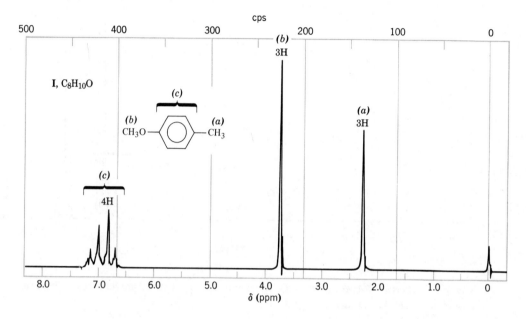

The pmr spectrum of compound **I** (problem 14.26). (Spectrum courtesy of Varian Associates, Palo Alto, Calif.)

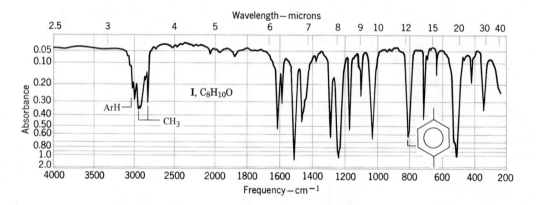

The infrared spectrum of compound **I** (problem 14.26). (Spectrum courtesy of Sadtler Research Laboratories, Philadelphia, Pa.)

14.27

Compound *J* is *cis*-1,2-dichloroethene,

We can make the following infrared assignments:

3125 cm^{-1}, alkene C—H stretching
1625 cm^{-1}, C=C stretching
695 cm^{-1}, out-of-plane bending of *cis* double bond.

14.28

(a) Compound K is,

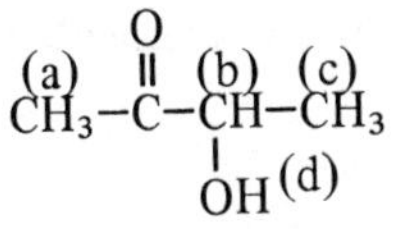

$$\underset{\substack{|\\ \underset{(d)}{OH}}}{\overset{O}{\underset{(a)}{CH_3}-\overset{\|}{C}-\underset{(b)}{CH}-\underset{(c)}{CH_3}}}$$

(a) Singlet $\delta 2.15$ (d) Singlet $\delta 3.75$
(b) Quartet $\delta 4.25$ C=O, 1720 cm^{-1}
(c) Doublet $\delta 1.35$

(b) When the compound is dissolved in D_2O, the —OH proton (d) is replaced by a deuteron and thus the pmr absorption peak disappears.

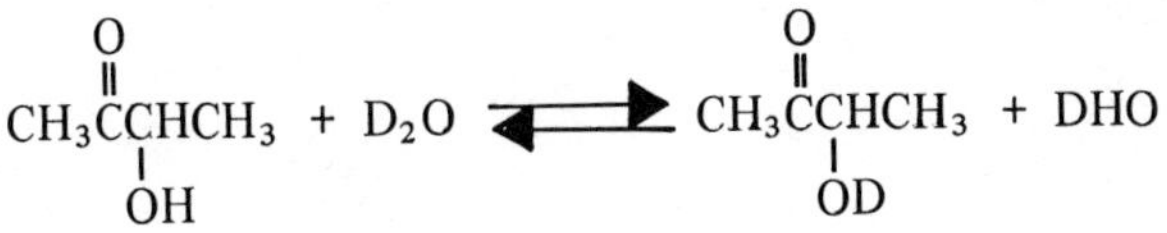

$$\underset{\substack{|\\ OH}}{\overset{O}{CH_3\overset{\|}{C}CHCH_3}} + D_2O \rightleftharpoons \underset{\substack{|\\ OD}}{\overset{O}{CH_3\overset{\|}{C}CHCH_3}} + DHO$$

14.29

Compound **L** is allylbenzene,

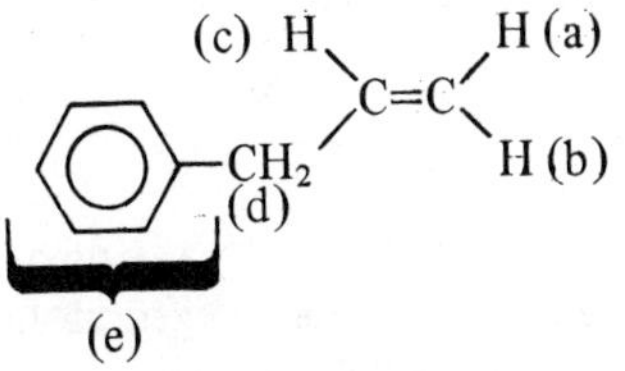

(d) Doublet $\delta 3.1$ (2H)
(a) or (b) Multiplet $\delta 4.8$
(a) or (b) Multiplet $\delta 5.1$
(c) Multiplet $\delta 5.8$
(e) Multiplet $\delta 7.1$ (5H)

The following infrared assignments can be made.

3035 cm^{-1}, C—H stretching of benzene ring
3020 cm^{-1}, C—H stretching of —CH=CH$_2$ group
2925 cm^{-1} and 2853 cm^{-1}, C—H stretching of —CH$_2$— group
1640 cm^{-1}, C=C stretching
990 cm^{-1} and 915 cm^{-1}, C—H bendings of —CH=CH$_2$ group
740 cm^{-1} and 695 cm^{-1}, C—H bendings of —C$_6$H$_5$ group

The ultraviolet absorbance maximum at 255 nm is indicative of a benzene ring that is not conjugated with a double bond.

14.30

Run the spectrum with the spectrometer operating at a different magnetic field strength (i.e., at 30 MHz or at 100 MHz). If the peaks are two singlets the distance between them—*when measured in Hertz*—will change because chemical shifts *expressed in Hertz* are proportional to the strength of the applied field (see p. 511). If, however, the two peaks represent a doublet then the distance that separates them, expressed in Hertz, will not change because this distance represents the magnitude of the coupling constant and coupling constants are independent of the applied magnetic field (p. 517).

14.31

Compound **M** is *m*-ethyltoluene. We can make the following assignments in the pmr spectrum.

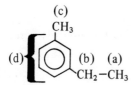

(a) Triplet, $\delta 1.3$

(b) Quartet, $\delta 2.6$

(c) Singlet, $\delta 2.4$

(d) Multiplet, $\delta 7.1$

Meta substitution is indicated by the strong peaks at 690 cm^{-1} and 780 cm^{-1} in the infrared spectrum.

14.32

Compound **N** is $C_6H_5CH=CHOCH_3$. The absence of absorption peaks due to O—H or C=O stretching in the infrared spectrum of **N** suggests that the oxygen atom is present as part of an ether linkage. The (5H) pmr multiplet at $\delta 7.3$ strongly suggests the presence of a monosubstituted benzene ring; this is confirmed by the strong peaks at ~690 cm^{-1} and ~770 cm^{-1} in the infrared spectrum.

We can make the following assignments in the pmr spectrum:

$$\underset{C_6H_5-CH=CH-OCH_3}{\overset{(a)\quad (b)\ (c)\quad (d)}{}}$$

(a) Multiplet $\delta 7.3$

(c) or (b) Doublet $\delta 6.05$

(b) or (c) Doublet $\delta 5.15$

(d) Singlet $\delta 3.7$

14.33

That the pmr spectrum shows only one signal indicates that all 12 protons of the carbocation are equivalent, and suggests very strongly that what is being observed is the bromonium ion:

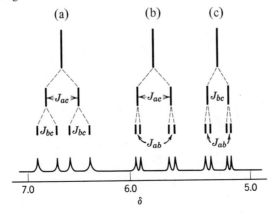

While this experiment does not prove that bromonium ions are intermediates in alkene additions, it does show that bromonium ions are capable of existence and thus makes postulating them as intermediates more plausible.

14.34
In the presence of SbF_5, **I** dissociates first to the cyclic allylic cation, **II**, and then to the aromatic dication, **III**.

14.35
The vinylic protons of *p*-chlorostyrene should give a spectrum approximately like the following:

15.1

(a) 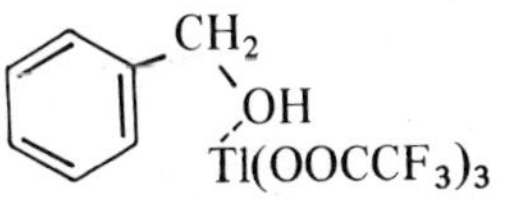 C_6H_5—CH_2—OH with $Tl(OOCCF_3)_3$

(b) C_6H_5—CH_2—OCH_3 with $Tl(OOCCF_3)_3$

(c) C_6H_5—CH_2—CH_2—O—CH_3 with $Tl(OOCCF_3)_3$

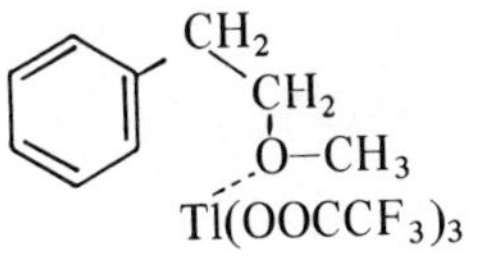

(d) 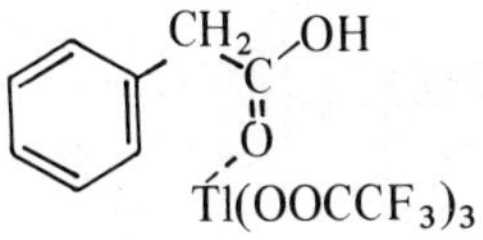 C_6H_5—CH_2—$\overset{O}{\underset{}{C}}$—OH with $Tl(OOCCF_3)_3$

(e) 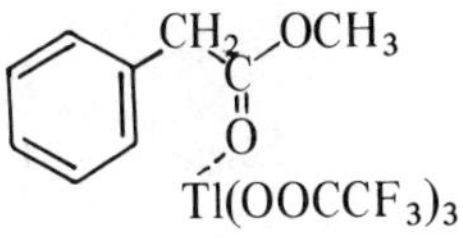 C_6H_5—CH_2—$\overset{O}{\underset{}{C}}$—$OCH_3$ with $Tl(OOCCF_3)_3$

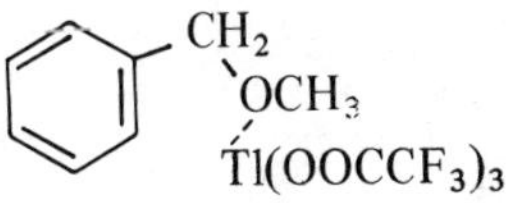

15.2

In the complex the Tl atom is too far from the ortho position:

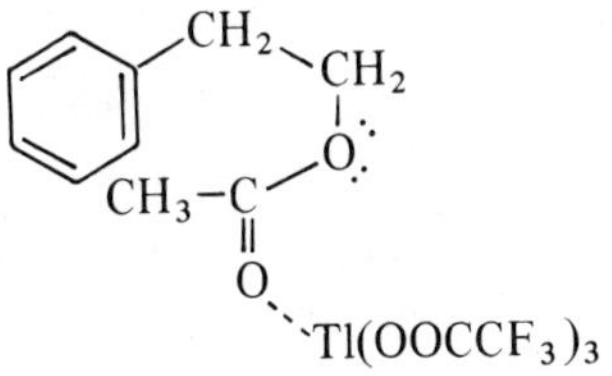

C_6H_5—CH_2—CH_2—O complexed with CH_3—C(=O) and $Tl(OOCCF_3)_3$

It must therefore react in the usual way at the para (least crowded) position.

15.3

(a) 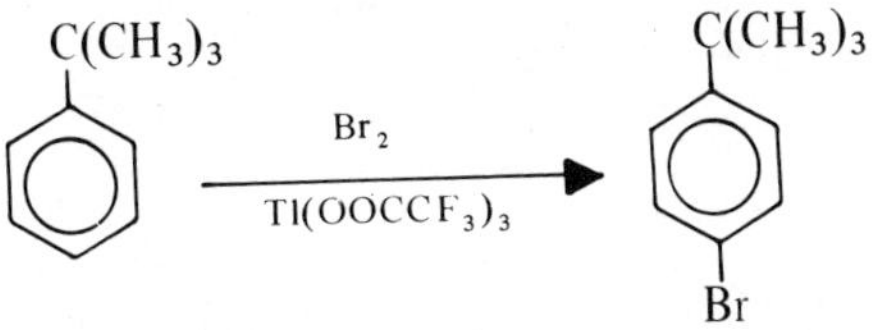 C_6H_5—$C(CH_3)_3$ $\xrightarrow[\text{Tl(OOCCF}_3)_3]{\text{Br}_2}$ para-Br—C_6H_4—$C(CH_3)_3$

(b) 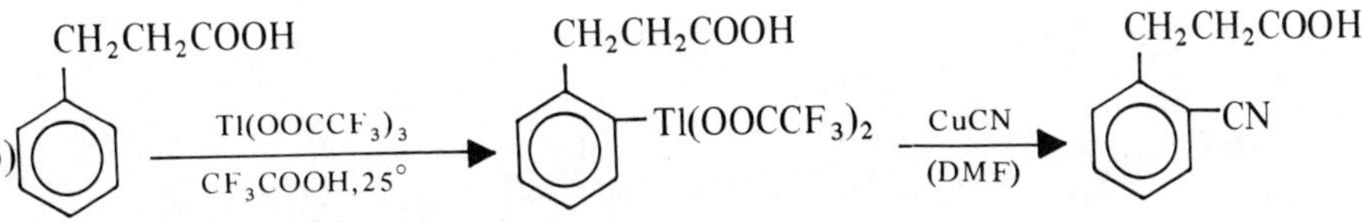 C_6H_5—CH_2CH_2COOH $\xrightarrow[\text{CF}_3\text{COOH},\,25°]{\text{Tl(OOCCF}_3)_3}$ ortho-$Tl(OOCCF_3)_2$—C_6H_4—CH_2CH_2COOH $\xrightarrow[\text{(DMF)}]{\text{CuCN}}$ ortho-CN—C_6H_4—CH_2CH_2COOH

(c)

$$\underset{\text{(1)Tl(OOCCF}_3)_3,\text{CF}_3\text{COOH,25}^\circ}{\overset{\text{(2)Pb(OOCCH}_3)_4,(\text{C}_6\text{H}_5)_3\text{P}}{\longrightarrow}} \quad \text{(3)OH}^-$$

(d)

$$\overset{\text{(1)Tl(OOCCF}_3)_3,\text{CF}_3\text{COOH,25}^\circ}{\underset{\text{(2)KI,H}_2\text{O}}{\longrightarrow}}$$

(e)

$$\overset{\text{(1)Tl(OOCCF}_3)_3,\text{CF}_3\text{COOH,73}^\circ}{\underset{\text{(2)KI,H}_2\text{O}}{\longrightarrow}}$$

(f)

$$\overset{\text{same as (d)}}{\underset{\text{above}}{\longrightarrow}}$$

(g)

$$\overset{\text{(1)Tl(OOCCF}_3)_3,\text{CF}_3\text{COOH,25}^\circ}{\underset{\text{(2)CuCN,DMF}}{\longrightarrow}}$$

15.4

(a) $CH_3CH_2CH_2CH_2 \overset{\delta-}{:} Li + \overset{\delta+}{H} : \overset{..}{\underset{..}{O}}H \longrightarrow CH_3CH_2CH_2CH_2-H + Li^+ : \overset{..}{\underset{..}{O}}H^-$
  (stronger base)   (stronger   (weaker acid)   (weaker
              acid)   base)

(b) $CH_3CH_2CH_2CH_2 \overset{\delta-}{:} Li + \overset{\delta+}{H} : \overset{..}{\underset{..}{O}}CH_2CH_3 \longrightarrow CH_3CH_2CH_2CH_2-H + Li^+ : \overset{..}{\underset{..}{O}}CH_2$
  (stronger base)   (stronger   (weaker acid CH_3
              acid)   (weaker
                                        base)

15.5

$$CH_3-\underset{\underset{CH_3}{|}}{\overset{\overset{CH_3}{|}}{C}}-Br + Mg \xrightarrow[35^\circ]{\text{ether}} CH_3-\underset{\underset{CH_3}{|}}{\overset{\overset{CH_3}{|}}{C}}-MgBr \xrightarrow{D_2O} CH_3-\underset{\underset{CH_3}{|}}{\overset{\overset{CH_3}{|}}{C}}-D$$

15.6

(a) $CH_3CH_2\underset{\underset{OH}{|}}{C}HCH_3 + PBr_3 \longrightarrow CH_3CH_2\underset{\underset{Br}{|}}{C}HCH_3 + P(OH)_3$

(b) $CH_3CH_2CH_2CH_2OH \xrightarrow{PBr_3} CH_3CH_2CH_2CH_2Br \xrightarrow{(CH_3)_3COK}$

$CH_3CH_2CH=CH_2 \xrightarrow[\text{(no peroxides)}]{HBr} CH_3CH_2\underset{Br}{CH}CH_3$

(c) See (b) above.

(d) $CH_3CH_2C\equiv CH \xrightarrow[\text{(no peroxides)}]{HBr} CH_3CH_2\underset{Br}{C}=CH_2 \xrightarrow{H_2}{Pt} CH_3CH_2\underset{Br}{CH}CH_3$

15.7

(a) $CH_3CH_2\underset{OH}{CH}CH_3 \xrightarrow{PBr_3} CH_3CH_2\underset{Br}{CH}CH_3 \xrightarrow{(CH_3)_3COK} CH_3CH_2CH=CH_2$
(+2-butenes)

$\xrightarrow[\text{peroxides}]{HBr} CH_3CH_2CH_2CH_2Br$

(b) $CH_3CH_2CH_2CH_2OH \xrightarrow{PBr_3} CH_3CH_2CH_2CH_2Br + P(OH)_3$

(c) $CH_3CH_2CH=CH_2 \xrightarrow[\text{(peroxides)}]{HBr} CH_3CH_2CH_2CH_2Br$

(d) $CH_3CH_2C\equiv CH \xrightarrow[\text{peroxides}]{HBr} CH_3CH_2CH=CHBr \xrightarrow[H_2]{Pt} CH_3CH_2CH_2CH_2Br$

15.8

(a) (cyclohexanol) $+ SOCl_2 \longrightarrow$ (chlorocyclohexane) $+ SO_2 + HCl$

(b) (cyclohexene) $+ HCl \longrightarrow$ (chlorocyclohexane)

(c) (1-methylcyclohexene) $+ HBr \xrightarrow[\text{peroxides}]{no}$ (1-bromo-1-methylcyclohexane)

(d) (1-methylcyclohexene) $+ HBr \xrightarrow{\text{peroxides}}$ (1-methyl-2-bromocyclohexane)

(e) (1-bromo-1-methylcyclohexane) $+ Mg \xrightarrow{ether}$ (Grignard) $\xrightarrow{D_2O}$ (1-deutero-1-methylcyclohexane)

15.9

(a) 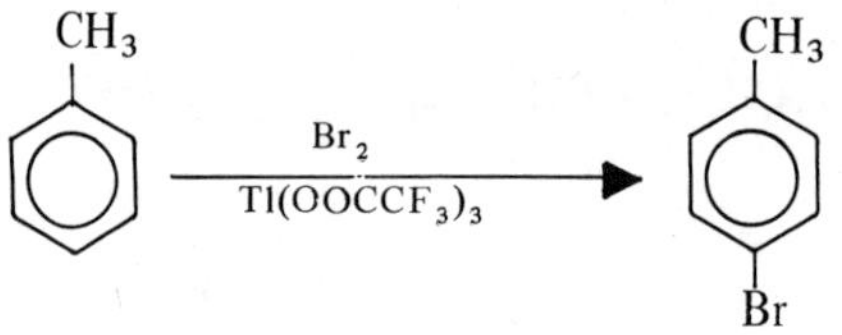
CH3
Br2
Tl(OOCCF3)3
CH3
Br

(b) 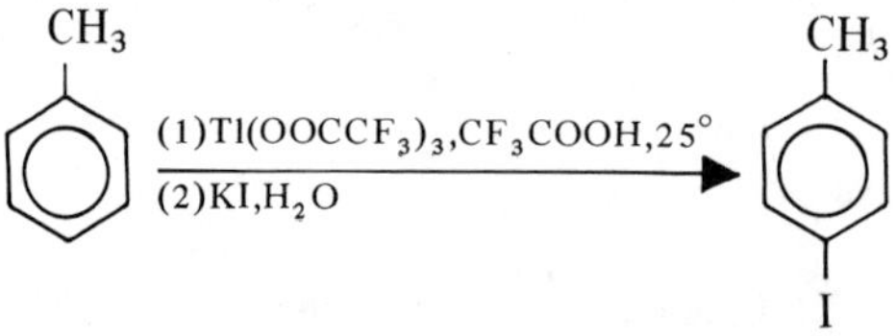
CH3
(1)Tl(OOCCF3)3,CF3COOH,25°
(2)KI,H2O
CH3
I

(c) 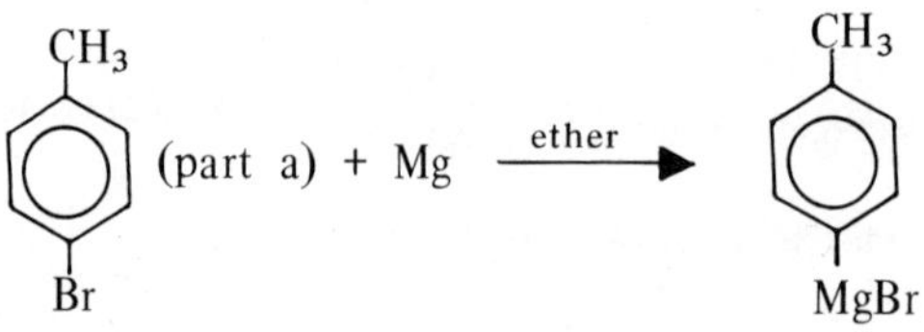
CH3
Br
(part a) + Mg
ether
CH3
MgBr

(d) 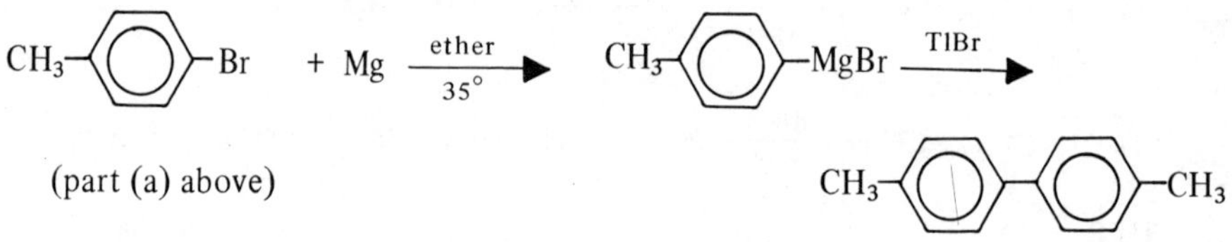
CH3
Br
+ Mg
ether
35°
CH3
MgBr
TlBr
(part (a) above)
CH3
CH3

(e) 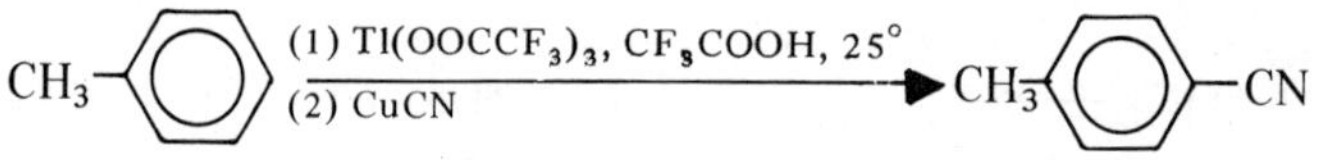
CH3
(1) Tl(OOCCF3)3, CF3COOH, 25°
(2) CuCN
CH3
CN

(f) 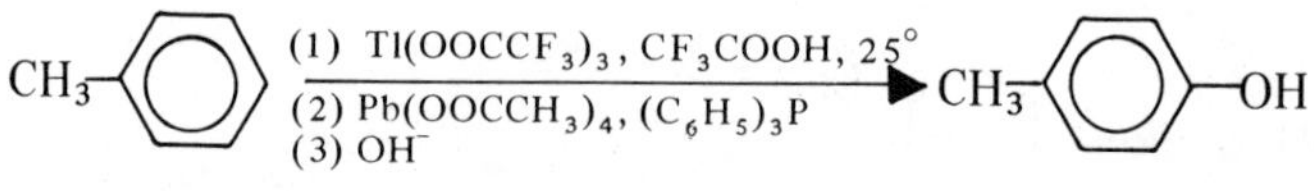
CH3
(1) Tl(OOCCF3)3, CF3COOH, 25°
(2) Pb(OOCCH3)4, (C6H5)3P
(3) OH⁻
CH3
OH

(g) 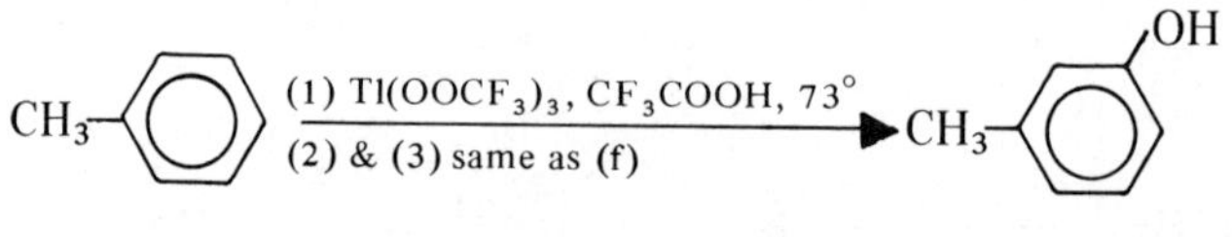
CH3
(1) Tl(OOCF3)3, CF3COOH, 73°
(2) & (3) same as (f)
CH3
OH

(h) 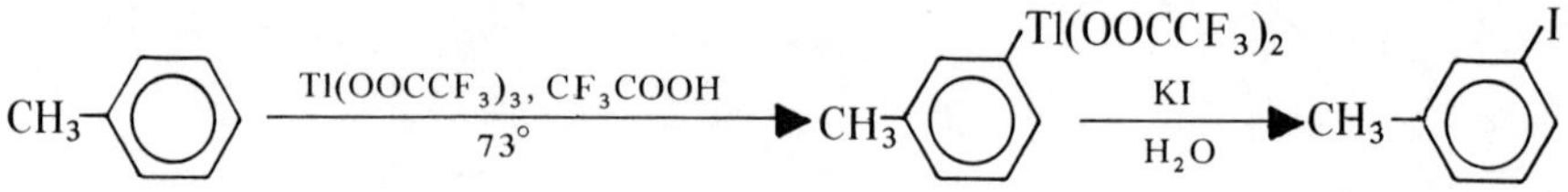
CH3
Tl(OOCCF3)3, CF3COOH
73°
CH3
Tl(OOCCF3)2
KI
H2O
CH3
I

(i) 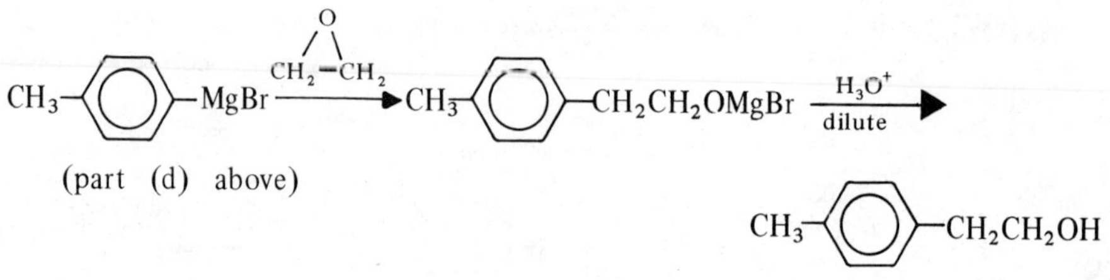
CH3
MgBr
O
CH2—CH2
CH3
CH2CH2OMgBr
H3O+
dilute
(part (d) above)
CH3
CH2CH2OH

(j) CH_3—⟨◯⟩—$MgBr$ $\xrightarrow{D_2O}$ CH_3—⟨◯⟩—D

15.10

The acetyl group is deactivating, thus the unsubstituted ring of the intermediate, mono-acetylated ferrocene is more reactive than the substituted one.

15.11

(a) [cyclopentane with Br] + $(CH_3)_2CuLi$ $\xrightarrow[\text{ether}]{0°}$ [cyclopentane with CH_3] + CH_3Cu + $LiBr$

(b) [cyclopentene with Br] + $(CH_3)_2CuLi$ $\xrightarrow[\text{ether}]{0°}$ [cyclopentene with CH_3] + CH_3Cu + $LiBr$

(c) $CH_2=CH-CH_2Br$ + $(CH_3CH_2)_2CuLi$ $\xrightarrow[\text{ether}]{0°}$ $CH_2=CH-CH_2-CH_2-CH_3$ + CH_3CH_2Cu + $LiBr$

(d)
$$\begin{matrix} CH_3 & & CH_3 \\ & C=C & \\ H & & I \end{matrix}$$
+ $(CH_3CH_2CH_2CH_2)_2CuLi$ $\xrightarrow[\text{ether}]{0°}$
$$\begin{matrix} CH_3 & & CH_3 \\ & C=C & \\ H & & CH_2CH_2CH_2CH_3 \end{matrix}$$
+ $CH_3CH_2CH_2CH_2Cu$ + LiI

15.12

(a) ⟨◯⟩ + $CH_3COO^- Li^+$ (b) ⟨◯⟩ + $CH_3O^- Li^+$

(c) CH_4 + $MgBrNH_2$ (d) $(CH_3)_4Si$ + $4MgBrCl$

(e) $\left[⟨◯⟩\right]_3 P$ + $3MgBrCl$ (f) $(CH_3CH_2)_2Cd$ + $2MgBrCl$

(g) ⟨◯⟩—$\overset{\overset{O}{\|}}{C}$—$OH$ + Mg^{++} (h) ⟨◯⟩—CH_2OH + Mg^{++}

15.13

(a) $CH_2=CH_2$

(b) $BrCH_2-CH_2-Br$ + Mg $\xrightarrow{\text{ether}}$ $\overset{\delta-}{Br}-CH_2-\overset{\delta+}{CH_2}-MgBr$ ⟶ $CH_2=CH_2$ + $MgBr_2$

15.14

(a) Allyl bromide decolorizes Br_2/CCl_4 solution; *n*-propyl bromide does not.

(b) Benzyl bromide gives a AgBr precipitate with $AgNO_3$ in alcohol; *p*-bromotoluene does not.

(c) Benzyl chloride gives an AgCl precipitate with $AgNO_3$ in alcohol; or vinyl chloride decolorizes Br_2/CCl_4 solution.

(d) Phenyllithium (a small amount) reacts vigorously with water to give benzene and a strongly basic aqueous solution (LiOH). Diphenylmercury does not react in this way.

(e) Aldrin decolorizes Br_2/CCl_4 solution rapidly. (The double bond of chlordan and the corresponding double bond of aldrin are not reactive.)

15.15

$$Cl_3C-\overset{\overset{\displaystyle O}{\|}}{C}H + H_2SO_4 \rightleftharpoons \left[Cl_3C-\overset{\overset{\displaystyle \overset{+}{O}H}{\|}}{C}H \longleftrightarrow Cl_3C-\overset{\overset{\displaystyle OH}{|}}{\underset{+}{C}}H \right] + HSO_4^-$$

$$Cl_3C-\overset{\overset{\displaystyle OH}{|}}{\underset{+}{C}}H + \text{(benzene-Cl)} \longrightarrow Cl_3C-\overset{\displaystyle OH}{\underset{\displaystyle H}{C}}\text{(ring-Cl)} \xrightarrow{-H^+} Cl_3C-\overset{\displaystyle OH}{\underset{\displaystyle H}{C}}\text{(ring-Cl)}$$

$$Cl_3C-\overset{\displaystyle OH}{C}H\text{(ring-Cl)} + H_2SO_4 \rightleftharpoons Cl_3C-\overset{\displaystyle \overset{+}{O}H_2}{C}H\text{(ring-Cl)} + HSO_4^- \rightleftharpoons Cl_3C-\overset{+}{C}H\text{(ring-Cl)} + H_2O$$

$$\text{(Cl-benzene)} + \overset{+}{C}H\text{(ring-Cl)} \longrightarrow \text{(structure)} \xrightarrow{-H^+}$$
with CCl_3 groups

$$Cl-\text{(ring)}-\overset{\displaystyle }{\underset{\displaystyle CCl_3}{C}H}-\text{(ring)}-Cl$$

15.16
An elimination reaction.

15.17
A is 3-chloro-2-chloromethyl-1-propene.

(b) H , (a) CH$_2$–Cl
     C=C
(b) H , (a) CH$_2$–Cl

(a) Singlet $\delta 4.25$ (4H)

(b) Singlet $\delta 5.35$ (2H)

B is 1,3-dichloro-2-butene

(a)   (c)   (b)
$CH_3-C=CH-CH_2Cl$
         |
         Cl

(a) Singlet $\delta 2.2$

(b) Doublet $\delta 4.15$

(c) Triplet $\delta 5.7$

15.18

Compound **C** is 1,3-dichlorobutane.

$$\text{(a)}\quad\text{(b)}\quad\text{(c)}\qquad\text{(d)}$$
$$CH_3-\underset{\underset{Cl}{|}}{CH}-CH_2-CH_2Cl$$

(a) Doublet $\delta 1.6$

(b) Multiplet $\delta 4.3$

(c) Multiplet $\delta 2.2$ (resembles a quartet)

(d) Multiplet $\delta 3.7$ (resembles a triplet)

Note: the protons labeled (c) are actually diastereotopic, but their chemical shifts are approximately the same. Coupling constants J_{bc} and J_{cd} are also approximately the same, and thus the spectrum is fortuitously simple.

15.19

D is 3-chloro-2-methylpropene.

$$\begin{array}{c} (c) \\ (a)\,H \qquad CH_2-Cl \\ \qquad\quad C{=}C \\ (b)\,H \qquad CH_3 \\ \qquad\qquad (d) \end{array}$$

(a) Singlet, $\delta 5.1$

(b) Singlet, $\delta 4.9$

(c) Singlet, $\delta 4.0$

(d) Singlet, $\delta 1.9$

15.20

E is 1,3-butadiene. **F** and **G** are the products of the 1,2- and 1,4-addition of chlorine. **H** is 1,2,3,4-tetrachlorobutane.

$$CH_2{=}CH-CH{=}CH_2 \xrightarrow{\;Cl_2\;} CH_2{=}CH-\underset{\underset{Cl}{|}}{C}HCH_2Cl \;+\; ClCH_2CH{=}CHCH_2Cl$$

$$\qquad\quad\text{E}\qquad\qquad\qquad\qquad\qquad\qquad \text{F}\qquad\qquad\qquad\text{G}$$

(or vice versa)

$$\text{F or G} \xrightarrow{\;Cl_2\;} \underset{(a)}{ClCH_2}-\underset{(b)}{\underset{\underset{Cl}{|}}{CH}}-\underset{(b)}{\underset{\underset{Cl}{|}}{CH}}-\underset{(a)}{CH_2Cl}$$

$$\qquad\qquad\qquad\qquad\qquad \text{H}$$

(a) Doublet $\delta 3.8$

(b) Triplet $\delta 4.6$

16 ALCOHOLS, PHENOLS, AND ETHERS

16.1

(a) Alcohols:

$CH_3CH_2CH_2CH_2CH_2OH$	1-pentanol

$$CH_3CH_2CH_2\underset{\underset{OH}{|}}{C}HCH_3 \quad \text{2-pentanol}$$

$$CH_3CH_2\underset{\underset{OH}{|}}{C}HCH_2CH_3 \quad \text{3-pentanol}$$

$$CH_3CH_2\overset{\overset{CH_3}{|}}{C}HCH_2OH \quad \text{2-methyl-1-butanol}$$

$$\overset{\overset{CH_3}{|}}{CH_3}CHCH_2CH_2OH \quad \text{3-methyl-1-butanol}$$

$$CH_3CH_2\overset{\overset{CH_3}{|}}{\underset{\underset{OH}{|}}{C}}CH_3 \quad \text{2-methyl-2-butanol}$$

$$\overset{\overset{CH_3}{|}}{CH_3}\underset{\underset{OH}{|}}{C}HCHCH_3 \quad \text{3-methyl-2-butanol}$$

$$CH_3-\overset{\overset{CH_3}{|}}{\underset{\underset{CH_3}{|}}{C}}-CH_2OH \quad \begin{array}{l}\text{2,2-dimethyl-1-propanol}\\ \text{(neopentyl alcohol)}\end{array}$$

Ethers:

$$CH_3CH_2CH_2CH_2-O-CH_3 \quad \begin{array}{l}\text{methyl } n\text{-butyl ether}\\ \text{(1-methoxybutane)}\end{array}$$

$$\overset{\overset{CH_3}{|}}{CH_3}CHCH_2-O-CH_3 \quad \begin{array}{l}\text{methyl isobutyl ether}\\ \text{(2-methyl-1-methoxypropane)}\end{array}$$

$$CH_3-\overset{\overset{CH_3}{|}}{\underset{\underset{CH_3}{|}}{C}}-O-CH_3 \quad \begin{array}{l}\text{methyl } tert\text{-butyl ether}\\ \text{(2-methyl-2-methoxypropane)}\end{array}$$

$$\underset{\underset{CH_3CH_2CH-O-CH_3}{|}}{\overset{CH_3}{|}}$$ methyl *sec*-butyl ether
(2-methoxybutane)

$CH_3CH_2CH_2-O-CH_2CH_3$ ethyl *n*-propyl ether
(1-ethoxypropane)

$$\underset{\underset{CH_3CH-O-CH_2CH_3}{|}}{\overset{CH_3}{|}}$$ ethyl isopropyl ether
(2-ethoxypropane)

(b) Alcohols: $CH_3CH_2CH=CHCH_2OH$ 2-penten-1-ol

$CH_3CH=CHCH_2CH_2OH$ 3-penten-1-ol

$CH_2=CHCH_2CH_2CH_2OH$ 4-penten-1-ol

$$\underset{\underset{CH_3CH=CHCHOH}{|}}{\overset{CH_3}{|}}$$ 3-penten-2-ol

$$\underset{\underset{CH_2=CHCH_2CHOH}{|}}{\overset{CH_3}{|}}$$ 4-penten-2-ol

$$\underset{\underset{OH}{|}}{CH_2=CHCHCH_2CH_3}$$ 1-penten-3-ol

$$\underset{\underset{CH_3CH_2CCH_2OH}{||}}{\overset{CH_2}{}}$$ 2-ethyl-2-propen-1-ol

$$\underset{\underset{CH_3CCH_2CH_2OH}{||}}{\overset{CH_2}{}}$$ 3-methyl-3-buten-1-ol

$$\underset{\underset{\underset{OH}{|}}{CH_3CCHCH_3}}{\overset{CH_2}{||}}$$ 3-methyl-3-buten-2-ol

$$\underset{\underset{CH_3}{|}}{CH_2=CHCHCH_2OH}$$ 3-methyl-3-buten-1-ol

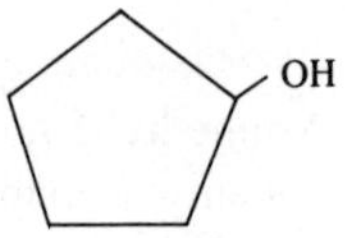

cyclopentanol

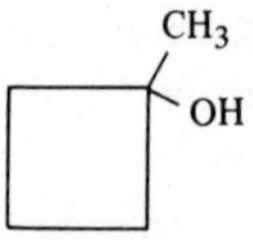

1-methylcyclobutanol

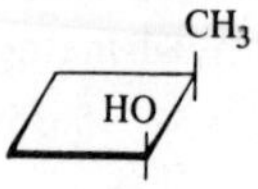

cis-2-methylcyclobutanol

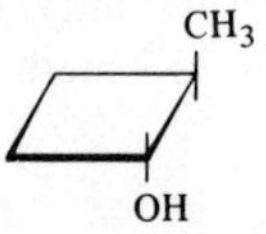

trans-2-methylcyclobutanol

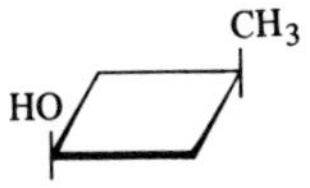

cis-3-methylcyclobutanol

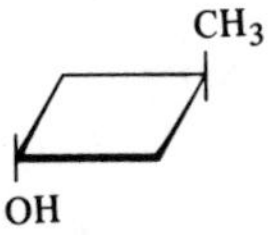

trans-3-methylcyclobutanol

There are also six isomers containing cyclopropane rings.

Ethers: $CH_2=CHCH_2CH_2-O-CH_3$ 4-methoxy-1-butene

$CH_2=CH-CH-O-CH_3$ 3-methoxy-1-butene
$\quad\quad\quad\ |$
$\quad\quad\quad CH_3$

$CH_2=C-OCH_3$ 2-methoxy-1-butene
$\quad\quad |$
$\quad\quad CH_2CH_3$

$CH_3-O-CH=CHCH_2CH_3$ 1-methoxy-1-butene

$CH_3CH=CHCH_2-O-CH_3$ 1-methoxy-2-butene

$CH_3CH=C-OCH_3$ 2-methoxy-2-butene
$\quad\quad\quad |$
$\quad\quad\quad CH_3$

$CH_2=CCH_2OCH_3$ 3-methoxy-2-methyl-propene
$\quad\quad |$
$\quad\quad CH_3$

$CH_2=CHCH_2-O-CH_2CH_3$ 3-ethoxypropene

$CH_2=C-O-CH_2CH_3$ 2-ethoxypropene
$\quad\quad |$
$\quad\quad CH_3$

$CH_3CH_2-O-CH=CHCH_2$ 1-ethoxypropene

$CH_3CH_2CH_2-O-CH=CH_2$ *n*-propyl vinyl ether

$\quad\quad CH_3$
$\quad\quad |$
$CH_3CH-O-CH=CH_2$ isopropyl vinyl ether

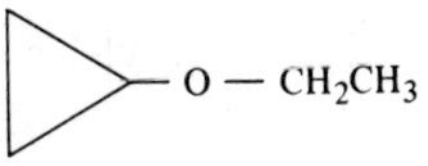

ethoxycyclopropane

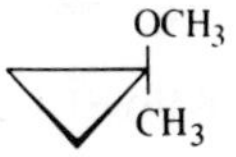

1-methoxy-1-methylcyclopropane

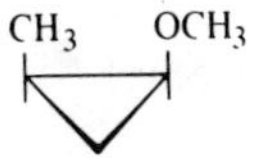

cis-1-methoxy-2-methylcyclopropane

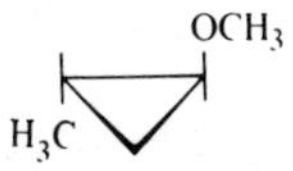

trans-1-methoxy-2-methylcyclopropane

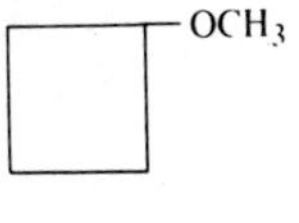

methoxycyclobutane

(c) Alcohols: $CH_3CH_2CH_2CH_2CH_2CH_2OH$ 1-hexanol

$$CH_3CH_2CH_2CH_2\overset{\overset{\displaystyle OH}{|}}{C}HCH_3$$ 2-hexanol

$$CH_3CH_2CH_2\overset{\overset{\displaystyle OH}{|}}{C}HCH_2CH_3$$ 3-hexanol

$$CH_3CH_2CH_2\overset{}{C}HCH_2OH \atop \underset{\displaystyle CH_3}{|}$$ 2-methyl-1-pentanol

$$CH_3CH_2\overset{}{C}HCH_2CH_2OH \atop \underset{\displaystyle CH_3}{|}$$ 3-methyl-1-pentanol

$$CH_3\overset{}{C}HCH_2CH_2CH_2OH \atop \underset{\displaystyle CH_3}{|}$$ 4-methyl-1-pentanol

$$CH_3CH_2CH_2-\overset{\overset{\displaystyle OH}{|}}{\underset{\underset{\displaystyle CH_3}{|}}{C}}-CH_3$$ 2-methyl-2-pentanol

$$CH_3CH_2\overset{}{C}H-\overset{}{C}HCH_3 \atop \underset{\displaystyle CH_3 \quad OH}{| \qquad |}$$ 3-methyl-2-pentanol

$$CH_3\overset{}{C}HCH_2\overset{}{C}HCH_3 \atop \underset{\displaystyle CH_3 \quad\; OH}{| \qquad\; |}$$ 4-methyl-2-pentanol

$$CH_3CH_2\overset{}{C}H-\overset{}{C}HCH_3 \atop \underset{\displaystyle OH \quad CH_3}{| \qquad |}$$ 2-methyl-3-pentanol

$$CH_3CH_2-\overset{\overset{\displaystyle OH}{|}}{\underset{\underset{\displaystyle CH_3}{|}}{C}}-CH_2CH_3$$ 3-methyl-3-pentanol

$$CH_3CH_2\overset{\overset{\displaystyle CH_3}{|}}{\underset{\underset{\displaystyle CH_3}{|}}{C}}CH_2OH$$ 2,2-dimethyl-1-butanol

$$CH_3\overset{\overset{\displaystyle CH_3}{|}}{C}H-\overset{\overset{\displaystyle CH_3}{|}}{C}HCH_2OH$$ 2,3-dimethyl-1-butanol

$$CH_3\overset{\overset{\displaystyle CH_3}{|}}{\underset{\underset{\displaystyle CH_3}{|}}{C}}CH_2CH_2OH$$ 3,3-dimethyl-1-butanol

$$CH_3\overset{\overset{\displaystyle CH_3}{|}}{C}H-\overset{\overset{\displaystyle CH_3}{|}}{\underset{\underset{\displaystyle OH}{|}}{C}}-CH_3$$ 2,3-dimethyl-2-butanol

$$CH_3-\underset{\underset{CH_3}{|}}{\overset{\overset{CH_3}{|}}{C}}\text{———}\underset{\underset{OH}{|}}{CH}-CH_3 \qquad \text{3,3-dimethyl-2-butanol}$$

Ethers: $CH_3CH_2CH_2CH_2CH_2-O-CH_3$ 1-methoxypentane

$$\underset{\underset{CH_3}{}}{CH_3CH_2CH_2\overset{\overset{CH_3}{|}}{CH}OCH_3} \qquad \text{2-methoxypentane}$$

$$CH_3CH_2\overset{\overset{OCH_3}{|}}{CH}CH_2CH_3 \qquad \text{3-methoxypentane}$$

$CH_3CH_2CH_2CH_2-O-CH_2CH_3$ 1-ethoxybutane

$$CH_3CH_2\overset{\overset{CH_3}{|}}{CH}-O-CH_2CH_3 \qquad \text{2-ethoxybutane}$$

$$CH_3CH_2\underset{\underset{CH_3}{|}}{CH}CH_2-O-CH_3 \qquad \text{1-methoxy-2-methylbutane}$$

$$CH_3CH_2\overset{\overset{CH_3}{|}}{\underset{\underset{CH_3}{|}}{C}}-O-CH_3 \qquad \text{2-methoxy-2-methylbutane}$$

$$CH_3\overset{\overset{CH_3}{|}}{CH}-\overset{\overset{OCH_3}{|}}{CH}CH_3 \qquad \text{2-methoxy-3-methylbutane}$$

$$CH_3\underset{\underset{CH_3}{|}}{CH}CH_2CH_2-O-CH_3 \qquad \text{1-methoxy-3-methylbutane}$$

$$CH_3\underset{\underset{CH_3}{|}}{CH}CH_2-O-CH_2CH_3 \qquad \text{1-ethoxy-2-methylpropane}$$

$$CH_3\overset{\overset{CH_3}{|}}{\underset{\underset{CH_3}{|}}{C}}-O-CH_2CH_3 \qquad \text{2-ethoxy-2-methylpropane}$$

$CH_3CH_2CH_2-O-CH_2CH_2CH_3$ di-*n*-propyl ether

$$CH_3CH_2CH_2-O-\overset{\overset{CH_3}{|}}{CH}CH_2 \qquad \textit{n}\text{-propyl isopropyl ether}$$

$$CH_3\overset{\overset{CH_3}{|}}{CH}-O-\overset{\overset{CH_3}{|}}{CH}CH_3 \qquad \text{diisopropyl ether}$$

16.2

The two hydroxyl groups in ethylene glycol allow the formation of more hydrogen bonds than in the monohydroxy alcohols. Thus a single diol molecule can be associated with many neighboring diol molecules.

16.3

(a) $CH_3\overset{\overset{\displaystyle CH_3}{|}}{C}=CH_2 + H_2O \xrightarrow{H^+} CH_3\overset{\overset{\displaystyle CH_3}{|}}{\underset{\underset{\displaystyle OH}{|}}{C}}CH_3$

(b) $CH_3CH_2CH_2CH_2CH=CH_2 + H_2O \xrightarrow{H^+} CH_3CH_2CH_2CH_2\overset{\overset{\displaystyle OH}{|}}{C}HCH_3$

(c) $+ H_2O \xrightarrow{H^+}$

(d) $+ H_2O \xrightarrow{H^+}$

16.4

Rearrangement of the secondary carbocation to the more stable tertiary carbocation,

$$CH_3\overset{\overset{\displaystyle CH_3}{|}}{\underset{\underset{\displaystyle CH_3}{|}}{C}}-CH=CH_2 \overset{H^+}{\rightleftharpoons} CH_3-\overset{\overset{\displaystyle CH_3}{|}}{\underset{\underset{\displaystyle CH_3}{|}}{C}}-\overset{+}{C}H-CH_3 \longrightarrow CH_3-\overset{+}{\underset{\underset{\displaystyle CH_3}{|}}{C}}-\overset{\overset{\displaystyle CH_3}{|}}{C}H-CH_3$$

followed by reaction of the resulting carbocation with water:

$$CH_3-\overset{\overset{\displaystyle CH_3}{|}}{\underset{\underset{\displaystyle CH_3}{|}}{\overset{+}{C}}}-CHCH_3 + H_2O \rightleftharpoons CH_3-\overset{\overset{\displaystyle ^+OH_2}{|}}{\underset{\underset{\displaystyle CH_3}{|}}{C}}-\overset{\overset{\displaystyle}{|}}{\underset{\underset{\displaystyle CH_3}{|}}{C}}HCH_3 \rightleftharpoons CH_3-\overset{\overset{\displaystyle OH}{|}}{\underset{\underset{\displaystyle CH_3}{|}}{C}}-\overset{}{\underset{\underset{\displaystyle CH_3}{|}}{C}}HCH_3$$
$$+ H^+$$

16.5

(a) $CH_3CH_2CH_2CH_2CH=CH_2 \xrightarrow[\text{THF-H}_2\text{O}]{\text{Hg(OAc)}_2} CH_3CH_2CH_2CH_2\overset{\overset{\displaystyle}{|}}{\underset{\underset{\displaystyle OH}{|}}{C}}HCH_2-HgOAc$

$\xrightarrow[\text{OH}^-]{\text{NaBH}_4} CH_3CH_2CH_2CH_2\overset{\overset{\displaystyle OH}{|}}{C}HCH_3$

(b) $\xrightarrow[\text{(2) NaBH}_4,\text{ OH}^-]{\text{(1) Hg(OAc)}_2,\text{ THF}-H_2O}$

(c) $CH_3\overset{\overset{\displaystyle CH_3}{|}}{\underset{\underset{\displaystyle CH_3}{|}}{C}}CH_2\overset{\overset{\displaystyle CH_3}{|}}{C}=CH_2 \xrightarrow{\text{same as (b)}} CH_3\overset{\overset{\displaystyle CH_3}{|}}{\underset{\underset{\displaystyle CH_3}{|}}{C}}CH_2\overset{\overset{\displaystyle CH_3}{|}}{\underset{\underset{\displaystyle OH}{|}}{C}}CH_3$

(d)

$$CH_3C=CH_2 \xrightarrow{\text{same as (b)}} CH_3\overset{\overset{\displaystyle OH}{|}}{C}CH_3$$

(with phenyl groups attached)

16.6

(a)

$$CH_3-\overset{\overset{\displaystyle CH_3}{|}}{\underset{\underset{\displaystyle CH_3}{|}}{C}}-CH=CH_2 + (BH_3)_2 \longrightarrow (CH_3-\overset{\overset{\displaystyle CH_3}{|}}{\underset{\underset{\displaystyle CH_3}{|}}{C}}-CH_2CH_2)_3B$$

$$\xrightarrow[OH^-,H_2O]{H_2O_2} CH_3-\overset{\overset{\displaystyle CH_3}{|}}{\underset{\underset{\displaystyle CH_3}{|}}{C}}-CH_2CH_2OH$$

(b) $CH_3CH_2CH_2CH_2CH=CH_2 \xrightarrow[(2)H_2O_2, OH^-, H_2O]{(1)(BH_3)_2} CH_3CH_2CH_2CH_2CH_2CH_2OH$

(c)

$$\text{(ring)}-CH=CH_2 \xrightarrow{\text{same as (b)}} \text{(ring)}-CH_2CH_2OH$$

(d)

same as (b)

+ enantiomer

16.7

(a) $LiAlH_4$ (b) $NaBH_4$ (c) $LiAlH_4$

16.8

(a) (1)$CH_3MgBr + CH_3\overset{\overset{\displaystyle O}{||}}{C}CH_3 \xrightarrow{(2)H_3O^+} CH_3-\overset{\overset{\displaystyle CH_3}{|}}{\underset{\underset{\displaystyle CH_3}{|}}{C}}-OH$

(2) $2CH_3MgBr\ CH_3\overset{\overset{\displaystyle O}{||}}{C}-OC_2H_5 \xrightarrow{(2)H_3O^+} CH_3-\overset{\overset{\displaystyle CH_3}{|}}{\underset{\underset{\displaystyle CH_3}{|}}{C}}-OH$

(b) (1)$CH_3MgBr + CH_3CH_2CH_2\overset{\overset{\displaystyle O}{||}}{CH} \xrightarrow{(2)H_3O^+} CH_3CH_2CH_2\overset{\overset{\displaystyle OH}{|}}{C}HCH_3$

(2)$CH_3CH_2CH_2MgBr + CH_3\overset{\overset{\displaystyle O}{||}}{CH} \xrightarrow{(2)H_3O^+} CH_3CH_2CH_2\overset{\overset{\displaystyle OH}{|}}{C}HCH_3$

(c) (1) C_6H_5MgBr + $CH_3\overset{\overset{O}{\|}}{C}CH_2CH_3$ $\xrightarrow{(2)H_3O^+}$ $C_6H_5\overset{\overset{CH_3}{|}}{\underset{\underset{OH}{|}}{C}}CH_2CH_3$

(2) CH_3MgBr + $C_6H_5\overset{\overset{O}{\|}}{C}CH_2CH_3$ $\xrightarrow{(2)H_3O^+}$ $C_6H_5\overset{\overset{CH_3}{|}}{\underset{\underset{OH}{|}}{C}}CH_2CH_3$

(3) CH_3CH_2MgBr + $C_6H_5\overset{\overset{O}{\|}}{C}CH_3$ $\xrightarrow{(2)H_3O^+}$ $C_6H_5\overset{\overset{CH_3}{|}}{\underset{\underset{OH}{|}}{C}}CH_2CH_3$

(d) (1) $CH_3CH_2CH_2CH_2MgBr$ + $\overset{O}{\overset{\diagup\diagdown}{CH_2-CH_2}}$ $\xrightarrow{(2)H_3O^+}$ $CH_3CH_2CH_2CH_2CH_2CH_2OH$

(2) $CH_3CH_2CH_2CH_2CH_2MgBr$ + CH_2O $\xrightarrow{(2)H_3O^+}$ $CH_3CH_2CH_2CH_2CH_2CH_2OH$

16.9

(a) $CH_3C\equiv\overset{-}{C}:$ + CH_3CH_2OH $\rightleftharpoons$ $CH_3C\equiv CH$ + $CH_3CH_2\overset{-}{O}$

    stronger   stronger   weaker   weaker
    base   acid   acid   base

(b) $CH_3CH_2CH_2\overset{\delta-}{C}\overset{\delta+}{H_2}:Li$ + CH_3CH_2OH $\rightleftharpoons$ $CH_3CH_2CH_2CH_3$ + $CH_3CH_2\overset{-}{O}\overset{+}{Li}$

    stronger   stronger   weaker   weaker
    base   acid   acid   base

(c) $CH_3\overset{\delta-}{C}\overset{\delta+}{H_2}:MgBr$ + CH_3CH_2OH $\rightleftharpoons$ CH_3CH_3 + $CH_3CH_2\overset{-}{O}\overset{+}{MgBr}$

    stronger   stronger   weaker   weaker
    base   acid   acid   base

16.10

(a) benzene + SO_3 $\xrightarrow{H_2SO_4}$ benzene-SO_3H $\xrightarrow{PCl_5}$ benzene-SO_2Cl $\xrightarrow{CH_3OH}$ benzene-SO_2-OCH_3

(b) toluene + SO_3 $\xrightarrow{H_2SO_4}$ toluene-SO_3H $\xrightarrow{PCl_5}$ toluene-SO_2Cl $\xrightarrow[OH^-]{CH_3\overset{\overset{OH}{|}}{C}HCH_3}$ toluene-$\overset{\overset{CH_3}{}}{\underset{\underset{\underset{CH_3}{|}}{OCHCH_3}}{O=S=O}}$

(c)

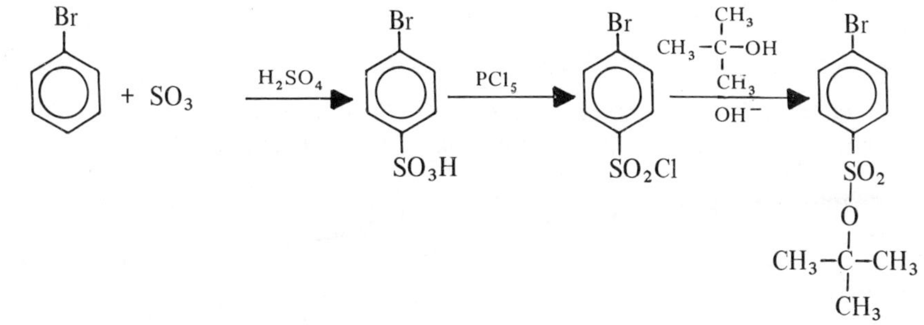

16.11

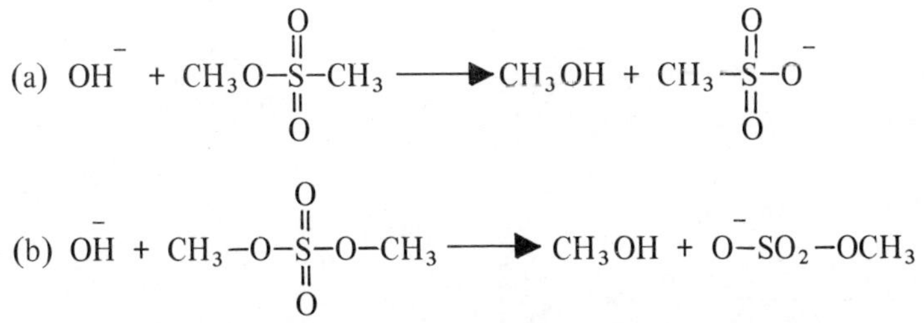

(a)
$$OH^- + CH_3O-\underset{\underset{O}{\|}}{\overset{\overset{O}{\|}}{S}}-CH_3 \longrightarrow CH_3OH + CII_3-\underset{\underset{O}{\|}}{\overset{\overset{O}{\|}}{S}}-O^-$$

(b)
$$OH^- + CH_3-O-\underset{\underset{O}{\|}}{\overset{\overset{O}{\|}}{S}}-O-CH_3 \longrightarrow CH_3OH + {}^-O-SO_2-OCH_3$$

$$OH^- + CH_3O-SO_2O^- \longrightarrow CH_3OH + {}^-O-SO_2-O^-$$

16.12

(a) *Trans*-2-pentene because it is more stable.

(b) A 1-phenylpropene is conjugated and thus is more stable than the unconjugated 3-phenylpropene, and *trans*-3-phenylpropene is more stable than the *cis* isomer.

16.13

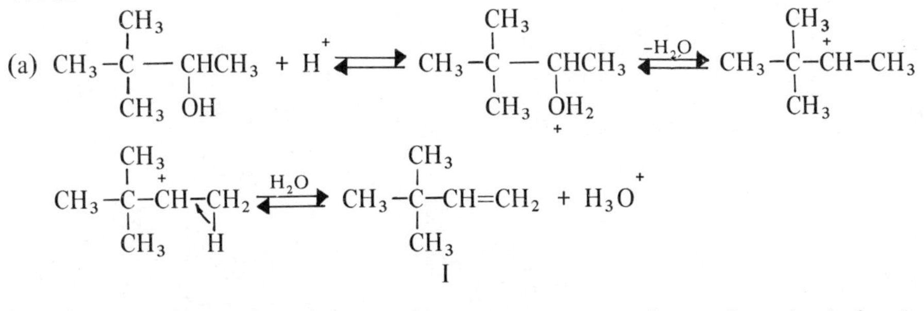

(a)

But this secondary carbocation can also rearrange to a tertiary carbocation before losing a proton:

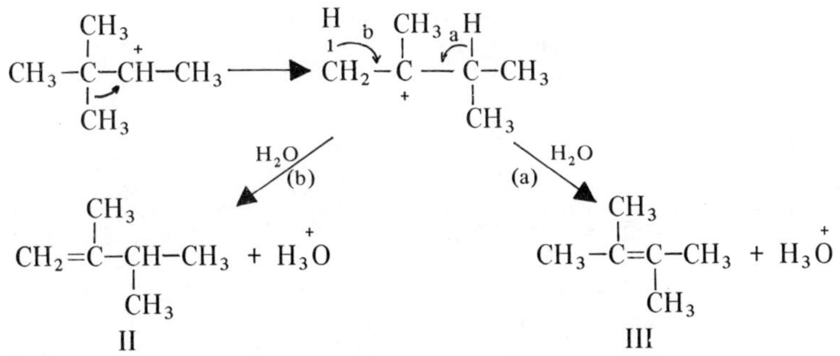

(b) 2,3-Dimethyl-2-butene (III) is the most substituted alkene, therefore it is most stable.

16.14

In structures **2 - 4** the carbon-oxygen bond is a double bond. Thus we would expect the carbon-oxygen bond of a phenol to be much stronger than that of an alcohol.

16.15

16.16

(g) $C_6H_5CH_3 \xrightarrow[\text{light}]{Br_2, CCl_4} C_6H_5CH_2Br \xrightarrow[\text{ether}]{Mg} C_6H_5CH_2MgBr$

$C_6H_5CH_2CH_2OH \xleftarrow[\text{(2) }H_3O^+]{\text{(1) }CH_2O}$

(h) $C_6H_6 \xrightarrow[\text{FeBr}_3]{Br_2} C_6H_5Br \xrightarrow[\text{ether}]{Mg} C_6H_5MgBr \xrightarrow[\text{(2) }H_3O^+]{\text{(1) }CH_2-CH_2 \text{ (epoxide)}} C_6H_5CH_2CH_2OH$

(i) $C_6H_5CH_2CO_2CH_3 \xrightarrow[\text{ether}]{LiAlH_4} C_6H_5CH_2CH_2OH$

16.17

(a) $CH_3CH_2CH_2CH_2OH \xrightarrow{PBr_3} CH_3CH_2CH_2CH_2Br$

$\xrightarrow{(CH_3)_3COK} CH_3CH_2CH=CH_2$

(b) $CH_3CH_2CH=CH_2 \xrightarrow[\text{(2) }NaBH_4, OH^-]{\text{(1) }Hg(OAc)_2, THF-H_2O} CH_3CH_2\overset{\displaystyle OH}{\underset{}{C}}HCH_3$

(c) $CH_3CH_2\overset{\displaystyle OH}{\underset{}{C}}HCH_3 \xrightarrow[H_2SO_4]{CrO_3} CH_3CH_2\overset{\displaystyle O}{\underset{}{C}}CH_3$

(d) $CH_3CH_2CH_2CH_2OH \xrightarrow{PBr_3} CH_3CH_2CH_2CH_2Br$

(e) $CH_3CH_2CH=CH_2 + HBr \xrightarrow{\text{(no peroxides)}} CH_3CH_2\overset{\displaystyle Br}{\underset{}{C}}HCH_3$

(f) $CH_3CH_2CH_2CH_2Br \xrightarrow[\text{ether}]{Mg} CH_3CH_2CH_2CH_2MgBr$

$\xrightarrow[\text{(2) }H_3O^+]{\text{(1) }CH_2O} CH_3CH_2CH_2CH_2CH_2OH$

(g) $CH_3CH_2CH_2CH_2MgBr \xrightarrow[\text{(2) }H_3O^+]{\text{(1) }CH_2-CH_2 \text{ (epoxide)}} CH_3CH_2CH_2CH_2CH_2CH_2OH$

$\xrightarrow{PBr_3} CH_3CH_2CH_2CH_2CH_2CH_2Br \xrightarrow{(CH_3)_3COK} CH_3CH_2CH_2CH_2CH=CH_2$

(h) $CH_3CH_2CH_2CH_2MgBr + CH_3\overset{\displaystyle O}{\underset{}{C}}CH_2CH_3 \xrightarrow{(2)H_3O^+} CH_3CH_2CH_2CH_2\overset{\displaystyle OH}{\underset{\displaystyle CH_3}{C}}CH_2CH_3$

(i) $CH_3CH_2CH_2CH_2OH \xrightarrow{CrO_3 \cdot 2C_5H_5N} CH_3CH_2CH_2\overset{\overset{\displaystyle O}{\|}}{C}H$

(j) $CH_3CH_2CH_2CH_2\,MgBr + CH_3CH_2CH_2\overset{\overset{\displaystyle O}{\|}}{C}H \xrightarrow{(2)\,H_3O^+}$

$$CH_3CH_2CH_2CH_2\overset{\overset{\displaystyle OH}{|}}{C}HCH_2CH_2CH_3$$

(k) $CH_3CH_2\overset{\overset{\displaystyle Br}{|}}{C}HCH_3 \xrightarrow[\text{ether}]{Mg} CH_3CH_2\overset{\overset{\displaystyle CH_3}{|}}{C}HMgBr$

$$\xrightarrow[(2)H_3O^+]{(1)CH_3CH_2CH_2\overset{\overset{\displaystyle O}{\|}}{C}H} CH_3CH_2\overset{\overset{\displaystyle CH_3}{|}}{C}H{-}\overset{\overset{\displaystyle OH}{|}}{C}HCH_2CH_2CH_3$$

(l) $CH_3CH_2CH_2CH_2MgBr + CO_2 \xrightarrow{(2)\,H_3O^+} CH_3CH_2CH_2CH_2COOH$

(m) (1)$CH_3CH_2\overset{\overset{\displaystyle CH_3}{|}}{C}HOH \xrightarrow{Na} CH_3CH_2\overset{\overset{\displaystyle CH_3}{|}}{C}HONa$

$$\xrightarrow{CH_3CH_2CH_2CH_2\,Br} CH_3CH_2\overset{\overset{\displaystyle CH_3}{|}}{C}H{-}O{-}CH_2CH_2CH_2CH_3$$

(2)$CH_3CH_2CH_2CH_2OH \xrightarrow{Na} CH_3CH_2CH_2CH_2ONa \xrightarrow{CH_3CH_2\overset{\overset{\displaystyle CH_3}{|}}{C}HBr}$

$CH_3CH_2CH_2CH_2O\overset{\overset{\displaystyle CH_3}{|}}{C}HCH_2CH_3$ (This method would give a lower yield than that of (1) or (3) because of competing elimination.)

(3)$CH_3CH_2CH=CH_2 + Hg(OAc)_2 \xrightarrow[CH_3CH_2CH_2CH_2OH]{THF}$

$$CH_3CH_2\overset{\overset{\displaystyle }{}}{C}H{-}CH_2{-}HgOAc$$
$$\overset{\overset{\displaystyle |}{}}{O}CH_2CH_2CH_2CH_3$$

$$CH_3CH_2\overset{\overset{\displaystyle CH_3}{|}}{C}H{-}O{-}CH_2CH_2CH_2CH_3 \xleftarrow[OH^-]{NaBH_4}$$

(n) (1)$2CH_3CH_2CH_2CH_2OH \xrightarrow[140°]{H_2SO_4} (CH_3CH_2CH_2CH_2)_2O$

(2)$CH_3CH_2CH_2CH_2OH + Na \longrightarrow CH_3CH_2CH_2CH_2ONa \xrightarrow{CH_3CH_2CH_2CH_2\,Br}$

$$(CH_3CH_2CH_2CH_2)_2O$$

(o) $CH_3CH_2CH_2CH_2Br + 2Li \longrightarrow CH_3CH_2CH_2CH_2Li + LiBr$

(p) $CH_3CH_2CH_2CH_2Li \xrightarrow{CuI} (CH_3CH_2CH_2CH_2)_2CuLi \xrightarrow{CH_3CH_2CH_2CH_2\,Br}$

$$CH_3CH_2CH_2CH_2CH_2CH_2CH_2CH_3$$

16.18

(a) $CH_3CH_2CH_2O^- Na^+$ Sodium propoxide

(b) $CH_3CH_2CH_2-O-CH_2CH_2CH_2CH_3$ propyl butyl ether

(c) $CH_3-\overset{\overset{O}{\|}}{\underset{\underset{O}{\|}}{S}}-OCH_2CH_2CH_3$ propyl methanesulfonate

(d) CH_3-⟨◯⟩$-SO_2-O-CH_2CH_2CH_3$ propyl p-toluenesulfonate (or propyl tosylate)

(e) $CH_3\overset{\overset{O}{\|}}{C}-O-CH_2CH_2CH_3$ propyl acetate

(f) $CH_3CH_2\overset{\overset{O}{\|}}{C}-O^- K^+$ potassium propanoate

(g) $CH_3CH_2CH_2Cl$ 1-chloropropane

(h) $CH_3CH_2CH_2Cl$ 1-chloropropane

(i) $CH_3CH_2CH_2-O-CH_2CH_2CH_3$ dipropyl ether

(j) $CH_3CH_2CH_2Br$ 1-bromopropane

(k) $CH_3-\overset{\overset{CH_3}{|}}{\underset{\underset{CH_3}{|}}{C}}-O-CH_2CH_2CH_3$ propyl $tert$-butyl ether

(l) ⟨◯⟩$-CH\overset{CH_3}{\underset{CH_3}{}}$ isopropylbenzene (major) +   ⟨◯⟩$-CH_2CH_2CH_3$ propylbenzene

16.19

(a) $CH_3\overset{\overset{CH_3}{|}}{CH}O^- Na^+$ sodium isopropoxide

(b) $CH_3\overset{\overset{CH_3}{|}}{CH}-O-CH_2CH_2CH_2CH_3$ isopropyl butyl ether

(c) $CH_3SO_2-O-\overset{\overset{CH_3}{|}}{CH}CH_3$ isopropyl methanesulfonate

(d) CH_3-⟨◯⟩$-SO_2-O-\overset{\overset{CH_3}{|}}{CH}CH_3$ isopropyl p-toluenesulfonate

(e) 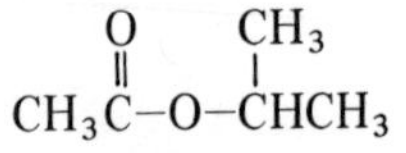 $CH_3\overset{O}{\overset{\|}{C}}-O-\overset{CH_3}{\overset{|}{C}HCH_3}$ isopropyl acetate

(f) 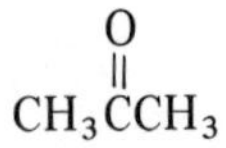 $CH_3\overset{O}{\overset{\|}{C}}CH_3$ acetone ($+ CH_3COOH$ and CO_2)

(g) 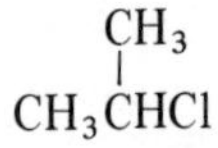 $CH_3\overset{CH_3}{\overset{|}{C}}HCl$ 2-chloropropane

(h) Same as (g)

(i) 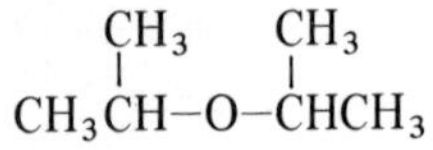 $CH_3\overset{CH_3}{\overset{|}{C}}H-O-\overset{CH_3}{\overset{|}{C}}HCH_3$ diisopropyl ether

(j) 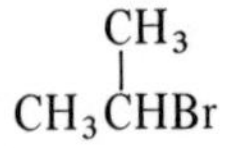 $CH_3\overset{CH_3}{\overset{|}{C}}HBr$ 2-bromopropane

(k) 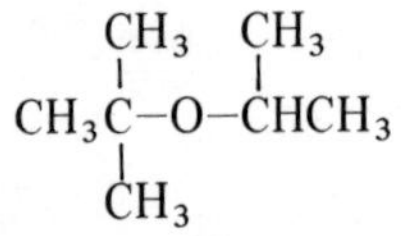 $CH_3\overset{CH_3}{\underset{\overset{|}{C}H_3}{\overset{|}{C}}}-O-\overset{CH_3}{\overset{|}{C}}HCH_3$ isopropyl *tert*-butyl ether

(l) 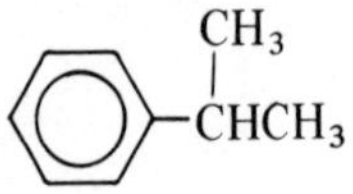 ⬡$-\overset{CH_3}{\overset{|}{C}}HCH_3$ isopropyl benzene

16.20

(a) 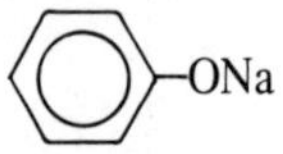 ⬡$-ONa + CH_3CH_2OH$ (e) ⬡$-ONa + H_2O,$

(b) 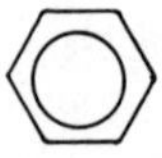 ⬡ $+ CH_3CH_2OMgBr$ (f) ⬡$-OH + NaCl$

(c) $CH_3CH_2SNa + H_2O$ (g) $CH_3CH_2OH + NaOH$

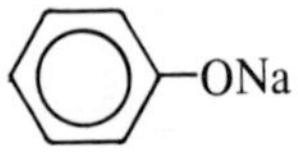

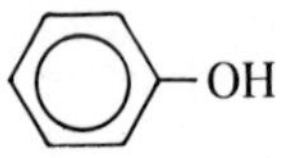

(d) $CH_3CH_2ONa + CH_3CH_2SH$

16.21
(d), (e), (f). (All are stronger acids than H_2CO_3, see page 612.)

16.22
(a) $CH_3Br + CH_3CH_2Br$ (c) $Br-CH_2CH_2CH_2CH_2-Br$

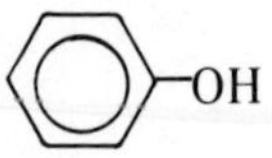

(b) ⬡$-OH + CH_3CH_2Br$ (d) $Br-CH_2CH_2-Br$ (2 moles)

16.23

(a) 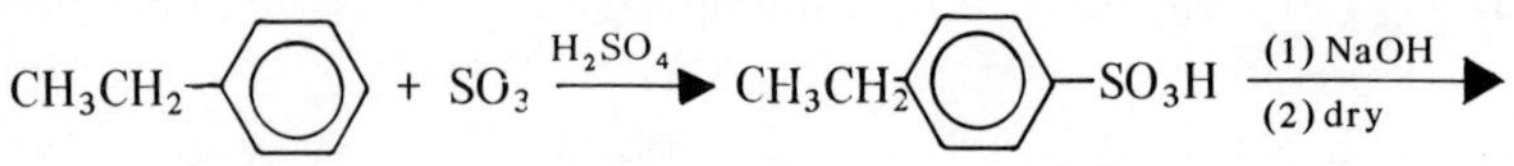 CH_3CH_2-⬡$ + SO_3 \xrightarrow{H_2SO_4} CH_3CH_2$⬡$-SO_3H \xrightarrow[(2)\,dry]{(1)\,NaOH}$

$$CH_3CH_2-\!\!\bigcirc\!\!-SO_3^-\overset{+}{Na} \xrightarrow[\text{(2)}H_3O^+]{\text{(1)}NaOH-KOH,330^\circ} CH_3CH_2-\!\!\bigcirc\!\!-OH$$

$$CH_3CH_2-\!\!\bigcirc \xrightarrow[\substack{CF_3COOH \\ 25^\circ}]{Tl(OOCCF_3)_3} CH_3CH_2-\!\!\bigcirc\!\!-Tl(OOCCF_3)_2 \xrightarrow[\text{(2) HO}^-]{\substack{\text{(1)}Pb(OOCCH_3)_4 \\ (C_6H_5)_3P}}$$

$$CH_3CH_2-\!\!\bigcirc\!\!-OH$$

(b) Thallation at 75° to produce the meta isomer, followed by reaction with $Pb(OOCCH_3)_4 + (C_6H_5)_3P$, then OH^-.

16.24

(a) $\underset{\quad\; OH\; OH}{CH_3CH\,CH_2}$

(b) $CH_3CH_2-O-CH_2CH_2OH$

(c) $C_6H_5O-CH_2CH_2OH$

(d) $CH_3-\!\!\bigcirc\!\!-O-SO_2-\!\!\bigcirc\!\!-CH_3$

(e) cyclohexyl-OOCCH$_3$

(f) phthalate: $\bigcirc$ with $C-O-\!\!\bigcirc$ and $C-OH$ (ortho diester/acid)

(g) $\underset{CH_3}{Br\text{-}\underset{OH}{\bigcirc}\text{-}Br}$

(h) $\bigcirc\!\!-CH_2-\overset{+}{S}=C\underset{NH_2}{\overset{NH_2}{<}}\ Br^-$

(i) $\bigcirc\!\!-CH_2SH$

(j) $\bigcirc\!\!-CH_2-S-S-CH_2-\!\!\bigcirc$

(k) $\bigcirc\!\!-CH_2-S^-Na^+$

(l) $\bigcirc\!\!-CH_2-S-CH_2-\!\!\bigcirc$

(m) $\bigcirc\!\!-\overset{O}{\overset{\|}{C}}-O^-K^+$

(n) $\bigcirc\!\!-\overset{O}{\overset{\|}{C}}H$

(o) $\bigcirc\!\!-CH_2O^-Na^+$

(p) $\bigcirc\!\!-CH_2OCH_3$

(q) $\underset{H}{CH_3CH_2}\!\!>\!\!\underset{O}{}\!\!<\!\!\underset{H}{CH_2CH_3}$

(r)

$$\underset{\substack{HO-\overset{|}{\underset{\underset{\underset{CH_3}{|}}{CH_2}}{C}}-H \\}}{\overset{\overset{\overset{CH_3}{|}}{CH_2}}{H-\overset{|}{C}-OH}}$$

+ enantiomer

16.25

(a) (1) CH_3CH_2MgBr + $\underset{\underset{CH_3}{|}}{\overset{\overset{CH_3}{|}}{C}}=O$ $\xrightarrow{\text{(2) } H_3O^+}$ $CH_3CH_2\underset{\underset{CH_3}{|}}{\overset{\overset{CH_3}{|}}{C}}-OH$

(2) CH_3MgBr + $CH_3CH_2\underset{\underset{CH_3}{|}}{C}=O$ $\xrightarrow{\text{(2) } H_3O^+}$ $CH_3CH_2\underset{\underset{CH_3}{|}}{\overset{\overset{CH_3}{|}}{C}}-OH$

(3) $CH_3CH_2\overset{\overset{O}{\|}}{C}-OCH_3$ + $2CH_3MgBr$ $\xrightarrow{\text{(2) } H_3O^+}$ $CH_3CH_2\underset{\underset{CH_3}{|}}{\overset{\overset{CH_3}{|}}{C}}-OH$

(b) (1) CH_3CH_2MgBr + $\underset{}{\bigcirc}\overset{\overset{O}{\|}}{C}-CH_2CH_3$ $\xrightarrow{\text{(2)}H_3O^+}$ $\underset{}{\bigcirc}\underset{\underset{CH_2CH_3}{|}}{\overset{\overset{OH}{|}}{C}}-CH_2CH_3$

(2) $\underset{}{\bigcirc}-MgBr$ + $CH_3CH_2\overset{\overset{O}{\|}}{C}CH_2CH_3$ $\xrightarrow{\text{(2) } H_3O^+}$ $\underset{}{\bigcirc}\underset{\underset{CH_2CH_3}{|}}{\overset{\overset{OH}{|}}{C}}-CH_2CH_3$

(3) $\underset{}{\bigcirc}\overset{\overset{O}{\|}}{C}-OCH_3$ + $2CH_3CH_2MgBr$ $\xrightarrow{\text{(2) } H_3O^+}$ $\underset{}{\bigcirc}\underset{\underset{CH_2CH_3}{|}}{\overset{\overset{OH}{|}}{C}}-CH_2CH_3$

(c) $\underset{}{\bigcirc}=O$ + C_6H_5-MgBr $\xrightarrow{\text{(2) } H_3O^+}$ $\underset{}{\bigcirc}\overset{OH}{\underset{C_6H_5}{<}}$

(d) $CH_3-\underset{}{\bigcirc}-MgBr$ + $CH_2\overset{O}{-}CH_2$ $\xrightarrow{\text{(2) } H_3O^+}$ $CH_3-\underset{}{\bigcirc}-CH_2CH_2OH$

(e) (1) $CH_3-\underset{}{\bigcirc}-\overset{\overset{O}{\|}}{C}H$ + CH_3MgBr $\xrightarrow{\text{(2) } H_3O^+}$ $CH_3-\underset{}{\bigcirc}-\underset{\underset{CH_3}{|}}{\overset{\overset{OH}{|}}{C}}H$

(2) $CH_3-\underset{}{\bigcirc}-MgBr$ + $CH_3\overset{\overset{O}{\|}}{C}H$ $\xrightarrow{\text{(2) } H_3O^+}$ $CH_3-\underset{}{\bigcirc}-\underset{\underset{CH_3}{|}}{\overset{\overset{OH}{|}}{C}}H$

16.26

(a) *p*-Cresol is soluble in aqueous NaOH; benzyl alcohol is not.

(b) Cyclohexanol is soluble in cold, concentrated H_2SO_4; cyclohexane is not. (Cyclohexanol also gives a positive test with CrO_3 in H_2SO_4, while cyclohexane does not.)

(c) Cyclohexene will decolorize Br_2/CCl_4 solution; cyclohexanol will not.

(d) Allyl propyl ether will decolorize Br_2/CCl_4 solution; dipropyl ether will not.

(e) *p*-Cresol is soluble in aqueous NaOH; anisole is not.

(f) Picric acid is soluble in aqueous $NaHCO_3$: 2,4,6-trimethylphenol is not (cf. Problem 16.21).

16.27

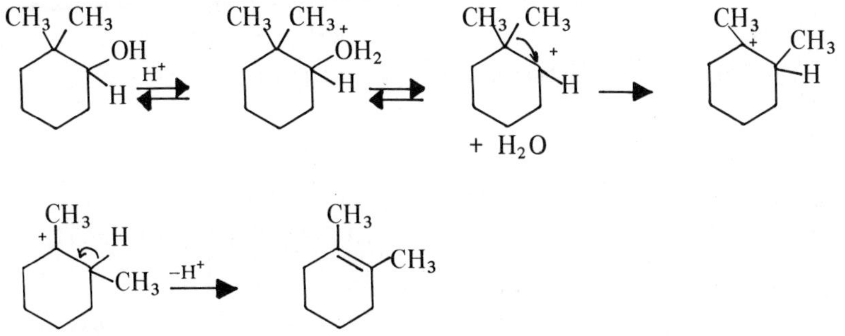

16.28

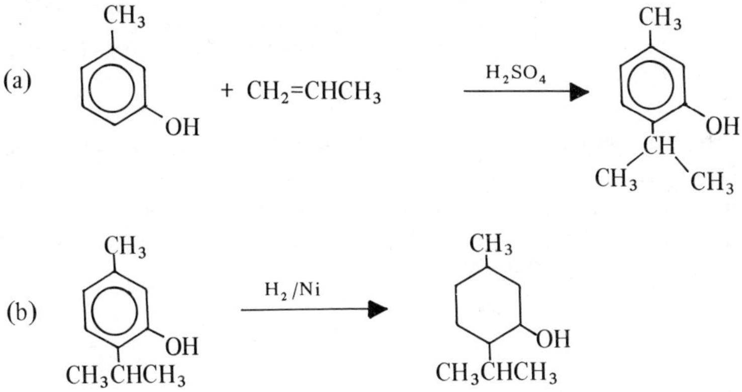

(a)

(b)

16.29

The position ortho to the isopropyl group is sterically more hindered than the position ortho to the methyl group:

16.30

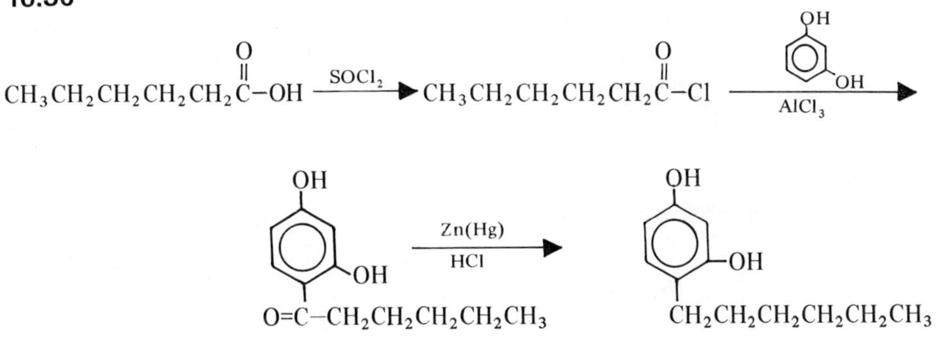

16.31

$$CH_3CH_2\overset{\overset{O}{\|}}{C}-OH \xrightarrow{SOCl_3} CH_3CH_2\overset{\overset{O}{\|}}{C}-Cl \xrightarrow[AlCl_3]{\overset{OCH_3}{\bigcirc}} $$

(para-methoxyphenyl propyl ketone with OCH$_3$ top and $\overset{}{C}CH_2CH_3$ / $\underset{O}{\|}$ below)

$$CH_3O-\bigcirc-\overset{\overset{O}{\|}}{C}CH_2CH_3 \xrightarrow[ether]{NaBH_4} CH_3O-\bigcirc-\overset{\overset{OH}{|}}{C}HCH_2CH_3 $$

$$\xrightarrow[heat]{H_2SO_4}$$

$$CH_3O-\bigcirc-CH=CH-CH_3$$

16.32

$$CH_2=CHCH_2Br + S=C\overset{\diagup NH_2}{\diagdown NH_2} \xrightarrow[(2)\ OH^-,\ H_2O]{(1)\ CH_3CH_2OH} CH_2=CHCH_2SH \xrightarrow{H_2O_2}$$

$$CH_2=CHCH_2-S-S-CH_2CH=CH_2$$

16.33

X is a phenol because it dissolves in aqueous NaOH but not in aqueous NaHCO$_3$. It gives a dibromo derivative, and must therefore be substituted in the ortho or para position. The broad infrared peak at 3250 cm^{-1} also suggests a phenol. The peak at 830 cm^{-1} indicates para substitution. The pmr singlet at $\delta 1.3$ (9H) suggests 9 methyl hydrogens which must be a *tert*-butyl group. The structure of X is:

(structure: para-tert-butylphenol — benzene ring with OH on top and CH$_3$–C(CH$_3$)–CH$_3$ below)

16.34

The broad infrared peak at 3400 cm^{-1} indicates a hydroxy group and the two bands at 720 and 770 cm^{-1} suggest a monosubstituted benzene ring. The presence of these groups is also indicated by the peaks at $\delta 2.7$ and $\delta 7.2$ in the pmr spectrum. The pmr spectrum also shows a triplet at $\delta 0.7$ indicating a –CH$_3$ group coupled with an adjacent –CH$_2$– group. What appears at first to be a quartet at $\delta 1.9$ actually shows further splitting. There is also a triplet at $\delta 4.35$ (1H). Putting these pieces together in the only way possible gives us the following structure for Y.

(structure: benzene ring with –CHCH$_2$CH$_3$ and OH below)

Analysed spectra are shown on p. 165 (Fig. 16.2 in text).

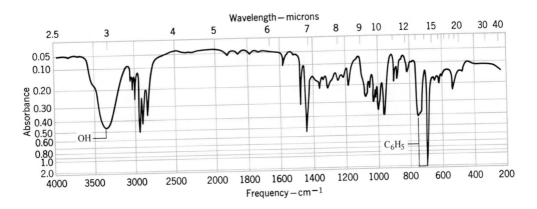

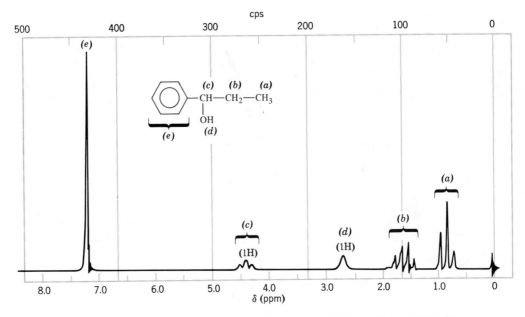

Fig. 16.2. The infrared and pmr spectra of compound Y, problem 16.34 (Spectra courtesy of Sadtler Research Laboratories Inc.)

16.35

$$CH_2=CHCH_2OH \xrightarrow{\ Br_2\ } CH_2BrCHBrCH_2OH \xrightarrow{\ NaSH\ } \underset{\underset{SH}{|}\ \underset{SH}{|}}{CH_2-CH-CH_2OH}$$

16.36

(a)

(b)

$$\text{phenol with OH and CH}_3 + 2\,CH_2=\overset{\underset{\displaystyle CH_3}{|}}{C}-CH_3 \xrightarrow{\;H^+\;} \text{BHT product}$$

BHT

Notice that both reactions are Friedel-Crafts alkylations.

16.37

Z is 3-methyl-2-buten-1-ol.

$$\text{(a) } CH_3 \diagdown \quad \overset{(c)}{\diagup} H$$
$$\qquad\qquad C=C$$
$$\text{(b) } CH_3 \diagup \quad \diagdown CH_2OH$$
$$\text{(d) (e)}$$

(a) or (b) Singlet $\delta 1.7$

(b) or (a) Singlet $\delta 1.8$

(c) Triplet $\delta 5.4$

(d) Doublet $\delta 4.1$

(e) Singlet $\delta 1.9$

16.38

(a) $ClCH_2CH_2\overset{\underset{\displaystyle \;}{\parallel}}{\overset{O}{C}}(CH_2)_4CO_2C_2H_5$ (this step is the Friedel-Crafts acylation of an alkene)

(b) $SOCl_2$

(c) $2\;C_6H_5CH_2SH$ and KOH

(d) H_3O^+

(e) $\begin{array}{c} \quad CH_2 \\ CH_2 \diagup \quad \diagdown CH(CH_2)_4COOH \\ \quad S\;\;\;S \\ \quad |\;\;\;\; | \\ \quad H\;\;H \end{array}$

16.34

(a) **A** is $CH_2=CH\overset{\underset{\displaystyle OH}{|}}{\overset{\displaystyle CH_3}{C}}C\equiv CH$,   **C** is $BrMgOCH_2CH=\overset{\underset{\displaystyle \;}{|}}{\overset{\displaystyle CH_3}{C}}C\equiv CMgBr$

(b) **A** is an allylic alcohol and thus forms a carbocation readily. **B** is a conjugated enyne and is therefore more stable than **A**.

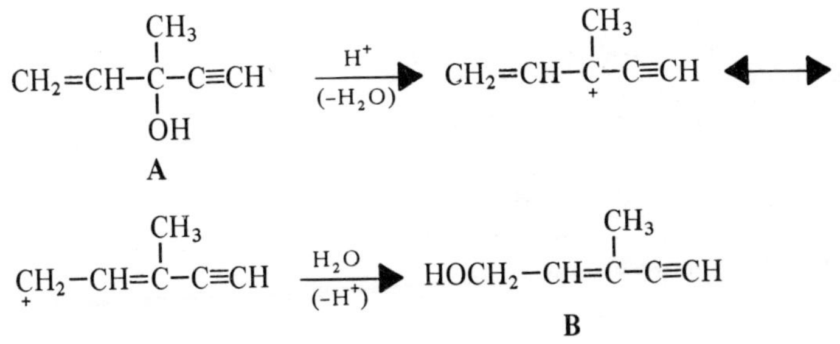

$$CH_2{=}CH{-}\underset{\underset{OH}{|}}{\overset{\overset{CH_3}{|}}{C}}{-}C{\equiv}CH \quad \xrightarrow[(-H_2O)]{H^+} \quad CH_2{=}CH{-}\underset{+}{\overset{\overset{CH_3}{|}}{C}}{-}C{\equiv}CH \quad \longleftrightarrow$$

A

$$\underset{+}{CH_2}{-}CH{=}\overset{\overset{CH_3}{|}}{C}{-}C{\equiv}CH \quad \xrightarrow[(-H^+)]{H_2O} \quad HOCH_2{-}CH{=}\overset{\overset{CH_3}{|}}{C}{-}C{\equiv}CH$$

B

16.40

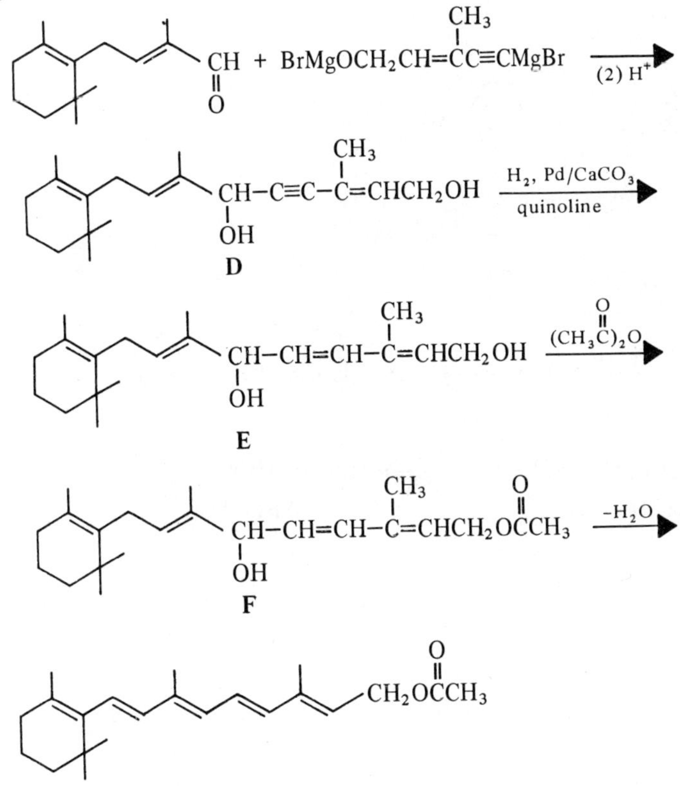

Vitamin A acetate

16.41

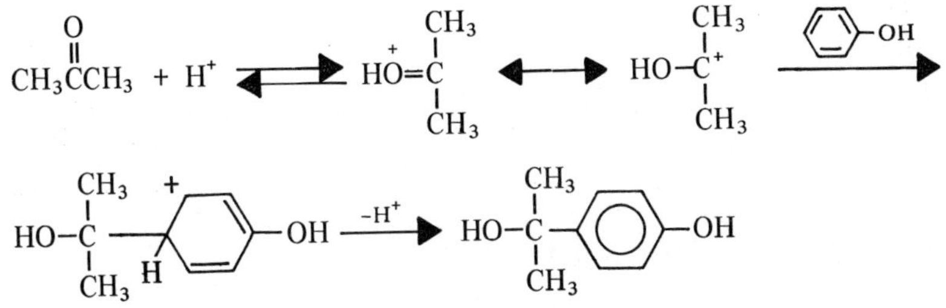

$$CH_3\overset{\overset{O}{\|}}{C}CH_3 + H^+ \rightleftharpoons HO\overset{+}{=}\underset{\underset{CH_3}{|}}{\overset{\overset{CH_3}{|}}{C}} \longleftrightarrow HO{-}\underset{\underset{CH_3}{|}}{\overset{\overset{CH_3}{|}}{\underset{+}{C}}} \quad \xrightarrow{\text{(phenol)}}$$

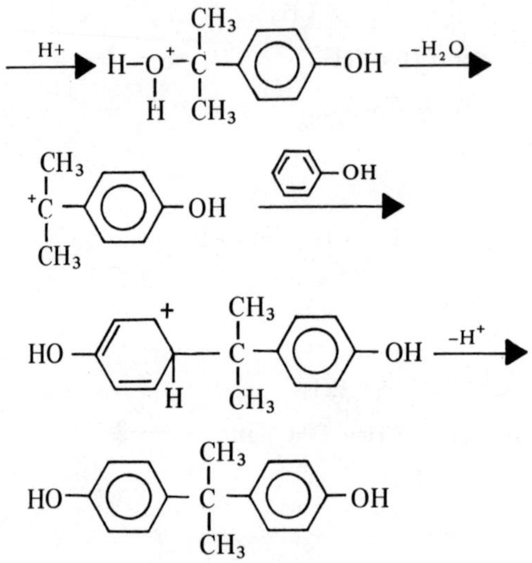

"Bisphenol A"

16.42

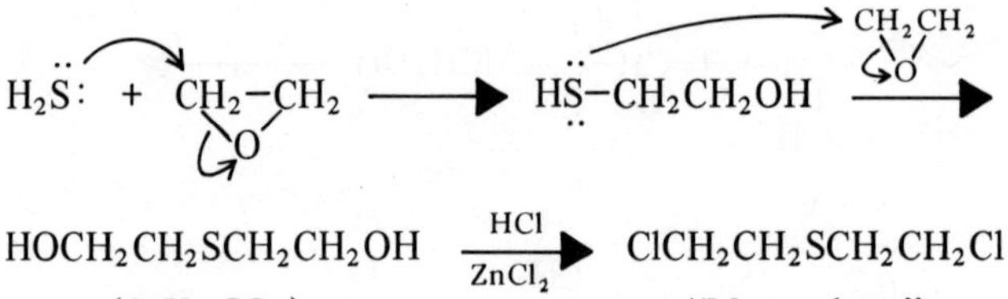

$$HOCH_2CH_2SCH_2CH_2OH \xrightarrow[ZnCl_2]{HCl} ClCH_2CH_2SCH_2CH_2Cl$$
$$(C_4H_{10}SO_2) \qquad\qquad \text{"Mustard gas"}$$

17 NUCLEOPHILIC SUBSTITUTION AND ELIMINATION REACTIONS

17.1

(a) $CH_3OH + NaO-\overset{\overset{\displaystyle O}{\|}}{\underset{\underset{\displaystyle O}{\|}}{S}}-CH_3$

(b) $C_6H_5CH_2-\overset{+}{N}(CH_3)_3 \ \overset{-}{Br}$

(c) $CH_3CH_2-N_3 + NaO-\overset{\overset{\displaystyle O}{\|}}{\underset{\underset{\displaystyle O}{\|}}{S}}-\!\!\left\langle\!\bigcirc\!\right\rangle\!\!-CH_3$

(d) $C_6H_5CH_2-I + NaCl$

(e) $CH_3\underset{\underset{\displaystyle OOCCH_3}{|}}{C}HCH_3 \quad + NaBr$

(f) $\begin{matrix} C_2H_5O\overset{\displaystyle O}{\overset{\displaystyle\|}{C}} \\ \diagdown \\ \qquad CH-CH_2CH_3 + NaBr \\ \diagup \\ C_2H_5O\underset{\displaystyle O}{\underset{\displaystyle\|}{C}} \end{matrix}$

(g) $CH_2=CHCH_2CH_2CH_3 + MgXBr$

17.2

$(C_6H_5)_2CH-Cl \underset{\xrightarrow{\hspace{1cm}}}{\overset{slow}{\longleftarrow}} (C_6H_5)_2\overset{+}{C}H + \overset{-}{Cl} \ \ (step \ 1)$

$(C_6H_5)_2\overset{+}{C}H + \overset{-}{F} \xrightarrow{\ fast\ } (C_6H_5)_2CH-F \ \ (step \ 2)$

Step 1 is rate-limiting, i.e., much slower than step 2. Only diphenylchloromethane is involved in step 1, therefore the reaction is first order with respect to diphenylchloromethane.

17.3

Compounds that undergo reactions by an S_N1 path must be capable of forming relatively stable carbocations. Primary halides of the type, $ROCH_2X$ form carbocations that are stabilized by resonance:

$$R-\overset{..}{\underset{..}{O}}-\overset{+}{C}H_2 \longleftrightarrow R-\overset{+}{\underset{..}{O}}=CH_2$$

17.4

The relative rates are in the order of the relative stabilities of the carbocations:

$$C_6H_5\overset{+}{C}H_2 < C_6H_5\overset{+}{C}HCH_3 < (C_6H_5)_2\overset{+}{C}H < (C_6H_5)_3\overset{+}{C}$$

The solvolysis reaction involves a carbocation intermediate or a solvent-separated ion-pair.

17.5

(a) Backside attack by the nucleophile is prevented by the cyclic structure. (Notice, too, that the carbon bearing the leaving group is tertiary.)

(b) The bridged cyclic structure prevents the carbon bearing the leaving group from assuming the planar trigonal conformation required of a carbocation.

17.6

(a) NH_2^- (b) Phenoxide ion

17.7

In quinuclidine the three groups attached to nitrogen are "pinned back" in an arrangement that makes approach to the carbon bearing the leaving group relatively unhindered. In triethylamine, on the other hand, the three ethyl groups partially block this approach.

17.8

In strong acid, ethers are protonated:

$$R-O-R + HA \rightleftarrows R-\overset{+}{\underset{\underset{H}{|}}{O}}-R + A^-$$

In weak acid or neutral media, protonation does not occur to any appreciable extent. Thus, in strong acid, the leaving group is an alcohol molecule,

$$X^- + R-\overset{+}{\underset{\underset{H}{|}}{O}}-R \longrightarrow X-R + HO-R$$

whereas in weak acid or neutral media, the leaving group would have to be an alkoxide ion,

$$X^- + R-O-R \longrightarrow X-R + \overset{-}{O}-R$$

$R-OH$ is a good leaving group; $R-O^-$ is a very poor one.

17.9

(a) Decrease   (b) Increase   (c) Decrease   (d) increase   (e)   decrease

(f) increase

17.10

Faster in dimethylsulfoxide. The reaction is S_N2, and dimethylsulfoxide is an aprotic solvent. Aprotic solvents accelerate the rates of all S_N2 reactions because the nucleophiles are unsolvated and, therefore, highly reactive.

17.11

(a) and (b)

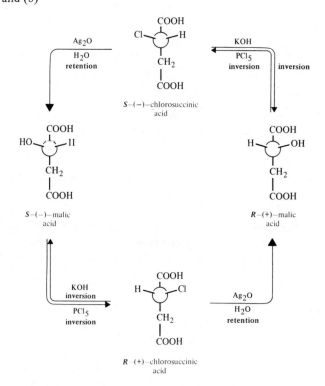

(c) The reaction takes place with retention of configuration.

(d)

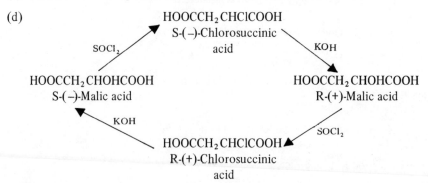

17.12

(a) Reaction of an alkene with halogen in water solution.

(b) 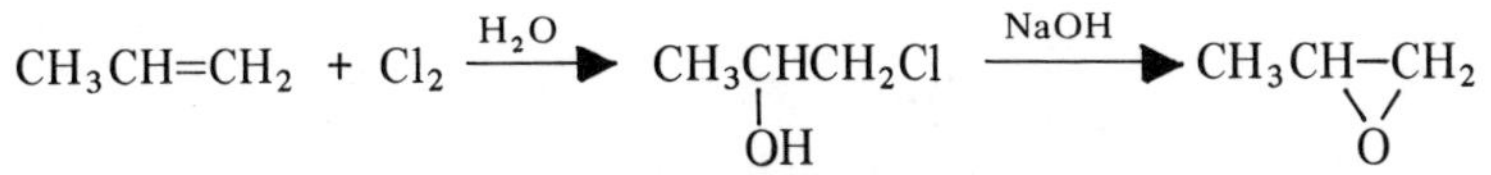 $CH_3CH=CH_2 + Cl_2 \xrightarrow{H_2O} CH_3CHCH_2Cl \xrightarrow{NaOH} CH_3CH-CH_2$

with OH below the first product and O (epoxide) below the second product.

(c) The $-\overset{..}{\underset{..}{O}}:^-$ group must displace the Cl$^-$ from the back side,

epoxide

Such displacement is impossible with the *cis*-isomer:

No epoxide

17.13
2-(p-hydroxyphenyl)-1-chloropropane $>$ 2-(p-tolyl)-1-chloropropane $>$ 2-phenyl-1-chloropropane $>$ 2-(p-nitrophenyl)-1-chloropropane

17.14
Reaction by a S_N2 pathway is hindered sterically by the large *tert*-butyl group. Reaction by a S_N1 pathway is very slow because a primary carbocation must be formed.

17.15
Only a proton or deuteron *anti* to the bromine can be eliminated. The two conformations of *erythro*-2-bromo-butane-3-*d* in which a proton or deuteron is *anti* to the bromine are I and II below.

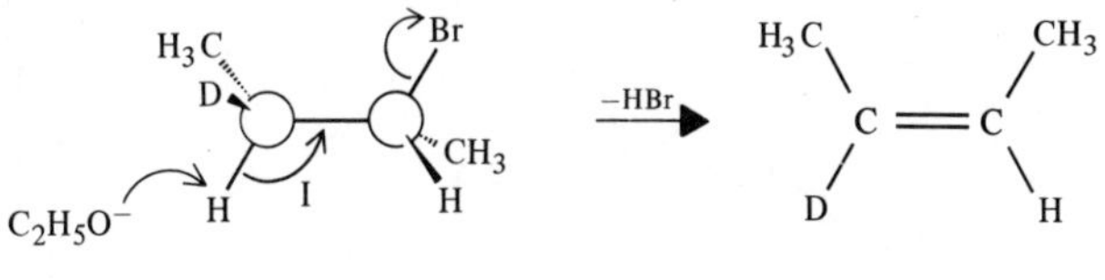

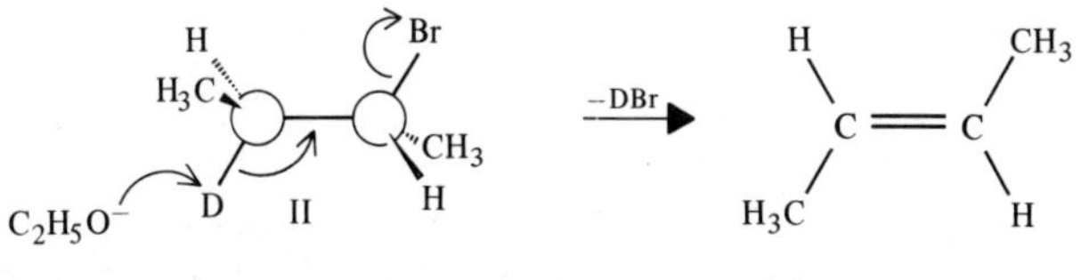

Conformation I can undergo loss of HBr to yield *cis*-2-butene-2-*d*. Conformation II can undergo loss of DBr to yield *trans*-2-butene.

17.16
(a) A $= CH_3CH_2-\overset{+}{N}(CH_3)_3 \ \overset{-}{Br}$ (b) B $= CH_3CH_2CH_2-\overset{+}{N}(CH_3)_3 \ \overset{-}{I}$

(c) C $=$ $-\overset{+}{N}(CH_3)_3 \ \overset{-}{Br}$

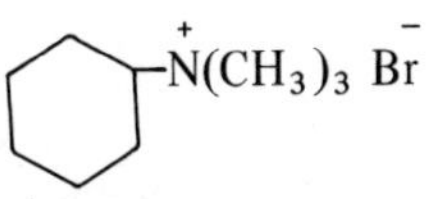

17.17

(a) $\underset{\underset{CH_3}{|}}{CH_3C}=CH_2 + CH_2=CH_2$ (major product)

(b) $CH_3CH=CH_2 + CH_2=CH_2$ (major product)

(c) $(CH_3)_2NCH_2CH_2CH_2CH=CH_2$

(d) $CH_2=CHCH_2CH=CH_2$

17.18

In each case a *syn* elimination takes place from a conformation in which the large phenyl substituents *are not eclipsed.*

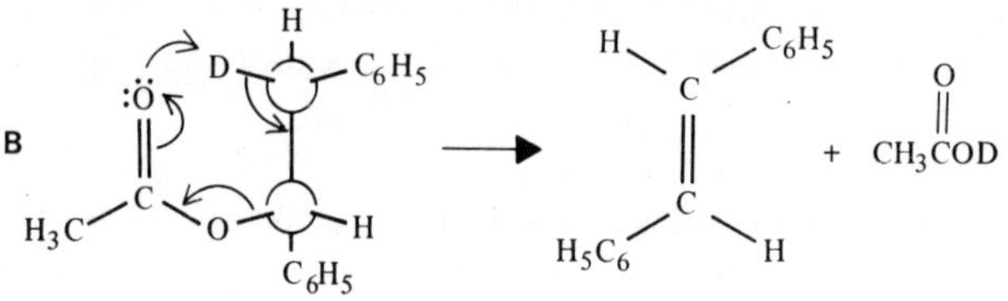

17.19

(a) 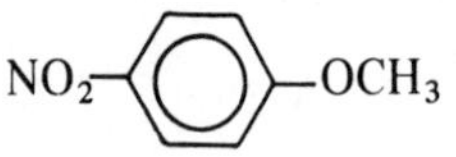 $NO_2\!-\!\langle\bigcirc\rangle\!-\!OCH_3$

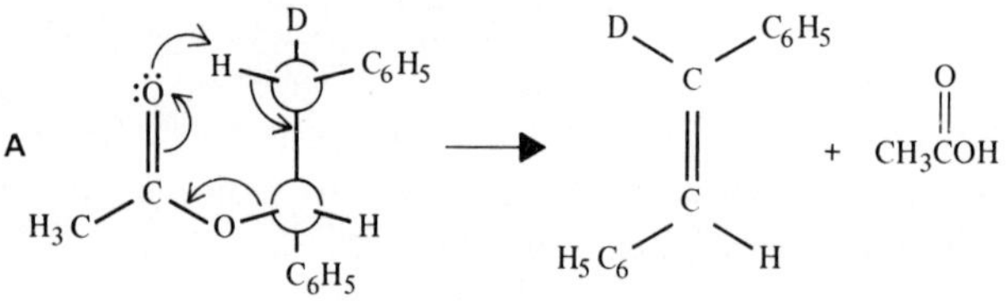

(b)

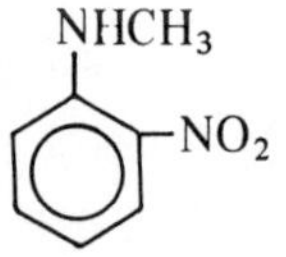

(c)

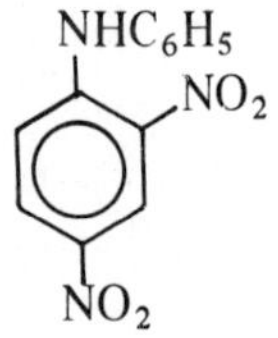

17.20

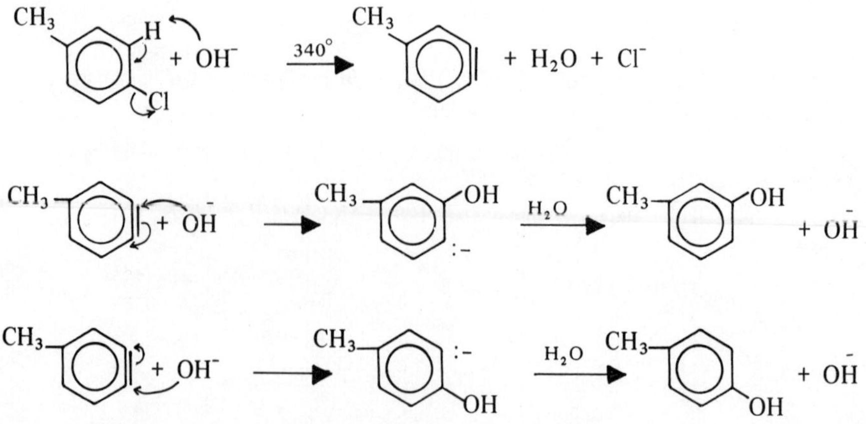

17.21

Since there are no hydrogens *ortho* to the halogen, elimination cannot take place. (Reaction by a bimolecular displacement is not possible either because the substrate lacks strong electron-withdrawing groups.) Thus the absence of a reaction must be due to the inability of 2-bromo-3-methylanisole to form a benzyne intermediate.

17.22

(a) If every substitution *involves an inversion*, then racemization will be complete when only *half* the substrate has incorporated radioactive iodine. (At this point there will be an equimolar mixture of the two enantiomers.) Thus, the rate of racemization *should be twice the rate of incorporation of radioactive iodine.*

(b) If an achiral intermediate such as a carbocation were involved, one would expect the rate of racemization to equal the rate of incorporation of radioactive iodine.

17.23

(a) The carbocation stability is in the order

(b) The methoxy group in the meta position cannot stabilize the carbocation by resonance:

no especially stable structure is possible

The small stabilization resulting from the release of electrons into the ring through resonance is cancelled by oxygen's electron-withdrawing inductive effect.

17.24

Reaction of the alcohol with K and then of the resulting salt with C_2H_5Br does not break bonds to the chiral carbon, and these reactions therefore occur with retention.

Reaction of the tosylate, $C_6H_5CH_2CHCH_3$, with C_2H_5OH in K_2CO_3 solution, however,

$$\underset{\text{OTs}}{|}$$

is an S_N2 reaction at the chiral carbon and thus it occurs with inversion.

17.25

Elimination is syn; i.e., the acetate group and the hydrogen that is eliminated must be on the same side of the ring. This is true for both ring hydrogens beta to the acetate group in but for only one ring hydrogen in

(In either case hydrogens of the α-CH_3 group can become syn to the acetate group.)

17.26

17.27

The acetate ion is a weaker base than OH^-, and elimination therefore occurs to a smaller extent. (The difference in nucleophilicity of OH^- and CH_3COO^- is not as large as the difference in basicity, however.)

17.28

17.29

(a) Use a strong, hindered base such as $(CH_3)_3COK$ in a solvent of low polarity in order to bring about an E2 reaction.

(b) Here we want an S_N1 reaction. We use ethanol as the solvent *and as the nucleophile*, and we carry out the reaction at a low temperature so that elimination will be minimized.

17.30

(a) $C_6H_5CH_2Br + NaCN \longrightarrow C_6H_5CH_2CN + NaBr$

(b) $CH_3CH_2CH_2Br + NaSH \longrightarrow CH_3CH_2CH_2SH + NaBr$

(c) $CH_3CH_2CH_2Br + NaSCH_3 \longrightarrow CH_3CH_2CH_2SCH_3 + NaBr$

(d) $C_6H_5CH_2Br + NaOCH_2CH_3 \longrightarrow C_6H_5CH_2OCH_2CH_3 + NaBr$

(e) $CH_3CH_2CH_2Br + NaN_3 \longrightarrow CH_3CH_2CH_2N_3 + NaBr$

(f) $CH_3CH_2\overset{\overset{O}{\|}}{C}ONa + C_6H_5CH_2Br \longrightarrow CH_3CH_2\overset{\overset{O}{\|}}{C}OCH_2C_6H_5 + NaBr$

(g) $BrCH_2CH_2CH_2CH_2Br + NaSH \longrightarrow BrCH_2CH_2CH_2CH_2SH \xrightarrow{NaOH}$

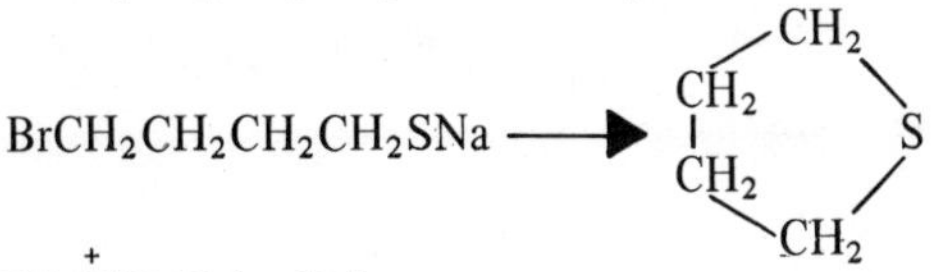

$BrCH_2CH_2CH_2CH_2SNa \longrightarrow$

(h) $C_6H_5CH_2Br + :P(C_6H_5)_3 \longrightarrow C_6H_5CH_2\overset{+}{P}(C_6H_5)_3 \ Br^-$

17.31

Ionization of $(C_6H_5)_2CHCl$ yields a carbocation that is highly resonance stabilized:

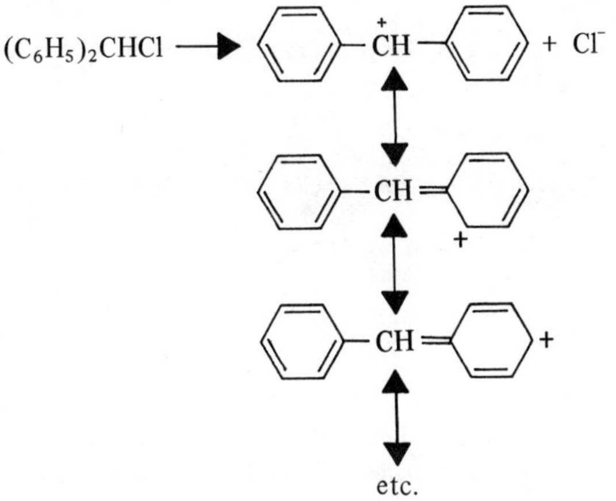

17.32

Iodide ion is a good nucleophile and a good leaving group; it can rapidly convert an alkyl chloride or alkyl bromide into an alkyl iodide, and the alkyl iodide can then react rapidly with another nucleophile. With methyl bromide in water, for example, the following reaction can take place:

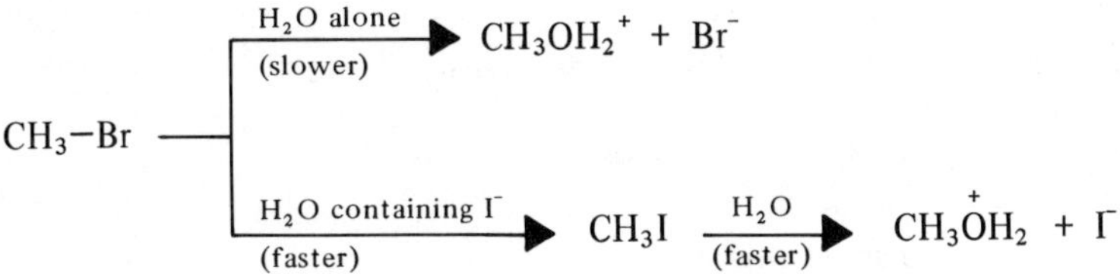

17.33

The Grignard reagent that forms early in the reaction reacts with the allyl halide in an S_N2 reaction:

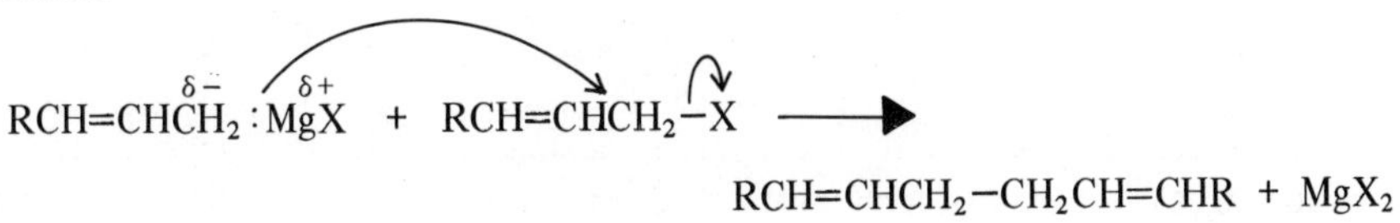

$RCH=CHCH_2-CH_2CH=CHR + MgX_2$

17.34

The reaction is an S_N2 reaction and thus nucleophilic attack takes place much more rapidly at the primary carbon than at the more hindered secondary carbon.

$$CH_3CH-CH_2 + C_2H_5O^- \xrightarrow[C_2H_5OH]{fast} CH_3CHCH_2OC_2H_5 \quad \text{Major}$$
$$\underset{O}{\diagup} \qquad\qquad\qquad\qquad\qquad \underset{OH}{|} \qquad\qquad \text{product}$$

$$CH_3CH-CH_2 + C_2H_5O^- \xrightarrow[C_2H_5OH]{slow} CH_3CHCH_2OH \quad \text{Minor}$$
$$\underset{O}{\diagup} \qquad\qquad\qquad\qquad\qquad \underset{OC_2H_5}{|} \qquad\qquad \text{product}$$

17.35

Ethoxide ion attacks the epoxide ring at the primary carbon because it is less hindered and the following reactions take place.

$$Cl-CH_2-CH-\overset{*}{C}H_2 + {}^-OC_2H_5 \longrightarrow Cl-CH_2-CH-\overset{*}{C}H_2OC_2H_5 \longrightarrow$$
$$\underset{O}{\diagdown\diagup} \qquad\qquad\qquad\qquad\qquad \underset{O}{|}$$

$$CH_2-CH-\overset{*}{C}H_2OC_2H_5$$
$$\underset{O}{\diagdown\diagup}$$

17.36

Both reactions are S_N1 reactions and thus the rate-limiting step for each is the formation of the carbocation. Since the carbocation formed in each instance is the same, the ratio of products is the same.

$$(C_6H_5)_2CHBr$$
$$\qquad\qquad \searrow slow$$
$$\qquad\qquad\qquad\qquad \longrightarrow (C_6H_5)_2\overset{+}{C}H \qquad \xrightarrow[fast]{H_2O} (C_6H_5)_2CHOH$$
$$\qquad\qquad \nearrow slower \qquad\qquad + \qquad\qquad \xrightarrow[fast]{N_3^-} (C_6H_5)_2CHN_3$$
$$(C_6H_5)_2CHCl \qquad\qquad\qquad Br^- \text{ or } Cl^-$$

Diphenylbromomethane reacts faster than diphenylchloromethane because Br^- is a better leaving group than Cl^-.

17.37

(a) Since the halides are all primary, this is almost certainly an S_N2 reaction with ethanol acting as the nucleophile.

$$C_2H_5\ddot{O}H \quad \underset{H}{\overset{R}{\underset{\diagup}{C}}}-Br \xrightarrow{-HBr} C_2H_5O-\underset{H}{\overset{R}{\underset{\diagup}{C}}}H$$

(b) Increasing the size of the R group increases steric hindrance to the approaching ethanol molecule and decreases the rate of reaction.

17.38

(a) This is another example of the relation between reactivity and selectivity that we first encountered in Chapter 4: generally speaking, highly reactive species are relatively unselective while less reactive species are more selective. In an S_N1 reaction the species that reacts with the nucleophile is a *carbocation*—a species that is electron deficient and thus is *highly reactive*. A carbocation, therefore, shows little tendency to discriminate between weak and strong nucleophiles—most often it simply reacts with the first nucleophile that it encounters. In S_N2 reactions, on the other hand, the species that reacts with the nucleophile is an alkyl halide or an alkyl tosylate. Such compounds are far less reactive toward nucleophiles than carbocations and they show much greater nucleophilic selectivities. An alkyl halide molecule, for example, might collide with a weak nucleophile thousands of times before a reaction takes place because few of the collisions will have sufficient energy to allow the weak nucleophile to displace the leaving group. On the other hand, an alkyl halide molecule might collide with a strong nucleophile only a few times before a collison leads to a reaction. This will be true because the strong nucleophile is better able to displace the leaving group and therefore a larger fraction of collisons will have sufficient energy to be fruitful.

(b) The reaction of $CH_3CH_2CH_2CH_2Cl$ is an S_N2 reaction and thus $CH_3CH_2CH_2CH_2Cl$ discriminates very effectively between the strongly nucleophilic CN^- ions and the weakly nucleophilic solvent molecules. By contrast, the reaction of $(CH_3)_3CCl$ is an S_N1 reaction and the carbocation that is formed shows little tendency to discriminate between solvent molecules and CN^- ions. Since solvent molecules are present in a much higher concentration the major product is $(CH_3)_3C-OCH_2CH_3$.

17.39

In each case, the reactions apparently involve the participation of the phenyl group and the formation of a phenonium ion as an intermediate. Solvolysis of **A** yields a chiral phenonium ion—one that reacts with solvent at either carbon to produce the same chiral (and thus optically active) acetate.

Chiral phenonium ion

Optically active

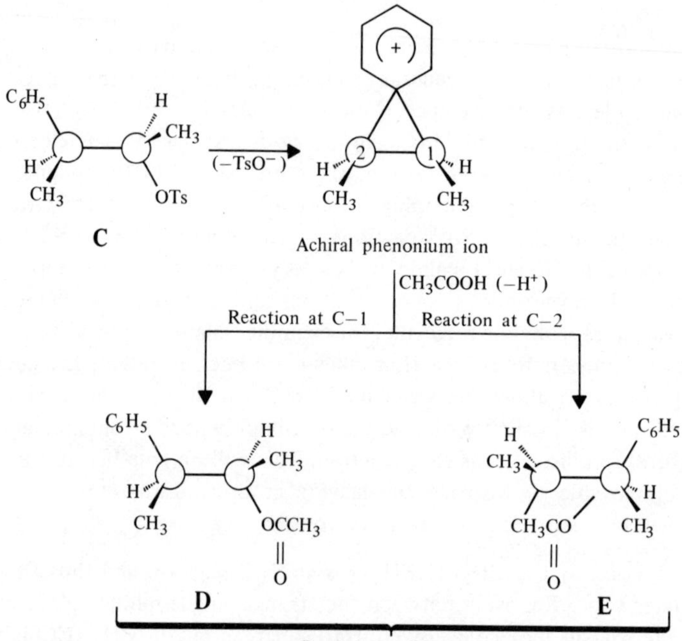

C

Achiral phenonium ion

$CH_3COOH\ (-H^+)$

Reaction at C–1 | Reaction at C–2

D

E

Racemic modification

(For an extensive discussion of this rearrangement, see J. M. Harris and C. C. Wamser, *Organic Reaction Mechanisms.* Wiley, New York, 1976, pp. 166-171.

18 ALKEHYDES AND KETONES

18.1

(a)

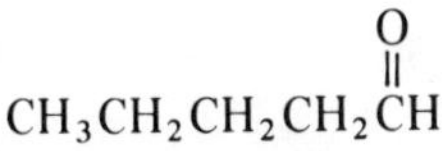

$$CH_3CH_2CH_2CH_2\overset{\overset{\displaystyle O}{\|}}{C}H$$
Pentanal

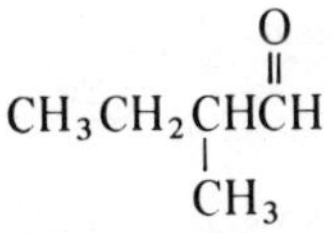

$$CH_3CH_2\overset{\overset{\displaystyle O}{\|}}{\underset{\underset{\displaystyle CH_3}{|}}{C}}HCH$$
2-Methylbutanal

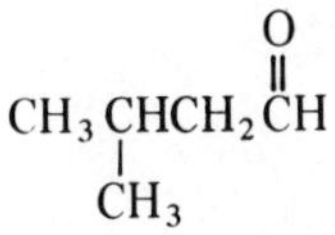

$$CH_3\overset{}{\underset{\underset{\displaystyle CH_3}{|}}{C}}HCH_2\overset{\overset{\displaystyle O}{\|}}{C}H$$
3-Methylbutanal

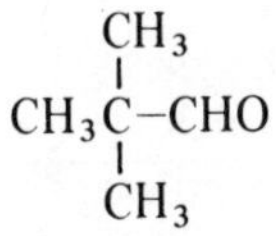

$$CH_3\overset{\overset{\displaystyle CH_3}{|}}{\underset{\underset{\displaystyle CH_3}{|}}{C}}-CHO$$
2,2-Dimethylpropanal

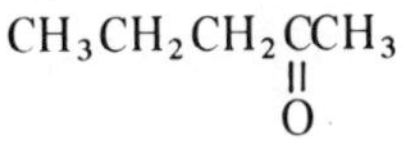

$$CH_3CH_2CH_2\overset{}{\underset{\underset{\displaystyle O}{\|}}{C}}CH_3$$
2-Pentanone

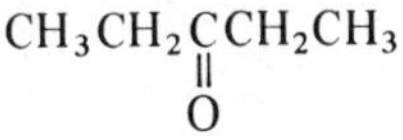

$$CH_3CH_2\overset{}{\underset{\underset{\displaystyle O}{\|}}{C}}CH_2CH_3$$
3-Pentanone

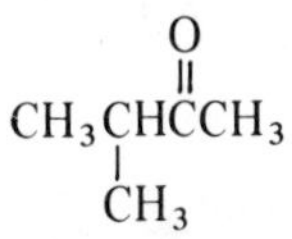

$$CH_3\overset{}{\underset{\underset{\displaystyle CH_3}{|}}{C}}H\overset{\overset{\displaystyle O}{\|}}{C}CH_3$$
3-Methyl-2-butanone

(b) and (c)

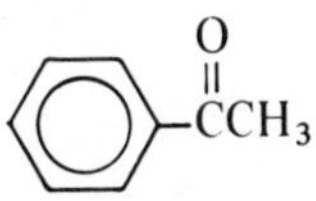

Acetophenone or
phenyl methyl ketone

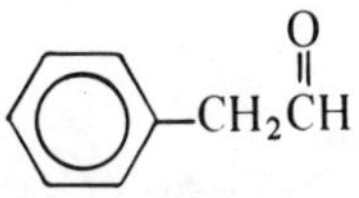

Phenylethanal or
phenylacetaldehyde

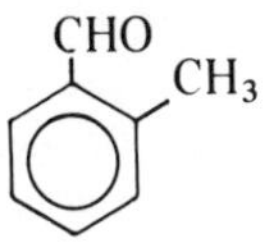

o-Tolualdehyde

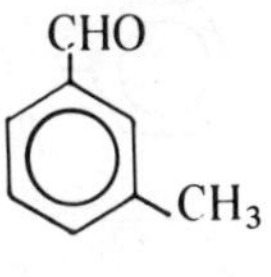

m-Tolualdehyde

p-Tolualdehyde

18.2

(a) 1-Pentanol, because its molecules form hydrogen bonds to each other.

(b) 2-Pentanol, because its molecules form hydrogen bonds to each other.

(c) Pentanal, because its molecules are more polar.

(d) 2-Phenylethanol, because its molecules form hydrogen bonds to each other.

(e) Benzyl alcohol because its molecules form hydrogen bonds to each other.

18.3

(a) $C_6H_6 \xrightarrow[\text{}]{Br_2,\ Fe} C_6H_5-Br \xrightarrow[\text{ether}]{Mg} C_6H_5-MgBr \xrightarrow[(2)\ H^+]{(1)\ HCHO}$

$C_6H_5-CH_2OH \xrightarrow[CH_2Cl_2]{CrO_3 \cdot C_5H_5N} C_6H_5-CHO$

(b) $C_6H_5-CH_3 \xrightarrow[(2)\,H^+]{(1)\ KMnO_4,\ OH^-,\ heat} C_6H_5-COOH \xrightarrow{SOCl_2}$

$C_6H_5-COCl \xrightarrow[\text{ether}]{LiAlH[OC(CH_3)_3]_3} C_6H_5-CHO$

$\xrightarrow[Pd(S)]{H_2} C_6H_5-CHO$

(c) $Br-C_6H_4-CH_3 \xrightarrow[CS_2]{CrO_2Cl_2} Br-C_6H_4-CH(OCrOHCl_2)_2$

$\xrightarrow{H_2O} Br-C_6H_4-CHO$

(d) $Cl-C_6H_4-CH_3 \xrightarrow[H_2SO_4]{\substack{CrO_3 \\ (CH_3CO)_2O}} Cl-C_6H_4-CH(OOCCH_3)_2 \xrightarrow[H_2O]{H^+} Cl-C_6H_4-CHO$

(e) $C_6H_5-CHCH_3 \xrightarrow[H_2SO_4]{CrO_3} C_6H_5-\underset{O}{\overset{\|}{C}}CH_3$
  $\overset{|}{OH}$

(f) $C_6H_6 \xrightarrow[AlCl_3]{CH_3COCl} C_6H_5-\underset{O}{\overset{\|}{C}}CH_3$

(g) $C_6H_5-\underset{O}{\overset{\|}{C}}Cl \xrightarrow[\substack{or \\ (CH_3)_2Cd}]{(CH_3)_2CuLi} C_6H_5-\underset{O}{\overset{\|}{C}}CH_3$

18.4

(a) The nucleophile is the negatively charged carbon of the Grignard reagent *acting as a carbanion*.

(b) The magnesium portion of the Grignard reagent acts as a Lewis acid and accepts an electron pair of the carbonyl oxygen. This makes the carbonyl carbon even more positive and, therefore, even more susceptible to nucleophilic attack.

$$\underset{\substack{\delta+\delta- \\ -\overset{|}{C}=\overset{..}{O} \quad \delta+ \\ \underset{R}{\nearrow} \quad Mg-X \\ \delta-}}{} \longrightarrow -\overset{|}{\underset{R}{C}}-\overset{..}{\underset{..}{O}}-MgX$$

(c) The product that forms initially (above) is a magnesium halide salt of an alcohol.

(d) On addition of water, the organic product that forms is an alcohol.

18.5
The nucleophile is a hydride ion.

18.6

(a) $\underset{\text{O}}{\overset{\text{O}}{\underset{\|}{CH_3CH}}} \xrightarrow[\text{H}_2\text{O}]{\text{NaHSO}_3} \underset{\text{OH}}{\overset{\text{OH}}{CH_3CHSO_3Na}} \xrightarrow[\text{H}_2\text{O}]{\text{NaCN}} \underset{\text{OH}}{\overset{\text{OH}}{CH_3CHCN}}$

$$\xrightarrow[\text{reflux}]{\text{HCl}} \underset{\substack{\text{OH} \\ \text{Lactic acid}}}{CH_3CHCOOH}$$

(b) A racemic modification.

18.7

(a) $\underset{\substack{\beta \quad \alpha \\ CH_3\ CH_2\overset{\text{O}}{\underset{\|}{CH}} + OH^-}}{} \rightleftharpoons \overset{\text{O}}{\underset{\|}{CH_3\overset{..}{\underset{..}{C}}HCH}} + H_2O$

$$CH_3CH_2\overset{\text{O}}{\underset{\|}{CH}} + \; ^-:\overset{\text{O}}{\underset{\substack{\| \\ CH_3}}{CHCH}} \rightleftharpoons CH_3CH_2\underset{\substack{| \\ CH_3}}{\overset{O^-}{CH}}\overset{\text{O}}{\underset{\|}{CHCH}}$$

$$CH_3CH_2\underset{\substack{| \\ CH_3}}{\overset{O^-}{CH}}CHCH + HOH \rightleftharpoons CH_3CH_2\underset{\substack{| \\ CH_3}}{\overset{OH}{CH}}\overset{\text{O}}{\underset{\|}{CHCH}}$$

(b) For $\underset{\substack{\text{OH} \\ CH_3CH_2CHCH_2CH_2\overset{\text{O}}{\underset{\|}{CH}}}}{}$ to form, a hydroxide ion would have to remove a β-proton in the first step. This does not happen because the anion that would be produced, i.e., $^-:CH_2CH_2CHO$, cannot be stabilized by resonance.

(c) $\underset{\substack{| \\ CH_3}}{CH_3CH_2CH=C\overset{\text{O}}{\underset{\|}{CH}}}$

18.8

(a) $2CH_3CH_2CH_2CHO \xrightarrow[H_2O]{OH^-}$ $CH_3CH_2CH_2\overset{\displaystyle OH}{\underset{\displaystyle \underset{\displaystyle CH_3}{\overset{\displaystyle |}{CH_2}}}{\overset{\displaystyle |}{C}}HCHCHO}$

(b) Product of (a) $\xrightarrow[(-H_2O)]{H^+}$ $CH_3CH_2CH_2CH=\underset{\displaystyle \underset{\displaystyle CH_3}{\overset{\displaystyle |}{CH_2}}}{C}CHO$

$\xrightarrow{NaBH_4}$ $CH_3CH_2CH_2CH=\underset{\displaystyle \underset{\displaystyle CH_3}{\overset{\displaystyle |}{CH_2}}}{C}CH_2OH$

(c) Product of (b) $\xrightarrow{\underset{Ni}{H_2}}$ $CH_3CH_2CH_2CH_2\underset{\displaystyle \underset{\displaystyle CH_3}{\overset{\displaystyle |}{CH_2}}}{C}HCH_2OH$

(d) Product of (a) $\xrightarrow{NaBH_4}$ $CH_3CH_2CH_2\overset{\displaystyle OH}{\underset{\displaystyle \underset{\displaystyle CH_3}{\overset{\displaystyle |}{CH_2}}}{\overset{\displaystyle |}{C}}HCHCH_2OH}$

18.9

(a) $CH_3\overset{\displaystyle O}{\overset{\displaystyle \|}{C}}CH_3 + OH^- \rightleftharpoons CH_3\overset{\displaystyle O}{\overset{\displaystyle \|}{C}}CH_2\!:^- + H_2O$

$CH_3\overset{\displaystyle O}{\overset{\displaystyle \|}{C}}CH_2\!:^- + CH_3\overset{\displaystyle O}{\overset{\displaystyle \|}{C}}CH_3 \rightleftharpoons CH_3\overset{\displaystyle O}{\overset{\displaystyle \|}{C}}CH_2\underset{\displaystyle CH_3}{\overset{\displaystyle O^-}{\underset{\displaystyle |}{\overset{\displaystyle |}{C}}}}CH_3$

$CH_3\overset{\displaystyle O}{\overset{\displaystyle \|}{C}}CH_2\underset{\displaystyle CH_3}{\overset{\displaystyle O^-}{\underset{\displaystyle |}{\overset{\displaystyle |}{C}}}}CH_3 + HOH \rightleftharpoons CH_3\overset{\displaystyle O}{\overset{\displaystyle \|}{C}}CH_2\underset{\displaystyle CH_3}{\overset{\displaystyle OH}{\underset{\displaystyle |}{\overset{\displaystyle |}{C}}}}CH_3 + OH^-$

(b) $CH_3\overset{\displaystyle O}{\overset{\displaystyle \|}{C}}CH=\underset{\displaystyle CH_3}{\overset{\displaystyle |}{C}}CH_3$

18.10

$CH_3CH_2\overset{\displaystyle O}{\overset{\displaystyle \|}{C}}H + OH^- \rightleftharpoons CH_3\overset{\displaystyle \cdot\cdot}{C}H\overset{\displaystyle O}{\overset{\displaystyle \|}{C}}H + H_2O$

$$CH_3\overset{\overset{\displaystyle O}{\|}}{C}H + CH_3\overset{..}{C}\overset{\overset{\displaystyle O}{\|}}{H} \rightleftharpoons CH_3\overset{\overset{\displaystyle O^-}{|}}{C}H\overset{}{C}H\overset{\overset{\displaystyle O}{\|}}{C}H$$
$$\underset{\displaystyle CH_3}{|}$$

$$CH_3\overset{\overset{\displaystyle O^-}{|}}{C}H\overset{}{C}H\overset{\overset{\displaystyle O}{\|}}{C}H + H_2O \rightleftharpoons CH_3\overset{\overset{\displaystyle OH}{|}}{C}H\overset{}{C}H\overset{\overset{\displaystyle O}{\|}}{C}H$$
$$\underset{\displaystyle CH_3}{|} \qquad\qquad \underset{\displaystyle CH_3}{|}$$

2-Methyl-3-
hydroxybutanal

$$CH_3\overset{\overset{\displaystyle O}{\|}}{C}H + OH^- \rightleftharpoons {}^-:CH_2\overset{\overset{\displaystyle O}{\|}}{C}H + H_2O$$

$$CH_3\,CH_2\overset{\overset{\displaystyle O}{\|}}{C}H + {}^-:CH_2\overset{\overset{\displaystyle O}{\|}}{C}H \rightleftharpoons CH_3CH_2\overset{\overset{\displaystyle O^-}{|}}{C}HCH_2\overset{\overset{\displaystyle O}{\|}}{C}H$$

$$CH_3CH_2\overset{\overset{\displaystyle O^-}{|}}{C}HCH_2\overset{\overset{\displaystyle O}{\|}}{C}H + H_2O \rightleftharpoons CH_3CH_2\overset{\overset{\displaystyle OH}{|}}{C}HCH_2\overset{\overset{\displaystyle O}{\|}}{C}H$$

3-Hydroxypentanal

18.11

Three successive aldol additions occur.

First
Aldol
Addition

$$CH_3\overset{\overset{\displaystyle O}{\|}}{C}H + OH^- \rightleftharpoons {}^-:CH_2\overset{\overset{\displaystyle O}{\|}}{C}H + H_2O$$

$$H\overset{\overset{\displaystyle O}{\|}}{C}H + {}^-:CH_2\overset{\overset{\displaystyle O}{\|}}{C}H \rightleftharpoons {}^-OCH_2CH_2\overset{\overset{\displaystyle O}{\|}}{C}H$$

$$^-OCH_2\overset{}{C}H\overset{\overset{\displaystyle O}{\|}}{C}H + H_2O \rightleftharpoons HOCH_2CH_2\overset{\overset{\displaystyle O}{\|}}{C}H$$

Second
Aldol
Addition

$$HOCH_2CH_2\overset{\overset{\displaystyle O}{\|}}{C}H + OH^- \rightleftharpoons HOCH_2\overset{..}{C}H\overset{\overset{\displaystyle O}{\|}}{C}H + H_2O$$

$$H\overset{\overset{\displaystyle O}{\|}}{C}H + HOCH_2\overset{..}{C}H\overset{\overset{\displaystyle O}{\|}}{C}H \rightleftharpoons HOCH_2\overset{\overset{\displaystyle CH_2O^-}{|}}{C}HCHO$$

$$HOCH_2\overset{\overset{\displaystyle CH_2O^-}{|}}{C}HCHO + H_2O \rightleftharpoons HOCH_2\overset{\overset{\displaystyle CH_2OH}{|}}{C}HCHO + OH^-$$

Third
Aldol
Addition

$$\left\{ \begin{array}{l} \underset{\displaystyle |}{\overset{\displaystyle CH_2OH}{}} \quad\quad\quad\quad\quad\quad \underset{\displaystyle |}{\overset{\displaystyle CH_2OH}{}} \\ HOCH_2CH{-}CHO + OH^- \;\rightleftharpoons\; HOCH_2\overset{}{\underset{-}{C}}{-}CHO \\[2mm] \quad O \quad\quad\quad\quad CH_2OH \quad\quad\quad\quad CH_2OH \\ \quad\| \quad\quad\quad\quad | \quad\quad\quad\quad\quad\quad | \\ H\overset{}{C}H + HOCH_2\overset{}{\underset{-}{C}}{-}CHO \;\rightleftharpoons\; HOCH_2{-}\overset{}{\underset{|}{C}}{-}CHO \\ \quad\quad\quad\quad\quad\quad\quad\quad\quad\quad\quad\quad\quad CH_2O^- \\[2mm] \quad\quad CH_2OH \quad\quad\quad\quad\quad CH_2OH \\ \quad\quad\quad | \quad\quad\quad\quad\quad\quad\quad\quad | \\ HOCH_2{-}\overset{}{\underset{|}{C}}{-}CHO + H_2O \;\rightleftharpoons\; HOCH_2{-}\overset{}{\underset{|}{C}}{-}CHO + OH^- \\ \quad\quad\quad CH_2O^- \quad\quad\quad\quad\quad CH_2OH \end{array} \right.$$

18.12

(a) $CH_3COOH + BF_3 \rightleftharpoons CH_3COOBF_3^- + H^+$

Pseudoionone

α-Ionone

β-Ionone

(b) In β-ionone the double bonds and the carbonyl group are conjugated, thus it is more stable.

(c) β-Ionone, because it is a conjugated unsaturated system.

18.13

(from Prob. 18.11)

$$H_2O + H-\underset{\underset{O}{\parallel}}{C}-O^- + H-\underset{\underset{H}{|}}{\overset{\overset{O^-}{|}}{C}}-\underset{\underset{CH_2OH}{|}}{\overset{\overset{CH_2OH}{|}}{C}}-CH_2OH \rightleftharpoons \longrightarrow$$

$$HOCH_2-\underset{\underset{CH_2OH}{|}}{\overset{\overset{CH_2OH}{|}}{C}}-CH_2OH + OH^- + H\underset{\underset{O}{\parallel}}{C}-O^-$$

Pentaerythritol

18.14

(a) $CH_3(CH_2)_4CH_2Br \xrightarrow{(C_6H_5)_3P} CH_3(CH_2)_4CH_2-P(C_6H_5)_3{}^+ Br^-$
$\xrightarrow{RLi} CH_3(CH_2)_4CH=P(C_6H_5)_3$

(b) $CH_3I \xrightarrow{(C_6H_5)_3P} CH_3-P(C_6H_5)_3{}^+ I^- \xrightarrow{RLi} CH_2=P(C_6H_5)_3$

(c) $BrCH_2CH_2CH_2CH_2Br + 2(C_6H_5)_3P \longrightarrow$

$$\underset{Br^-}{(C_6H_5)_3\overset{+}{P}-CH_2CH_2CH_2CH_2}-\underset{Br^-}{\overset{+}{P}(C_6H_5)_3} \xrightarrow{2\ RLi}$$

$$(C_6H_5)_3P=CHCH_2CH_2CH=P(C_6H_5)_3$$

18.15

(a) $C_6H_5CH_2Br \xrightarrow[\text{(2) RLi}]{\text{(1) } (C_6H_5)_3P} C_6H_5CH=P(C_6H_5)_3 \xrightarrow{CH_3\overset{\overset{O}{\parallel}}{C}CH_3}$

$$C_6H_5CH=\underset{\underset{CH_3}{|}}{C}CH_3$$

(b) $CH_3I \xrightarrow[\text{(2) RLi}]{\text{(1) } (C_6H_5)_3P} CH_2=P(C_6H_5)_3 \xrightarrow{C_6H_5\overset{\overset{O}{\parallel}}{C}CH_3} C_6H_5\underset{\underset{CH_3}{|}}{C}=CH_2$

(c) $CH_3CH_2Br \xrightarrow[\text{(2) RLi}]{\text{(1) } (C_6H_5)_3P} CH_3CH=P(C_6H_5)_3 \xrightarrow{C_6H_5\overset{\overset{O}{\parallel}}{C}CH_3}$

$$C_6H_5\underset{\underset{CH_3}{|}}{C}=CHCH_3$$

(d) $CH_2=P(C_6H_5)_3 \xrightarrow{CH_3\overset{\overset{O}{\parallel}}{C}CH_3} \underset{CH_3}{\overset{CH_3}{}}C=CH_2$
   (from part b)

(e) $CH_2=P(C_6H_5)_3 \longrightarrow$
   (from part b)

(f) $CH_3CH_2CH_2Br \xrightarrow[\text{(2) RLi}]{\text{(1) }(C_6H_5)_3P} CH_3CH_2CH=P(C_6H_5)_3$

$\xrightarrow{CH_3\overset{\overset{\displaystyle O}{\|}}{C}CH_2CH_3} CH_3CH_2CH=\overset{\overset{\displaystyle CH_3}{|}}{C}CH_2CH_3$

(g) $CH_2=CHCH_2Br \xrightarrow[\text{(2) RLi}]{\text{(1) }(C_6H_5)_3P} CH_2=CHCH=P(C_6H_5)_3$

$\xrightarrow{C_6H_5\overset{\overset{\displaystyle O}{\|}}{C}H} C_6H_5CH=CHCH=CH_2$

(h) $C_6H_5CH=P(C_6H_5)_3 \xrightarrow{C_6H_5\overset{\overset{\displaystyle O}{\|}}{C}H} C_6H_5CH=CHC_6H_5$
 (from part (a))

18.16

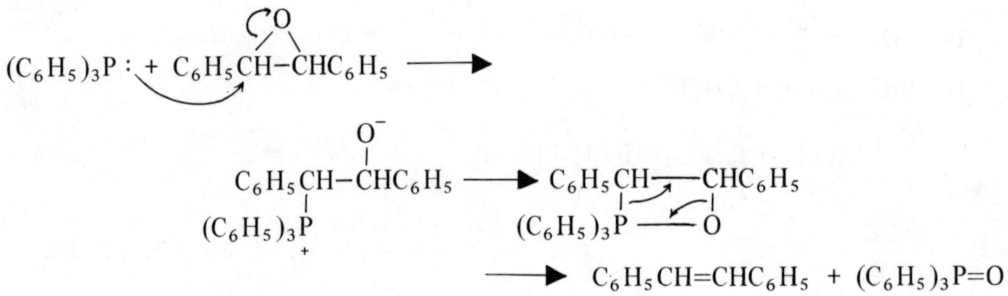

18.17

(a) $CH_3OCH_2Br + (C_6H_5)_3P \xrightarrow{\text{(2) base}} CH_3OCH=P(C_6H_5)_3$

(b) Hydrolysis of the ether yields a hemiacetal (see the example in part (c) below) that then goes on to form an aldehyde.

(c)

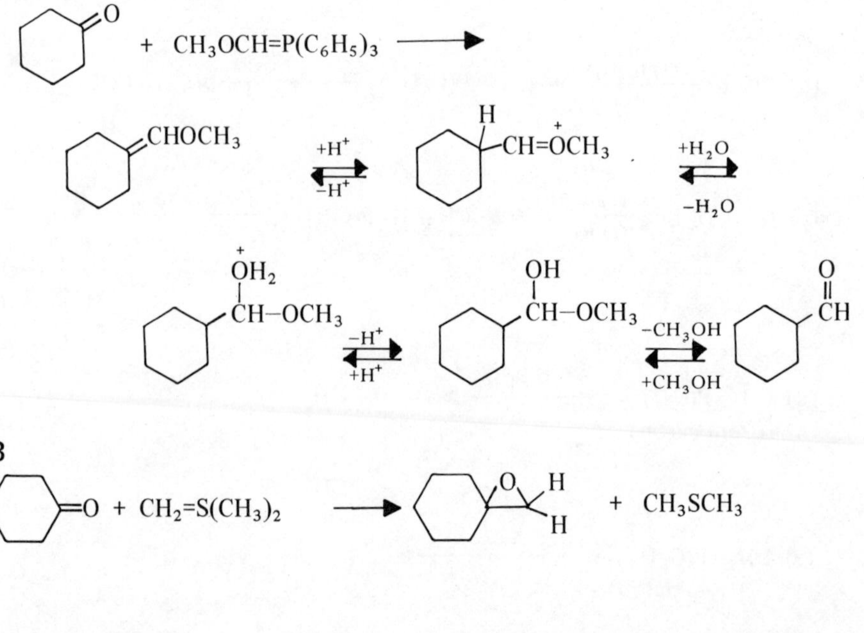

18.18

(a)

(b)

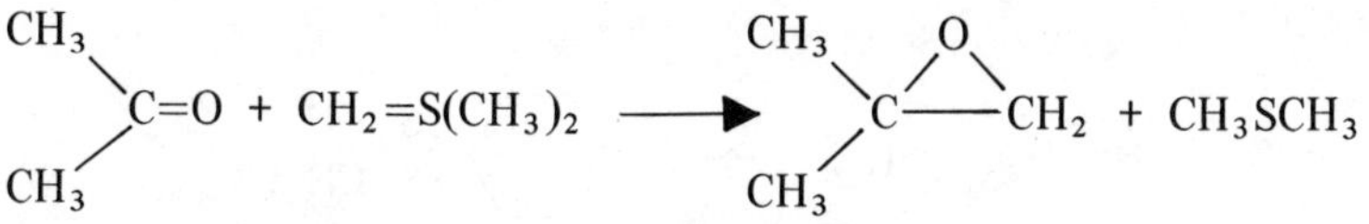

$$\underset{CH_3}{\overset{CH_3}{>}}C=O \ + \ CH_2=S(CH_3)_2 \ \longrightarrow \ \underset{CH_3}{\overset{CH_3}{>}}C\overset{O}{-\!\!\!\triangle\!\!\!-}CH_2 \ + \ CH_3SCH_3$$

18.19

(a) 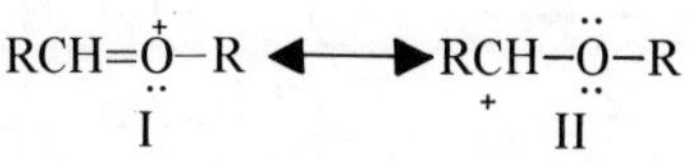 $RCH=\overset{+}{\underset{..}{O}}-R \ \longleftrightarrow \ RCH-\underset{+}{\overset{..}{O}}-R$

$\quad\quad\quad\quad\quad$ I $\quad\quad\quad\quad\quad\quad\quad$ II

(b) and (c) Structure I should make a greater contribution because in it both the carbon atom and the oxygen atom have an octet of electrons and because it has one more bond.

18.20

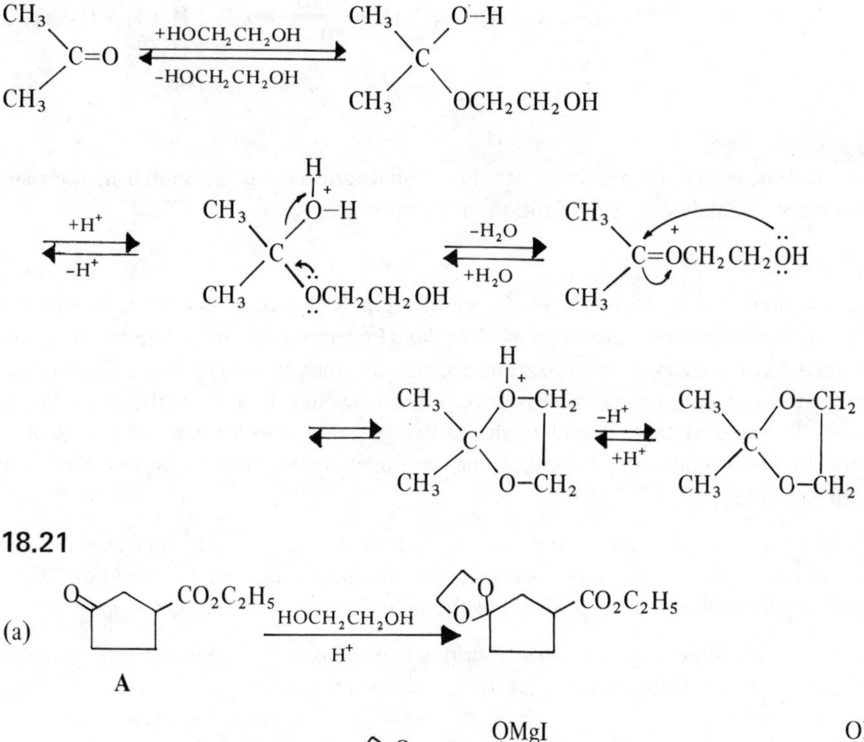

18.21

(a)

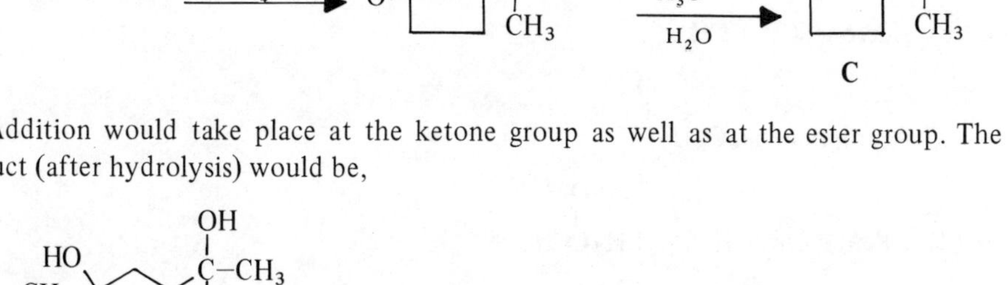

(b) Addition would take place at the ketone group as well as at the ester group. The product (after hydrolysis) would be,

18.22

(a)

(b)

$$CH_3\overset{O}{\overset{\|}{C}}CH_2CH_2CO_2C_2H_5 + HSCH_2CH_2SH \xrightarrow{BF_3}$$

18.23

The reaction is said to be "base promoted" because base is consumed as the reaction takes place. A catalyst is, by definition, not consumed.

18.24

(a) The slow step in base-catalyzed racemization is the same as that in base-promoted halogenation—*the formation of an enolate ion*. (Formation of an enolate ion from phenyl *sec*-butyl ketone leads to racemization because the enolate ion is achiral. When it accepts a proton it yields a racemic modification.) The slow step in acid-catalyzed racemization is also the same as that in acid-catalyzed halogenation—*the formation of an enol*. (The enol, like the enolate ion, is achiral and tautomerizes to yield a racemic modification of the ketone.)

(b) According to the mechanism given, the slow step for acid-catalyzed iodination (formation of the enol) is the same as that for acid-catalyzed bromination. Thus we would expect both reactions to occur at the same rate.

(c) Again, the slow step for both reactions (formation of the enolate ion) is the same, and consequently, both reactions take place at the same rate.

18.25

(a) Acetone, $CH_3\overset{O}{\overset{\|}{C}}CH_3$

(b) Acetophenone, $C_6H_5\overset{O}{\overset{\|}{C}}CH_3$

(d) 2-Pentanone, $CH_3CH_2CH_2\overset{O}{\overset{\|}{C}}CH_3$

(f) 1-Phenylethanol, $C_6H_5\overset{OH}{\overset{|}{C}}HCH_3$

(h) 2-Butanol, $CH_3CH_2\overset{\overset{\displaystyle OH}{|}}{C}HCH_3$

(i) 1-Acetylnaphthalene,

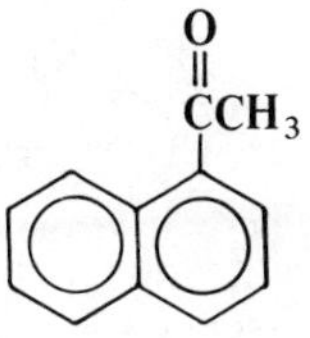

18.26

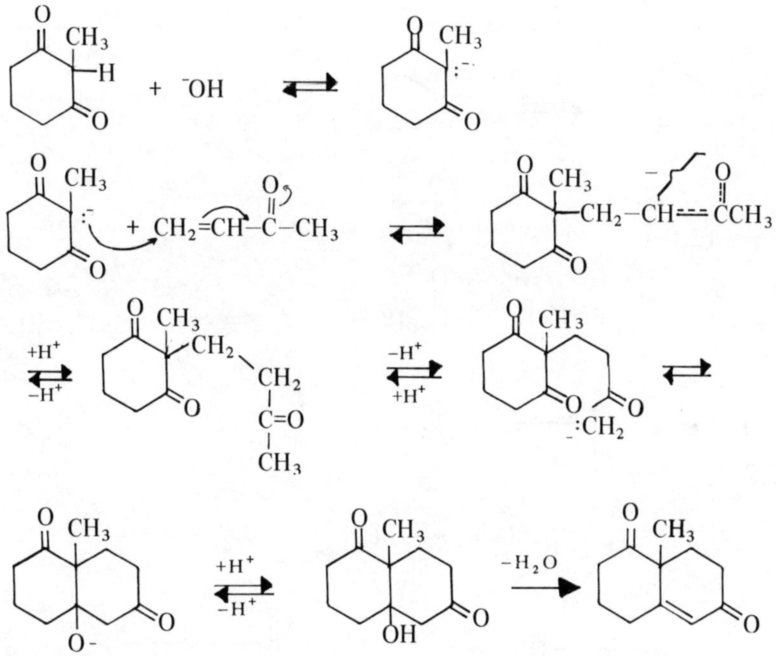

(b) 2-Methyl-1,3-cyclohexanedione is more acidic because its enolate ion is stabilized by an additional resonance structure.

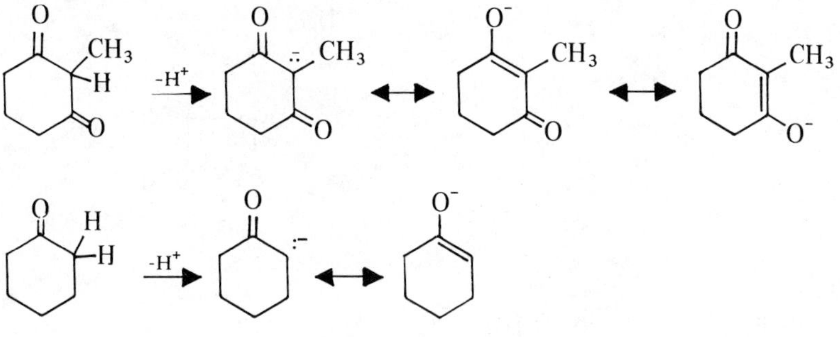

18.27

(a) $C_6H_5\overset{\overset{\displaystyle O}{\|}}{C}CH_3 \underset{+H^+}{\overset{-H^+}{\rightleftharpoons}} C_6H_5\overset{\overset{\displaystyle O}{\|}}{C}CH_2{:}^-$

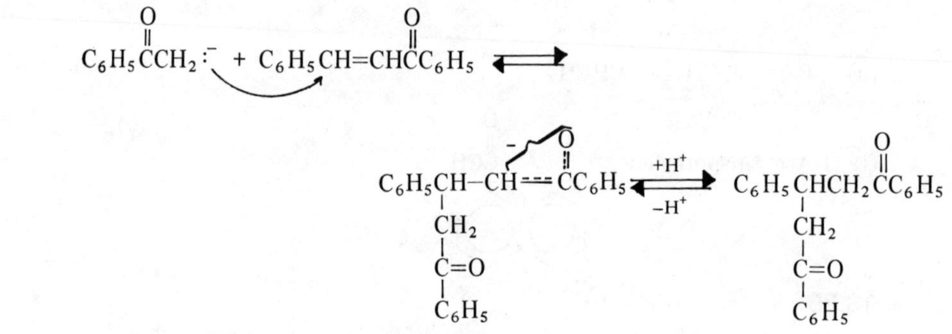

$$C_6H_5\overset{O}{\overset{\|}{C}}CH_2{:}^- \;+\; C_6H_5CH{=}CH\overset{O}{\overset{\|}{C}}C_6H_5 \;\rightleftharpoons$$

$$C_6H_5CH{-}CH{=\!=\!=}\overset{-}{\overset{O}{\overset{\|}{C}}}CC_6H_5 \;\underset{-H^+}{\overset{+H^+}{\rightleftharpoons}}\; C_6H_5CHCH_2\overset{O}{\overset{\|}{C}}CC_6H_5$$
$$\quad\quad\;\; |\qquad\qquad\qquad\qquad\qquad\qquad\quad |$$
$$\quad\quad\; CH_2 \qquad\qquad\qquad\qquad\qquad\quad CH_2$$
$$\quad\quad\; C{=}O \qquad\qquad\qquad\qquad\qquad\quad C{=}O$$
$$\quad\quad\; C_6H_5 \qquad\qquad\qquad\qquad\qquad\; C_6H_5$$

(b)

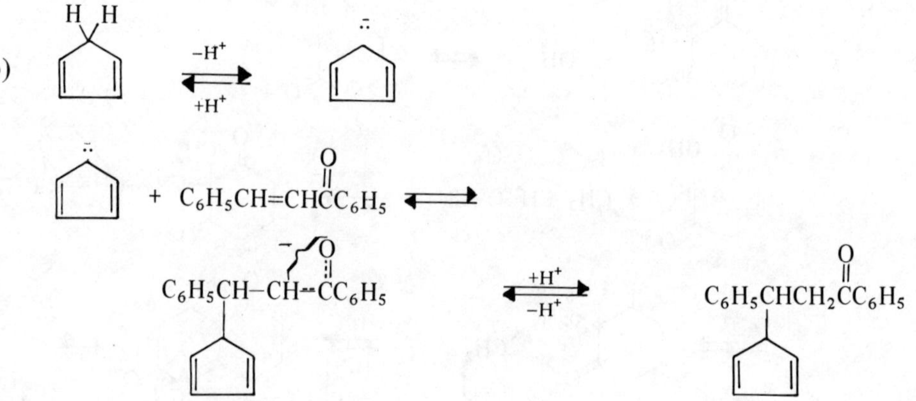

18.28

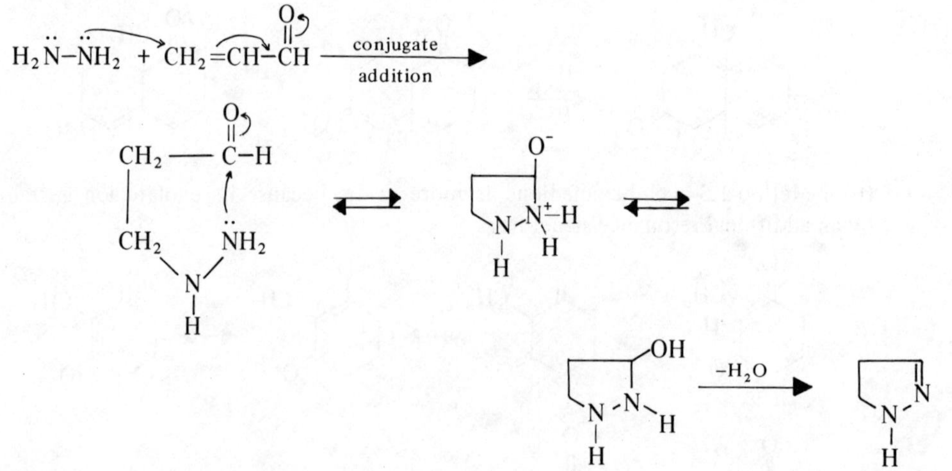

18.29

(a) HCHO

(b) CH$_3$CHO

(c) C$_6$H$_5$CH$_2$CHO

(d) CH$_3$COCH$_3$

(e) CH$_3$COCH$_2$CH$_3$

(f) CH$_3$COC$_6$H$_5$

(g) $C_6H_5CH=CHCOCH_3$

(j)

(h) $C_6H_5CH=CHCOC_6H_5$

(k)

(i) $C_6H_5COC_6H_5$

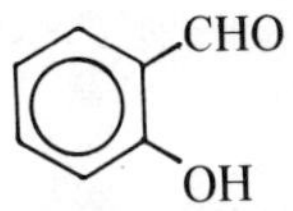

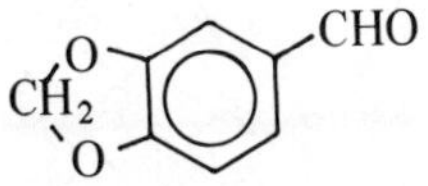

(l)

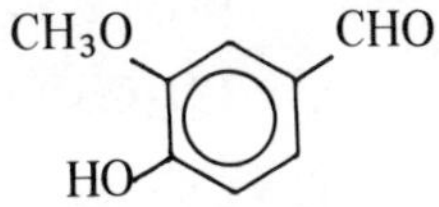

18.30

(a) $CH_3CH_2CH_2OH$

(b) $CH_3CH_2CHOHC_6H_5$

(c) $CH_3CH_2CH_2OH$

(d) $CH_3CH_2CHOHSO_3Na$

(e) $CH_3CH_2CHOHCN$

(f) $CH_3CH_2CHOHCH(CH_3)CHO$

(g) $CH_3CH_2CH=C(CH_3)CHO$

(h) $CH_3CH_2CH_2OH$

(i)

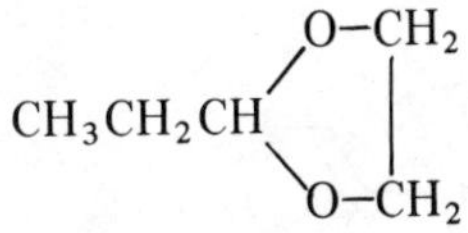

(j) $CH_3CH_2CH=CHCH_3$

(k) $CH_3CHBrCHO$

(l) $CH_3CH_2COO^- + Ag\downarrow$

(m) $CH_3CH_2CH=NOH$

(n) $CH_3CH_2CH=NNHCONH_2$

(o) $CH_3CH_2CH=NNHC_6H_5$

(p) CH_3CH_2COOH

(q)

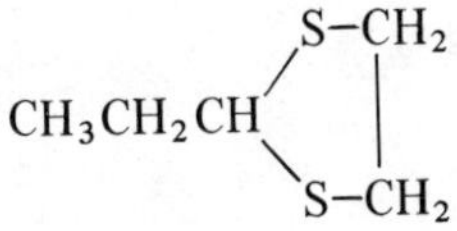

(r) $CH_3CH_2CH_3 + CH_3CH_3 + NiS$

18.31

(a) $CH_3CHOHCH_3$

(b) $C_6H_5\underset{\underset{CH_3}{|}}{C}OHCH_3$

(c) $CH_3CHOHCH_3$

(d) $CH_3\underset{\underset{CH_3}{|}}{\overset{\overset{OH}{|}}{C}}SO_3Na$

(e) $CH_3\underset{\underset{CH_3}{|}}{\overset{\overset{OH}{|}}{C}}CN$

(f) $CH_3COCH_2\underset{\underset{CH_3}{|}}{\overset{\overset{OH}{|}}{C}}CH_3$

(g) $CH_3COCH=\underset{\underset{CH_3}{|}}{C}CH_3$

(h) $CH_3CHOHCH_3$

(i)

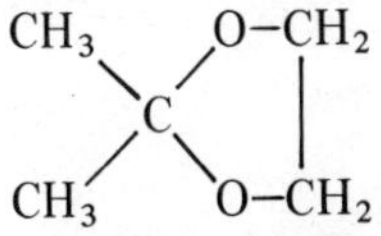

(j) $CH_3CH=C(CH_3)_2$

(k) CH_3COCH_2Br

(o) $CH_3\underset{\overset{|}{CH_3}}{C}=NNHC_6H_5$

(l) No reaction

(p) No reaction

(m) $CH_3\underset{\overset{|}{CH_3}}{C}=NOH$

(q) $CH_3\underset{CH_3}{\overset{}{\diagdown}}C\underset{S-CH_2}{\overset{S-CH_2}{\diagup\diagdown}}|$

(n) $CH_3\underset{\overset{|}{CH_3}}{C}=NNHCONH_2$

(r) $CH_3CH_2CH_3 + CH_3CH_3 + NiS$

18.32

(a) $CH_3-\langle\bigcirc\rangle-CH=CHCHO$

(d) $CH_3-\langle\bigcirc\rangle-COOH$

(b) $CH_3-\langle\bigcirc\rangle-COO^- + CH_3-\langle\bigcirc\rangle-CH_2OH$

(e) $HOOC-\langle\bigcirc\rangle-COOH$

(c) $CH_3-\langle\bigcirc\rangle-CH_2OH + HCOO^-$

(f) $CH_3-\langle\bigcirc\rangle-CH=CH_2$

18.33

(a) 3-nitroacetophenone structure

(d) $\langle\bigcirc\rangle-\underset{\overset{|}{OH}}{C}HCH_3$

(b) $\langle\bigcirc\rangle-COO^- + CHCl_3$

(e) $\langle\bigcirc\rangle-\underset{\overset{|}{OH}}{\overset{\overset{CH_3}{|}}{C}}-\langle\bigcirc\rangle$

(c) $\langle\bigcirc\rangle-\underset{CH_3}{\overset{CH_2}{C}}$

18.34

(a) $\langle\bigcirc\rangle + CH_3CH_2CH_2COCl \xrightarrow{AlCl_3} \langle\bigcirc\rangle-\overset{\overset{O}{\|}}{C}CH_2CH_2CH_3$

$\langle\bigcirc\rangle + (CH_3CH_2CH_2CO)_2O \xrightarrow{AlCl_3} \langle\bigcirc\rangle-\overset{\overset{O}{\|}}{C}CH_2CH_2CH_3$

$\langle\bigcirc\rangle \xrightarrow[Fe]{Br_2} \langle\bigcirc\rangle-Br \xrightarrow[(2)\ CdCl_2]{(1)\ Mg,\ ether} \left[\langle\bigcirc\rangle\right]_2 Cd$

$$CH_3CH_2CH_2\overset{\overset{\displaystyle O}{\|}}{C}Cl \longrightarrow \text{⟨⟩}-\overset{\overset{\displaystyle O}{\|}}{C}CH_2CH_2CH_3$$

(b)

$$\text{⟨⟩}-\overset{\overset{\displaystyle O}{\|}}{C}CH_2CH_2CH_3 \longrightarrow$$

$\xrightarrow[\text{HCl}]{\text{Zn(Hg)}}$ ⟨⟩-$CH_2CH_2CH_2CH_3$

$\xrightarrow[\text{OH}^-]{NH_2NH_2}$ ⟨⟩-$CH_2CH_2CH_2CH_3$

$\xrightarrow[\text{H}^+]{HSCH_2CH_2SH}$ ⟨⟩-C$\begin{array}{c} S-CH_2 \\ | \\ S-CH_2 \\ CH_2 \\ CH_2 \\ CH_3 \end{array}$

$\xrightarrow[\text{Ni}]{\text{Raney}}$ ⟨⟩-$CH_2CH_2CH_2CH_3$

18.35

(a) ⟨⟩-CH_2OD + ⟨⟩-COO^-

(b) ⟨⟩-CH_2OH + ⟨⟩-COO^-

(c) Yes. In both reactions a hydride ion (rather than a deuteride ion) is transferred to benzaldehyde. This shows that the hydride ion is transferred from one benzaldehyde molecule to another (as shown on page 705) and not from the solvent to benzaldehyde.

18.36

(a) $Ag(NH_3)_2{}^+OH^-$ (positive test with benzaldehyde)

(b) $Ag(NH_3)_2{}^+OH^-$ (positive test with hexanal)

(c) I_2 in NaOH (Iodoform, from 2-hexanone)

(d) CrO_3 in H_2SO_4 (positive test with 2-hexanol)

(e) I_2 in NaOH (Iodoform from 2-hexanol)

(f) CrO_3 in H_2SO_4 (positive test with 3-hexanol)

(g) Br_2 in CCl_4 (decolorization with benzalacetophenone)

(h) I_2 in NaOH (Iodoform from 1-phenylethanol)

(i) $Ag(NH_3)_2{}^+OH^-$ (positive test with pentanal)

(j) Br_2 in CCl_4 (immediate decolorization occurs with enol form)

(k) $Ag(NH_3)_2{}^+OH^-$ (positive test with cyclic hemiacetal)

18.37

(a) In simple addition the carbonyl peak (1665-1780 cm^{-1} region) does not appear in the product; in conjugate addition it does.

(b) As the reaction takes place the long wavelength absorption arising from the conjugated system should disappear. One could follow the rate of the reaction by following the rate at which this absorption peak disappears.

18.38

(a) The conjugate base is a hybrid of the following structures:

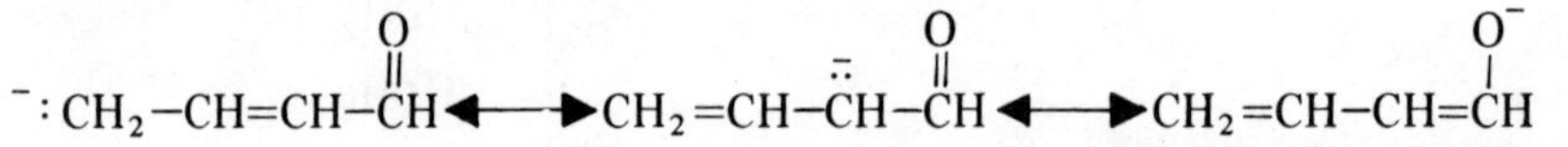

$$^-:CH_2-CH=CH-\overset{\overset{O}{\|}}{CH} \longleftrightarrow CH_2=CH-\overset{\overset{\cdot\cdot}{\overset{O}{\|}}}{CH}-\overset{\cdot\cdot}{CH} \longleftrightarrow CH_2=CH-CH=\overset{\overset{O^-}{|}}{CH}$$

This structure is especially stable because the negative charge is on oxygen

(b) $CH_3CH=CHCHO \underset{+H^+}{\overset{-H^+}{\rightleftarrows}} {}^-:CH_2CH=CHCHO$

$$C_6H_5CH=CH\overset{\overset{O}{\|}}{CH} + {}^-:CH_2CH=CHCHO \rightleftarrows$$

$$C_6H_5\overset{\overset{O^-}{|}}{CH}=CHCH-CH_2CH=CHCHO \underset{-H^+}{\overset{+H^+}{\rightleftarrows}} C_6H_5\overset{\overset{OH}{|}}{CH}=CHCH-CH_2CH=CHCHO$$

$$\xrightarrow{-H_2O} C_6H_5CH=CHCH=CHCH=CHCHO$$

18.39

(a)

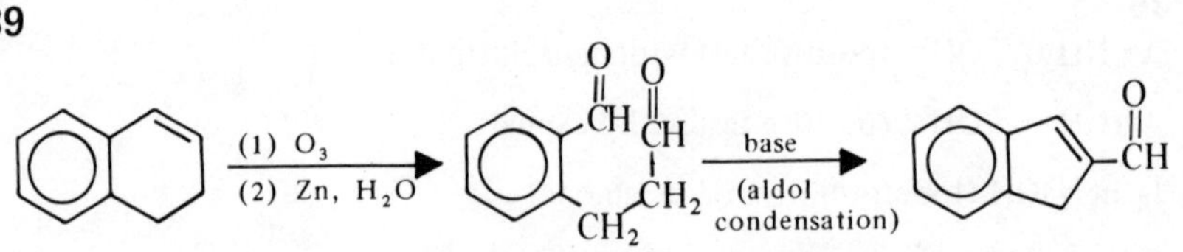

(b)

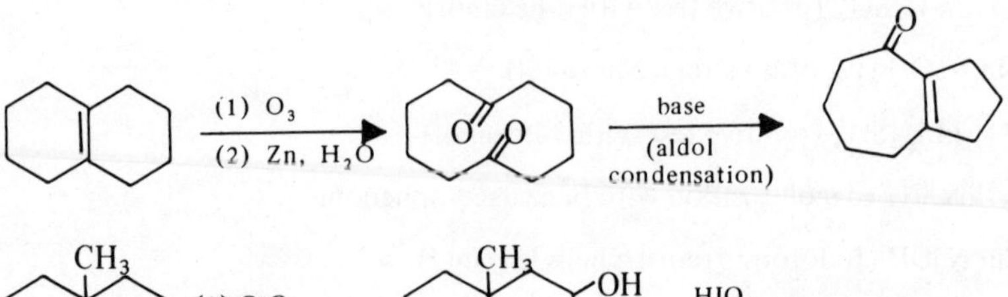

(c)

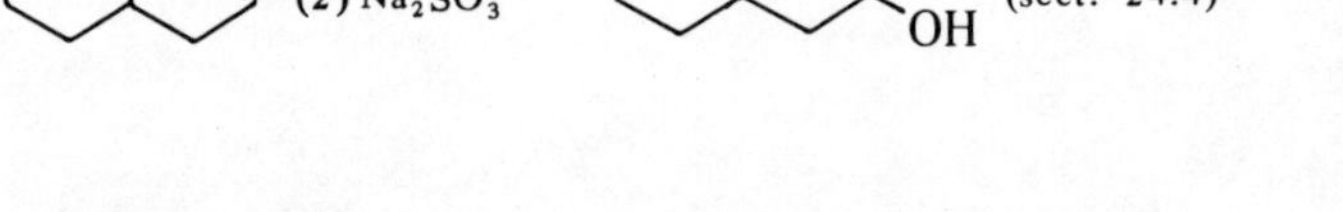

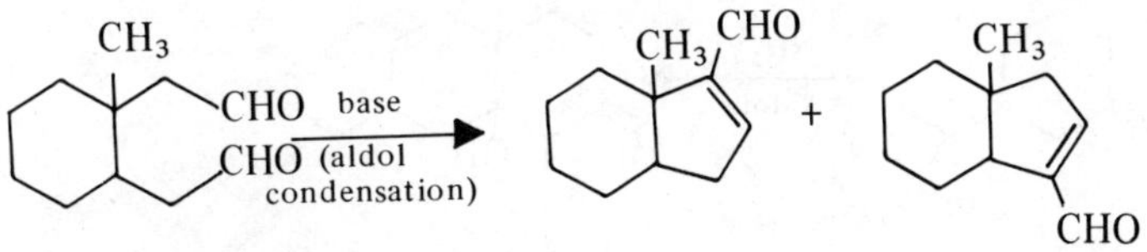

(d)

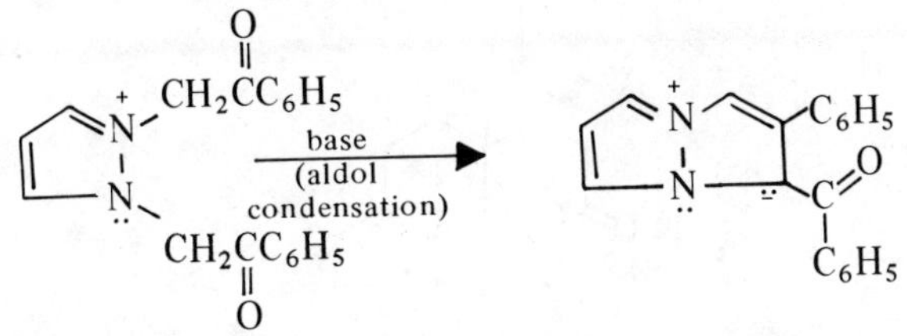

18.40

(a)

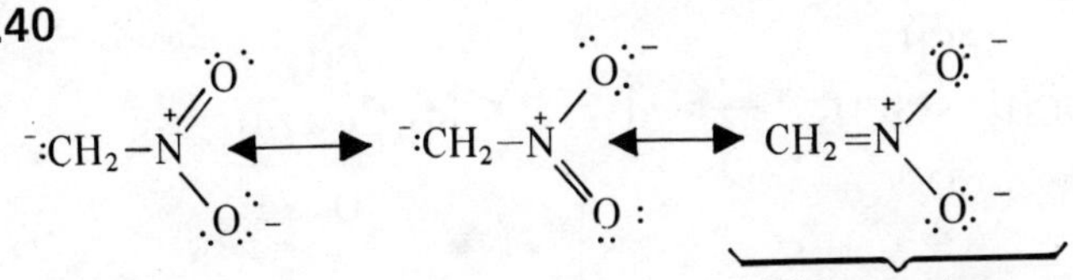

This structure is especially stable because both negative charges are on oxygen

(b) All of these syntheses are variations of a crossed aldol addition or condensation.

$$\underset{\substack{\parallel \\ O}}{HCH} + CH_3NO_2 \xrightarrow{base} HOCH_2CH_2NO_2$$

$$C_6H_5CHO + CH_3NO_2 \xrightarrow[(-H_2O)]{base} C_6H_5CH=CHNO_2$$

$$C_6H_5CHO + CH_3CH_2NO_2 \xrightarrow[(-H_2O)]{base} \underset{\substack{| \\ CH_3}}{C_6H_5CH=CNO_2}$$

18.41

First an elimination takes place,

$$R_3\overset{+}{N}CH_2CH_2\underset{\substack{\parallel \\ O}}{C}CH_2CH_3 + NH_2^- \longrightarrow CH_2=CH\underset{\substack{\parallel \\ O}}{C}CH_2CH_3 + R_3N + NH_3$$

then a conjugate addition occurs, followed by an aldol addition:

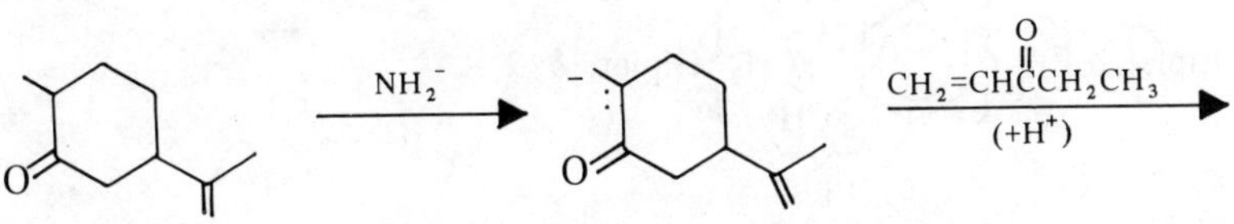

18.42

18.43

(a) Compound U is phenyl ethyl ketone:

$\delta 7.7$ $\delta 3.0, \delta 1.2$

(b) Compound V is benzyl methyl ketone:

$\delta 7.1$ $\delta 3.5$ $\delta 2.0$

18.44

Compound W is:

multiplet, δ 7.3   singlet δ 3.4

infrared peak near 1715 cm^{-1}

Compound X is:

multiplet, δ 7.5   triplet, δ 2.5

triplet
δ 3.1

$\xrightarrow[\text{(2) H}_3\text{O}^+]{\substack{\text{heat} \\ \text{(1)KMnO}_4, \text{OH}^-}}$ Phthalic aicd

18.45

The pmr spectra (Figures 18.2 and 18.3) each have a five hydrogen peak near $\delta 7.1$, suggesting that Y and Z each have a C_6H_5- group. The infrared spectrum of each compound shows a strong peak near 1705 cm^{-1}. This indicates that each compound has a C=O group. We have, therefore, the following pieces,

$$\text{and} \quad -\overset{\overset{\displaystyle O}{\|}}{C}-$$

If we subtract the atoms of these pieces from the molecular formula,

$$\begin{array}{r} C_{10}H_{12}O \\ - C_7\ H_5\ O\ (C_6H_5\ +\ C{=}O) \\ \hline \end{array}$$

We are left with, $C_3\ H_7$

In the pmr spectrum of Y we see an ethyl group [triplet, $\delta 1.0$ (3H) and quartet, $\delta 2.3$ (2H)] and an unsplit $-CH_2-$ group [singlet, $\delta 3.7$ (2H)]. This means that Y must be,

$$-CH_2\overset{\overset{\displaystyle O}{\|}}{C}CH_2CH_3$$

1-Phenyl-2-butanone

In the pmr spectrum of Z, we see an unsplit $-CH_3$ group [singlet, $\delta 2.0$ (3H)] and a multiplet (actually two superimposed triplets) at $\delta 2.8$. This means Z must be,

$$-CH_2CH_2\overset{\overset{\displaystyle O}{\|}}{C}CH_3$$

4-Phenyl-2-butanone

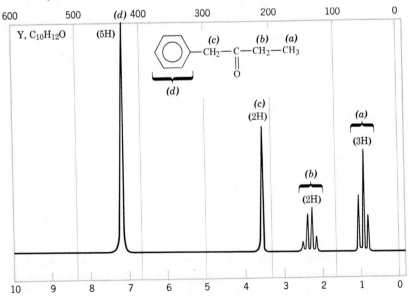

FIG. 18.2. The pmr spectrum of compound Y, problem 18.45. (Spectrum courtesy of Aldrich Chemical Co., Milwaukee, Wis.)

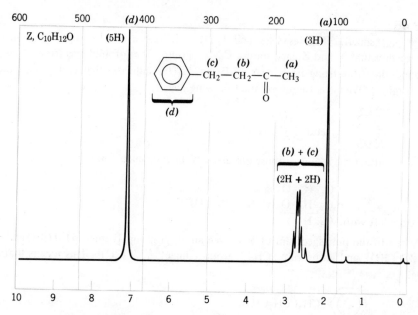

FIG. 18.3. The pmr spectrum of compound Z, problem 18.45. (Spectrum courtesy of Aldrich Chemical Co., Milwaukee, Wis.)

18.46

$$\underset{\qquad}{\overset{O}{\underset{\parallel}{}}}$$
A is CH$_3$CCH$_2$CH(OCH$_3$)$_2$

$$CH_3-\overset{O}{\underset{\parallel}{C}}-CH_2-CH(OCH_3)_2 \quad \xrightarrow[\text{NaOH}]{\text{I}_2} \quad CHI_3\downarrow$$

$$\xrightarrow[]{\text{Ag(NH}_3)_2{}^+OH^-} \quad \text{No reaction}$$

$$\downarrow \text{H}^+, \text{H}_2\text{O}$$

$$CH_3\overset{O}{\underset{\parallel}{C}}CH_2\overset{O}{\underset{\parallel}{C}}H \quad \xrightarrow{\text{Ag(NH}_3)_2{}^+OH^-} \quad Ag\downarrow + CH_3\overset{O}{\underset{\parallel}{C}}CH_2\overset{O}{\underset{\parallel}{C}}O^-$$

(a) $\overset{O}{\underset{\parallel}{}}$ (b) (d) (c)
CH$_3$–C–CH$_2$–CH(OCH$_3$)$_2$

    (a) Singlet δ2.1

    (b) Doublet δ2.6

    (c) Singlet δ3.2

    (d) Triplet δ4.7

18.47

Reasonance contributions such as those shown below render the carbonyl groups of 2-hydroxybenzaldehyde and 4-hydroxybenzaldehyde highly unreactive toward nucleophilic attack. Hence no hydride transfer takes place.

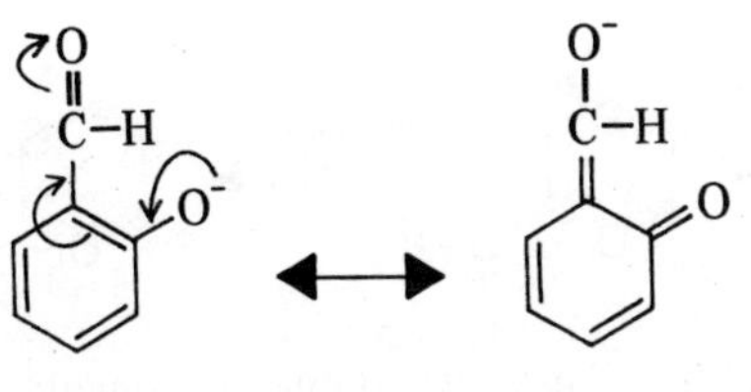

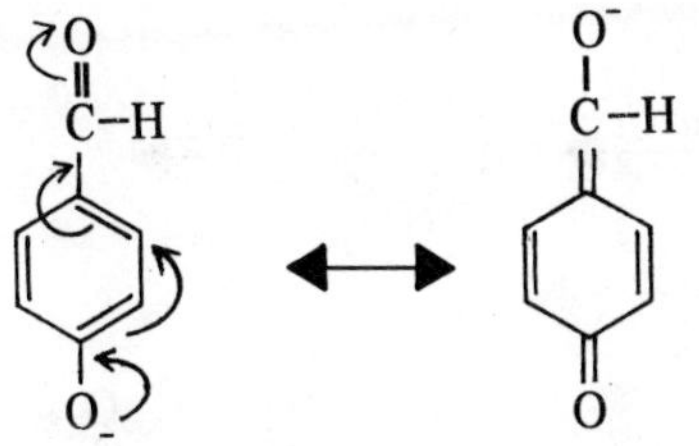

18.48

The two nitrogens of semicarbazide that are adjacent to the C=O group bear partial positive charges because of resonance contributions made by the second and third structures below,

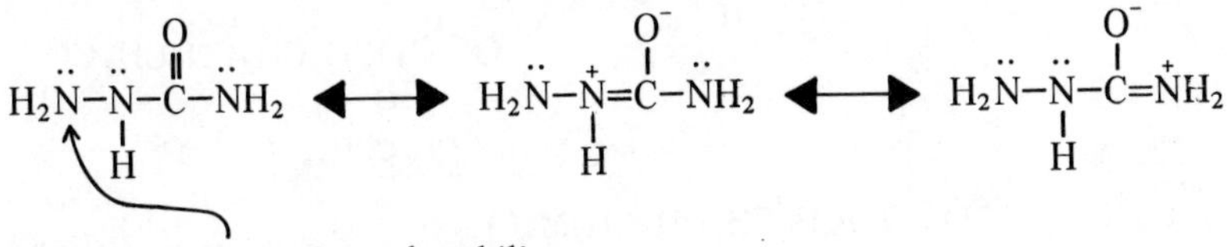

Only this nitrogen is nucleophilic.

18.49

Abstraction of an α-hydrogen at the ring junction yields an enolate ion that can then accept a proton to form either *trans*-1-decalone or *cis*-1-decalone. Since *trans*-1-decalone is more stable, it predominates at equilibrium.

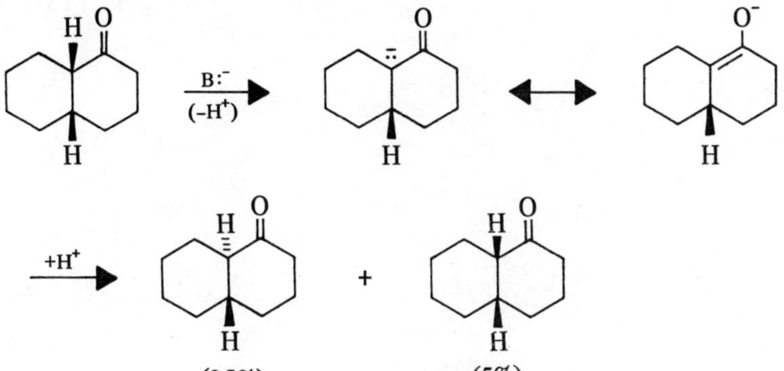

(95%)
trans-1-Decalone
(more stable)

(5%)
cis-1-Decalone
(less stable)

18.50

(a)

$$\text{(dihydropyran)} \xrightarrow{+H^+} \text{(oxocarbenium ion)} \xrightarrow{+ROH} \underset{\overset{|}{\text{H}}}{\text{(tetrahydropyran–}O\overset{+}{-}R)} \xrightarrow{-H^+} \text{(tetrahydropyran–}O\text{–}R)$$

(b) Tetrahydropyranyl ethers are acetals; thus they are stable in aqueous base and hydrolyze readily in aqueous acid.

$$\text{(pyran }O\text{–}R) \xrightarrow{H^+} \underset{\overset{|}{\text{H}}}{\text{(pyran }O\overset{+}{-}R)} \xrightarrow{-ROH} \text{(oxocarbenium ion)} \xrightarrow{H_2O}$$

$$\underset{\overset{|}{\text{H}}}{\text{(pyran }O\overset{+}{-}H)} \xrightarrow{-H^+} \text{(pyran OH)} \rightleftharpoons \underset{\text{5-Hydroxybutanal}}{HOCH_2CH_2CH_2CH_2\overset{\overset{\text{O}}{\|}}{C}H}$$

(c) $HOCH_2CH_2CH_2CH_2Cl \xrightarrow[H^+]{\text{(dihydropyran)}} \text{(pyran)}OCH_2CH_2CH_2CH_2Cl$

$$\xrightarrow[\text{ether}]{Mg} \text{(pyran)}OCH_2CH_2CH_2CH_2MgCl \xrightarrow{CH_3\overset{\overset{\text{O}}{\|}}{C}CH_3}$$

$$\text{(pyran)}OCH_2CH_2CH_2CH_2\underset{\overset{|}{CH_3}}{\overset{\overset{|}{CH_3}}{C}}OMgCl \xrightarrow[H_2O]{H^+} HOCH_2CH_2CH_2CH_2\underset{\overset{|}{CH_3}}{\overset{\overset{|}{CH_3}}{C}}OH$$

$$(+ \; HOCH_2CH_2CH_2CH_2\overset{\overset{\text{O}}{\|}}{C}H)$$

19

CARBOXYLIC ACIDS AND THEIR DERIVATIVES: NUCLEOPHILIC SUBSTITUTION AT ACYL CARBON

19.1

(a) Carbon dioxide is an acid; it converts an aqueous solution of the strong base, NaOH, into an aqueous solution of the weaker base, $NaHCO_3$.

$$NaOH_{(aq)} + CO_2 \longrightarrow NaHCO_{3\,(aq)}$$

In this new solution, the more strongly basic p-cresoxide ion accepts a proton and becomes p-cresol,

$$p\text{-}CH_3C_6H_4O^- + HCO_3^- \xrightleftharpoons{} \underset{\text{water-insoluble}}{p\text{-}CH_3C_6H_4OH} + CO_3^=$$

The more weakly basic benzoate ion remains in solution.

(b) Dissolve all three compounds in an organic solvent such as CH_2Cl_2, then extract with aqueous NaOH. The organic layer will contain cyclohexanol, which can be separated by distillation. The aqueous layer will contain the benzoic acid, as sodium benzoate, and the p-cresol, as sodium p-cresoxide.

Now pass CO_2 into the aqueous layer; this will cause p-cresol to separate (it can then be extracted into an organic solvent and purified by distillation). After separation of the p-cresol, the aqueous phase can be acidified with aqueous HCl to yield benzoic acid as a precipitate.

19.2

(a) CH_2FCOOH (F− is more electronegative than H−)

(b) CH_2FCOOH (F− is more electronegative than Cl−)

(c) $CH_2ClCOOH$ (Cl− is more electronegative than Br−)

(d) $CH_3CHClCH_2COOH$ (Cl− is closer to −COOH)

(e) $CH_3CH_2CHClCOOH$ (Cl− is closer to −COOH)

(f) $(CH_3)_3\overset{+}{N}$—⟨O⟩—COOH [$(CH_3)_3\overset{+}{N}$− is more electronegative than H−]

(g) CF_3—⟨O⟩—COOH (CF_3− is more electronegative than CH_3−)

19.3

(a) The carboxyl group is an electron-withdrawing group; thus in a dicarboxylic acid such as those in Table 19.3, one carboxyl group increases the acidity of the other.

(b) As the distance between the carboxyl groups increases the acid-strengthening, inductive effect decreases.

19.4

These syntheses are easy to see if we work backward.

(a) $C_6H_5CH_2COOH \xleftarrow[\text{(2) H}^+]{\text{(1) CO}_2} C_6H_5CH_2MgBr$

$\Big\uparrow$ Mg, ether

$C_6H_5CH_2Br$

(b) $CH_3CH_2CH_2\overset{\overset{\displaystyle CH_3}{\displaystyle |}}{\underset{\underset{\displaystyle CH_3}{\displaystyle |}}{C}}COOH \xleftarrow[\text{(2) H}^+]{\text{(1) CO}_2} CH_3CH_2CH_2\overset{\overset{\displaystyle CH_3}{\displaystyle |}}{\underset{\underset{\displaystyle CH_3}{\displaystyle |}}{C}}MgBr$

$\Big\uparrow$ Mg, ether

$CH_3CH_2CH_2\overset{\overset{\displaystyle CH_3}{\displaystyle |}}{\underset{\underset{\displaystyle CH_3}{\displaystyle |}}{C}}Br$

(c) $CH_2{=}CHCH_2COOH \xleftarrow[\text{(2) H}^+]{\text{(1) CO}_2} CH_2{=}CHCH_2MgBr$

$\Big\uparrow$ Mg, ether

$CH_2{=}CHCH_2Br$

(d) $CH_3{-}\langle\bigcirc\rangle{-}COOH \xleftarrow[\text{(2) H}^+]{\text{(1) CO}_2} CH_3{-}\langle\bigcirc\rangle{-}MgBr$

$\Big\uparrow$ Mg, ether

$CH_3{-}\langle\bigcirc\rangle{-}Br$

(e) $CH_3CH_2CH_2CH_2CH_2COOH \xleftarrow[\text{(2) H}_2\text{O}]{\text{(1) CO}_2} CH_3CH_2CH_2CH_2CH_2MgBr$

$\Big\uparrow$ Mg, ether

$CH_3CH_2CH_2CH_2CH_2Br$

19.5

(a) $C_6H_5CH_2COOH \xleftarrow[\text{(2) H}^+, \text{H}_2\text{O, heat}]{\text{(1) CN}^-} C_6H_5CH_2Br$

$$CH_2=CHCH_2COOH \xleftarrow[\text{(2) H}^+\text{, H}_2\text{O, heat}]{\text{(1) CN}^-} CH_2=CHCH_2Br$$

$$CH_3CH_2CH_2CH_2COOH \xleftarrow[\text{(2) H}^+\text{, H}_2\text{O, heat}]{\text{(1) CN}^-} CH_3CH_2CH_2CH_2Br$$

(b) A nitrile synthesis. Preparation of a Grignard reagent from $HOCH_2CH_2CH_2CH_2Br$ would not be possible because of the presence of the acidic hydroxyl group.

19.6

(a) $CH_3COOH + C_6H_5COCl \xrightarrow{\text{pyridine}}$ $\underset{\qquad\qquad}{CH_3\overset{\overset{\displaystyle O}{\|}}{C}O\overset{\overset{\displaystyle O}{\|}}{C}C_6H_5}$

(b) $CH_3(CH_2)_4COOH + (CH_3CO)_2O \xrightarrow{\text{heat}}$

$$CH_3(CH_2)_4\overset{\overset{\displaystyle O}{\|}}{C}O\overset{\overset{\displaystyle O}{\|}}{C}(CH_2)_4CH_3 \;+\; 2CH_3COOH$$
(remove by
distillation)

(c)

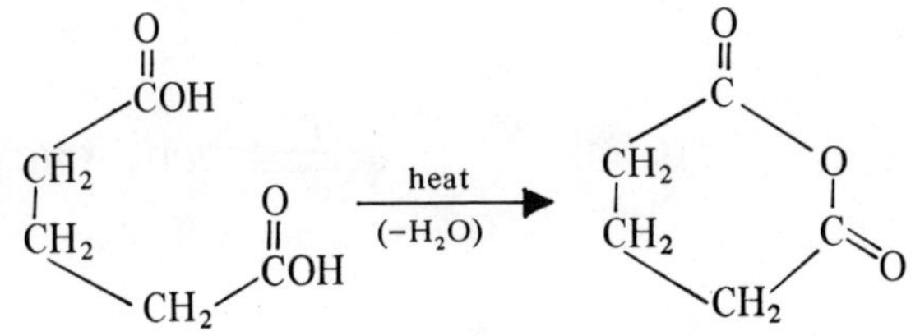

19.7

Since maleic acid is a *cis*-dicarboxylic acid, dehydration occurs readily:

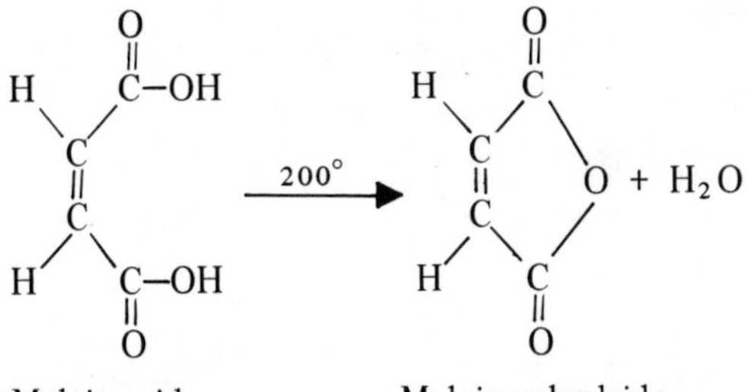

Maleic acid   Maleic anhydride

Being a *trans*-dicarboxylic acid, fumaric acid must undergo isomerization to maleic acid first. This requires a higher temperature.

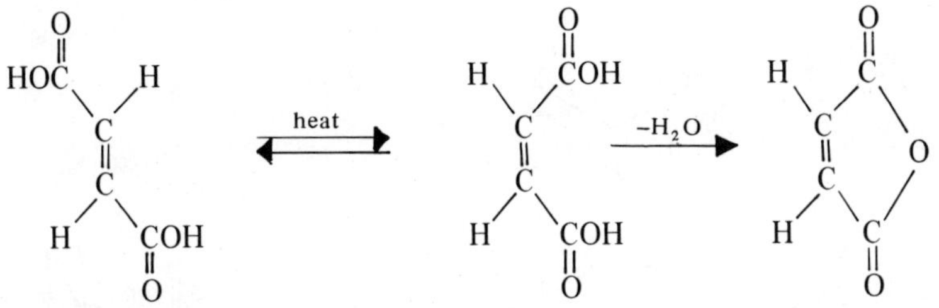

Fumaric acid

19.8

The labeled oxygen should appear in the carboxyl group of the acid. (Follow the reverse steps of the mechanism on page 756 of the text using $H_2{}^{18}O$.)

19.9

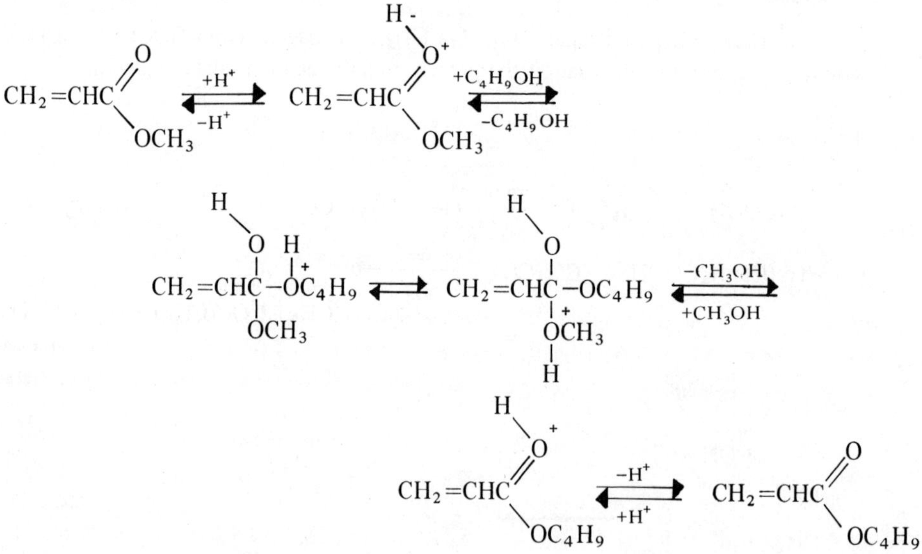

$$CH_2=CHC\underset{OCH_3}{\overset{O}{\big\|}} \underset{-H^+}{\overset{+H^+}{\rightleftharpoons}} CH_2=CHC\underset{OCH_3}{\overset{\overset{H\cdot}{\underset{\big|}{O^+}}}{\big\|}} \underset{-C_4H_9OH}{\overset{+C_4H_9OH}{\longrightarrow}}$$

$$CH_2=CHC\underset{OCH_3}{\overset{\overset{H}{\underset{\big|}{O}}\ \overset{H}{\underset{\big|}{\ }}}{\underset{\big|}{-}}OC_4H_9} \rightleftharpoons CH_2=CHC\underset{\underset{H}{\overset{\big|}{OCH_3}}}{\overset{\overset{H}{\underset{\big|}{O}}}{-}}OC_4H_9 \underset{+CH_3OH}{\overset{-CH_3OH}{\longrightarrow}}$$

$$CH_2=CHC\underset{OC_4H_9}{\overset{\overset{H}{\underset{\big|}{O}}^+}{\big\|}} \underset{+H^+}{\overset{-H^+}{\rightleftharpoons}} CH_2=CHC\underset{OC_4H_9}{\overset{O}{\big\|}}$$

19.10

(a)

(1)

$A \xrightarrow[\text{(inversion)}]{OH^-,\ heat}$

$B \quad + \quad C_6H_5SO_3^-$

(2)

$\xrightarrow[\text{(retention)}]{OH^-,\ heat}$ D $\quad + \quad C_6H_5CO_2^-$

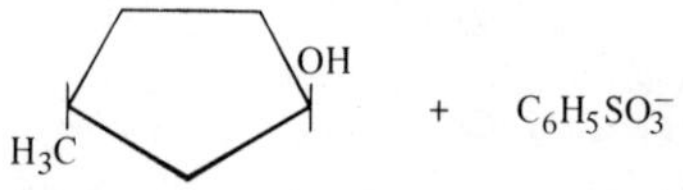

(3)

$+ \quad CH_3\overset{O}{\overset{\big\|}{C}}O^-Na^+ \xrightarrow{\text{(inversion)}}$ E

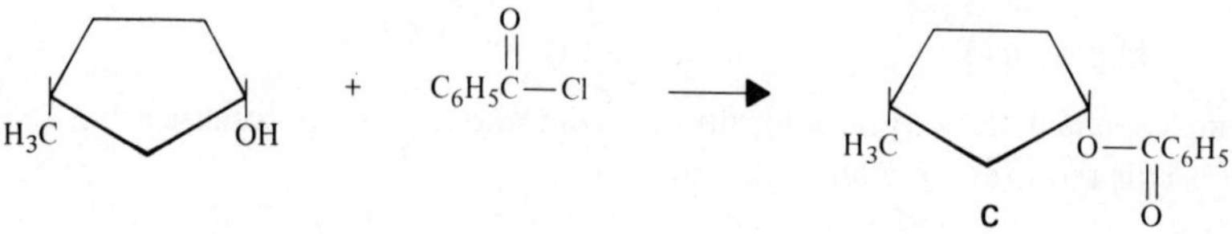

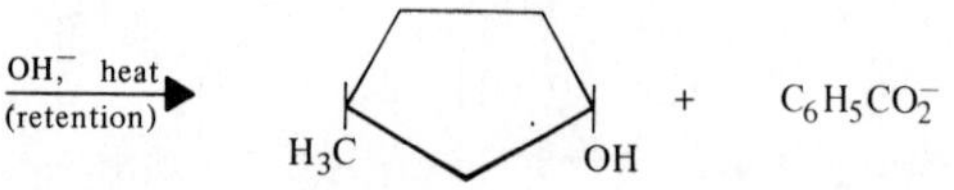

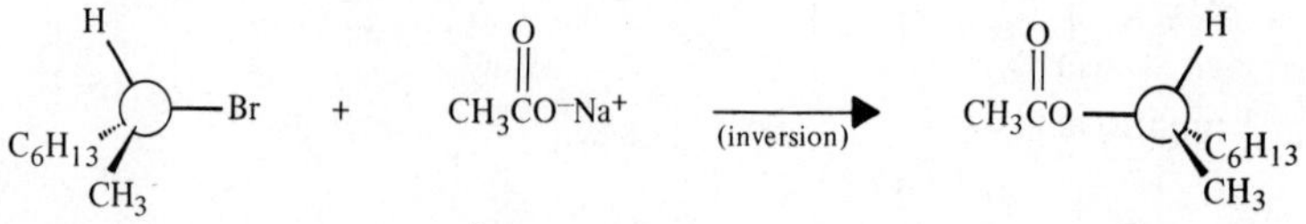

(b) Method (3) should give a higher yield of **F** than method (4). Since the hydroxide ion is a strong base and since the alkyl halide is secondary, method (4) is likely to be accompanied by considerable elimination. Method (3), on the other hand, employs a weaker base, acetate ion, in the S_N2 step and is less likely to be complicated by elimination. Hydrolysis of the ester **E** that results should also proceed in high yield.

19.11

(a) Steric hindrance presented by the di-*ortho* methyl groups of methyl mesitoate prevents formation of the tetrahedral intermediate that must accompany attack at the acyl carbon.

(b) Carry out hydrolysis with labeled OH$^-$ in labeled H_2O. The label should appear in the methanol.

19.12

19.13

(a) $(CH_3)_3CCOOH \xrightarrow{SOCl_2} (CH_3)_3CCOCl$

$\xrightarrow{NH_3} (CH_3)_3CCONH_2 \xrightarrow[\text{heat}]{P_2O_5} (CH_3)_3CC\equiv N$

(b) An elimination reaction would take place.

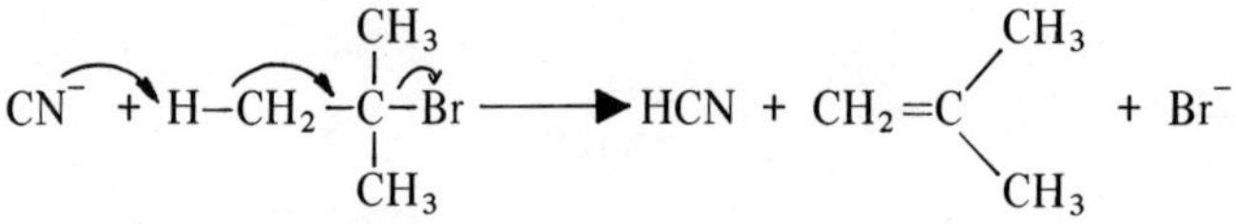

$$CN^- + H-CH_2-\overset{\overset{\displaystyle CH_3}{|}}{\underset{\underset{\displaystyle CH_3}{|}}{C}}-Br \longrightarrow HCN + CH_2=\overset{\overset{\displaystyle CH_3}{}}{\underset{\underset{\displaystyle CH_3}{}}{C}} + Br^-$$

19.14

(a) $CH_3(CH_2)_4COOH$

(b) $CH_3(CH_2)_4CONH_2$

(c) $CH_3(CH_2)_4CONHC_2H_5$

(d) $CH_3(CH_2)_4CON(C_2H_5)_2$

(e) $CH_3CH_2CH=CHCH_2COOH$

(f) $CH_3CH=CHCH_2\overset{\displaystyle CHCOOH}{\underset{\displaystyle CH_3}{|}}$

(g) $HOOCCH_2CH_2CH_2CH_2COOH$

(h)

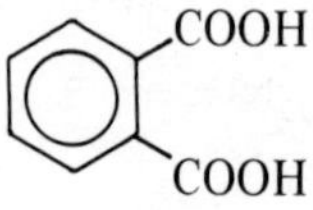

(i)

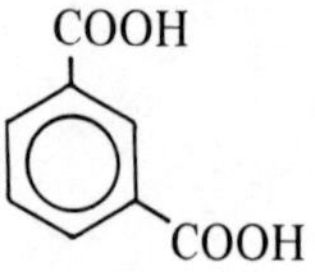

(j)

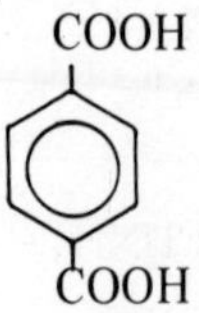

(k) $C_2H_5OOC-COOC_2H_5$

(l) $C_2H_5OOC(CH_2)_4COOC_2H_5$

(m) $CH_3CH_2COOCH_2CH(CH_3)_2$

(n)

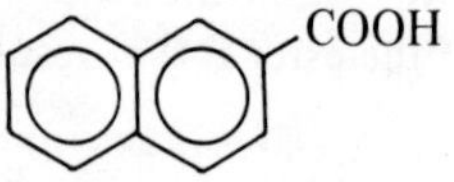

(o)

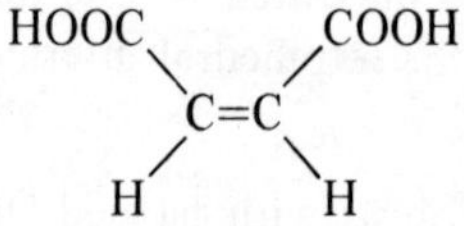

$$\overset{HOOC}{}\,\underset{H}{\overset{}{C}}=\underset{H}{\overset{COOH}{C}}$$

(p) $HOOCCHOHCH_2COOH$

(q)

$$\overset{HOOC}{}\,\underset{H}{\overset{}{C}}=\underset{COOH}{\overset{H}{C}}$$

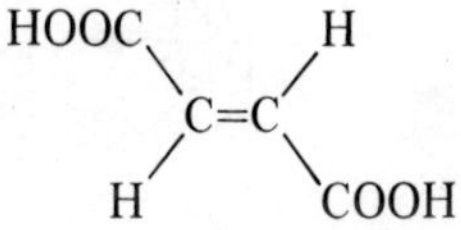

(r) $HOOCCH_2CH_2COOH$

(s)

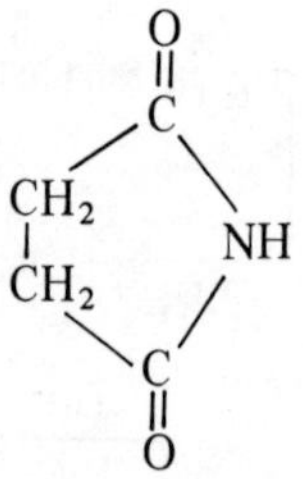

(t) $HOOCCH_2COOH$

(u) $C_2H_5OOCCH_2COOC_2H_5$

19.15

(a) Benzoic acid

(b) Benzoyl chloride

(c) Benzamide

(d) Benzoic anhydride

(e) Benzyl benzoate

(f) Phenyl benzoate

(g) Isopropyl acetate

(h) N,N-Dimethylacetamide

(i) Acetonitrile

(j) Maleic anhydride

(k) Phthalic anhydride

(l) Phthalimide

(m) Glyceryl tripalmitate

(n) α-Ketosuccinic acid
(oxaloacetic acid)

(o) Methyl salicylate

19.16

(a)

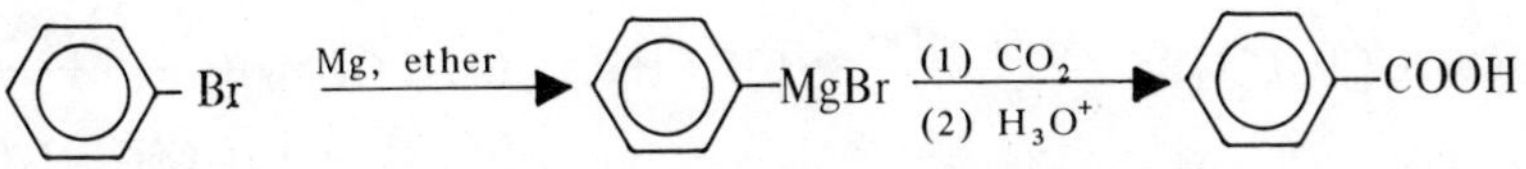

(b)

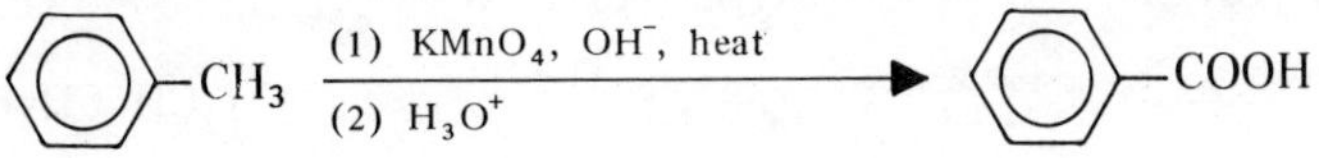

(c)

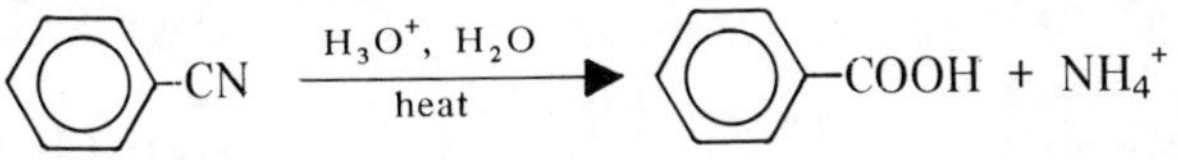

(d)

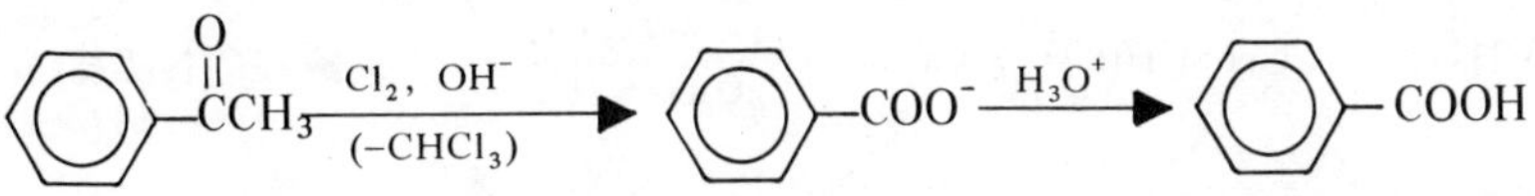

(e)

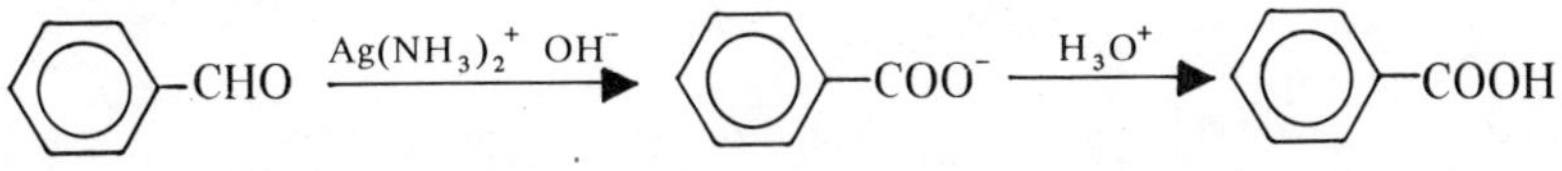

(f)

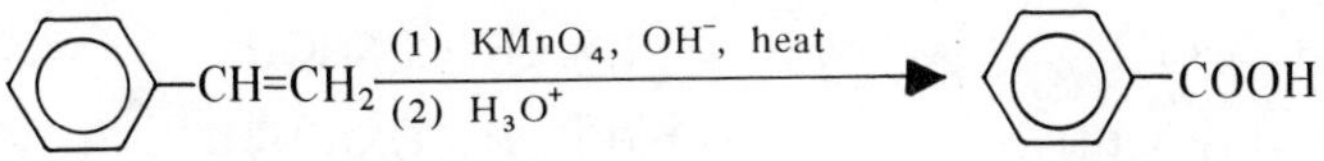

(g)

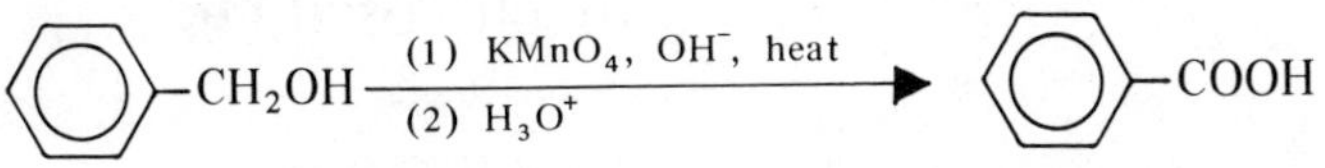

19.17

(a)

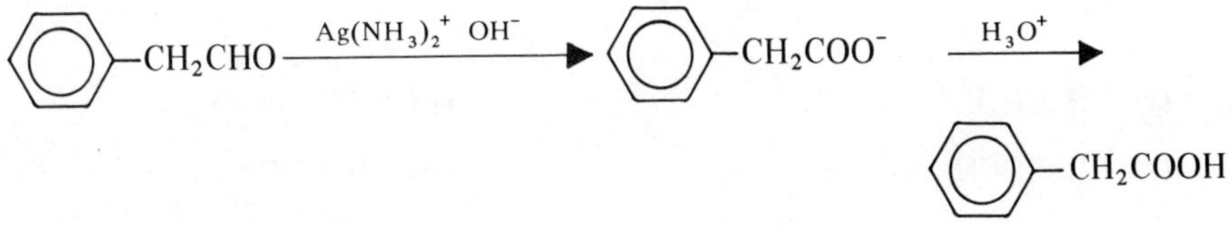

(b) $C_6H_5-CH_2Br \xrightarrow[\text{(2) CO}_2]{\text{(1) Mg, ether}} C_6H_5-CH_2COOMgBr \xrightarrow{H_3O^+}$

$C_6H_5-CH_2COOH$

$C_6H_5-CH_2Br \xrightarrow{CN^-} C_6H_5-CH_2CN \xrightarrow[\text{heat}]{H_3O^+,\ H_2O}$

$C_6H_5-CH_2COOH$

19.18

(a) $CH_3CH_2CH_2CH_2CH_2OH \xrightarrow[\text{(2) H}_3O^+]{\text{(1) KMnO}_4,\ OH^-,\ heat} CH_3CH_2CH_2CH_2COOH$

(b) $CH_3CH_2CH_2CH_2Br \xrightarrow[\text{(2) CO}_2]{\text{(1) Mg, ether}} CH_3CH_2CH_2CH_2COOMgBr \xrightarrow{H_3O^+}$

$CH_3CH_2CH_2CH_2COOH$

$CH_3CH_2CH_2CH_2Br \xrightarrow{CN^-} CH_3CH_2CH_2CH_2CN \xrightarrow[\text{heat}]{H_3O^+,\ H_2O}$

$CH_3CH_2CH_2CH_2COOH$

(c) $CH_3CH_2CH_2CH_2\underset{\underset{O}{\|}}{C}CH_3 \xrightarrow[\text{(-CHCl}_3)]{\text{Cl}_2,\ OH^-} CH_3CH_2CH_2CH_2COO^- \xrightarrow{H_3O^+}$

$CH_3CH_2CH_2CH_2COOH$

(d) $CH_3(CH_2)_3CH=CH(CH_2)_3CH_3 \xrightarrow[\text{(2) H}_3O^+]{\text{(1) KMnO}_4,\ OH^-,\ heat} 2CH_3(CH_2)_3COOH$

(e) $CH_3CH_2CH_2CH_2CHO \xrightarrow[\text{(2) H}_3O^+]{\text{(1) Ag(NH}_3)_2{}^+OH^-} CH_3CH_2CH_2CH_2COOH$

19.19

(a) $CH_3COOH + HCl$

(b) $CH_3COOH + AgCl$

(c) $CH_3COOCH_2(CH_2)_2CH_3$

(d) CH_3CONH_2

(e)

$+ \ CH_3-C_6H_4-\underset{\underset{}{\overset{\overset{O}{\|}}{}}}{C}CH_3$

(f) CH_3CHO

(g) CH_3COCH_3

(h) $CH_3COCH_2CH_3$

(i) $CH_3CONHCH_3$

(j) $CH_3CONHC_6H_5$

(k) $CH_3CON(CH_3)_2$

(l) $CH_3COSCH_2CH_3$

(m) $(CH_3CO)_2O$

(n) $(CH_3CO)_2O$

(o) $CH_3COOC_6H_5$

19.20

(a) CH_3CONH_2 + CH_3COONH_4

(b) $2CH_3COOH$

(c) $CH_3COOCH_2CH_2CH_3$ + CH_3COOH

(d) $C_6H_5COCH_3$ + CH_3COOH

(e) $CH_3CONHCH_2CH_3$ + $CH_3COO^-\ CH_3CH_2NH_3^+$

(f) $CH_3CON(CH_2CH_3)_2$ + $CH_3COO^-\ (CH_3CH_2)_2NH_2^+$

19.21

(a) 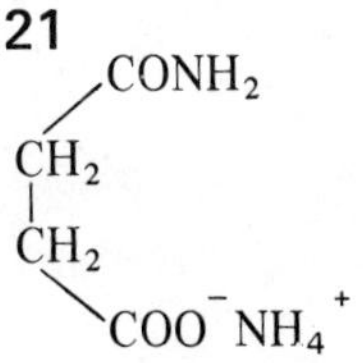
$$\begin{array}{c} CONH_2 \\ CH_2 \\ | \\ CH_2 \\ COO^-\ NH_4^+ \end{array}$$

(b)
$$\begin{array}{c} COOH \\ CH_2 \\ | \\ CH_2 \\ COOH \end{array}$$

(c)
$$\begin{array}{c} COOCH_2CH_2CH_3 \\ CH_2 \\ | \\ CH_2 \\ COOH \end{array}$$

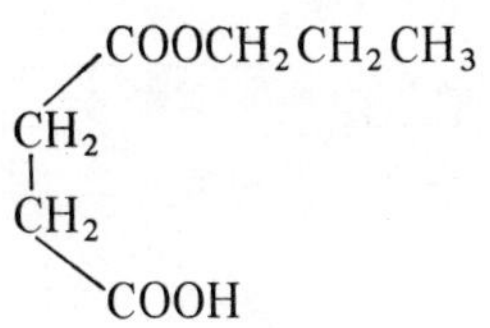

(d)

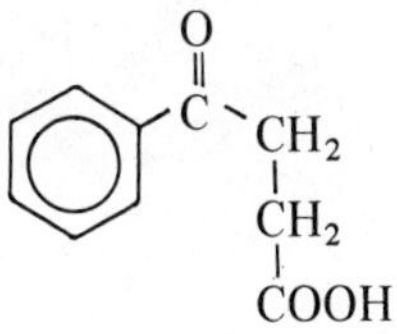

(e) 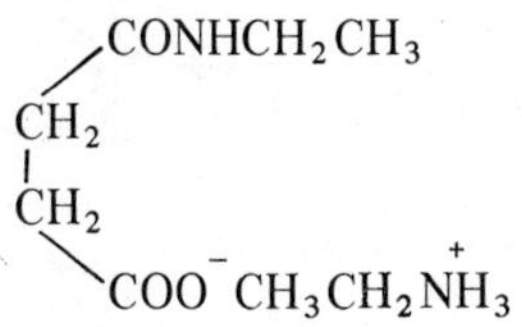
$$\begin{array}{c} CONHCH_2CH_3 \\ CH_2 \\ | \\ CH_2 \\ COO^-\ CH_3CH_2NH_3^+ \end{array}$$

(f) 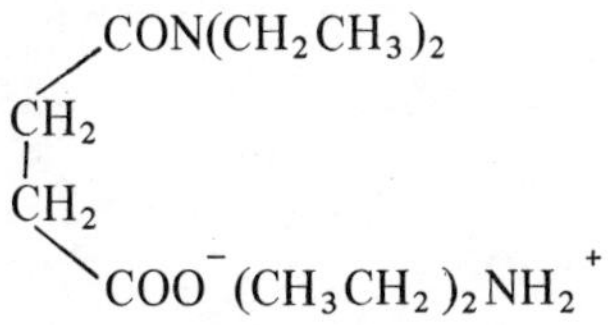
$$\begin{array}{c} CON(CH_2CH_3)_2 \\ CH_2 \\ | \\ CH_2 \\ COO^-\ (CH_3CH_2)_2NH_2^+ \end{array}$$

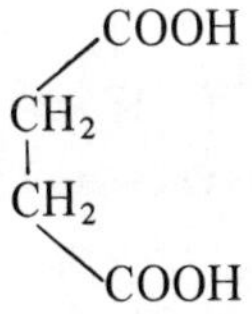

19.22

(a)

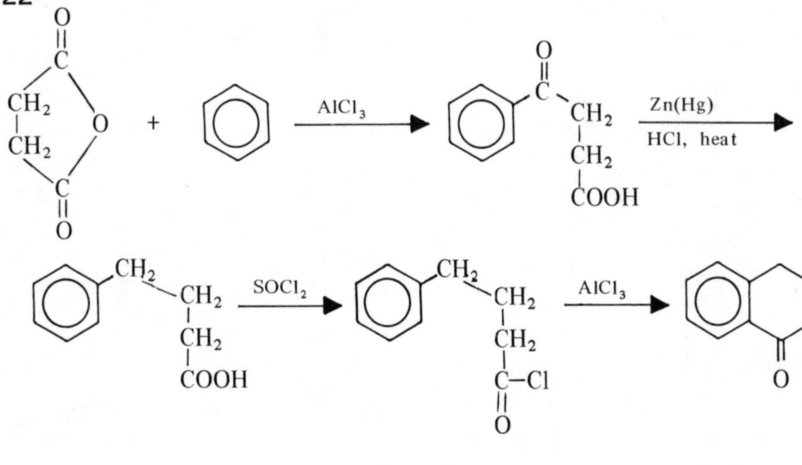

(b)

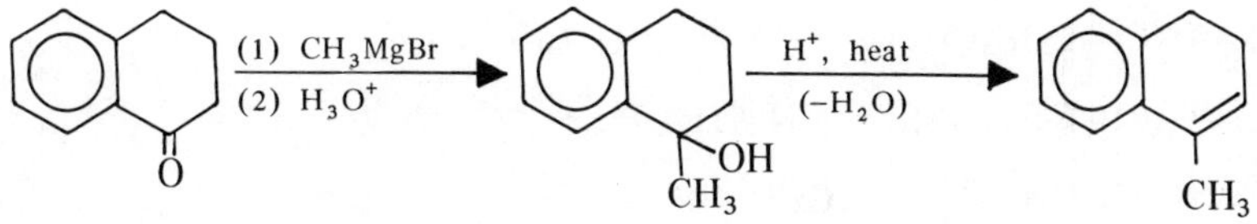

(c)

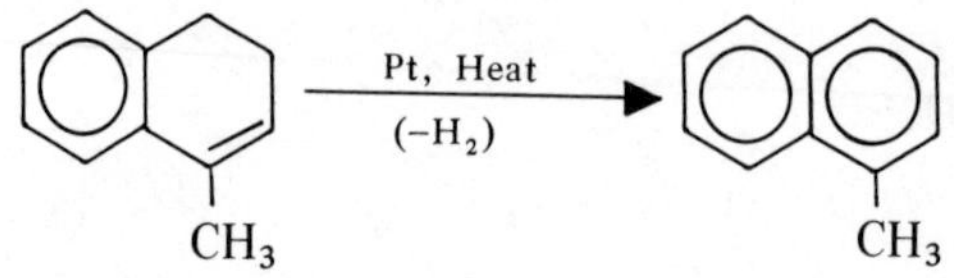

(d)

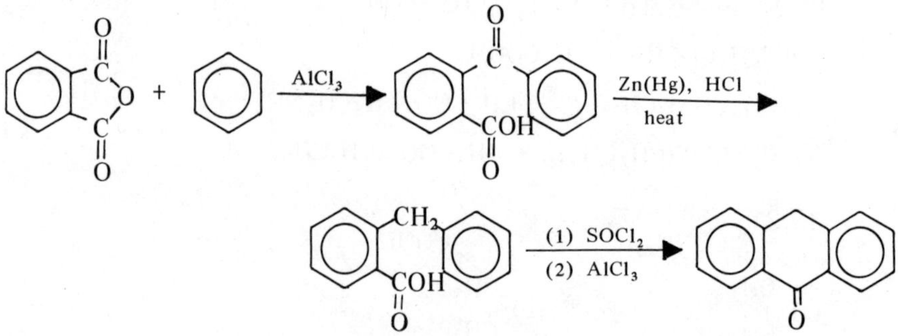

(e)

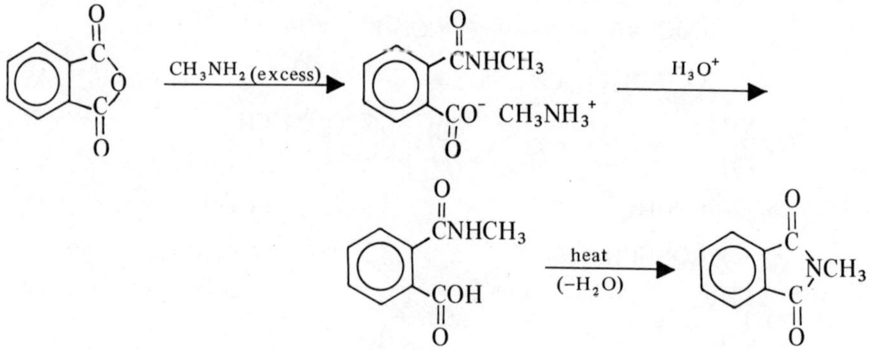

(f)

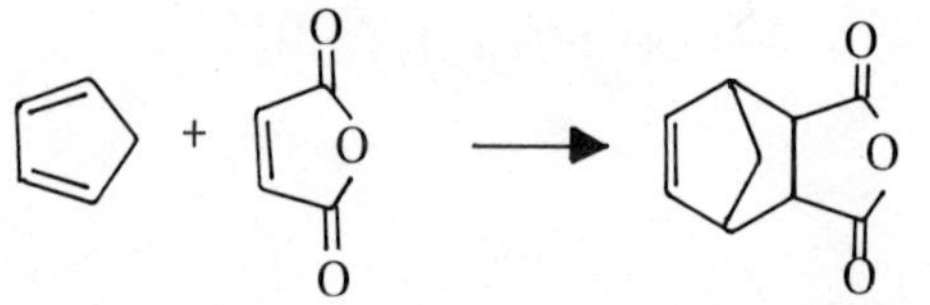

(g)

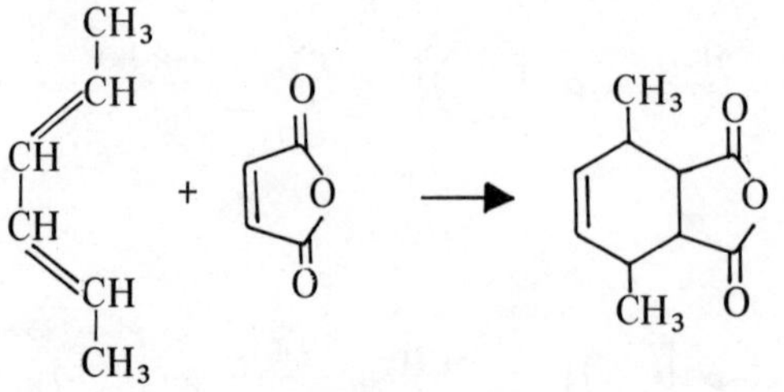

19.23

(a) $CH_3CH_2COOH + CH_3CH_2OH$

(b) $CH_3CH_2COO^- + CH_3CH_2OH$

(c) $CH_3CH_2COO(CH_2)_7CH_3 + CH_3CH_2OH$

(d) $CH_3CH_2CONHCH_3 + CH_3CH_2OH$

(e) $CH_3CH_2CH_2OH + CH_3CH_2OH$

$$\overset{\displaystyle C_6H_5}{\underset{\displaystyle OH}{CH_3CH_2\overset{|}{\underset{|}{C}}-C_6H_5}} + CH_3CH_2OH$$

(f)

19.24

(a) $CH_3CH_2COOH + NH_4{}^{+}$

(b) $CH_3CH_2COO^{-} + NH_3$

(c) CH_3CH_2CN

19.25

(a) Benzoic acid dissolves in aqueous $NaHCO_3$. Methyl benzoate does not.

(b) Benzoyl chloride gives a precipitate (AgCl) when treated with alcoholic $AgNO_3$. Benzoic acid does not.

(c) Benzoic acid dissolves in aqueous $NaHCO_3$. Benzamide does not.

(d) Benzoic acid dissolves in aqueous $NaHCO_3$. p-Cresol does not.

(e) Refluxing benzamide with aqueous NaOH liberates NH_3 which can be detected in the vapors with moist red litmus paper. Ethyl benzoate does not liberate NH_3.

(f) Cinnamic acid, because it has a double bond, decolorizes Br_2 in CCl_4. Benzoic acid does not.

(g) Benzoyl chloride gives a precipitate (AgCl) when treated with alcoholic $AgNO_3$. Ethyl benzoate does not.

(h) α-Chlorobutanoic acid gives a precipitate (AgCl) when treated with alcoholic silver nitrate. Butanoic acid does not.

19.26

(a)
$$\begin{array}{c} CH_2-CH_2 \\ | \qquad\qquad \diagdown \\ \qquad\qquad\qquad C=O \\ | \qquad\qquad \diagup \\ CH_2-O \end{array}$$

(b) $CH_3CH=CHCOOH$

(c)
$$CH_3CH_2CH \underset{O-C}{\overset{C-O}{\diagup\diagdown}} CHCH_2CH_3$$
with $C=O$ at top and bottom

(d)
$$\begin{array}{c} CH_2-C\diagup^{O} \\ \diagup \qquad\qquad \diagdown \\ CH_2 \qquad\qquad O \\ \diagdown \qquad\qquad \diagup \\ CH_2-C\diagdown_{O} \end{array}$$

(e)
$$\begin{array}{c} CH_2-C\diagup^{O} \\ \diagup \qquad\qquad \diagdown \\ CH_2 \qquad\qquad NH \\ \diagdown \qquad\qquad \diagup \\ CH_2-CH \\ \diagdown CH_3 \end{array}$$

(f) benzene ring fused with five-membered ring containing C=O, NH, and CH$_2$

19.27

(a)

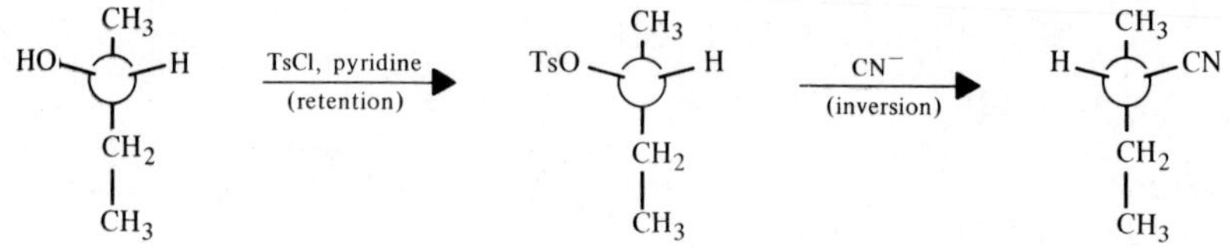

$R-(-)-2-$butanol → A → B

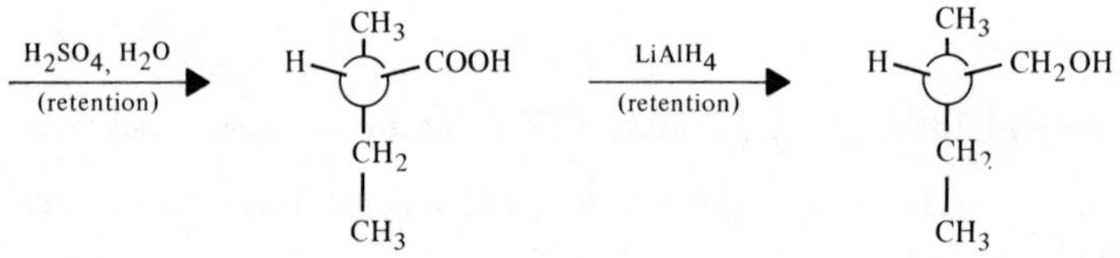

$R-(-)-2-$butanol —TsCl, pyridine (retention)→ A —CN⁻ (inversion)→ B

—H₂SO₄, H₂O (retention)→ (+) – C —LiAlH₄ (retention)→ (−) – D

(b)

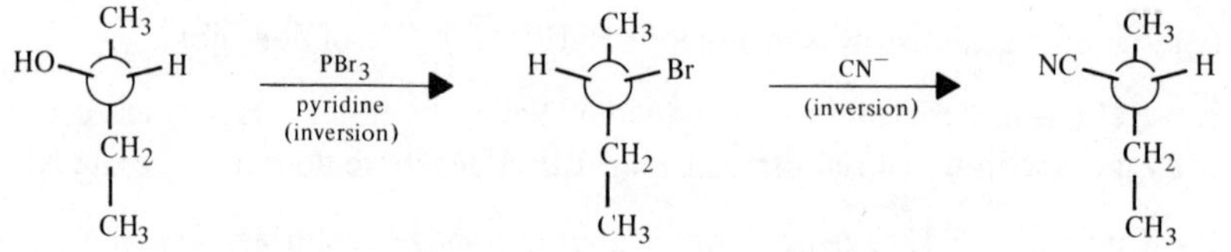

$R-(-)-2-$butanol —PBr₃ pyridine (inversion)→ E —CN⁻ (inversion)→ F

—H₂SO₄, H₂O (retention)→ (−) – C —LiAlH₄→ (+) – D

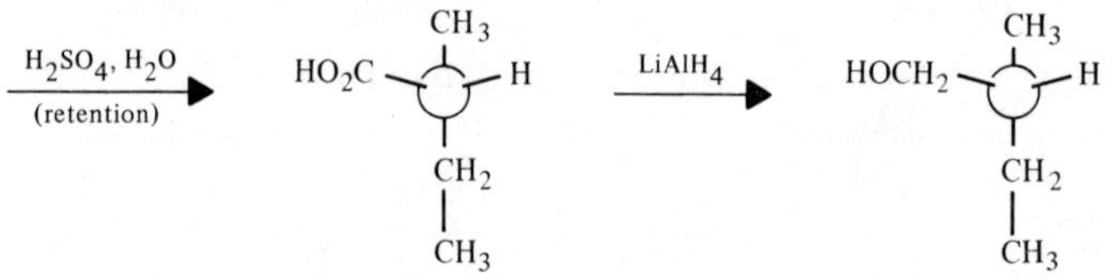

(c)

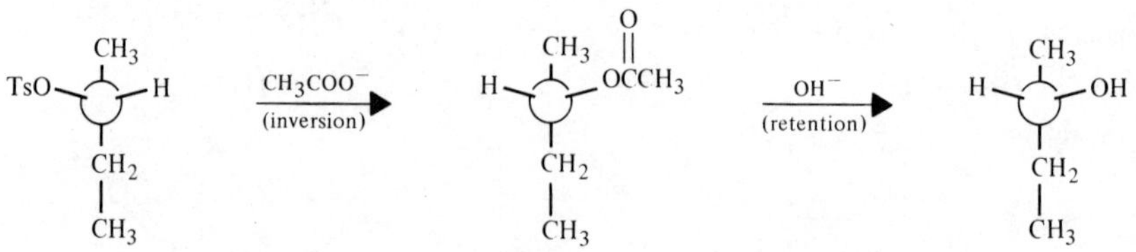

A —CH₃COO⁻ (inversion)→ G —OH⁻ (retention)→ (+) – H
$S-(+)-2-$butanol

(d)

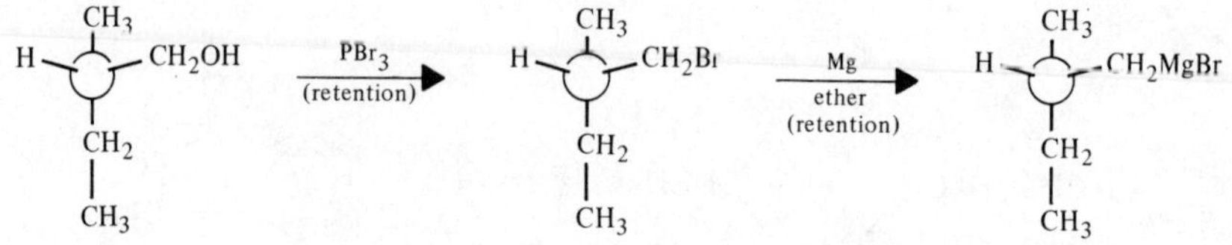

(−) – D —PBr₃ (retention)→ J —Mg ether (retention)→ K

$$\xrightarrow[\substack{(2)\ H^+ \\ (\text{retention})}]{(1)\ CO_2}$$

CH₃ — (H, CH₂COOH) — CH₂ — CH₃

L

(e)

R–(+)–glyceraldehyde (H, CHO, OH, CH₂OH) $\xrightarrow{HCN}$ **M** (CN, H, OH, CH₂OH) + **N** (CN, HO, H, CH₂OH)

(f) **M** $\xrightarrow[\substack{H_2O \\ \text{heat}}]{H_2SO_4}$ **P** (COOH, H, OH, CH₂OH) $\xrightarrow[HNO_3]{(O)}$ *meso*–tartaric acid (COOH, H, OH, COOH)

(g) **N** $\xrightarrow[\substack{H_2O \\ \text{heat}}]{H_2SO_4}$ (COOH, HO, H, CH₂OH) $\xrightarrow[HNO_3]{(O)}$ (–)–tartaric acid (COOH, HO, H, COOH)

19.28

(a) $CH_3\overset{CH_3}{\underset{}{CHCHO}}$ + $\overset{O}{\underset{}{HCH}}$ $\xrightarrow[H_2O]{K_2CO_3}$ $CH_3\overset{CH_3}{\underset{CH_2OH}{CCHO}}$ $\xrightarrow[(2)\ KCN]{(1)\ NaHSO_3}$

A

$CH_3\overset{CH_3}{\underset{CH_2OH}{C}}\!\!-\!\!\overset{OH}{\underset{}{CHCN}}$ $\xrightarrow{H_3O^+}$ $\left[CH_3\overset{CH_3}{\underset{CH_2OH}{C}}\!\!-\!\!\overset{OH}{\underset{}{CHCOOH}} \right]$ $\xrightarrow{-H_2O}$

(±)-**B** (±)-**C**

$CH_3\overset{CH_3}{\underset{CH_2\ \ C=O \atop \diagdown O \diagup}{C}}\!\!-\!\!\overset{OH}{\underset{}{CH}}$ $\xrightarrow{H_2NCH_2CH_2\overset{O}{\overset{\|}{C}}OH}$ (±)-Pantothenic acid

(±)-**D**

$\xrightarrow{H_2NCH_2CH_2\overset{O}{\overset{\|}{C}}NHCH_2CH_2SH}$ (±)-Pantetheine

(b)

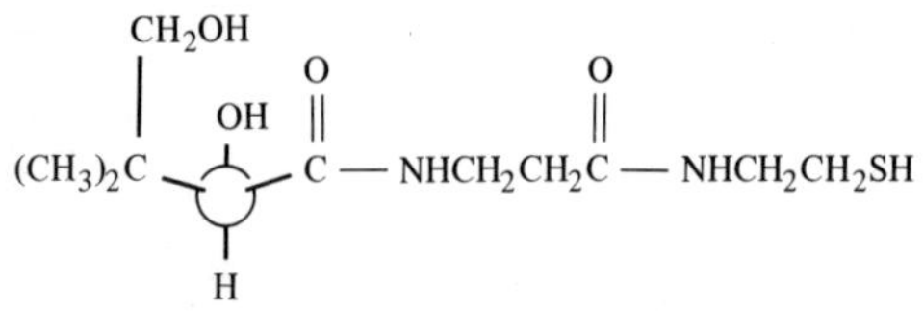

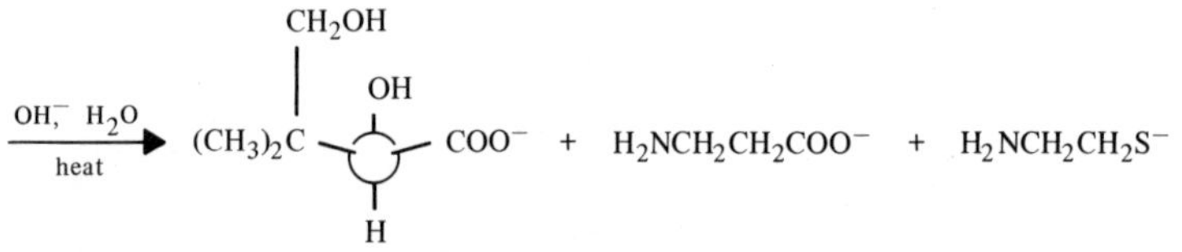

$$(CH_3)_2C \underset{H}{\overset{CH_2OH}{|}} \overset{OH}{\underset{}{}} \overset{O}{\overset{||}{C}} - NHCH_2CH_2\overset{O}{\overset{||}{C}} - NHCH_2CH_2SH$$

$$\xrightarrow[\text{heat}]{OH^-,\ H_2O} (CH_3)_2C \underset{H}{\overset{CH_2OH}{|}} \overset{OH}{\underset{}{}} COO^- + H_2NCH_2CH_2COO^- + H_2NCH_2CH_2S^-$$

19.29

$$CH_3CH_2O-\bigcirc-NH-\overset{O}{\overset{||}{C}}-CH_3 \xrightarrow[\substack{H_2O \\ reflux}]{OH^-} CH_3CH_2O-\bigcirc-NH_2$$

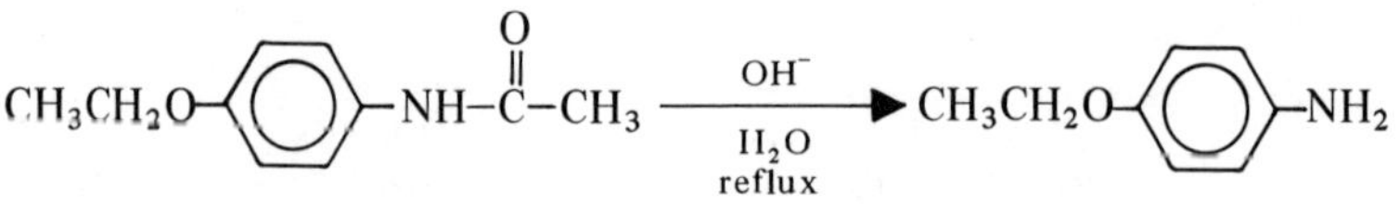

Phenacetin   Phenetidine
+
CH_3COO^-

An interpretation of the spectral data for phenacetin is given in Fig. 19.1.

19.30

(a) $CH_3CH_2-O-\overset{O}{\overset{||}{C}}-CH_2CH_2-\overset{O}{\overset{||}{C}}-O-CH_2CH_3$
    a   c   b   b   c   a

Interpretation:

a Triplet δ1.2 (6H) $2-\overset{O}{\overset{||}{C}}-$, 1740 cm⁻¹

b Singlet δ2.5 (4H)

c Quartet δ4.1 (4H)

(b)

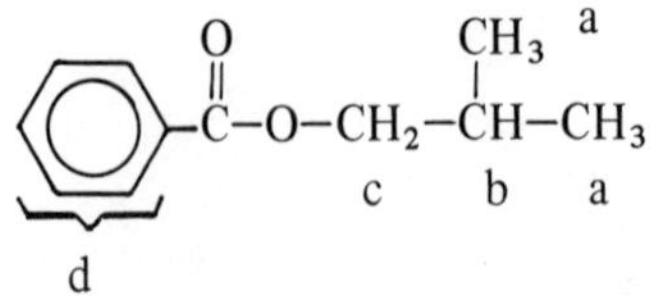

$$\bigcirc-\overset{O}{\overset{||}{C}}-O-CH_2-\overset{\overset{\displaystyle CH_3\ a}{|}}{CH}-CH_3$$
              c   b   a
d

Interpretation:

a Doublet δ1.0 (6H) $-\overset{O}{\overset{||}{C}}-$, 1720 cm⁻¹

b Multiplet δ2.1 (1H)

c Doublet δ4.1 (2H)

d Multiplet δ7.8 (5H)

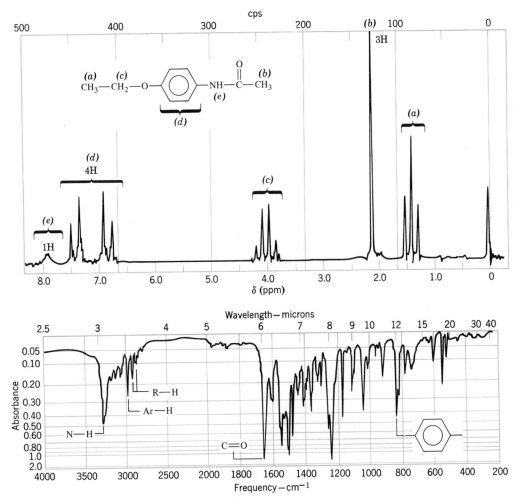

FIG. 19.1. The pmr and infrared spectra of phenacetin. (pmr spectrum courtesy of Varian Associates. IR spectrum courtesy of Sadtler Inc.)

(c)

$$\text{C}_6\text{H}_5\text{—CH}_2\text{—}\overset{\overset{\displaystyle O}{\|}}{\text{C}}\text{—O—CH}_2\text{CH}_3$$

$$\quad\quad\quad\quad\quad\;\; b \quad\quad\quad\;\; c \quad a$$

$$\underbrace{}_{d}$$

Interpretation:

a Triplet δ 1.2 (3H) $\quad -\overset{\overset{\displaystyle O}{\|}}{\text{C}}-$, 1740 cm^{-1}

b Singlet δ 3.5 (2H)

c Quartet δ 4.1 (2H)

d Multiplet δ 7.3 (5H)

(d)
$$\underset{\underset{a}{}}{Cl-}\underset{}{\overset{\overset{Cl}{|}}{C}H}-\underset{b}{COOH}$$

Interpretation:

    a Singlet $\delta 6.0$ $-OH$, $2500\text{-}2700$ cm^{-1}

    b Singlet $\delta 11.70$ $-\overset{\overset{O}{\|}}{C}-$, 1705 cm^{-1}

(e)
$$\underset{b}{Cl-CH_2}-\overset{\overset{O}{\|}}{\underset{c}{C}}-\underset{a}{OCH_2CH_3}$$

Interpretation:

    a Triplet $\delta 1.3$ $-\overset{\overset{O}{\|}}{C}-$, 1745 cm^{-1}

    b Singlet $\delta 4.0$

    c Quartet $\delta 4.2$

19.31

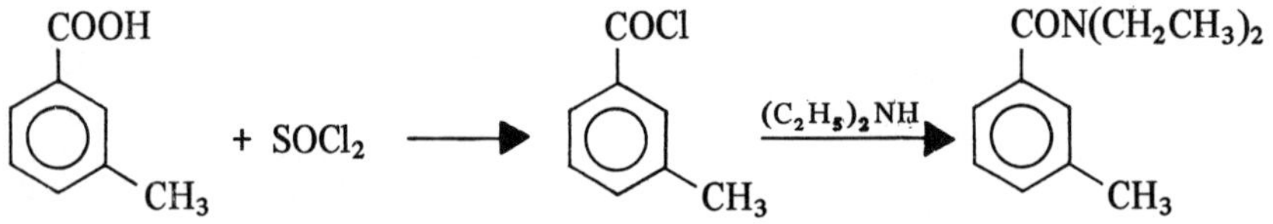

19.32

Alkyl groups are electron releasing; they help disperse the positive charge of an alkyl-ammonium salt and thereby help to stabilize it.

$$R\overset{..}{N}H_2 + H_3O^+ \longrightarrow R\rightarrow NH_3^+ + H_2O$$

*Stabilized by
electron-
releasing
alkyl group.*

Alkylamines, consequently, are somewhat stronger bases than ammonia.

    Amides, on the other hand, have acyl groups, $R-\overset{\overset{O}{\|}}{C}-$, attached to nitrogen, and acyl groups are electron withdrawing. They are especially electron withdrawing because of resonance contributions of the kind shown below,

$$R-\overset{\overset{:\overset{..}{O}}{\|}}{C}-\overset{..}{N}H_2 \longleftrightarrow R-\overset{\overset{:\overset{..}{O}:^-}{|}}{C}=\overset{+}{N}H_2$$

This kind of resonance also *stabilizes* the amide. The tendency of the acyl group to be electron withdrawing, however, *destabilizes* the conjugate acid of an amide and reactions such as the following do not take place to an appreciable extent.

$$\underset{\substack{\text{Stabilized}\\ \text{by}\\ \text{resonance.}}}{\overset{\displaystyle\overset{O}{\underset{\|}{}}}{R\ddot{C}-\overset{..}{N}H_2}} + H_3O^+ \xrightarrow{\quad X \quad} \underset{\substack{\text{Destabilized}\\ \text{by electron-}\\ \text{withdrawing}\\ \text{acyl group.}}}{\overset{\displaystyle\overset{O}{\underset{\|}{}}}{R\ddot{C}-NH_3^+}} + H_2O$$

19.33

(a) The conjugate base of an amide is stabilized by resonance.

$$R-\overset{\displaystyle :\overset{..}{O}}{\underset{\|}{C}}-\overset{..}{N}H_2 + :B^- \rightleftharpoons R-\overset{\displaystyle :\overset{..}{O}}{\underset{\|}{C}}-\overset{..}{\ddot{N}}H^- + BH$$

$$\updownarrow$$

$$R-\overset{\displaystyle :\overset{..}{O}:^-}{\underset{|}{C}}=\overset{..}{N}H$$

*This structure
is especially
stable because the
negative charge is
on oxygen.*

(b) The conjugate base of an imide is stabilized by an additional resonance structure,

$$R\overset{\displaystyle :\overset{..}{O}}{\underset{\|}{C}}-\overset{..}{N}H-\overset{\displaystyle \overset{..}{O}:}{\underset{\|}{C}}R + OH^- \longrightarrow R\overset{\displaystyle :\overset{..}{O}}{\underset{\|}{C}}-\overset{..}{\ddot{N}}-\overset{\displaystyle \overset{..}{O}:}{\underset{\|}{C}}R + H_2O$$

An imide

$$\updownarrow$$

$$R\overset{\displaystyle ^-:\overset{..}{O}:}{\underset{|}{C}}=\overset{..}{N}-\overset{\displaystyle \overset{..}{O}:}{\underset{\|}{C}}R$$

$$\updownarrow$$

$$R\overset{\displaystyle :\overset{..}{O}}{\underset{\|}{C}}-\overset{..}{N}=\overset{\displaystyle :\overset{..}{O}:^-}{\underset{|}{C}}R$$

19.34

(a) (1) $RO-OR \longrightarrow 2\,RO\cdot$ } Chain-initiating steps

(2) $CH_3\overset{O}{\underset{\|}{C}}SH + RO\cdot \longrightarrow RC\overset{O}{\underset{\|}{}}S\cdot + ROH$

(3) $CH_3\overset{O}{\underset{\|}{C}}S\cdot + CH_2=CHR \longrightarrow CH_3\overset{O}{\underset{\|}{C}}SCH_2\overset{\cdot}{C}HR$ } Chain-propagating steps

(4) $CH_3\overset{O}{\underset{\|}{C}}SCH_2\overset{\cdot}{C}HR + CH_3\overset{O}{\underset{\|}{C}}SH \longrightarrow CH_3\overset{O}{\underset{\|}{C}}SCH_2CH_2R + CH_3\overset{O}{\underset{\|}{C}}S\cdot$

(b)

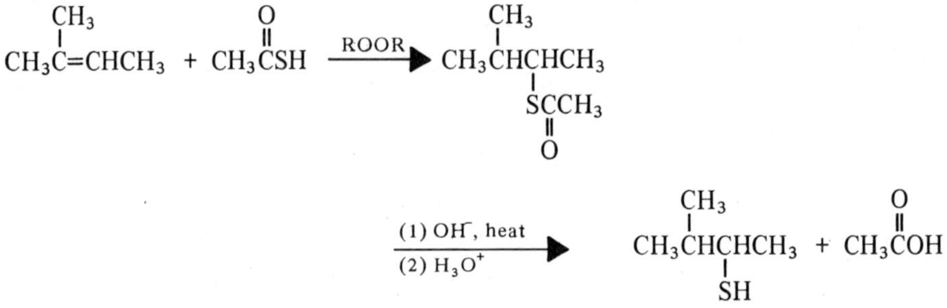

$$\text{CH}_3\text{C}=\text{CHCH}_3 + \text{CH}_3\overset{\text{O}}{\underset{\|}{\text{C}}}\text{SH} \xrightarrow{\text{ROOR}} \text{CH}_3\overset{\text{CH}_3}{\underset{\underset{\|}{\text{O}}}{\underset{|}{\text{CHCHCH}_3}}} $$

$$\xrightarrow[\text{(2) } \text{H}_3\text{O}^+]{\text{(1) OH}^-,\text{ heat}} \text{CH}_3\text{CHCHCH}_3 + \text{CH}_3\text{COH}$$

19.35

cis-4-Hydroxycyclohexane carboxylic acid can assume a boat conformation that permits lactone formation.

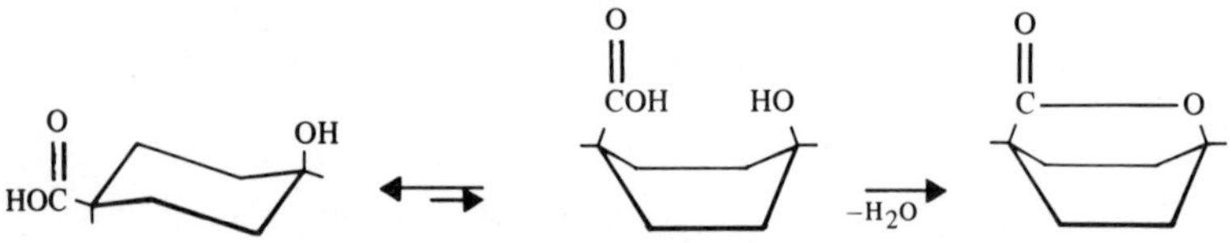

Neither of the chair conformations nor the boat form of *trans*-4-hydroxycyclohexane carboxylic acid places the −OH group and the −COOH group close enough together to permit lactonization.

19.36

(a)

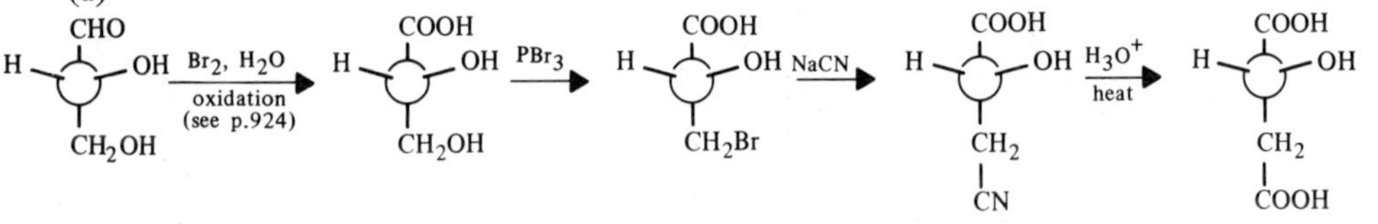

(−)−glyceric acid   (−)−3−bromo−2−hydroxypropanoic acid   ($C_4H_5NO_3$)   R−(+)−malic acid

19.37

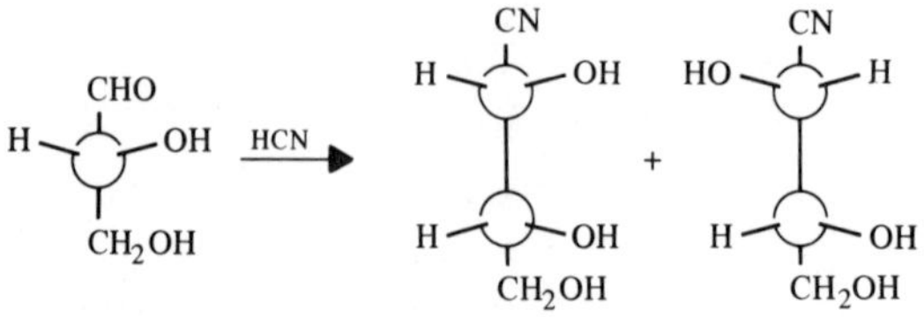

R−(+)−glyceraldehyde   **M**   **N**

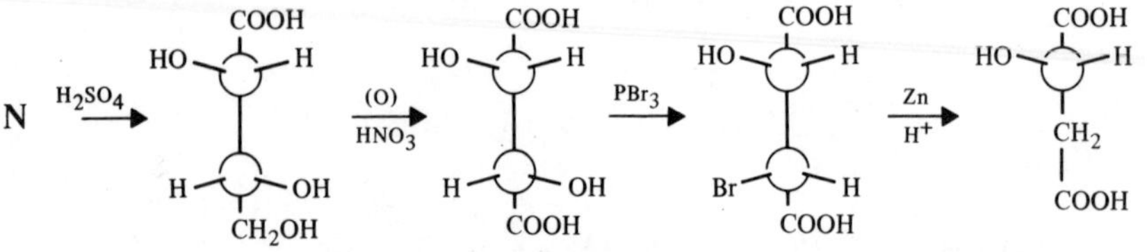

(−)−tartaric acid

(b) Replacement of either alcoholic —OH by a reaction that proceeds with inversion produces the same stereoisomer.

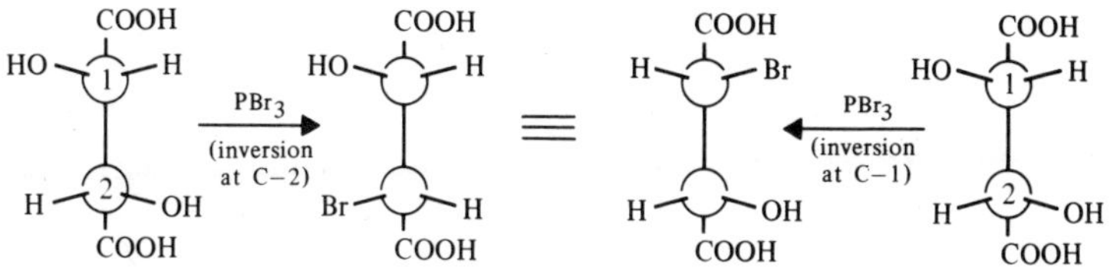

(c) Two. The stereoisomer given in (b) above and the one given below.

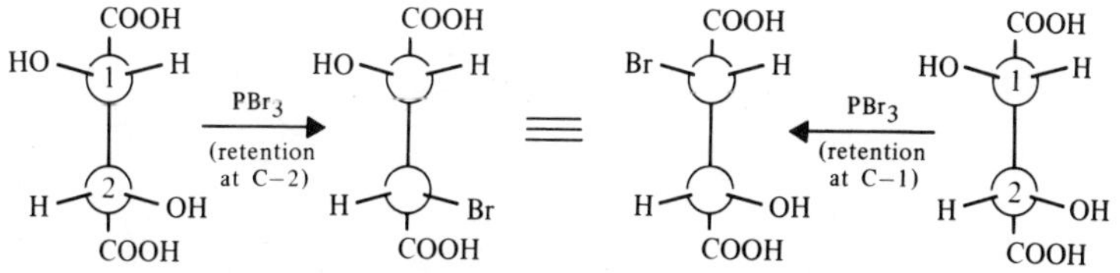

(d) It would have made no difference because treating either isomer (or both together) with zinc and acid produces (−)-malic acid.

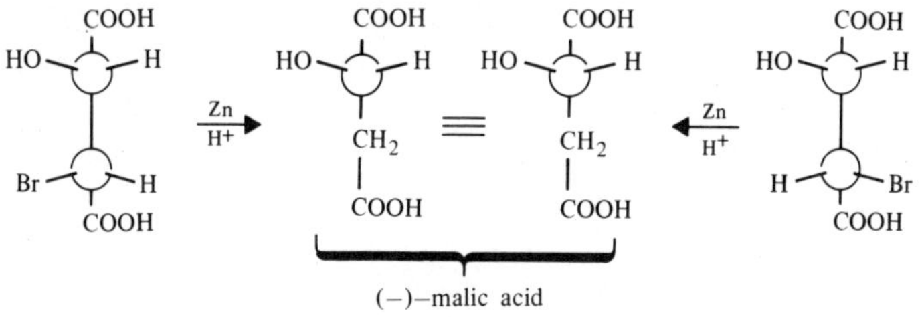

$(-)$-malic acid

19.38

(a) $CH_3O_2C-C{\equiv}C-CO_2CH_3$. This is a Diels-Alder reaction.

(b) H_2, Pd. The disubstituted double bond is less hindered than the tetrasubstituted double bond and hence is more reactive.

(c) $CH_2{=}CH-CH{=}CH_2$. Another Diels-Alder reaction.

(d) $LiAlH_4$

(e) $CH_3\overset{\overset{\displaystyle O}{\|}}{\underset{\underset{\displaystyle O}{\|}}{S}}-Cl$ and pyridine

(f) $CH_3CH_2S^-$

(g) OsO_4

(h) Raney Ni

(i) Base. This is an aldol condensation.

(j) C_6H_5Li (or C_6H_5MgBr) followed by H_3O^+

(k) H_3O^+. This is an acid-catalyzed rearrangement of an allylic alcohol.

(l) 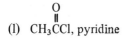 CH₃CCl, pyridine

(m) Heat. This is an acetate pyrolysis.

(n) O₃ followed by oxidation.

(o) Heat

19.39

(a)

Furan

Dimethylmaleic
anhydride

Cantharadin

(b) Cantharidin apparently undergoes dehydrogenation to the Diels-Alder adduct shown above and then the adduct spontaneously decomposes through a reverse Diels-Alder reaction to furan and dimethylmaleic anhydride. These results suggest that the attempted Diels-Alder synthesis fails because the position of equilibrium favors reactants rather than products.

20

SYNTHESIS AND REACTIONS OF β- DICARBONYL COMPOUNDS

20.1

(a) Step 1 $CH_3\underset{\underset{H}{|}}{CH}-\overset{\overset{O}{\|}}{C}OC_2H_5 \quad {}^-OC_2H_5 \quad \rightleftarrows \quad CH_3\ddot{\underset{..}{CH}}-\overset{\overset{O}{\|}}{C}OC_2H_5 \quad + \quad C_2H_5OH$

$$CH_3CH=\overset{\overset{O^-}{|}}{C}OC_2H_5$$

Step 2 $CH_3CH_2\overset{\nearrow O}{\underset{\searrow OC_2H_5}{C}} \quad + \quad {}^-:\underset{\underset{CH_3}{|}}{CH}COC_2H_5 \quad \rightleftarrows \quad CH_3CH_2\overset{\overset{O^-}{|}}{\underset{\underset{C_2H_5O}{|}}{C}}-\underset{\underset{CH_3}{|}}{CH}-\overset{\overset{O}{\|}}{C}OC_2H_5$

$$C_2H_5O^- \quad + \quad CH_3CH_2\overset{\overset{O}{\|}}{C}-\underset{\underset{CH_3}{|}}{CH}-\overset{\overset{O}{\|}}{C}OC_2H_5$$

Step 3 $CH_3CH_2\overset{\overset{O}{\|}}{C}-\underset{\underset{CH_3}{|}}{\overset{\overset{H}{|}}{C}}-\overset{\overset{O}{\|}}{C}OC_2H_5 \quad + \quad {}^-OC_2H_5 \quad \rightleftarrows \quad CH_3CH_2\overset{\overbrace{O \qquad O}^{-}}{\underset{\underset{CH_3}{|}}{C}=C=C}OC_2H_5$

$$+ \quad C_2H_5OH$$

(b) $CH_3CH_2\overset{\overset{O}{\|}}{C}\underset{\underset{CH_3}{|}}{CH}\overset{\overset{O}{\|}}{C}OC_2H_5 \quad + \quad CH_3CH_2\overset{\overset{OH}{|}}{C}=\underset{\underset{CH_3}{|}}{C}\overset{\overset{O}{\|}}{C}OC_2H_5$

20.2

(a) $C_2H_5O\overset{\overset{O}{\|}}{C}CH_2CH_2CH_2CH_2\overset{\overset{O}{\|}}{C}OC_2H_5 \quad \underset{+H^+}{\overset{-H^+}{\rightleftharpoons}} \quad \underset{\underset{CH_2\underline{\quad}CH_2}{|}}{\overset{C_2H_5O\overset{\curvearrowright}{\underset{CH_2}{C}}\overset{O}{\diagup}}{}}\,{}^-:CHCO_2C_2H_5$

$$C_2H_5O \overset{O^-}{\underset{}{\underset{\displaystyle}{}}} \overset{H}{\underset{}{}} \overset{O}{\overset{\|}{C}}OC_2H_5 \quad \rightleftharpoons \quad \overset{O}{\underset{}{}} \overset{H}{\underset{}{}} \overset{O}{\overset{\|}{C}}OC_2H_5 \; + \; C_2H_5O^-$$

$$\rightleftharpoons \quad \overset{\bar{}}{\overset{O \cdots O}{\underset{}{}}} COC_2H_5 \; + \; C_2H_5OH \quad \overset{H^+}{\longrightarrow} \quad \overset{O}{\underset{}{}} \overset{O}{\overset{\|}{C}}OC_2H_5 \; + \; \text{enol form}$$

(b)

$$\overset{O}{\underset{}{}} \overset{O}{\overset{\|}{C}}OC_2H_5$$

(c) To undergo a Dieckmann condensation, diethyl glutarate would have to form a highly strained four-membered ring.

20.3

$$CH_3\overset{O}{\overset{\|}{C}}OC_2H_5 \; + \; C_2H_5O^- \; \rightleftharpoons \; {:}CH_2\overset{O}{\overset{\|}{C}}OC_2H_5 \; + \; C_2H_5OH$$

$$C_6H_5\overset{O}{\overset{\|}{C}}OC_2H_5 \; + \; {:}CH_2\overset{O}{\overset{\|}{C}}OC_2H_5 \; \rightleftharpoons \; C_6H_5\overset{O^-}{\underset{OC_2H_5}{C}}-CH_2\overset{O}{\overset{\|}{C}}OC_2H_5$$

$$\rightleftharpoons \; C_6H_5\overset{O}{\overset{\|}{C}}CH_2\overset{O}{\overset{\|}{C}}OC_2H_5 \; + \; C_2H_5O^- \; \rightleftharpoons \; C_6H_5\overset{O \quad O}{\overset{\bar{}}{C}}{=}CH{=}\overset{}{C}OC_2H_5 \; + \; C_2H_5OH$$

$$\overset{H^+}{\longrightarrow} C_6H_5\overset{O}{\overset{\|}{C}}CH_2\overset{O}{\overset{\|}{C}}OC_2H_5$$

$$C_6H_5CH_2\overset{O}{\overset{\|}{C}}OC_2H_5 \; + \; C_2H_5O^- \; \rightleftharpoons \; C_6H_5\overset{\cdot\cdot}{C}H\overset{O}{\overset{\|}{C}}OC_2H_5 \; + \; C_2H_5OH$$

$$C_6H_5\overset{\cdot\cdot}{C}H\overset{O}{\overset{\|}{C}}OC_2H_5 \; + \; C_2H_5O\overset{O}{\overset{\|}{C}}OC_2H_5 \; \rightleftharpoons \; C_6H_5\overset{}{C}H \begin{matrix} \overset{O^-}{\underset{}{}}C_2H_5O-\overset{O^-}{\underset{}{C}}-OC_2H_5 \\ \overset{}{\underset{O}{C}}OC_2H_5 \end{matrix}$$

$$\rightleftharpoons \; C_6H_5\overset{}{C}H \begin{matrix} \overset{O}{\overset{\|}{C}}OC_2H_5 \\ \overset{}{\underset{O}{\overset{\|}{C}}}OC_2H_5 \end{matrix} \; + \; C_2H_5O^- \; \longrightarrow \; C_6H_5\overset{}{C}{:} \begin{matrix} \overset{O}{\overset{\|}{C}}OC_2H_5 \\ \overset{}{\underset{O}{\overset{\|}{C}}}OC_2H_5 \end{matrix} \; + \; C_2H_5OH$$

resonance
stabilized

$$\xrightarrow{\;H^+\;} C_6H_5\underset{\underset{O}{\overset{\|}{C}OC_2H_5}}{\overset{\overset{O}{\overset{\|}{C}OC_2H_5}}{\underset{|}{C}H}}$$

20.4

(a) $CH_3CH_2\overset{O}{\overset{\|}{C}}OC_2H_5 \; + \; C_2H_5O\overset{O}{\overset{\|}{C}}-\overset{O}{\overset{\|}{C}}OC_2H_5 \xrightarrow[\text{(2) H}^+]{\text{(1) NaOC}_2\text{H}_5} \; CH_3\underset{\underset{O\;\;O}{\overset{\|\;\;\|}{C-COC_2H_5}}}{\overset{\overset{O}{\overset{\|}{C}}OC_2H_5}{\underset{|}{C}H}}$

(b) $CH_3\overset{O}{\overset{\|}{C}}OC_2H_5 \; + \; H\overset{O}{\overset{\|}{C}}OC_2H_5 \xrightarrow[\text{(2) H}^+]{\text{(1) NaOC}_2\text{H}_5} \; H\overset{O}{\overset{\|}{C}}CH_2\overset{O}{\overset{\|}{C}}OC_2H_5$

20.5

(a) $+ \; H\overset{O}{\overset{\|}{C}}OC_2H_5 \xrightarrow[\text{(2) H}^+]{\text{(1) NaOC}_2\text{H}_5}$

(b) $CH_3CH_2\overset{O}{\overset{\|}{C}}CH_2CH_2CH_2\overset{O}{\overset{\|}{C}}OC_2H_5 \xrightarrow[\text{(2) H}^+]{\text{(1) NaOC}_2\text{H}_5}$

(c) $C_2H_5O_2CCH_2\underset{\underset{CH_3}{|}}{\overset{\overset{CH_3}{|}}{C}}CH_2CO_2C_2H_5 \; + \; C_2H_5O\overset{O}{\overset{\|}{C}}-\overset{O}{\overset{\|}{C}}OC_2H_5 \xrightarrow[\text{(2) H}^+]{\text{(1) NaOC}_2\text{H}_5}$

$\xrightarrow[\text{(2) H}^+]{\text{(1) NaOC}_2\text{H}_5}$

20.6

(a) $CH_3\overset{O}{\overset{\|}{C}}CH_2CH_2CH_2CH_2\overset{O}{\overset{\|}{C}}OC_2H_5 \; + \; {}^-OC_2H_5 \underset{(-C_2H_5OH)}{\rightleftarrows}$

$+ \; C_2H_5O^-$ (cont. on p. 212)

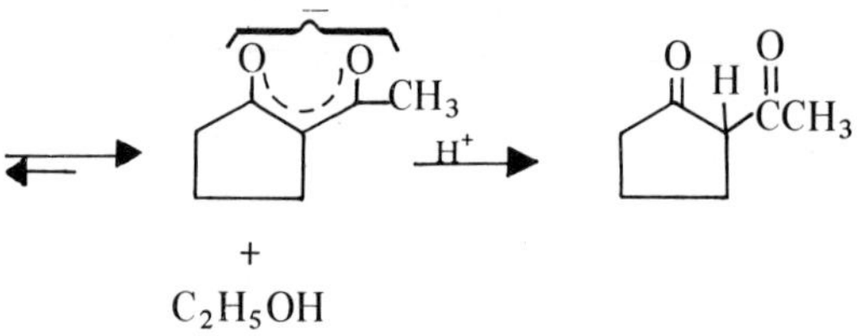

+

C_2H_5OH

(b) $CH_3\overset{O}{\overset{\|}{C}}CH_2CH_2CH_2\overset{O}{\overset{\|}{C}}OC_2H_5$ + $^-OC_2H_5$ $\rightleftharpoons$ (−C_2H_5OH)

$+$ $^-OC_2H_5$

$+$ C_2H_5OH

H^+

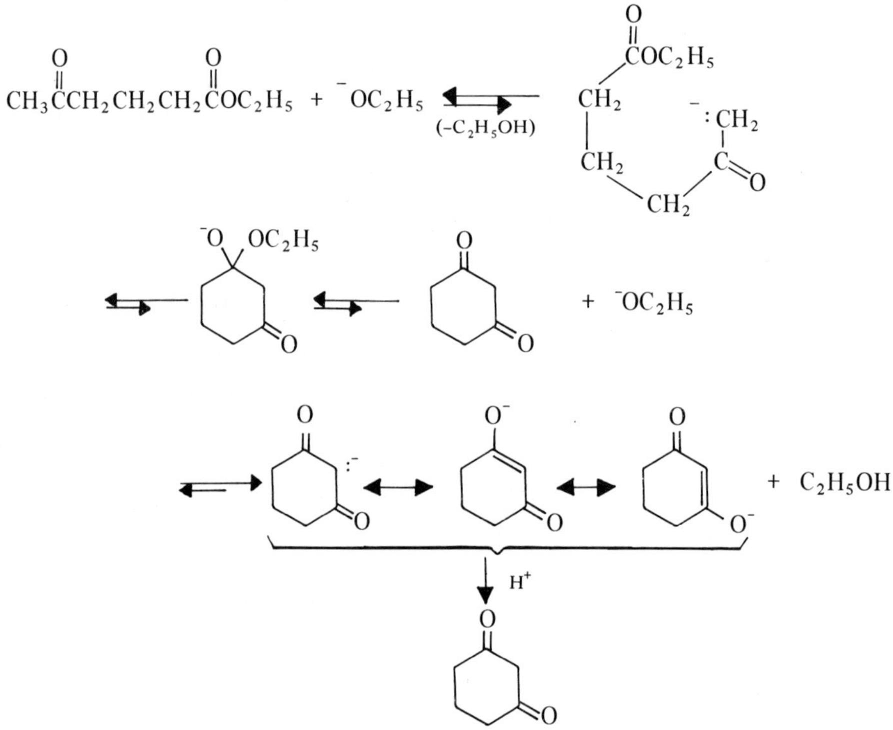

Ionization of a hydrogen alpha to the ketone group might, in the first step, also yield $CH_3\overset{O}{\overset{\|}{C}}\overset{..}{C}HCH_2CH_2\overset{O}{\overset{\|}{C}}OC_2H_5$. Cyclization of this enolate anion does not occur to any appreciable extent, however, because to do so it would yield a (highly-strained) four-membered ring.

20.7

(a) The iminium salt transfers hydrogen chloride to the enamine in the following way.

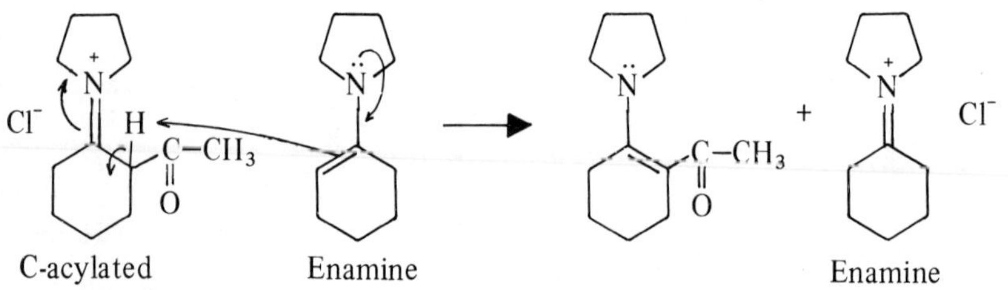

Cl^- $\longrightarrow$ $+$ Cl^-

C-acylated   Enamine   Enamine
iminium   hydrochloride
salt

(b) Since the enamine hydrochloride formed in this acid-base reaction is not susceptible to acylation, the overall yield of the enamine synthesis will be decreased.

20.8
These syntheses are easy to see if we work backward.

(a)

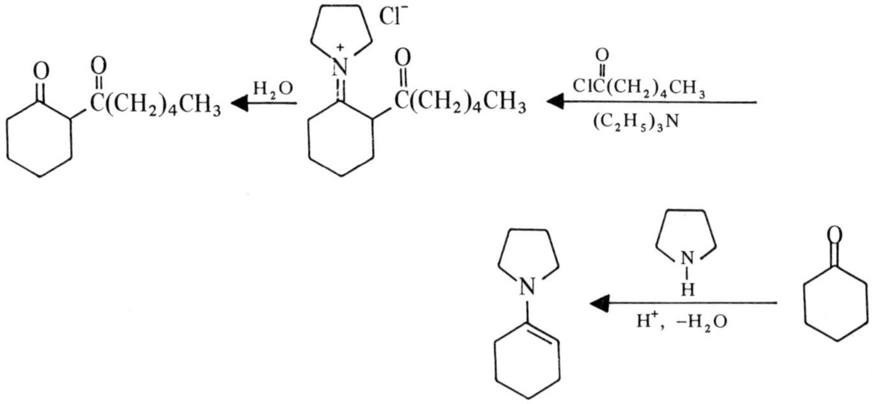

(b)

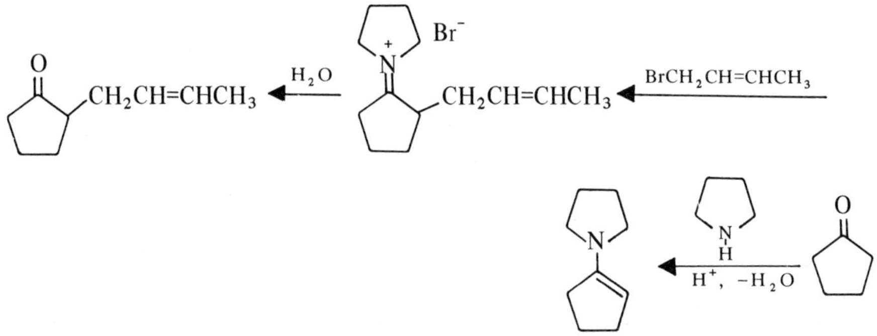

(c)

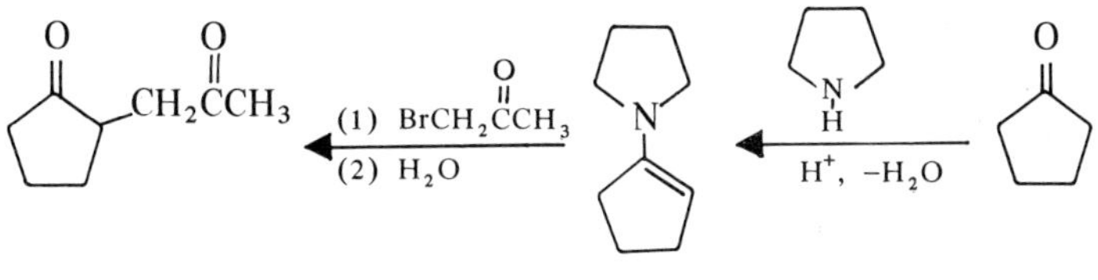

(d)

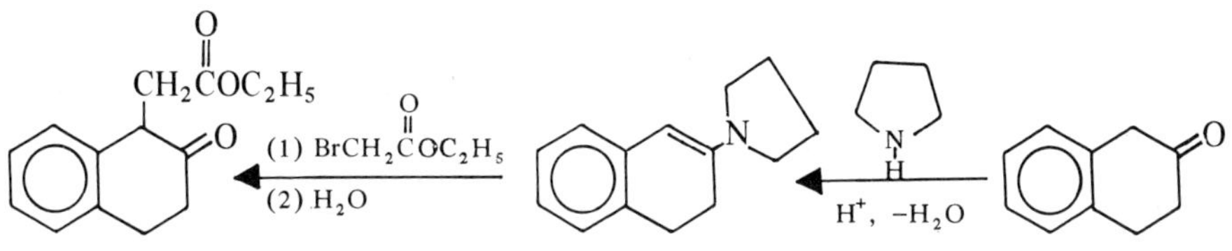

20.9
(a) Nucleophilic substitution, S_N2.

(b) The partially negative oxygen of sodioacetoacetic ester acts as the nucleophile.

$$\underset{\underset{CH_3C-CH-C-OC_2H_5}{}}{\overset{\overset{O\quad\quad O}{\parallel\quad\ddots\quad\parallel}}{}} \longleftrightarrow \underset{\underset{CH_3C=CH-C-OC_2H_5}{}}{\overset{\overset{O^-\quad\quad O}{\mid\quad\quad\parallel}}{}}$$

20.10

Again, working backward,

(a) $\underset{\underset{CH_3CCH_2CH_2CH_3}{}}{\overset{\overset{O}{\parallel}}{}} \xleftarrow[-CO_2]{heat} \underset{\underset{\underset{CH_3}{\mid}}{\underset{CH_2}{\mid}}}{\underset{CH_3CCH-COH}{\overset{\overset{O\quad\ O}{\parallel\ \ \parallel}}{}}} \xleftarrow[(2)\ H_3O^+]{(1)\ dil.\ NaOH,\ heat}$

$\underset{\underset{\underset{CH_3}{\mid}}{\underset{CH_2}{\mid}}}{\underset{CH_3CCHCOC_2H_5}{\overset{\overset{O\ \ O}{\parallel\ \ \parallel}}{}}} \xleftarrow[(2)\ CH_3CH_2Br]{(1)\ NaOC_2H_5} \underset{CH_3CCH_2COC_2H_5}{\overset{\overset{O\ \ \ \ O}{\parallel\ \ \ \ \parallel}}{}}$

(b) $\underset{\underset{\underset{CH_3}{\mid}}{\underset{\underset{CH_2}{\mid}}{\underset{CH_2}{\mid}}}}{\underset{CH_3CCHCH_2CH_2CH_3}{\overset{\overset{O}{\parallel}}{}}} \xleftarrow[-CO_2]{heat} \underset{\underset{\underset{CH_3}{\mid}}{\underset{\underset{CH_2}{\mid}}{\underset{CH_2}{\mid}}}}{\overset{\overset{\overset{CH_3}{\mid}}{\overset{CH_2}{\mid}}}{\underset{CH_3C-C-COOH}{\overset{\overset{O\ \ CH_2}{\parallel\ \ \mid}}{}}}} \xleftarrow[(2)\ H_3O^+]{(1)\ dil.\ NaOH,\ heat}$

$\underset{\underset{\underset{CH_3}{\mid}}{\underset{\underset{CH_2}{\mid}}{\underset{CH_2}{\mid}}}}{\overset{\overset{\overset{CH_3}{\mid}}{\overset{CH_2}{\mid}}}{\underset{CH_3C-C-COOC_2H_5}{\overset{\overset{O\ \ CH_2}{\parallel\ \ \mid}}{}}}} \xleftarrow[(2)\ CH_3CH_2CH_2Br]{(1)\ (CH_3)_3COK} \underset{\underset{\underset{CH_3}{\mid}}{\underset{CH_2}{\mid}}}{\underset{CH_3C-CHCOC_2H_5}{\overset{\overset{O\ \ \ \ O}{\parallel\ \ \ \ \parallel}}{}}}$

$\xleftarrow[(2)\ CH_3CH_2CH_2Br]{(1)\ NaOC_2H_5} \underset{CH_3CCH_2COC_2H_5}{\overset{\overset{O\ \ \ \ O}{\parallel\ \ \ \ \parallel}}{}}$

(c) $\underset{CH_3CCH_2CH_2C_6H_5}{\overset{\overset{O}{\parallel}}{}} \xleftarrow[-CO_2]{heat} \underset{\underset{\underset{C_6H_5}{\mid}}{\underset{CH_2}{\mid}}}{\underset{CH_3CCHCOH}{\overset{\overset{O\ \ O}{\parallel\ \ \parallel}}{}}} \xleftarrow[(2)\ H_3O^+]{(1)\ NaOH,\ heat} \underset{\underset{\underset{C_6H_5}{\mid}}{\underset{CH_2}{\mid}}}{\underset{CH_3CCHCOC_2H_5}{\overset{\overset{O\ \ O}{\parallel\ \ \parallel}}{}}}$

$\xleftarrow[(2)\ C_6H_5CH_2Br]{(1)\ NaOC_2H_5} \underset{CH_3CCH_2COC_2H_5}{\overset{\overset{O\ \ \ \ O}{\parallel\ \ \ \ \parallel}}{}}$

20.11

(a) The first alkylation produces ethyl methylacetoacetate; it then reacts with sodium ethoxide to produce an anion that can undergo a second alkylation.

$$CH_3\overset{\overset{\displaystyle O}{\|}}{C}-\overset{\overset{\displaystyle H}{|}}{\underset{\underset{\displaystyle CH_3}{|}}{C}}-\overset{\overset{\displaystyle O}{\|}}{C}OC_2H_5 \xrightarrow{\text{NaOC}_2\text{H}_5} CH_3\overset{\overset{\displaystyle O}{\|}}{C}-\overset{\overset{\displaystyle \cdot\cdot}{C}}{\underset{\underset{\displaystyle CH_3}{|}}{}}-\overset{\overset{\displaystyle O}{\|}}{C}OC_2H_5 \xrightarrow{\text{CH}_3\text{I}} CH_3\overset{\overset{\displaystyle O}{\|}}{C}-\overset{\overset{\displaystyle CH_3}{|}}{\underset{\underset{\displaystyle CH_3}{|}}{C}}-COOC_2H_5$$

Ethyl methyl-
acetoacetate

Ethyl dimethyl-
acetoacetate

(b) It favors monoalkylation.

(c) Methyl and ethyl halides present less steric hindrance to the attacking anion.

(d) Carry out the first alkylation with the larger alkyl halide, that is, with propyl bromide. This will minimize dialkylation in the first step.

20.12

(a) Reactivity is the same as with any S_N2 reaction. With primary halides substitution is highly favored, with secondary halides elimination competes with substitution, and with tertiary halides elimination is the exclusive course of reaction.

(b) Acetoacetic ester and 2-methylpropene.

(c) Bromobenzene is unreactive toward nucleophilic substitution (Cf. Section 17.11 of the text).

20.13

$$CH_3CH_2CH_2\overset{\overset{\displaystyle O}{\|}}{C}OC_2H_5 \xrightarrow[\text{(2) H}^+]{\text{(1) NaOC}_2\text{H}_5} CH_3CH_2CH_2\overset{\overset{\displaystyle O}{\|}}{C}\overset{\overset{\displaystyle O}{\|}}{\underset{\underset{\underset{\displaystyle CH_3}{|}}{CH_2}}{C}H}COC_2H_5$$

$$\xrightarrow[\text{(2) H}_3\text{O}^+]{\text{(1) NaOH, H}_2\text{O, heat}} CH_3CH_2CH_2\overset{\overset{\displaystyle O}{\|}}{C}\overset{\overset{\displaystyle O}{\|}}{\underset{\underset{\underset{\displaystyle CH_3}{|}}{CH_2}}{C}H}COH \xrightarrow[-\text{CO}_2]{\text{heat}} CH_3CH_2CH_2\overset{\overset{\displaystyle O}{\|}}{C}CH_2CH_2CH_3$$

20.14

The carboxyl group that is lost most readily is the one that is *beta* to the keto group (cf. page 797 of the text).

20.15

$$CH_3\overset{\overset{\displaystyle O}{\|}}{C}CH_2CH_2\overset{\overset{\displaystyle O}{\|}}{C}C_6H_5 \xleftarrow[-\text{CO}_2]{\text{heat}} CH_3\overset{\overset{\displaystyle O}{\|}}{C}\overset{\overset{\displaystyle O}{\|}}{\underset{\underset{\underset{\underset{\displaystyle C_6H_5}{|}}{C=O}}{CH_2}}{C}H}COH \xleftarrow[\text{(2) H}_3\text{O}^+]{\text{(1) OH}^-,\text{ H}_2\text{O, heat}}$$

$$\underset{\substack{\text{CH}_2\\|\\\text{C=O}\\|\\\text{C}_6\text{H}_5}}{\text{CH}_3\overset{\text{O}}{\overset{||}{\text{C}}}\text{CHC}\overset{\text{O}}{\overset{||}{\text{C}}}\text{OC}_2\text{H}_5} \xleftarrow[\text{(2)C}_6\text{H}_5\text{COCH}_2\text{Br}]{\text{(1) NaOC}_2\text{H}_5} \text{CH}_3\overset{\text{O}}{\overset{||}{\text{C}}}\text{CH}_2\overset{\text{O}}{\overset{||}{\text{C}}}\text{OC}_2\text{H}_5$$

20.16

$$\text{CH}_3\overset{\text{O}}{\overset{||}{\text{C}}}\text{CH}_2\overset{\text{O}}{\overset{||}{\text{C}}}\text{C}_6\text{H}_5 \xleftarrow[-\text{CO}_2]{\text{heat}} \underset{\substack{\text{C=O}\\|\\\text{C}_6\text{H}_5}}{\text{CH}_3\overset{\text{O}}{\overset{||}{\text{C}}}\text{CHC}\overset{\text{O}}{\overset{||}{\text{OH}}}} \xleftarrow[\text{(2) H}_3\text{O}^+]{\text{(1) OH}^-,\text{ H}_2\text{O, heat}}$$

$$\underset{\substack{\text{C=O}\\|\\\text{C}_6\text{H}_5}}{\text{CH}_3\overset{\text{O}}{\overset{||}{\text{C}}}\text{CHC}\overset{\text{O}}{\overset{||}{\text{C}}}\text{OC}_2\text{H}_5} \xleftarrow[\text{(2) C}_6\text{H}_5\text{COCl}]{\text{(1) NaH}} \text{CH}_3\overset{\text{O}}{\overset{||}{\text{C}}}\text{CH}_2\overset{\text{O}}{\overset{||}{\text{C}}}\text{OC}_2\text{H}_5$$

20.17
Here we acylate the dianion,

$$\text{CH}_3-\overset{\text{O}}{\overset{||}{\text{C}}}-\text{CH}_2-\overset{\text{O}}{\overset{||}{\text{C}}}\text{OC}_2\text{H}_5 \xrightarrow[\text{liq NH}_3]{2\text{KNH}_2} \text{:CH}_2-\overset{\text{O}}{\overset{||}{\text{C}}}-\overset{\cdot\cdot}{\text{CH}}-\overset{\text{O}}{\overset{||}{\text{C}}}\text{OC}_2\text{H}_5$$

$$\xrightarrow[\text{(2) NH}_4\text{Cl}]{\text{(1) C}_6\text{H}_5\overset{\text{O}}{\overset{||}{\text{C}}}\text{Cl}} \text{C}_6\text{H}_5\overset{\text{O}}{\overset{||}{\text{C}}}\text{CH}_2\overset{\text{O}}{\overset{||}{\text{C}}}\text{CH}_2\overset{\text{O}}{\overset{||}{\text{C}}}\text{OC}_2\text{H}_5$$

20.18
Working backward,

(a) $\text{CH}_3\text{CH}_2\text{CH}_2\text{CH}_2\text{COOH} \xleftarrow[-\text{CO}_2]{\text{heat}} \underset{\text{COOH}}{\overset{\text{COOH}}{\text{CH}_3\text{CH}_2\text{CH}_2\text{CH}}} \xleftarrow[\text{(2) H}_3\text{O}^+]{\text{(1) OH}^-,\text{ H}_2\text{O, heat}}$

$$\underset{\text{COOC}_2\text{H}_5}{\overset{\text{COOC}_2\text{H}_5}{\text{CH}_3\text{CH}_2\text{CH}_2\text{CH}}} \xleftarrow[\text{CH}_3\text{CH}_2\text{CH}_2\text{Br}]{\text{NaOC}_2\text{H}_5} \underset{\text{COOC}_2\text{H}_5}{\overset{\text{COOC}_2\text{H}_5}{\overset{|}{\underset{|}{\text{CH}_2}}}}$$

(b) $\underset{\text{CH}_3}{\overset{}{\text{CH}_3\text{CH}_2\text{CH}_2\text{CHCOOH}}} \xleftarrow[-\text{CO}_2]{\text{heat}} \underset{\text{CH}_3}{\overset{\text{CH}_3\text{CH}_2\text{CH}_2}{\text{C}}}\underset{\text{COOH}}{\overset{\text{COOH}}{}} \xleftarrow[\text{(2) H}_3\text{O}^+]{\text{(1) OH}^-,\text{ H}_2\text{O, heat}}$

$$\underset{\text{CH}_3}{\overset{\text{CH}_3\text{CH}_2\text{CH}_2}{\text{C}}}\underset{\text{COOC}_2\text{H}_5}{\overset{\text{COOC}_2\text{H}_5}{}} \xleftarrow[\text{NaOC}_2\text{H}_5]{\text{CH}_3\text{I}} \underset{\text{COOC}_2\text{H}_5}{\overset{\text{COOC}_2\text{H}_5}{\text{CH}_3\text{CH}_2\text{CH}_2\text{CH}}}$$

$$\xleftarrow[\text{NaOC}_2\text{H}_5]{\text{CH}_3\text{CH}_2\text{CH}_2\text{Br}} \quad \begin{array}{c} \text{COOC}_2\text{H}_5 \\ | \\ \text{CH}_2 \\ | \\ \text{COOC}_2\text{H}_5 \end{array}$$

(c) $\text{CH}_3\text{CHCH}_2\text{CH}_2\text{COOH} \xleftarrow[-\text{CO}_2]{\text{heat}} \text{CH}_3\text{CHCH}_2\text{CH} \xleftarrow[\text{(2) H}_3\text{O}^+]{\text{(1) OH}^-, \text{H}_2\text{O, heat}}$

with CH_3 below first and CH_3 / COOH, COOH on second

$$\text{CH}_3\text{CHCH}_2\text{CH} \begin{array}{c}\text{COOC}_2\text{H}_5 \\ \text{COOC}_2\text{H}_5\end{array} \xleftarrow[\text{CH}_3\text{CHCH}_2\text{Br}]{\text{NaOC}_2\text{H}_5} \begin{array}{c}\text{COOC}_2\text{H}_5 \\ | \\ \text{CH}_2 \\ | \\ \text{COOC}_2\text{H}_5\end{array}$$

20.19

$$\begin{array}{c}\text{CH}_3 \quad \text{O} \\ | \quad \quad || \\ \text{CH}_3-\text{CCH}_2\text{COC}_2\text{H}_5 \\ | \\ \text{CH} \\ \diagup \quad \diagdown \\ \text{O}=\text{C} \quad \quad \text{C}=\text{O} \\ | \quad \quad \quad | \\ \text{OC}_2\text{H}_5 \quad \text{OC}_2\text{H}_5 \end{array} \xrightarrow[\text{(2) H}_3\text{O}^+]{\text{(1) OH}^-, \text{H}_2\text{O, heat}} \begin{array}{c}\text{CH}_3 \quad \text{O} \\ | \quad \quad || \\ \text{CH}_3-\text{CCH}_2\text{COH} \\ | \\ \text{CH} \\ \diagup \quad \diagdown \\ \text{O}=\text{C} \quad \quad \text{C}=\text{O} \\ | \quad \quad \quad | \\ \text{OH} \quad \quad \text{OH} \end{array}$$

A malonic acid

$$\xrightarrow[-\text{CO}_2]{\text{heat}} \quad \begin{array}{c}\text{O} \quad \quad \text{CH}_3 \quad \text{O} \\ || \quad \quad \quad | \quad \quad || \\ \text{HOCCH}_2\text{CCH}_2\text{COH} \\ \quad \quad \quad | \\ \quad \quad \quad \text{CH}_3 \end{array}$$

20.20

(a) $\begin{array}{c}\text{H} \\ \diagdown \\ \quad \text{C}=\text{O} \\ \diagup \\ \text{H}\end{array} + \text{HN(CH}_3)_2 \underset{}{\overset{\text{H}^+}{\rightleftharpoons}} \text{CH}_2=\overset{+}{\text{N}}\begin{array}{c}\text{CH}_3 \\ \diagdown \\ \diagup \\ \text{CH}_3\end{array} + \text{H}_2\text{O}$

(b) $\begin{array}{c}\text{H} \\ \diagdown \\ \quad \text{C}=\text{O} \\ \diagup \\ \text{H}\end{array} + \text{H}-\text{N}\begin{array}{c}\\ \\ \end{array} \underset{}{\overset{\text{H}^+}{\rightleftharpoons}} \text{CH}_2=\overset{+}{\text{N}}\begin{array}{c}\\ \\ \end{array} + \text{H}_2\text{O}$

(c)
$$\begin{array}{c}H\\ \\ H\end{array}C{=}O + HN\begin{array}{c}CH_3\\ \\ CH_3\end{array} \underset{}{\overset{H^+}{\rightleftharpoons}} CH_2{=}\overset{+}{N}\begin{array}{c}CH_3\\ \\ CH_3\end{array} + H_2O$$

20.21

(a) $CH_3CH_2CH_2\overset{O}{\overset{\|}{C}}\overset{O}{\overset{\|}{CH}COC_2H_5}$ $\underset{(2)\,H^+}{\overset{(1)\,NaOC_2H_5}{\longleftarrow}}$ $CH_3CH_2CH_2\overset{O}{\overset{\|}{C}}OC_2H_5$
with substituent $\overset{}{\underset{CH_3}{CH_2}}$

(b) $CH_3CH_2CH_2\overset{O}{\overset{\|}{C}}CH_2CH_2CH_3$ $\underset{-CO_2}{\overset{heat}{\longleftarrow}}$ $CH_3CH_2CH_2\overset{O}{\overset{\|}{C}}\overset{O}{\overset{\|}{CH}COH}$
with substituent $\overset{}{\underset{CH_3}{CH_2}}$

$\underset{(2)\,H_3O^+}{\overset{(1)\,OH^-,\,H_2O,\,heat}{\longleftarrow}}$ Product of (a)

(c) $C_6H_5\overset{CH_3}{\underset{}{CH}COOH}$ $\underset{-CO_2}{\overset{heat}{\longleftarrow}}$ $\begin{array}{c}CH_3\quad COOH\\ \diagdown\;\diagup\\ C\\ \diagup\;\diagdown\\ C_6H_5\quad COOH\end{array}$ $\underset{(2)\,H_3O^+}{\overset{(1)\,OH^-,\,H_2O,\,heat}{\longleftarrow}}$

$\begin{array}{c}CH_3\quad COOC_2H_5\\ \diagdown\;\diagup\\ C\\ \diagup\;\diagdown\\ C_6H_5\quad COOC_2H_5\end{array}$ $\underset{CH_3I}{\overset{NaOC_2H_5}{\longleftarrow}}$ $C_6H_5{-}\overset{COOC_2H_5}{\underset{COOC_2H_5}{CH}}$

$\underset{(2)\,H^+}{\overset{(1)\,C_2H_5O\overset{O}{\overset{\|}{C}}OC_2H_5}{\underset{NaOC_2H_5}{\longleftarrow}}}$ $C_6H_5CH_2\overset{O}{\overset{\|}{C}}OC_2H_5$

(d) $CH_3CH_2\overset{O}{\overset{\|}{CH}COC_2H_5}$ $\underset{(2)\,H^+}{\overset{(1)\,C_2H_5O\overset{O}{\overset{\|}{C}}{-}\overset{O}{\overset{\|}{C}}OC_2H_5}{\underset{NaOC_2H_5}{\longleftarrow}}}$ $CH_3CH_2CH_2\overset{O}{\overset{\|}{C}}OC_2H_5$
with substituent $\overset{O}{\overset{\|}{C}}{-}\overset{O}{\overset{\|}{C}}OC_2H_5$

(e)
$$CH_3CH_2CH_2\overset{O}{\underset{\|}{C}}-\overset{O}{\underset{\|}{C}}OC_2H_5 \xleftarrow[C_2H_5OH]{H^+} CH_3CH_2CH_2\overset{O}{\underset{\|}{C}}-\overset{O}{\underset{\|}{C}}OH$$

$$\xleftarrow[-CO_2]{heat} CH_3CH_2\underset{\underset{O}{\overset{\|}{C}}-\overset{O}{\underset{\|}{C}}OH}{CHCOOH} \xleftarrow[(2)\,H_3O^+]{(1)\,OH^-,\,H_2O,\,heat} \text{Product of (d)}$$

(f)
$$C_6H_5\underset{\underset{O}{\overset{\|}{C}}H}{CHCOC_2H_5} \xleftarrow[\substack{NaOC_2H_5 \\ (2)\,H^+}]{(1)\,H\overset{O}{\overset{\|}{C}}OC_2H_5} C_6H_5CH_2\overset{O}{\overset{\|}{C}}OC_2H_5$$

(g)

(h)

Product of (g)

(i)

20.22

(a)
$$CH_3\overset{O}{\overset{\|}{C}}-\underset{\underset{CH_3}{|}}{\overset{CH_3}{\overset{|}{C}}}-CH_3 \xleftarrow{Zn,\,H^+} CH_3\overset{O}{\overset{\|}{C}}-\underset{\underset{CH_3}{|}}{\overset{CH_3}{\overset{|}{C}}}-CH_2Br \xleftarrow{PBr_3} CH_3\overset{O}{\overset{\|}{C}}-\underset{\underset{CH_3}{|}}{\overset{CH_3}{\overset{|}{C}}}-CH_2OH$$

$$CH_3\overset{O}{\overset{\|}{C}}-\underset{\underset{CH_3}{|}}{\overset{CH_3}{\overset{|}{C}}}-COOC_2H_5 \xleftarrow[NaOC_2H_5]{CH_3I} CH_3\overset{O}{\overset{\|}{C}}CH_2\overset{O}{\overset{\|}{C}}OC_2H_5$$

is also present...

$$CH_3\overset{O}{\overset{\|}{C}}-\underset{\underset{CH_3}{|}}{\overset{}{C}H}-COOC_2H_5 \xleftarrow[NaOC_2H_5]{CH_3I} CH_3\overset{O}{\overset{\|}{C}}CH_2\overset{O}{\overset{\|}{C}}OC_2H_5$$

(b) $CH_3\overset{O}{\overset{\|}{C}}CH_2CH_2CH_2CH_3 \xleftarrow[-CO_2]{heat} CH_3\overset{O}{\overset{\|}{C}}\overset{O}{\overset{\|}{C}}HCOH \xleftarrow[(2)H_3O^+]{(1)\ OH^-,\ H_2O,\ heat}$
with the middle structure bearing $CH_2-CH_2-CH_3$ branch.

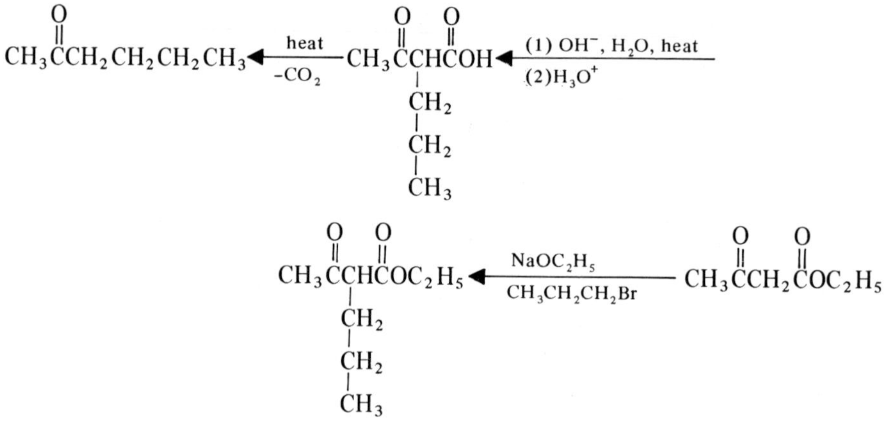

$CH_3\overset{O}{\overset{\|}{C}}\overset{O}{\overset{\|}{C}}HCOC_2H_5 \xleftarrow[CH_3CH_2CH_2Br]{NaOC_2H_5} CH_3\overset{O}{\overset{\|}{C}}CH_2\overset{O}{\overset{\|}{C}}OC_2H_5$
with $CH_2-CH_2-CH_3$ branch.

(c) $CH_3\overset{O}{\overset{\|}{C}}CH_2CH_2\overset{O}{\overset{\|}{C}}CH_3 \xleftarrow[-CO_2]{heat} CH_3\overset{O}{\overset{\|}{C}}\overset{O}{\overset{\|}{C}}HCOH \xleftarrow[(2)\ H_3O^+]{(1)\ OH^-,\ H_2O,\ heat}$
with $CH_2-C(=O)-CH_3$ branch.

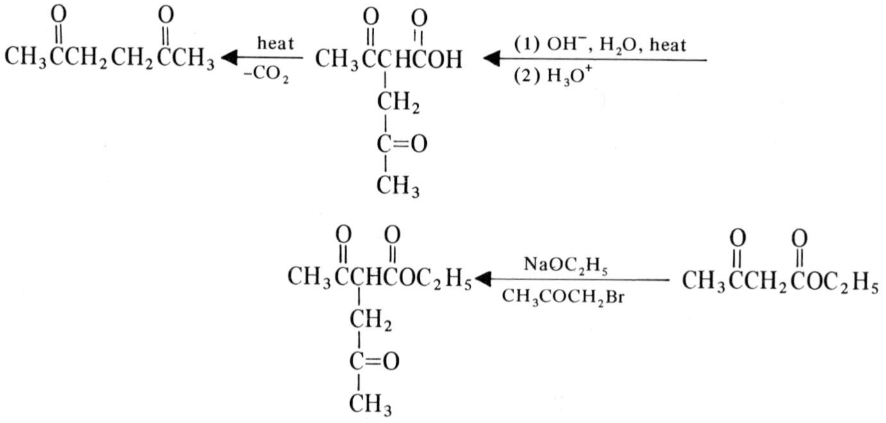

$CH_3\overset{O}{\overset{\|}{C}}\overset{O}{\overset{\|}{C}}HCOC_2H_5 \xleftarrow[CH_3COCH_2Br]{NaOC_2H_5} CH_3\overset{O}{\overset{\|}{C}}CH_2\overset{O}{\overset{\|}{C}}OC_2H_5$
with $CH_2-C(=O)-CH_3$ branch.

(d) $CH_3\overset{OH}{\overset{|}{C}}HCH_2CH_2COOH \xleftarrow{NaBH_4} CH_3\overset{O}{\overset{\|}{C}}CH_2CH_2\overset{O}{\overset{\|}{C}}OH \xleftarrow[-CO_2]{heat}$

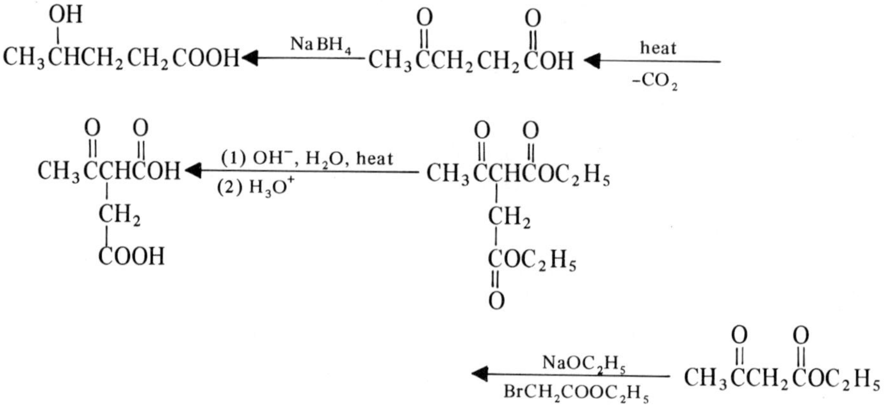

$CH_3\overset{O}{\overset{\|}{C}}\overset{O}{\overset{\|}{C}}HCOH \xleftarrow[(2)\ H_3O^+]{(1)\ OH^-,\ H_2O,\ heat} CH_3\overset{O}{\overset{\|}{C}}\overset{O}{\overset{\|}{C}}HCOC_2H_5$
left structure with CH_2-COOH branch; right structure with $CH_2-C(=O)OC_2H_5$ branch.

$\xleftarrow[BrCH_2COOC_2H_5]{NaOC_2H_5} CH_3\overset{O}{\overset{\|}{C}}CH_2\overset{O}{\overset{\|}{C}}OC_2H_5$

(e) $CH_3\overset{OH}{\overset{|}{C}}HCHCH_2OH \xleftarrow[(2)\ H^+]{(1)\ LiAlH_4} CH_3\overset{O}{\overset{\|}{C}}\overset{O}{\overset{\|}{C}}HCOC_2H_5$
left structure with C_2H_5 branch; right structure with C_2H_5 branch.

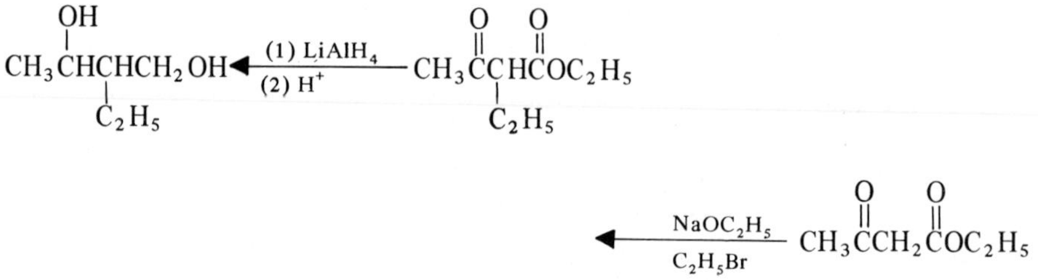

$\xleftarrow[C_2H_5Br]{NaOC_2H_5} CH_3\overset{O}{\overset{\|}{C}}CH_2\overset{O}{\overset{\|}{C}}OC_2H_5$

(f) $\underset{\underset{\text{OH}}{|}}{CH_3CH}\underset{}{CH_2}\underset{\underset{\text{OH}}{|}}{CH}C_6H_5 \xleftarrow{\text{NaBH}_4} \underset{\underset{O}{||}}{CH_3C}CH_2\underset{\underset{O}{||}}{C}C_6H_5 \longleftarrow$ cf. Problem 20.16

(g) $C_6H_5\underset{\underset{\text{OH}}{|}}{CH}CH_2\underset{\underset{\text{OH}}{|}}{CH}CH_2CH_2OH \xleftarrow[\text{(2) H}^+]{\text{(1) LiAlH}_4} C_6H_5\underset{\underset{O}{||}}{C}CH_2\underset{\underset{O}{||}}{C}CH_2\underset{\underset{O}{||}}{C}OC_2H_5$

$\longleftarrow$ cf. Problem 20.17

20.23

(a) $CH_3CH_2\underset{\underset{\text{CH}_3}{|}}{CH}COOH \xleftarrow{-CO_2} \underset{CH_3}{\overset{CH_3CH_2}{>}}C\underset{COOH}{\overset{COOH}{<}} \xleftarrow[\text{(2) H}_3\text{O}^+]{\text{(1) OH}^-,\ \text{H}_2\text{O, heat}}$

$\underset{CH_3}{\overset{CH_3CH_2}{>}}C\underset{COOC_2H_5}{\overset{COOC_2H_5}{<}} \xleftarrow[\text{NaOC}_2\text{H}_5]{\text{CH}_3\text{I}} CH_3CH_2CH\overset{COOC_2H_5}{\underset{COOC_2H_5}{<}}$

$\xleftarrow[\text{NaOC}_2\text{H}_5]{\text{CH}_3\text{CH}_2\text{Br}} \underset{\underset{\text{COOC}_2\text{H}_5}{|}}{\overset{\overset{\text{COOC}_2\text{H}_5}{|}}{CH_2}}$

(b) $CH_3\underset{\underset{\text{CH}_3}{|}}{CH}CH_2CH_2CH_2OH \xleftarrow[\text{(2) H}^+]{\text{(1) LiAlH}_4} CH_3\underset{\underset{\text{CH}_3}{|}}{CH}CH_2CH_2COOH$

(from Problem 20.18 c)

(c) $CH_3CH_2\underset{\underset{\text{CH}_2\text{OH}}{|}}{CH}CH_2OH \xleftarrow[\text{(2) H}^+]{\text{(1) LiAlH}_4} CH_3CH_2CH\overset{COOC_2H_5}{\underset{COOC_2H_5}{<}} \longleftarrow$ cf. p. 804

(d) $HOCH_2CH_2CH_2CH_2OH \xleftarrow[\text{(2) H}^+]{\text{(1) LiAlH}_4} HOOCCH_2CH_2COOH \xleftarrow[-CO_2]{\text{heat}}$

$\underset{HOOC}{\overset{HOOC}{>}}CHCH_2COOH \xleftarrow{\text{HCl, heat}} \underset{C_2H_5OOC}{\overset{C_2H_5OOC}{>}}CHCH_2COOC_2H_5$

$\longleftarrow \underset{\underset{\text{COOC}_2\text{H}_5}{|}}{\overset{\overset{\text{COOC}_2\text{H}_5}{|}}{CH_2}} + NaOC_2H_5 + BrCH_2COOC_2H_5$

20.24

The following reaction took place,

$$CH_3\overset{O}{\overset{\|}{C}}CH_2\overset{O}{\overset{\|}{C}}OC_2H_5 + BrCH_2CH_2CH_2Br \xrightarrow{NaOC_2H_5} BrCH_2CH_2CH_2CH\begin{smallmatrix} CH_3 \\ | \\ C=O \\ \\ COC_2H_5 \\ \| \\ O \end{smallmatrix}$$

$$\xrightarrow[(-H^+)]{NaOC_2H_5} C_2H_5OOC-C\overset{CH_3}{\underset{CH_2-CH_2}{\overset{|}{\underset{\|}{C}}}}\overset{\overset{|}{C}}{\overset{O^-}{\diagdown}}CH_2-Br \longrightarrow$$

Perkin's ester

$$C_2H_5O\overset{O}{\overset{\|}{C}}-C\overset{CH_3}{\underset{CH_2-CH_2}{\overset{|}{\underset{}{C}}}}\overset{C-O}{\diagup}CH_2$$

$$\xrightarrow[(2)\ H_3O^+]{(1)\ OH^-,\ H_2O,\ heat} HOOC-C\overset{CH_3}{\underset{CH_2-CH_2}{\overset{|}{\underset{}{C}}}}\overset{C-O}{\diagup}CH_2$$

Perkin's acid

20.25

(a) $BrCH_2CH_2Br + \overset{COOC_2H_5}{\underset{COOC_2H_5}{\overset{|}{CH_2}}} + NaOC_2H_5 \longrightarrow$

$$\left[BrCH_2CH_2-\overset{COOC_2H_5}{\underset{COOC_2H_5}{\overset{|}{CH}}}\right] \xrightarrow[(-H^+)]{NaOC_2H_5} \left[BrCH_2CH_2-\overset{COOC_2H_5}{\underset{COOC_2H_5}{\overset{|}{C}}}\bar{:}\right]$$

$$\longrightarrow \begin{smallmatrix} CH_2 \\ | \\ | \\ CH_2 \end{smallmatrix}\!\!\!\diagup\!\!\!C\!\!\!\diagup\!\!\!\overset{CO_2C_2H_5}{\underset{CO_2C_2H_5}{}} \xrightarrow[\substack{(2)\ H_3O^+ \\ (3)\ heat,\ -CO_2}]{(1)\ OH^-,\ H_2O,\ heat} \triangleright\!\!-COOH$$

(b) $2NaCH(CO_2C_2H_5)_2 + BrCH_2CH_2CH_2Br$

$$\longrightarrow \overset{C_2H_5O_2C}{\underset{C_2H_5O_2C}{}}\!\!\!\diagdown\!\!\!H-\overset{}{C}CH_2CH_2CH_2\overset{}{C}-H\!\!\!\diagup\!\!\!\overset{CO_2C_2H_5}{\underset{CO_2C_2H_5}{}}$$

A

$$\xrightarrow[Br_2]{NaOC_2H_5} \left[\overset{C_2H_5O_2C}{\underset{C_2H_5O_2C}{}}\!\!\!\diagdown\!\!\!H-\overset{}{C}CH_2CH_2CH_2\overset{}{C}-Br\!\!\!\diagup\!\!\!\overset{CO_2C_2H_5}{\underset{CO_2C_2H_5}{}}\right] \xrightarrow{NaOC_2H_5}$$

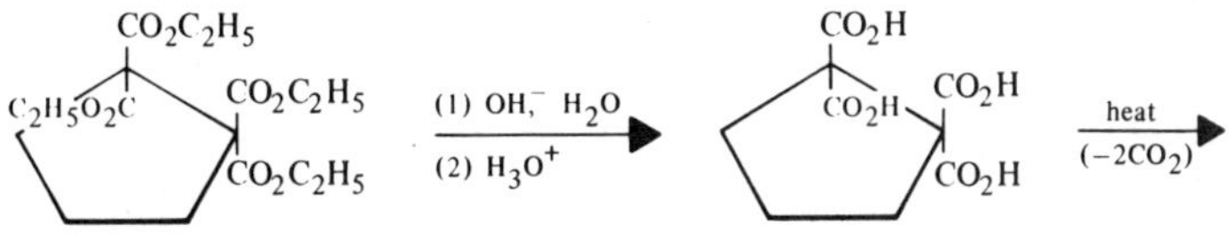

B

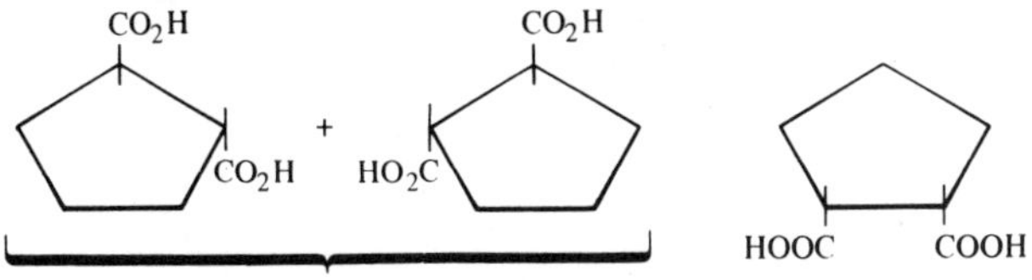

D racemic modification

E *meso*-compound

(c) $BrCH_2CH_2CH_2CH_2Br \xrightarrow{NaCH(CO_2C_2H_5)_2} BrCH_2CH_2CH_2CH_2CH(CO_2C_2H_5)_2$

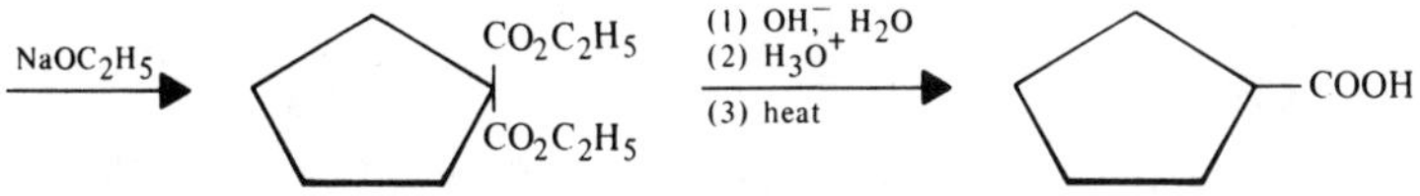

20.26

The α-hydrogens of a thiol ester are more acidic than those of an ordinary ester, and the RS^- group is a better leaving group than RO^-. (See pages 772 and 773 of the text for further explanation.)

20.27

(a) $CH_2(COOC_2H_5)_2 + {}^-OC_2H_5 \rightleftharpoons {}^-:CH_2(COOC_2H_5)_2 + C_2H_5OH$

$$C_6H_5CH{=}CH{-}\overset{O}{\overset{\|}{C}}OC_2H_5 + {}^-:CH(COOC_2H_5)_2 \rightleftharpoons C_6H_5\underset{CH(COOC_2H_5)_2}{CH}CH{=}\overset{O^-}{C}OC_2H_5$$

$$\xrightleftharpoons{+H^+} C_6H_5\underset{CH(COOC_2H_5)_2}{CH}CH_2\overset{O}{\overset{\|}{C}}OC_2H_5$$

(b) $CH_3NH_2 + CH_2{=}CH{-}COCH_3 \rightleftharpoons CH_3{-}\overset{H}{\underset{H}{\overset{+}{N}}}{-}CH_2{-}CH{=}COCH_3$

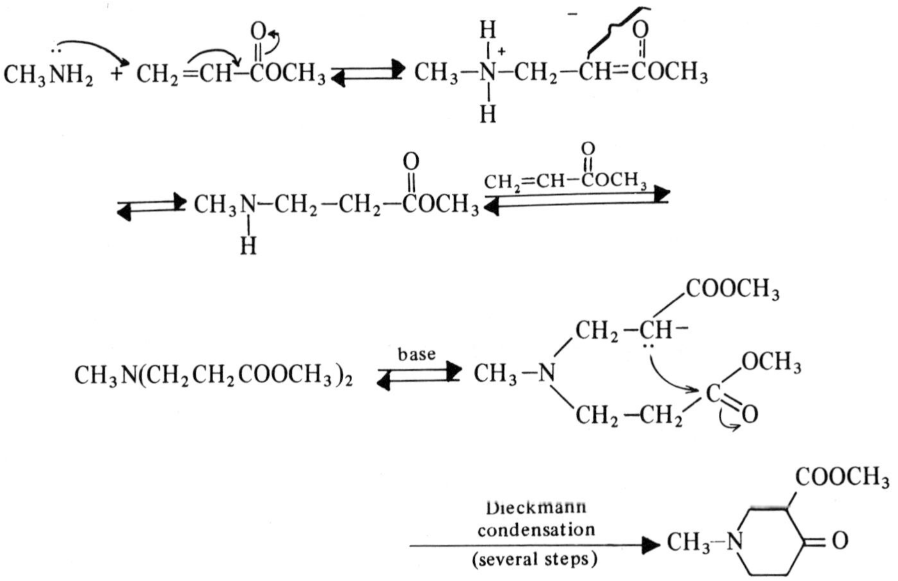

$\rightleftharpoons CH_3\underset{H}{N}{-}CH_2{-}CH_2{-}COCH_3 \xleftarrow{CH_2{=}CH{-}COCH_3}$

$CH_3N(CH_2CH_2COOCH_3)_2 \xrightleftharpoons{\text{base}} CH_3{-}N \begin{cases} CH_2{-}CH{-}COOCH_3 \\ CH_2{-}CH_2{-}C{\overset{OCH_3}{\underset{O}{}}} \end{cases}$

$\xrightarrow[\substack{\text{Dieckmann} \\ \text{condensation} \\ \text{(several steps)}}]{} CH_3{-}N \quad \text{(ring)} \quad COOCH_3, {=}O$

(c) $\underset{\underset{CH(CO_2C_2H_5)_2}{\overset{CH_3}{|}}}{CH_3{-}C}{-}CH_2{-}COC_2H_5 + C_2H_5O^- \rightleftharpoons \underset{\underset{CH(CO_2C_2H_5)_2}{\overset{CH_3}{|}}}{CH_3{-}C}{-}CH{=}COC_2H_5$

$+ C_2H_5OH$

$\underset{\underset{CH(CO_2C_2H_5)_2}{\overset{CH_3}{|}}}{CH_3{-}C}{-}CH{=}COC_2H_5 \rightleftharpoons \underset{\overset{CH_3}{|}}{CH_3{-}C}{=}CH{-}COC_2H_5 + {:}CH(CO_2C_2H_5)_2$

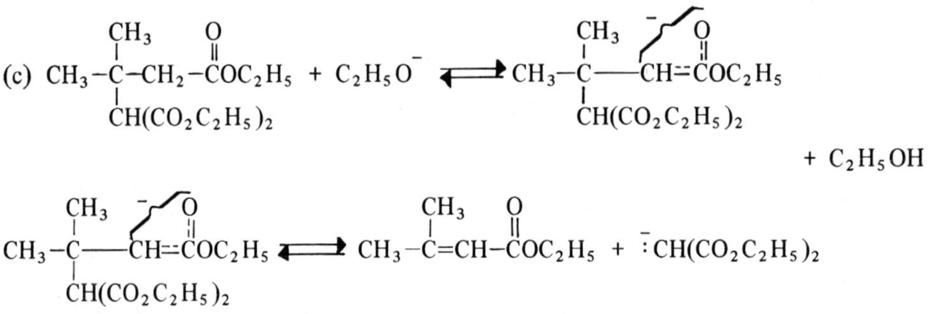

The Michael reaction is reversible and the reaction just given is an example of a reverse Michael reaction.

(d)

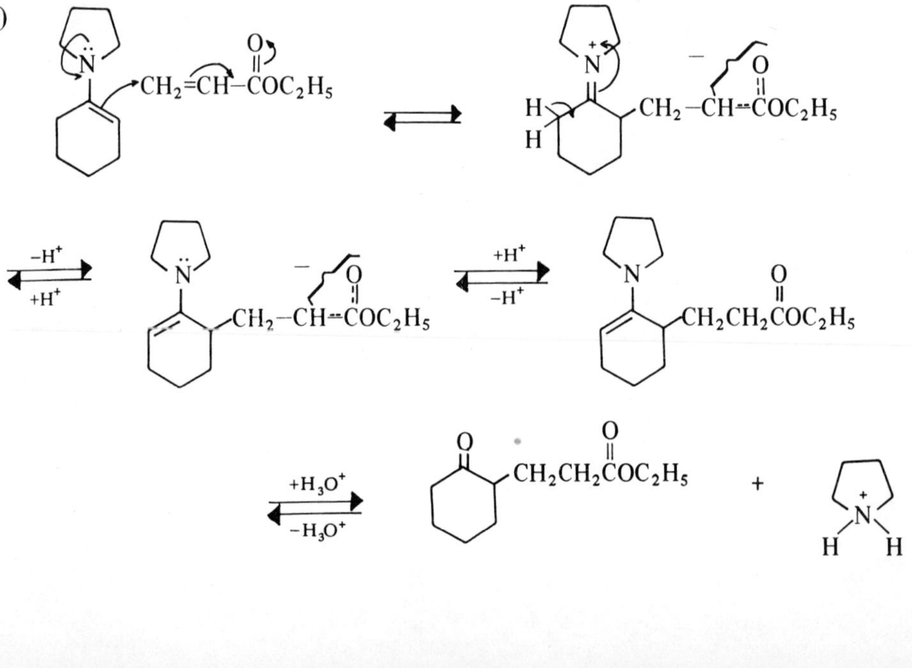

$\xrightleftharpoons[+H^+]{-H^+}$

$\xrightleftharpoons[-H^+]{+H^+}$

$\xrightleftharpoons[-H_3O^+]{+H_3O^+}$

(e) This one is a real challenge.

20.28

Two reactions take place. The first is a normal Knoevenagel condensation,

Then the α,β-unsaturated diketone reacts with a second mole of the active methylene compound in a Michael addition.

20.29

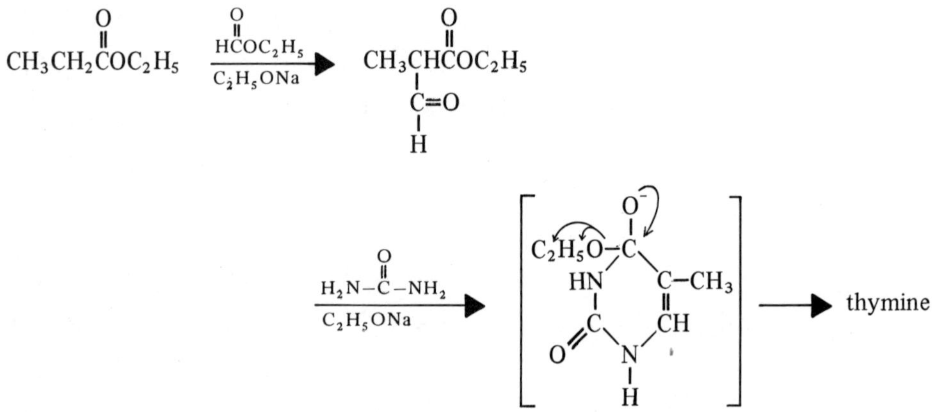

$$CH_3CH_2\overset{O}{\overset{\|}{C}}OC_2H_5 \xrightarrow[C_2H_5ONa]{H\overset{O}{\overset{\|}{C}}OC_2H_5} CH_3\underset{\underset{H}{\overset{|}{\underset{C=O}{|}}}}{CH}\overset{O}{\overset{\|}{C}}OC_2H_5$$

$$\xrightarrow[C_2H_5ONa]{H_2N-\overset{O}{\overset{\|}{C}}-NH_2} \left[\begin{array}{c} C_2H_5O-C \\ HN \\ C \\ O \end{array} \right] \longrightarrow \text{thymine}$$

20.30

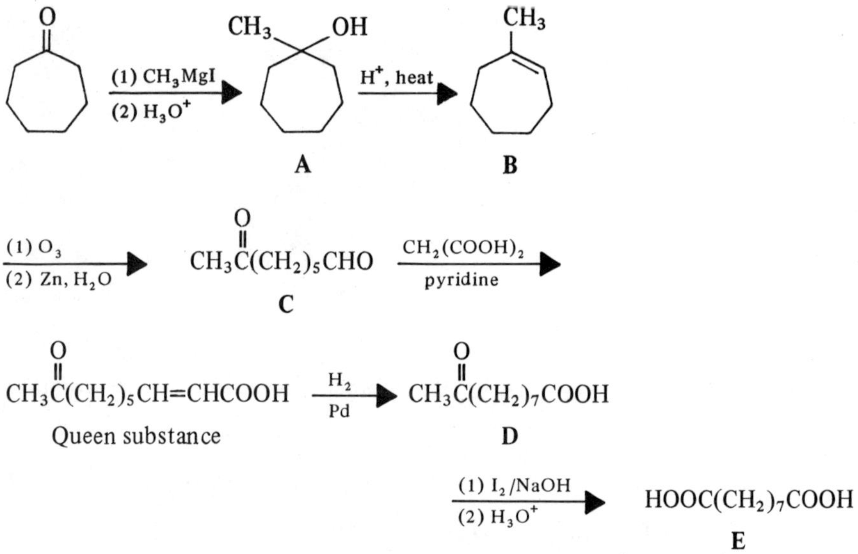

$$\xrightarrow[(2)\ Zn,\ H_2O]{(1)\ O_3} CH_3\overset{O}{\overset{\|}{C}}(CH_2)_5CHO \xrightarrow[\text{pyridine}]{CH_2(COOH)_2}$$

C

$$CH_3\overset{O}{\overset{\|}{C}}(CH_2)_5CH=CHCOOH \xrightarrow[Pd]{H_2} CH_3\overset{O}{\overset{\|}{C}}(CH_2)_7COOH$$

Queen substance   D

$$\xrightarrow[(2)\ H_3O^+]{(1)\ I_2/NaOH} HOOC(CH_2)_7COOH$$

E

20.31

$$CH_2=\underset{CH_3}{\overset{|}{C}}-CH=CH_2 + HBr \longrightarrow CH_3\underset{CH_3}{\overset{|}{C}}=CHCH_2Br$$

F

$$Na^+ \ ^-:\underset{CO_2C_2H_5}{\overset{\overset{O}{\overset{\|}{C}}CH_3}{CH}} \longrightarrow CH_3\underset{CH_3}{\overset{|}{C}}=CHCH_2\underset{CO_2C_2H_5}{\overset{\overset{O}{\overset{\|}{C}}CH_3}{CH}} \xrightarrow[(2)\ H_3O^+,\ (3)\ heat]{(1)\ dil.\ NaOH}$$

G

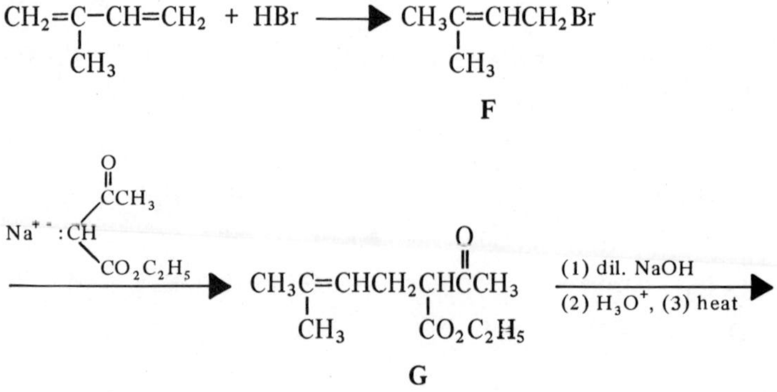

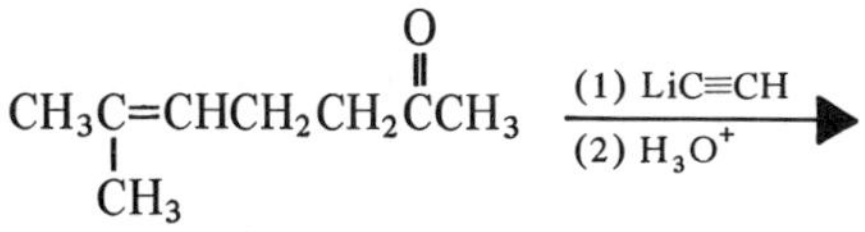

$$CH_3C=CHCH_2CH_2\overset{\overset{O}{\|}}{C}CH_3 \xrightarrow[\text{(2) } H_3O^+]{\text{(1) LiC} \equiv \text{CH}}$$

with CH₃ substituent below the first carbon

H

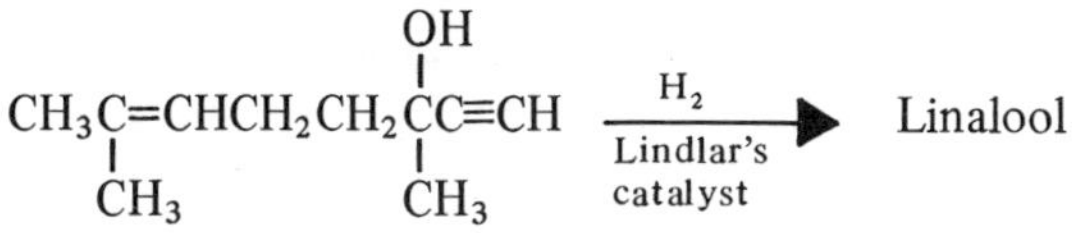

$$CH_3C=CHCH_2CH_2\overset{\overset{OH}{|}}{C}C \equiv CH \xrightarrow[\substack{\text{Lindlar's} \\ \text{catalyst}}]{H_2} \text{Linalool}$$

with CH₃ substituents below

I

20.32

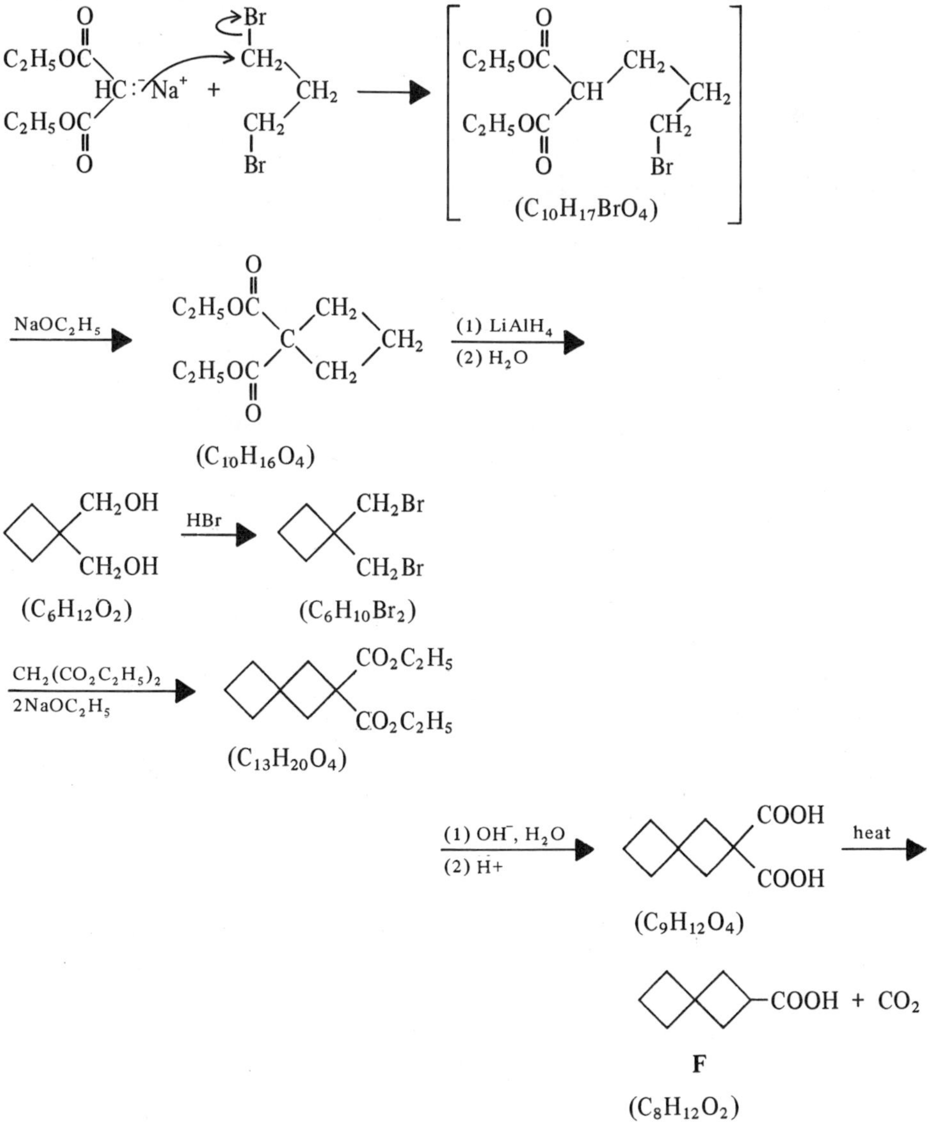

$(C_{10}H_{17}BrO_4)$

$(C_{10}H_{16}O_4)$

$(C_6H_{12}O_2)$ → $(C_6H_{10}Br_2)$

$$\xrightarrow[\text{2NaOC}_2\text{H}_5]{\text{CH}_2(\text{CO}_2\text{C}_2\text{H}_5)_2}$$

$(C_{13}H_{20}O_4)$

$$\xrightarrow[\text{(2) H+}]{\text{(1) OH}^-, H_2O} $$

$(C_9H_{12}O_4)$ $\xrightarrow{\text{heat}}$

F

$(C_8H_{12}O_2)$

20.33

(a) $ClCH_2COOC_2H_5$ + $C_2H_5O^-$ $\rightleftharpoons$ $Cl-\overset{..}{\underset{..}{C}}HCOOC_2H_5$ + C_2H_5OH

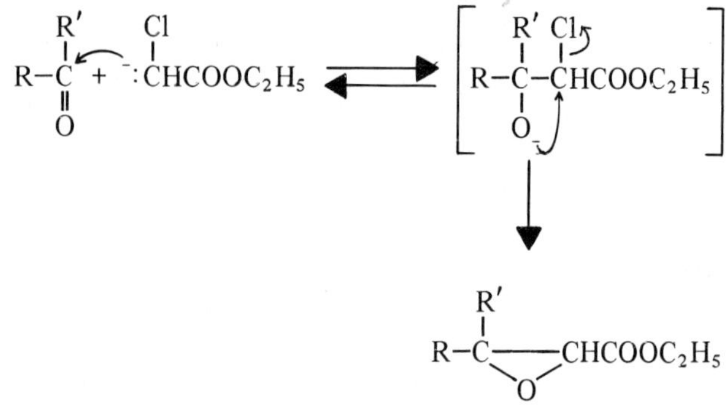

(b) Decarboxylation of the epoxy acid gives an enol anion which, on protonation, gives an aldehyde.

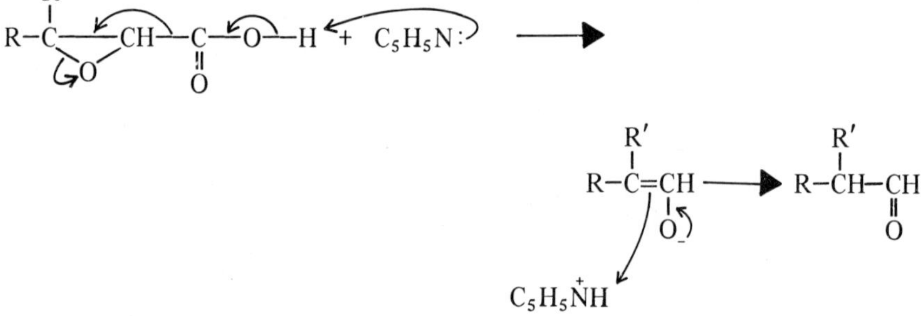

(c)

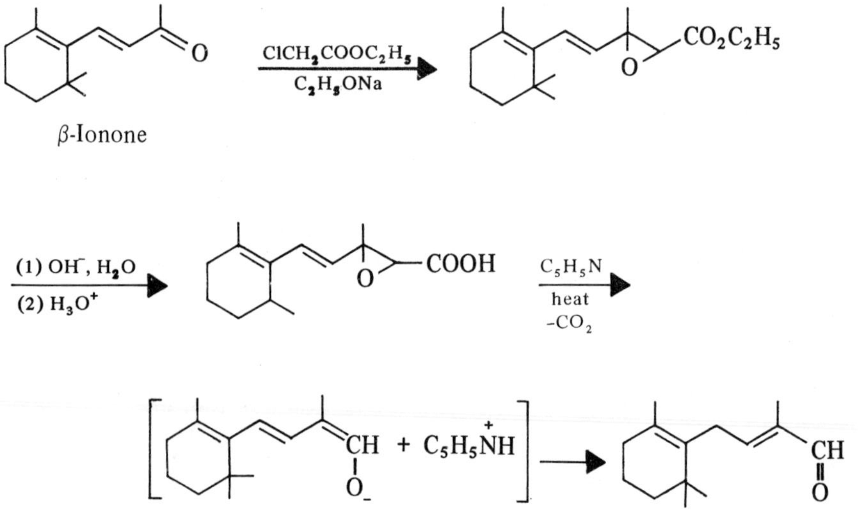

β-Ionone

20.34

(a) 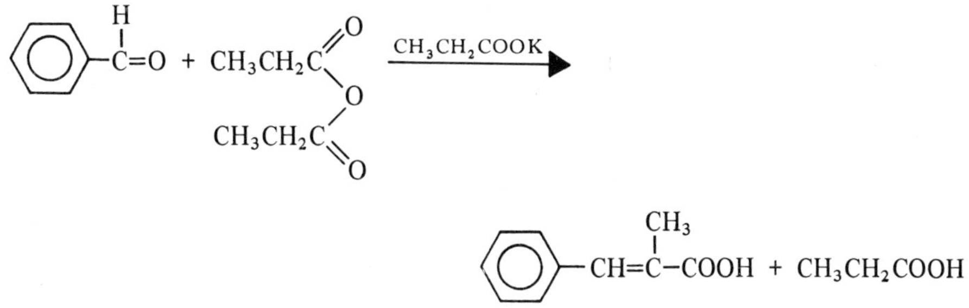 C₆H₅—CH(H)=O + CH₃CH₂C(=O)—O—C(=O)CH₂CH₃ →[CH₃CH₂COOK]→

$$\text{C}_6\text{H}_5-\text{CH}=\overset{\displaystyle \text{CH}_3}{\underset{}{\text{C}}}-\text{COOH} + \text{CH}_3\text{CH}_2\text{COOH}$$

(b) 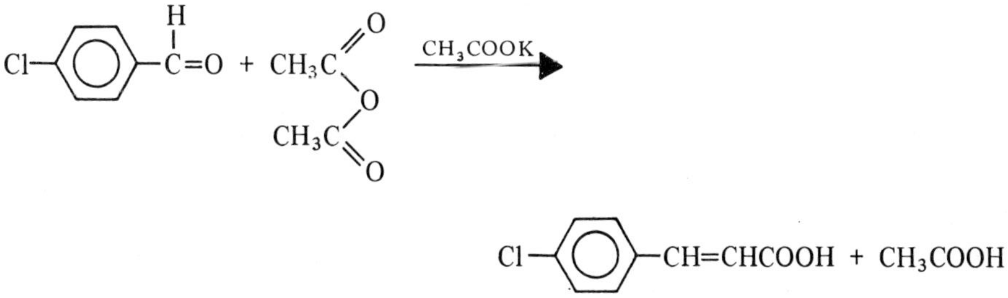 Cl—C₆H₄—CH(H)=O + CH₃C(=O)—O—C(=O)CH₃ →[CH₃COOK]→

$$\text{Cl}-\text{C}_6\text{H}_4-\text{CH}=\text{CHCOOH} + \text{CH}_3\text{COOH}$$

20.35

(a) $\text{CH}_2=\overset{\displaystyle \text{CH}_3}{\underset{}{\text{C}}}-\text{COOCH}_3$

(b) KMnO_4, OH^-, then H_3O^+

(c) CH_3OH, H^+

(d) CH_3ONa, then H^+

(e) and (f) 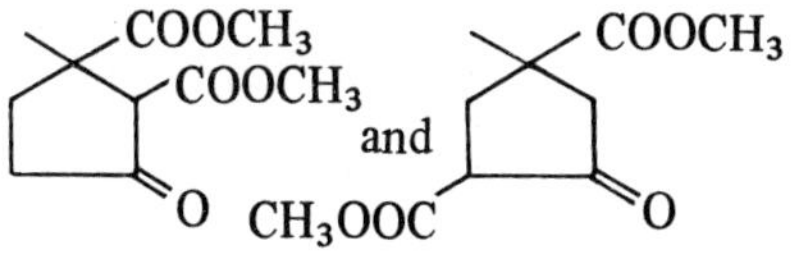 and

(g) OH^-, H_2O, then H_3O^+

(h) heat ($-\text{CO}_2$)

(i) CH_3OH, H^+

(j)

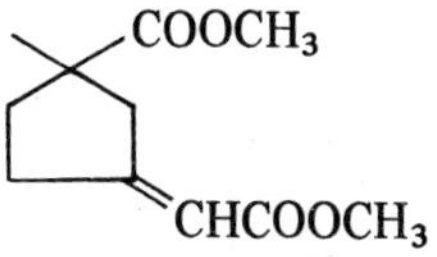

(k) H_2, Pt

(l) CH_3ONa, then H^+

(m) $2 \text{ NaNH}_2 + 2 \text{ CH}_3\text{I}$

20.36

(a) A Grignard synthesis.

(b) An organozinc compound, CH_3OOCCH_2ZnBr.

(c) The organozinc compound adds to the carbonyl group of a ketone (or aldehyde) but it does not add to the carbonyl group of an ester. Hence the Reformatsky reagent does not react with itself (or with other ester groups).

(d) Aldehydes and ketones are more reactive toward nucleophilic attack than esters are because esters are stabilized by the type of resonance shown below.

$$\underset{\substack{\| \\ O}}{R-C-OR} \longleftrightarrow \underset{\substack{| \\ O^-}}{R-C=O^+R}$$

Organozinc compounds are less reactive than organomagnesium compounds because zinc is less electropositive. Hence organozinc compounds (Reformatsky reagents) add to the more reactive carbonyl groups of aldehydes and ketones, but unlike organomagnesium compounds (Grignard reagents) they do not react at the less reactive carbonyl groups of esters.

(e) $C_6H_5CHO + CH_3CHCOOCH_3 \xrightarrow[\text{(2) } H_3O^+]{\text{(1) Zn, ether}}$
                         |
                        Br

$$\underset{\substack{| \\ OH}}{C_6H_5\underset{\substack{| \\ }}{CH}CHCOOCH_3} \xrightarrow[\text{heat}]{H^+} C_6H_5CH=\underset{\substack{| \\ CH_3}}{C}COOCH_3 \xrightarrow[Pt]{H_2}$$

where the central carbon bears a CH_3 group above.

$$C_6H_5CH_2\underset{\substack{| \\ CH_3}}{CH}COOCH_3$$

20.37

If we look at the molecule as consisting of the following pieces,

we can begin to see how it might be constructed from acetone, malonic ester, and two moles of benzaldehyde.

We can begin by carrying out a double Claisen-Schmidt condensation using acetone, and two moles of benzaldehyde.

$$2\ C_6H_5\underset{\substack{| \\ H}}{C}=O + CH_3-\underset{\substack{\| \\ O}}{C}-CH_3 \xrightarrow{OH^-} C_6H_5CH=CH-\underset{\substack{\| \\ O}}{C}-CH=CHC_6H_5$$

Then we carry out a double Michael addition, followed by hydrolysis and decarboxylation.

21 AMINES

21.1

Because it has two phenyl groups attached to the amine nitrogen, diphenylamine is stabilized by resonance to a much greater extent than aniline. In addition to structures involving different Kekulé structures for the phenyl groups, structures such as the following delocalize the nitrogen electron pair into both benzene rings:

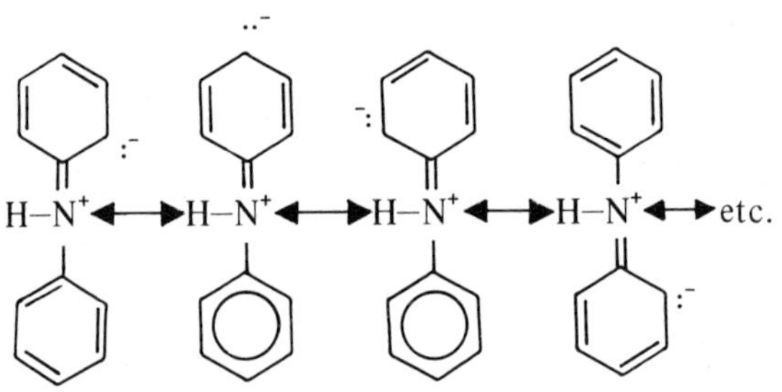

The diphenylammonium ion does not have this type of resonance stabilization. This greater resonance stabilization of diphenylamine means that the reaction,

$$(C_6H_5)_2\overset{..}{N}H + H_2O \longrightarrow (C_6H_5)_2NH_2^+ + OH^-$$

is even more endothermic than the reaction,

$$C_6H_5\overset{..}{N}H_2 + H_2O \longrightarrow C_6H_5NH_3^+ + OH^-$$

and that diphenylamine, consequently, is a weaker base.

21.2

The electron-releasing groups (CH_3- and CH_3O-) of p-toluidine and p-anisidine stabilize the cations that are produced:

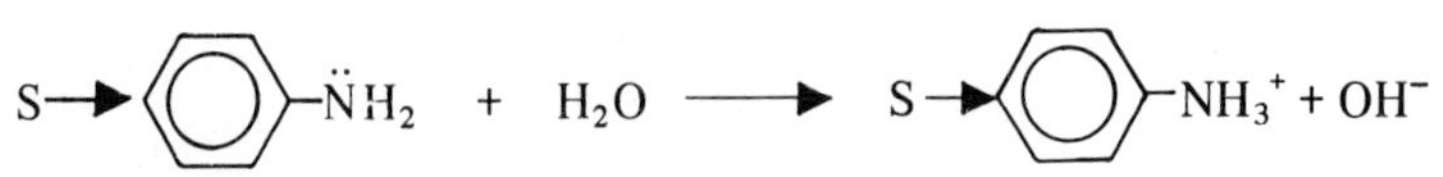

(S = CH_3- or CH_3O-)

S releases electrons and stabilizes the product of the acid-base reaction.

Greater stabilization of the product ($S-C_6H_4NH_3^+$) relative to the reactant ($S-C_6H_4NH_2$) means that these reactions will be less endothermic than the corresponding reaction of aniline. *p*-Toluidine and *p*-anisidine, consequently, are stronger bases.

With *p*-chloroaniline and *p*-nitroaniline the substituents (Cl– and NO_2-) are electron-withdrawing. These groups destabilize the cations that are the products of an acid-base reaction:

$$S \longleftarrow \langle\bigcirc\rangle - \ddot{N}H_2 \ + \ H_2O \ \longrightarrow \ S \longleftarrow \langle\bigcirc\rangle - NH_3^+ \ + \ OH^-$$

(S = Cl or NO_2)

S withdraws electrons and destabilizes the product of the acid-base reaction.

The reactions, therefore, are more endothermic than the corresponding reaction of aniline, and *p*-chloroaniline and *p*-nitroaniline are weaker bases.

21.3

Dissolve both compounds in ether and extract with aqueous HCl. This gives an ether layer that contains cyclohexane and an aqueous layer that contains cyclohexylammonium chloride. Cyclohexane may then be recovered from the ether layer by distillation. Cyclohexylamine may be recovered from the aqueous layer by adding aqueous NaOH (to convert cyclohexylammonium chloride to cyclohexylamine) and then by ether extraction and distillation

$$C_6H_{12} \ + \ C_6H_{11}NH_2$$
(in ether)

$\Big| H_3O^+Cl^-/H_2O$

ether layer
C_6H_{12}
(evaporate ether
and distill)

aqueous layer
$C_6H_{11}NH_3^+ \ Cl^-$

$\Big| OH^-$

$C_6H_{11}NH_2$
(extract into ether
and distill)

21.4

We begin by dissolving the mixture in a water-immiscible organic solvent such as CH_2Cl_2 or ether. Then, extractions with aqueous acids and bases allow us to separate the components. (We separate *p*-cresol from benzoic acid by taking advantage of benzoic acid's solubility in the more weakly basic aqueous $NaHCO_3$, whereas, *p*-cresol requires the more strongly basic, aqueous NaOH (cf., Problem 16.21, page 625 of the text)).

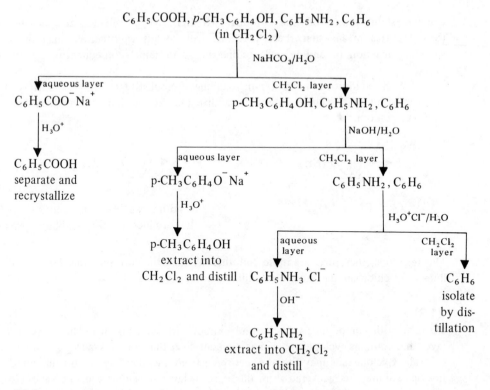

$C_6H_5COOH, p\text{-}CH_3C_6H_4OH, C_6H_5NH_2, C_6H_6$
(in CH_2Cl_2)

NaHCO$_3$/H$_2$O

aqueous layer
$C_6H_5COO^-Na^+$

CH$_2$Cl$_2$ layer
$p\text{-}CH_3C_6H_4OH, C_6H_5NH_2, C_6H_6$

H$_3$O$^+$

NaOH/H$_2$O

C_6H_5COOH
separate and
recrystallize

aqueous layer
$p\text{-}CH_3C_6H_4O^-Na^+$

CH$_2$Cl$_2$ layer
$C_6H_5NH_2, C_6H_6$

H$_3$O$^+$

H$_3$O$^+$Cl$^-$/H$_2$O

$p\text{-}CH_3C_6H_4OH$
extract into
CH_2Cl_2 and distill

aqueous
layer
$C_6H_5NH_3^+Cl^-$

CH$_2$Cl$_2$
layer
C_6H_6
isolate
by dis-
tillation

OH$^-$

$C_6H_5NH_2$
extract into CH_2Cl_2
and distill

21.5

(a) Neglecting Kekulé forms of the ring, we can write the following resonance structures for the phthalimide anion.

(b) Phthalimide is more acidic than benzamide because its anion is stabilized by resonance to a greater extent than the anion of benzamide. (Benzamide has only one carbonyl group attached to the nitrogen and thus fewer resonance contributors are possible.)

21.6

21.7

(a) $CH_3(CH_2)_3CHO + NH_3 \xrightarrow{H_2,\ Ni} CH_3(CH_2)_3CH_2NH_2$

(b) $C_6H_5\overset{\underset{\|}{O}}{C}CH_3 + NH_3 \xrightarrow{H_2,\ Ni} C_6H_5\underset{\underset{NH_2}{|}}{C}HCH_3$

(c) $CH_3(CH_2)_4CHO + C_6H_5NH_2 \xrightarrow[CH_3OH]{LiBH_3CN} CH_3(CH_2)_4CH_2NHC_6H_5$

21.8

The reaction of a secondary halide with ammonia would inevitably be accompanied by considerable elimination thus decreasing the yield.

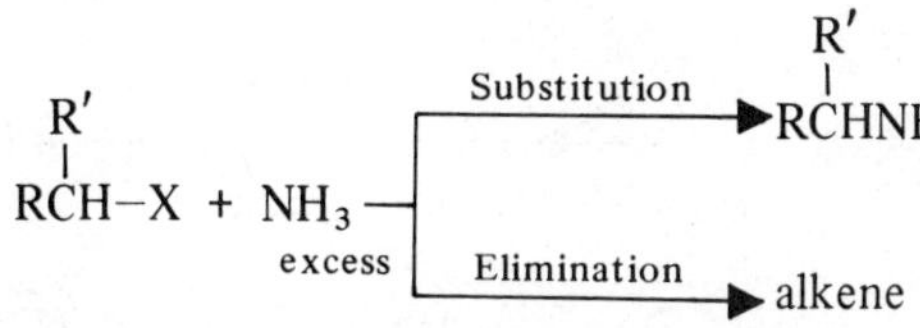

21.9

(a) $C_6H_5COOH \xrightarrow{SOCl_2} C_6H_5COCl \xrightarrow{CH_3CH_2NH_2}$

$C_6H_5CONHCH_2CH_3 \xrightarrow{LiAlH_4} C_6H_5CH_2NHCH_2CH_3$

(b) $CH_3CH_2CH_2CH_2CH_2Br \xrightarrow{NaCN} CH_3CH_2CH_2CH_2CH_2CN$

$\xrightarrow{LiAlH_4} CH_3CH_2CH_2CH_2CH_2CH_2NH_2$

(c) $CH_3CH_2COOH \xrightarrow{SOCl_2} CH_3CH_2COCl \xrightarrow{(CH_3CH_2CH_2)_2NH}$

$CH_3CH_2CON(CH_2CH_2CH_3)_2 \xrightarrow{LiAlH_4} (CH_3CH_2CH_2)_3N$

21.10

(a)

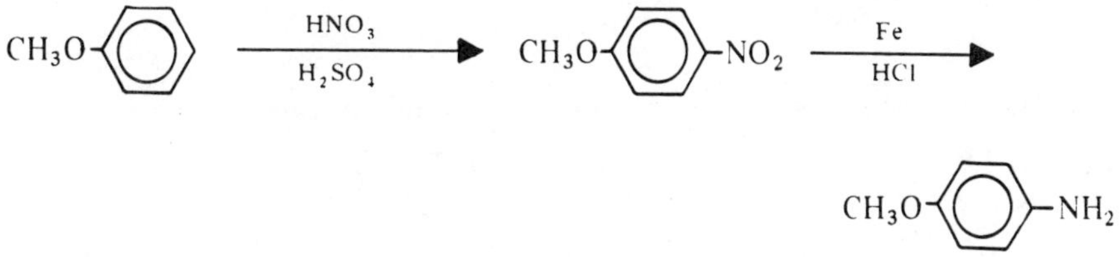

(b)

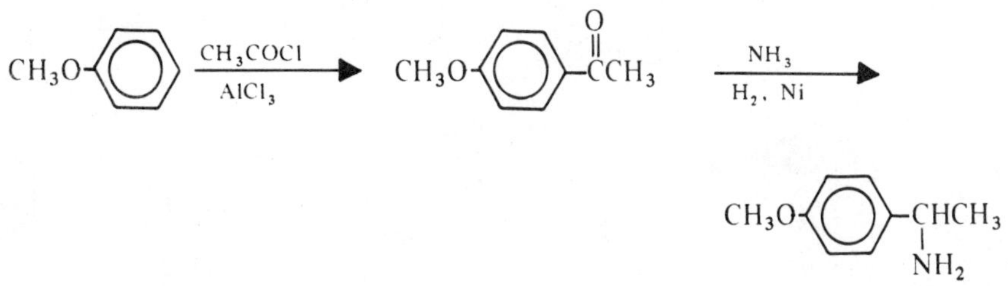

(c) 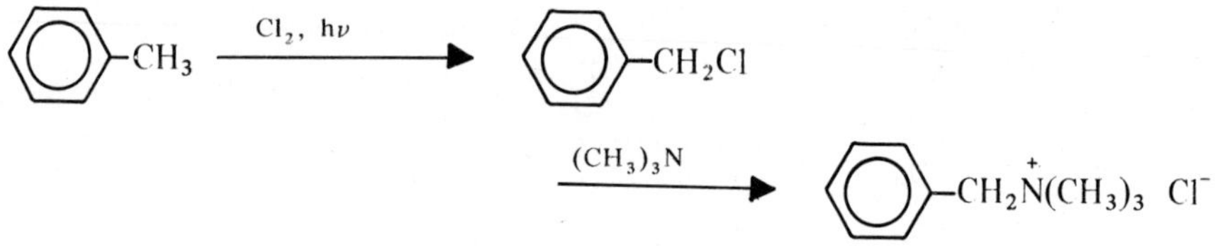

(d) 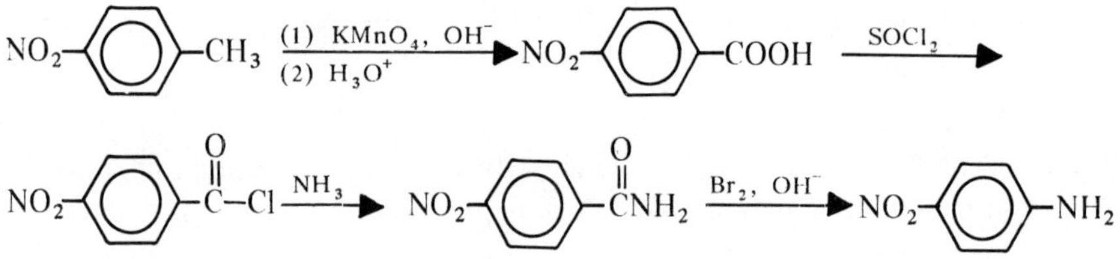

(e) 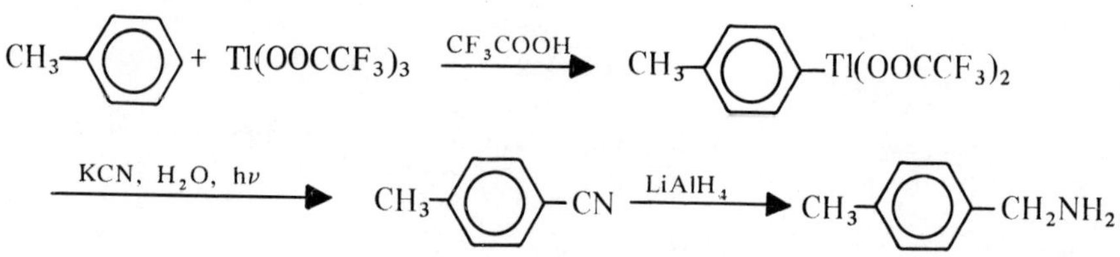

21.11

An amine acting as a base.

$$CH_3CH_2\ddot{N}H_2 + H_3O^+ \rightleftharpoons CH_3CH_2NH_3^+ + H_2O$$

An amine acting as a nucleophile in an alkylation reaction.

$$(CH_3CH_2)_3N: + CH_3{-}I \longrightarrow (CH_3CH_2)_3\overset{+}{N}{-}CH_3\ I^-$$

An amine acting as a nucleophile in an acylation reaction.

$$\underset{\text{(excess)}}{(CH_3)_2\ddot{N}H} + CH_3C(=O)Cl \longrightarrow (CH_3)_2N\overset{O}{\overset{\|}{C}}CH_3 + (CH_3)_2NH_2Cl$$

An enamine acting as a carbon nucleophile in an acylation reaction.

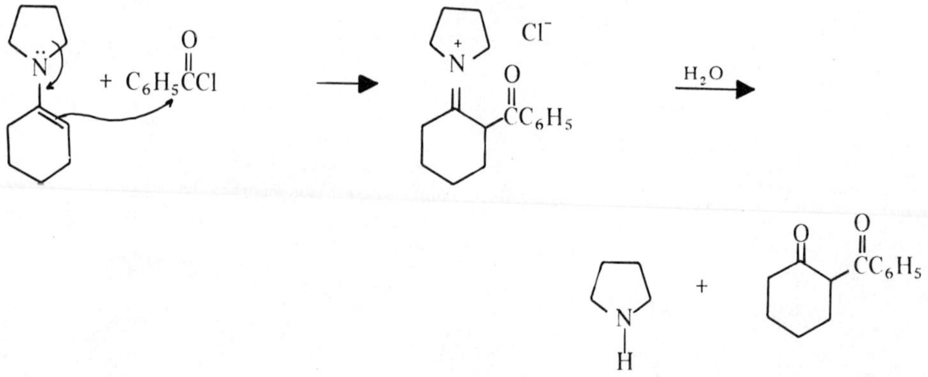

An enamine acting as a carbon nucleophile in an alkylation reaction.

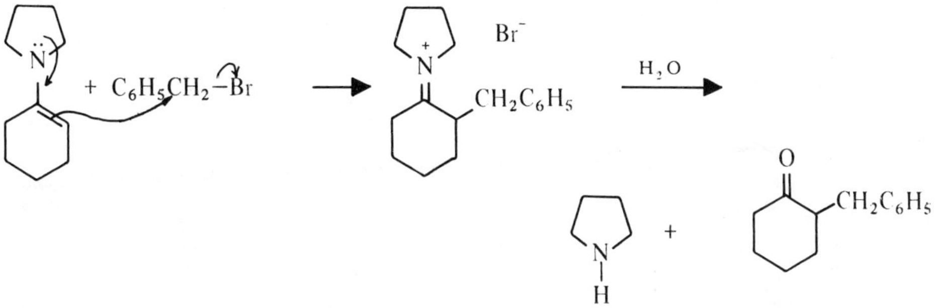

An amino group acting as an activating group and as an ortho-para director in electrophilic aromatic substituion.

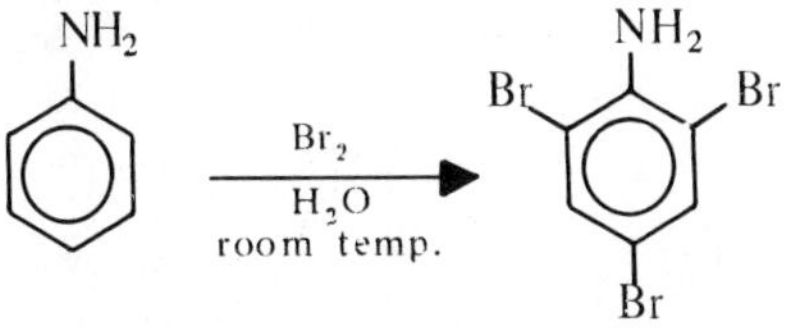

21.12

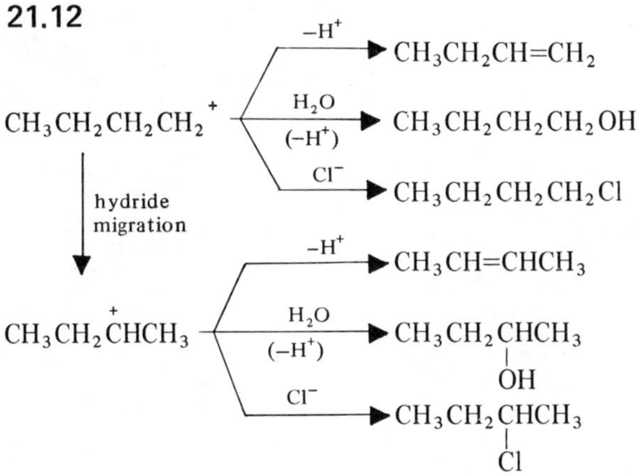

21.13

(a) $^-O-N=O + H_3O^+ \rightleftharpoons HO-N=O$

$HO-N=O + H_3O^+ \rightleftharpoons HO^+-N=O$
$\qquad\qquad\qquad\qquad\qquad\quad |$
$\qquad\qquad\qquad\qquad\qquad\;\; H$

$HO^+-N=O \rightleftharpoons H_2O + \overset{+}{N}=O$
$\;\; |$
$\;\; H$

(b)

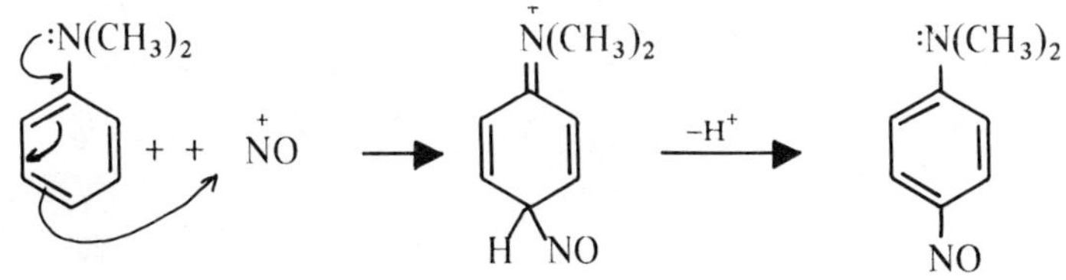

(c) The $\overset{+}{N}O$ ion is a weak electrophile. For it to react with an aromatic ring, the ring must have a powerful activating group such as —OH or —NR$_2$.

21.14

(a)

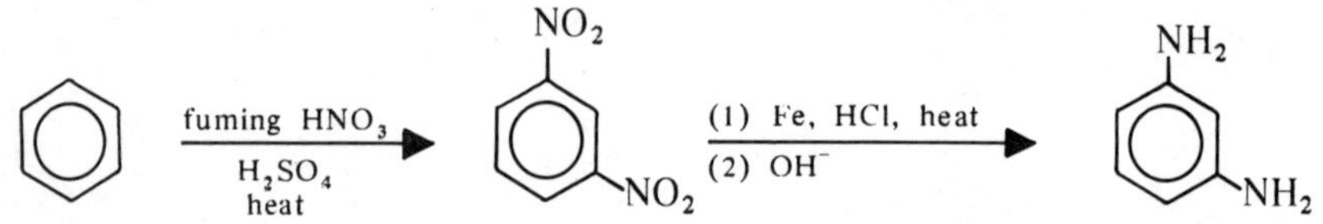

(b)

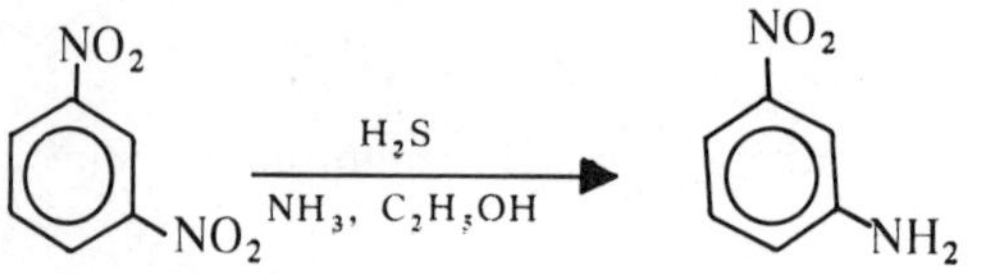

(from part a)

(c)

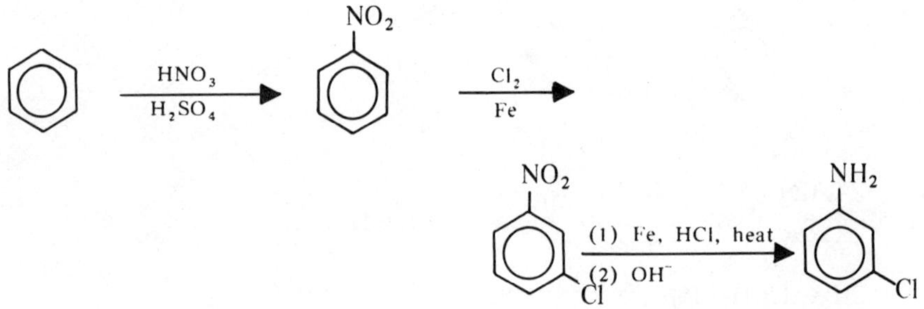

(d)

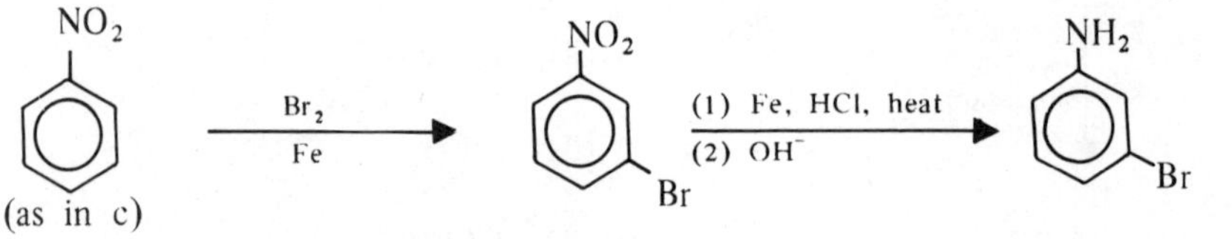

(as in c)

(e)

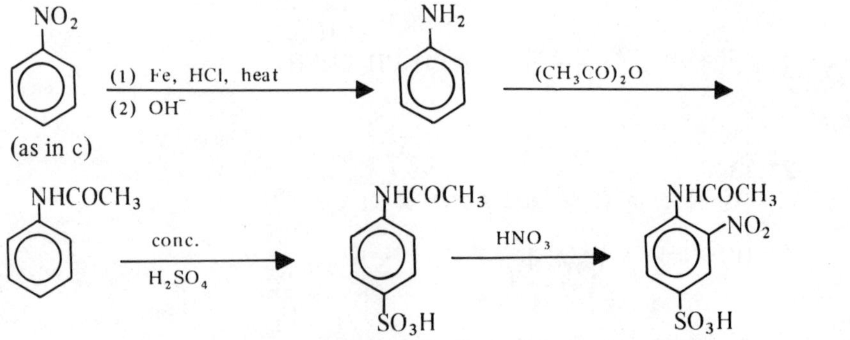

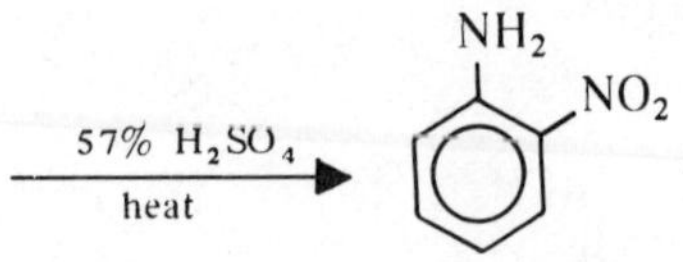

(as in c)

(56% yield from acetanilide cf., page 498 of text)

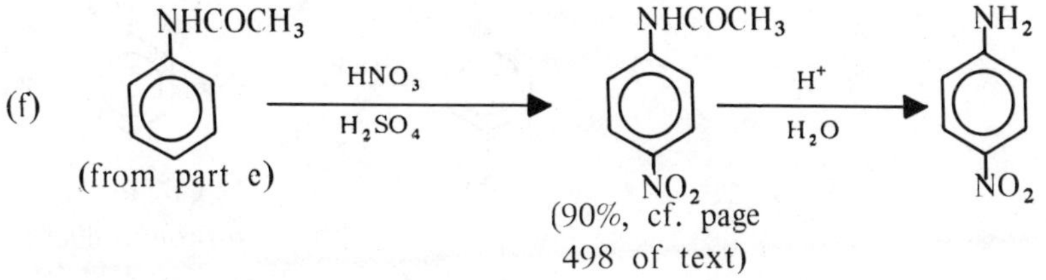
(f)
NHCOCH3
HNO3
H2SO4
NHCOCH3
NO2
H+
H2O
NH2
NO2
(from part e)
(90%, cf. page 498 of text)

21.15

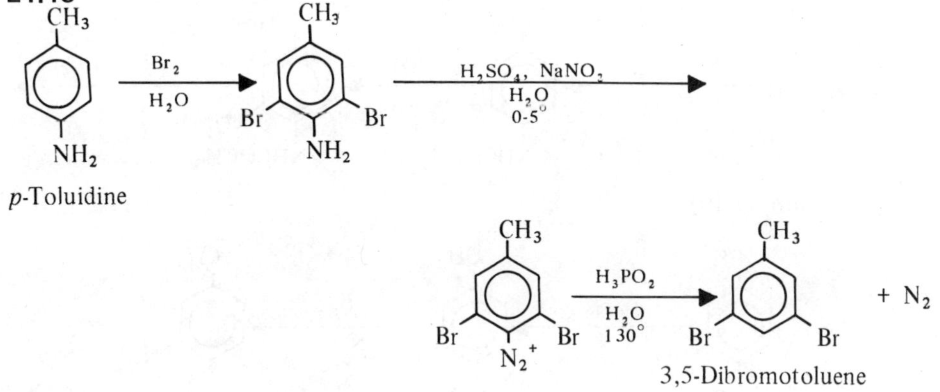
CH3
NH2
p-Toluidine
Br2
H2O
CH3
Br
Br
NH2
H2SO4, NaNO2
H2O
0-5°
CH3
Br
Br
N2+
H3PO2
H2O
130°
CH3
Br
Br
+ N2
3,5-Dibromotoluene

21.16

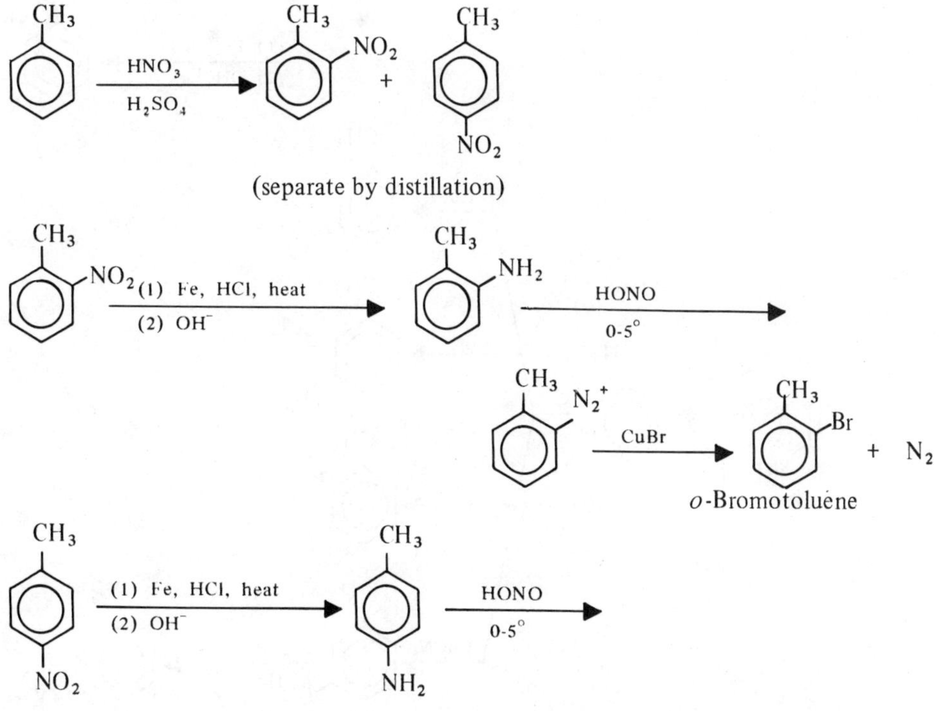
CH3
HNO3
H2SO4
CH3
NO2
+
CH3
NO2
(separate by distillation)
CH3
NO2
(1) Fe, HCl, heat
(2) OH−
CH3
NH2
HONO
0-5°
CH3
N2+
CuBr
CH3
Br
+ N2
o-Bromotoluene
CH3
NO2
(1) Fe, HCl, heat
(2) OH−
CH3
NH2
HONO
0-5°

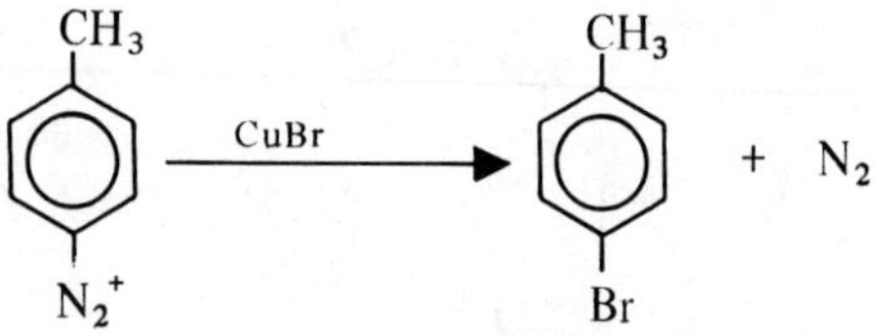

p-Bromotoluene

21.17

(a)

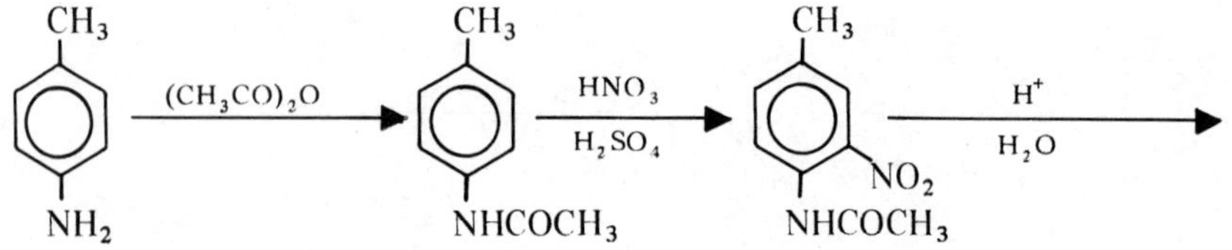

(cf. Prob. 21.16)

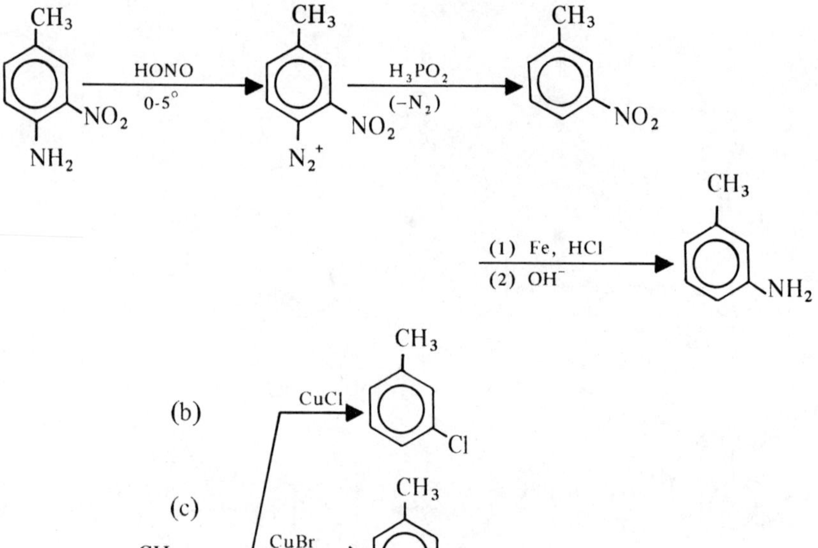

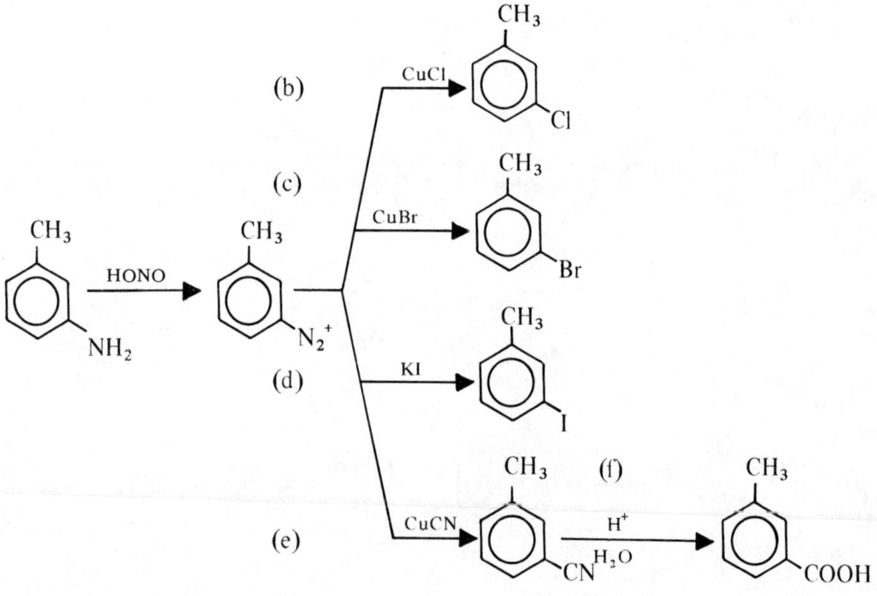

(b)

(c)

(d)

(e)

(f)

21.18

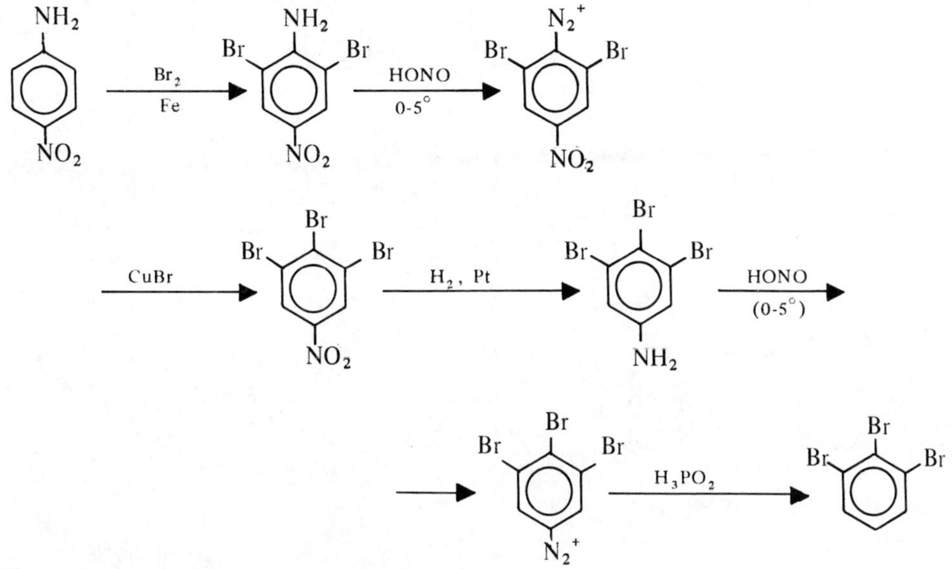

21.19

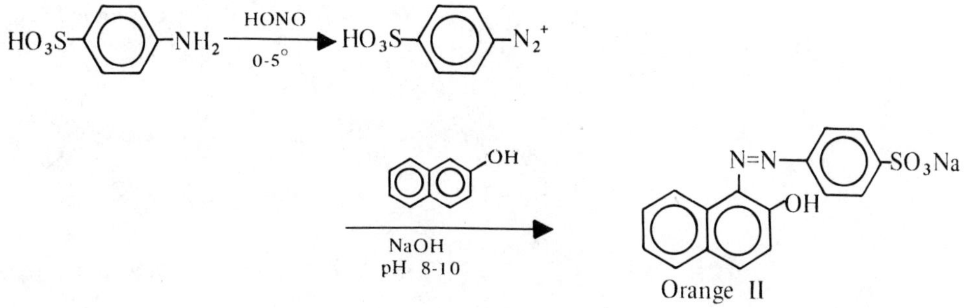

Orange II

21.20

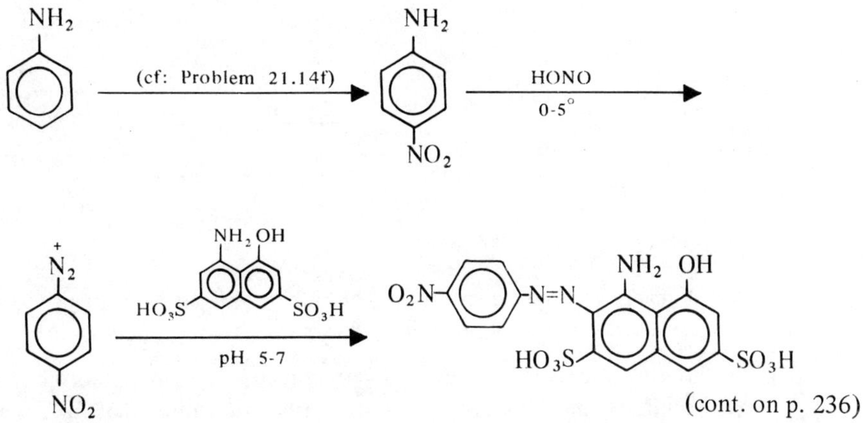

(cont. on p. 236)

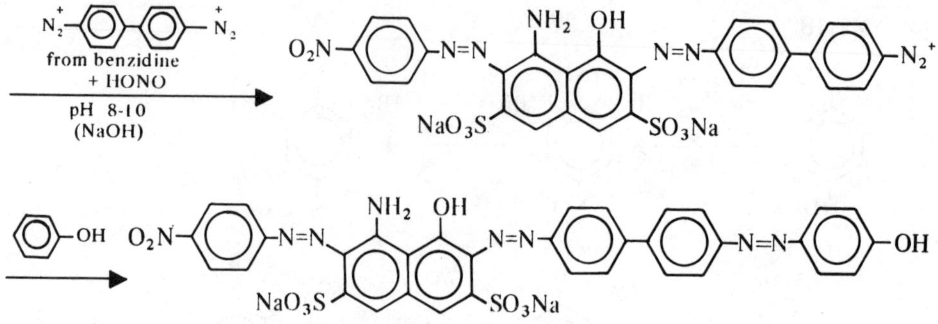

Diamine Green B

21.21

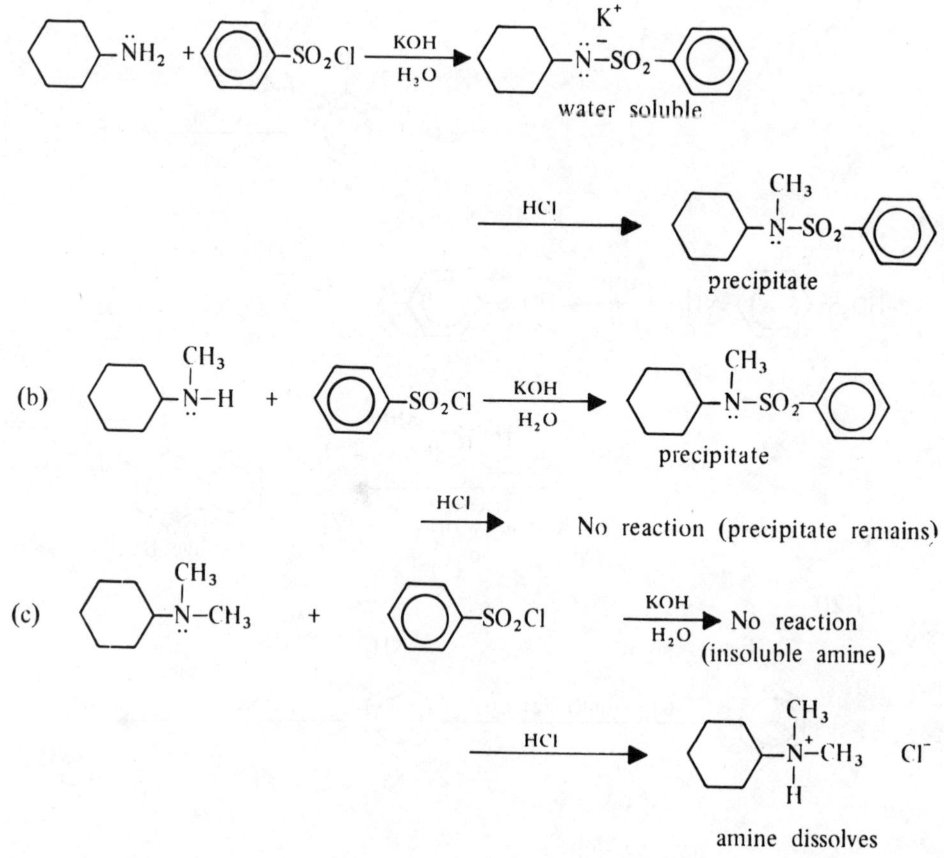

water soluble

precipitate

(b)

precipitate

No reaction (precipitate remains)

(c)

No reaction (insoluble amine)

amine dissolves

21.22

(1) That A reacts with benzenesulfonyl chloride in aqueous KOH to give a clear solution which on acidification yields a precipitate shows that A is a primary amine.

(2) That diazotization of A followed by treatment with 2-naphthol gives an intensely colored precipitate shows that A is a primary aromatic amine, that is, A is a substituted aniline.

(3) Consideration of the molecular formula of A leads us to conclude that A is a toluidine

$$
\begin{array}{c}
\mathrm{C_7H_9N} \\
-\ \underline{\mathrm{C_6H_6N}} \\
\mathrm{CH_3}
\end{array}
=
\raisebox{-0.5em}{\text{[benzene ring]}}-\mathrm{NH_2}
$$

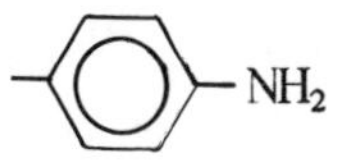

But is A *o*-toluidine, *m*-toluidine, or *p*-toluidine?

(4) This question is answered by the infrared data. A single absorption peak in the 680-840 cm^{-1} region at 815 cm^{-1} is indicative of a *para* substituted benzene. Thus A is *p*-toluidine.

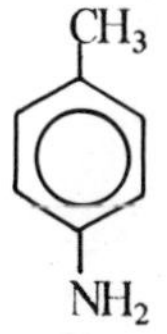

A

21.23

(a)

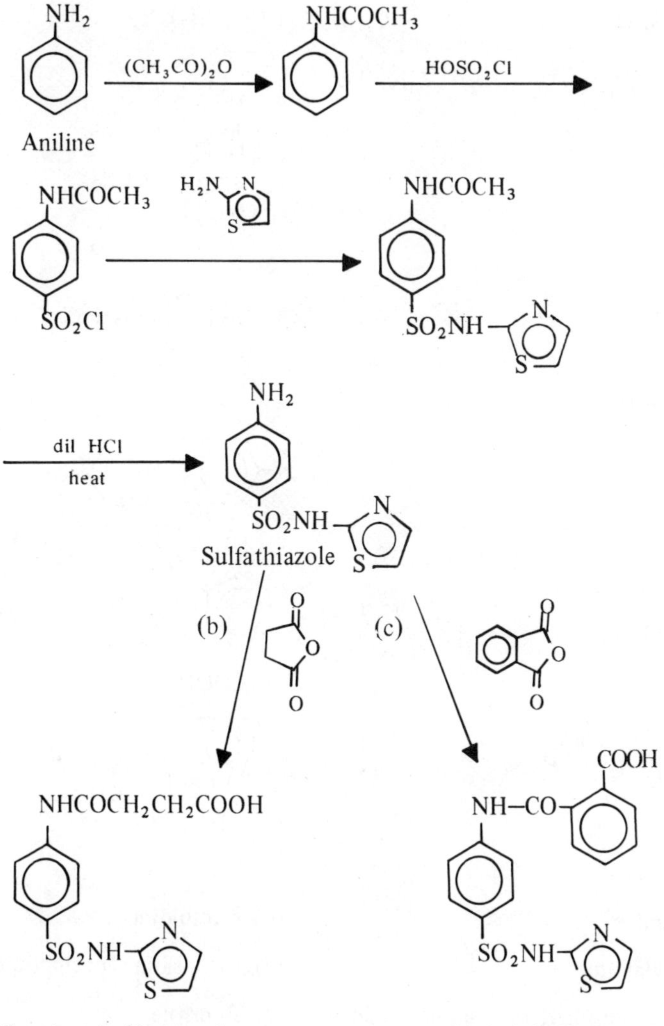

Aniline

Sulfathiazole

(b)   (c)

Succinoylsulfathiazole   Phthalylsulfathiazole

21.24

(a) $C_6H_5CH_2NHCH_3$

(b) $(CH_3CH)_3N$
$\quad\quad\;\; |$
$\quad\quad\;\; CH_3$

(c) [benzene ring]$-N$
$\quad\quad\;\;\; \diagup CH_3$
$\quad\quad\;\;\; \diagdown CH_2CH_3$

(d) [benzene ring with CH_3 and NH_2 substituents]

(e) [pyrrole ring with CH_3, N–H]

(f) [piperidine ring with N–CH_2CH_3]

(g) [pyridinium ring, $\overset{+}{N}$–CH_2CH_3, Br^-]

(h) [pyridine ring with $COOH$]

(i) [indole ring, N–H]

(j) [benzene ring]$-NH\overset{\overset{\displaystyle O}{\|}}{C}CH_3$

(k) $CH_3-\overset{\overset{\displaystyle H}{|}}{\underset{\underset{\displaystyle CH_3}{|}}{N}}-\overset{+}{H}\;\; Cl^-$

(l) [imidazole ring with CH_3, N–H]

(m) [benzene ring with NH_2 and SO_2NH–pyridine]

(n) $(CH_3CH_2CH_2)_4N^+\quad Cl^-$

(o) [pyrrolidine ring, N–H]

(p) CH_3–[benzene ring]$-N$
$\quad\quad\quad\quad\quad\quad\;\; \diagup CH_3$
$\quad\quad\quad\quad\quad\quad\;\; \diagdown CH_3$

(q) CH_3O–[benzene ring]$-NH_2$

(r) H_2N–[benzene ring]–[benzene ring]$-NH_2$

(s) [benzene ring with NH_2 and $COOH$]

(t) [imidazole ring]$-CH_2\underset{\underset{\displaystyle NH_2}{|}}{CH}COOH$

21.25

(a) Propylamine

(b) N-methylaniline

(c) Trimethylisopropylammonium iodide

(d) *o*-toluidine

(e) *o*-Anisidine (or *o*-methoxyaniline)

(f) Pyrazole

(g) 2-Aminopyrimidine

(h) Benzylammonium chloride

(i) N,N-Dipropylaniline

(j) Benzenesulfonamide

(k) Methylammonium acetate

(l) 3-Aminopropanol

(m) Purine

(n) N-Methylpyrrole

21.26

(a)

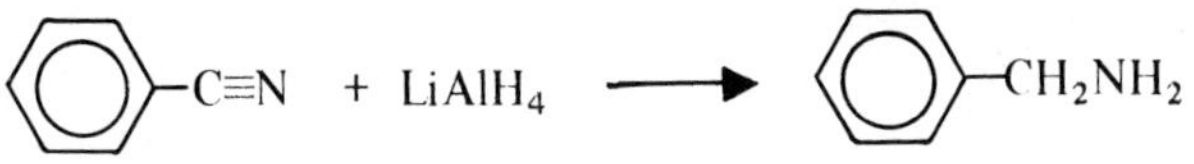

(b)

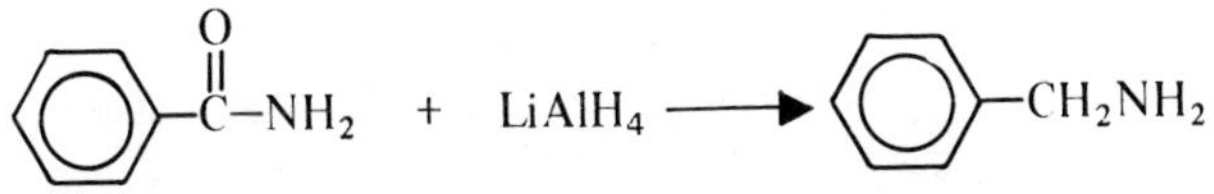

(c)

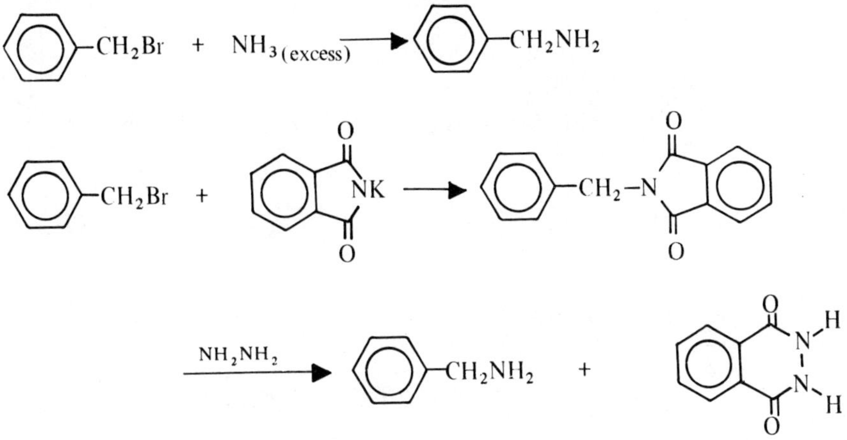

(d)

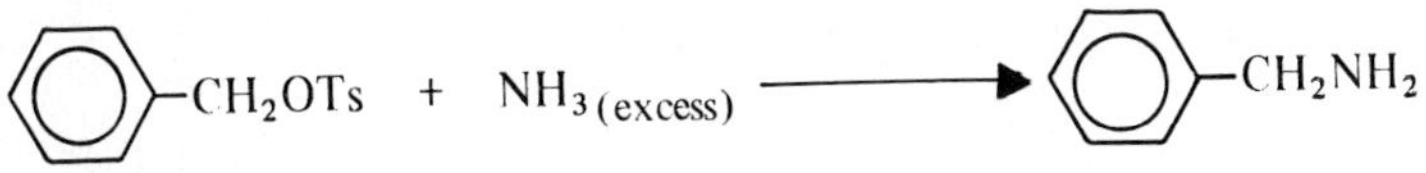

(e)

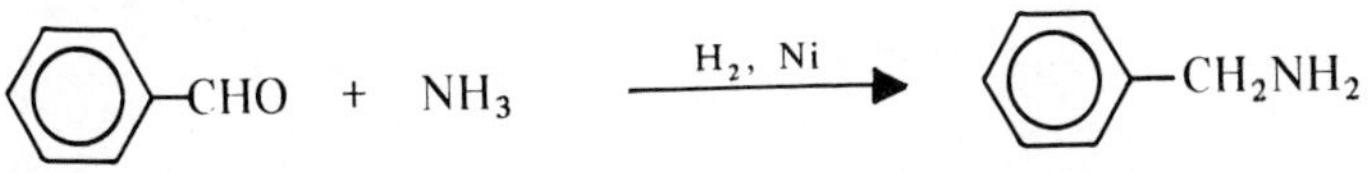

(f)

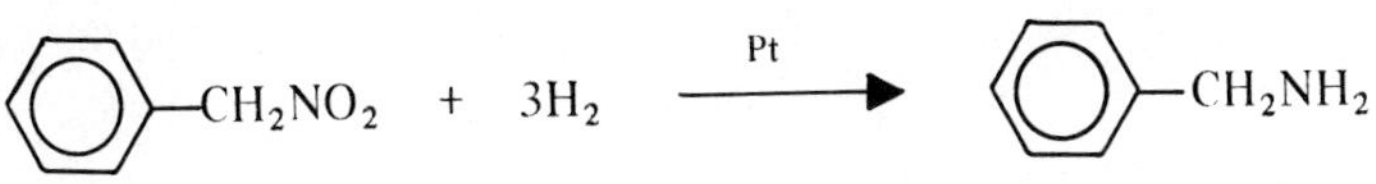

(g)

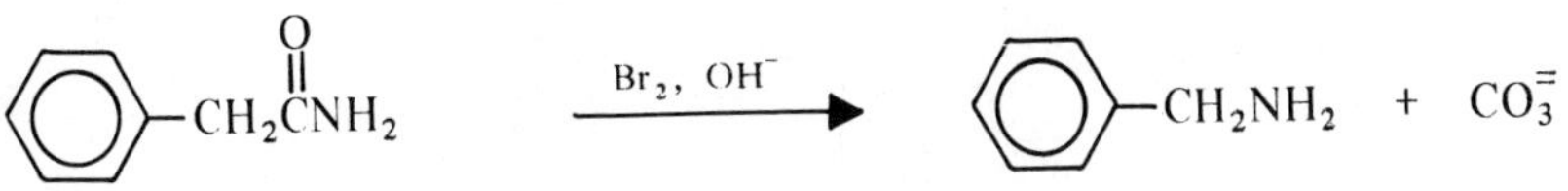

21.27

(a)

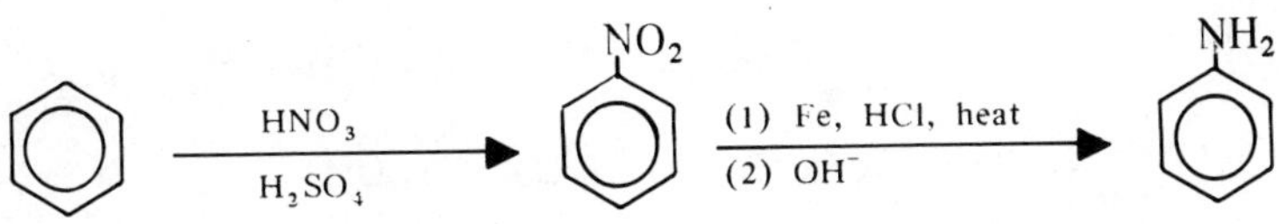

(b)

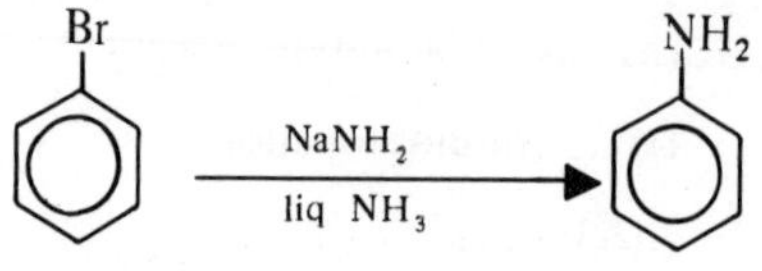

$$\text{(bromobenzene)} \xrightarrow[\text{liq } NH_3]{NaNH_2} \text{(aniline)}$$

(c)

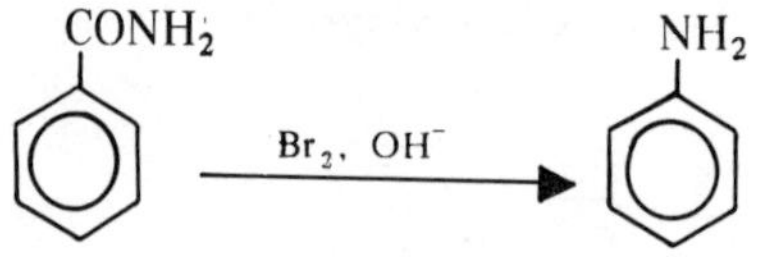

$$\text{(benzamide)} \xrightarrow{Br_2, OH^-} \text{(aniline)}$$

21.28

(a) 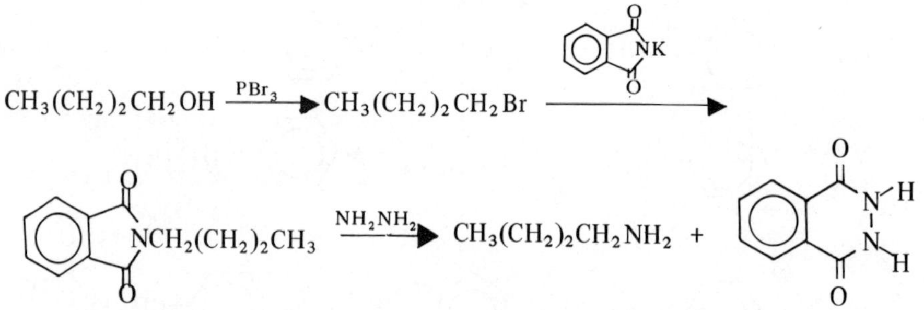 $CH_3(CH_2)_2CH_2OH \xrightarrow{PBr_3} CH_3(CH_2)_2CH_2Br \longrightarrow$

$$\xrightarrow{NH_2NH_2} CH_3(CH_2)_2CH_2NH_2 \; + \;$$

(b) $CH_3(CH_2)_2CH_2Br \xrightarrow{NaCN} CH_3(CH_2)_3CN \xrightarrow{LiAlH_4} CH_3(CH_2)_3CH_2NH_2$
   (from part a)

(c) $CH_3(CH_2)_2CH_2OH \xrightarrow[\text{(2) } H_3O^+]{\text{(1) } KMnO_4, \; OH^-} CH_3CH_2CH_2COOH$

$$\xrightarrow[\text{(2) } NH_3]{\text{(1) } SOCl_2} CH_3CH_2CH_2CONH_2 \xrightarrow{Br_2, \; OH^-} CH_3CH_2CH_2NH_2$$

(d) $CH_3CH_2CH_2CH_2OH \xrightarrow[C_5H_5N]{CrO_3} CH_3CH_2CH_2CHO \xrightarrow[H_2, \; Ni]{CH_3NH_2}$

$$CH_3CH_2CH_2CH_2NHCH_3$$

21.29

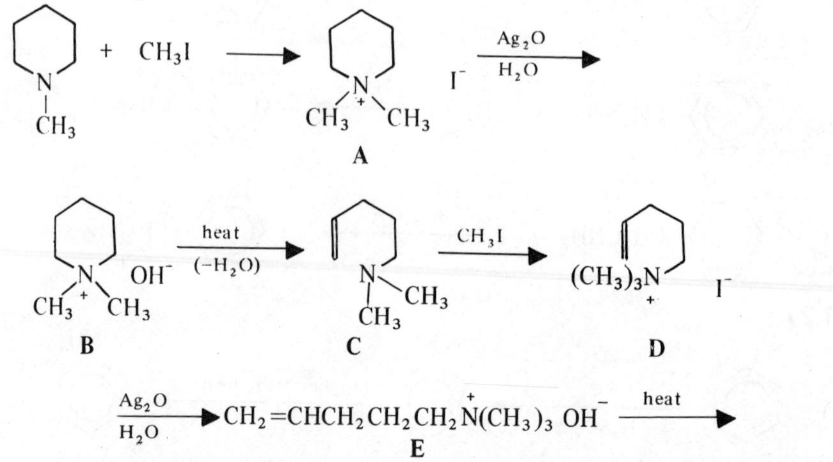

$$\xrightarrow[H_2O]{Ag_2O} CH_2=CHCH_2CH_2CH_2\overset{+}{N}(CH_3)_3 \; OH^- \xrightarrow{heat}$$
                               E

$$CH_2=CHCH_2CH=CH_2 + H_2O + (CH_3)_3N$$
$$F$$

21.30

(a)

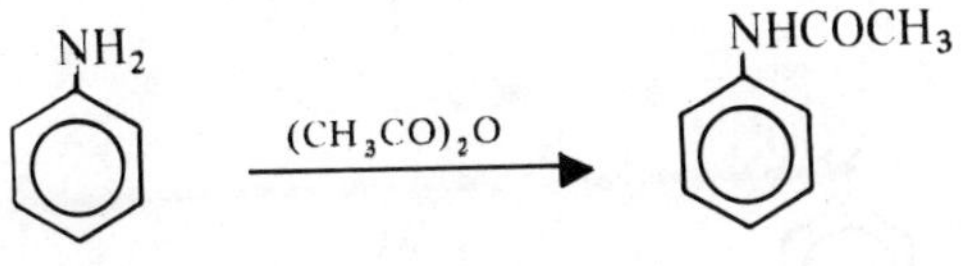

$$\xrightarrow{(CH_3CO)_2O}$$

(b)

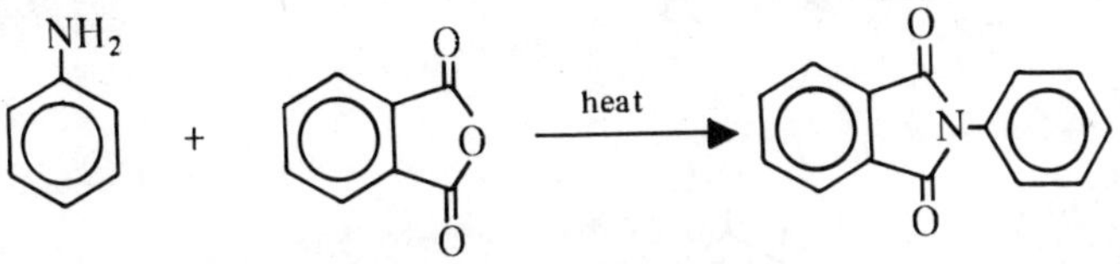

$$\xrightarrow{heat}$$

(c) 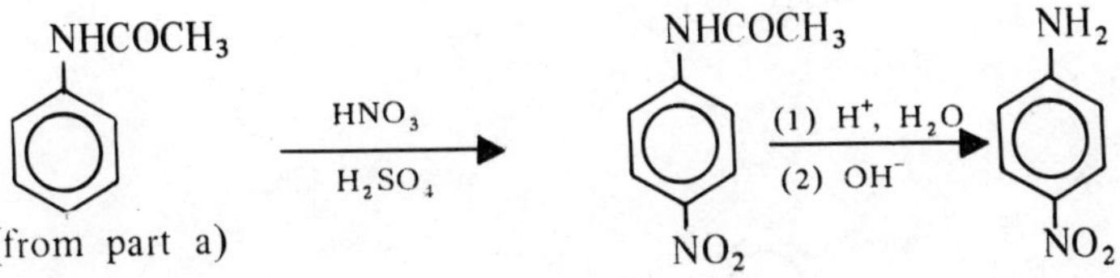 (from part a)

$$\xrightarrow[H_2SO_4]{HNO_3}$$

$$\xrightarrow[(2)\ OH^-]{(1)\ H^+,\ H_2O}$$

(d) (from part a)

$$\xrightarrow{HOSO_2Cl}$$

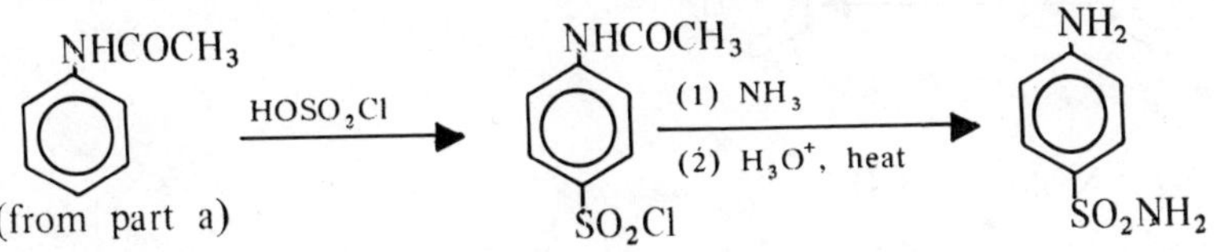

$$\xrightarrow[(2)\ H_3O^+,\ heat]{(1)\ NH_3}$$

(e)

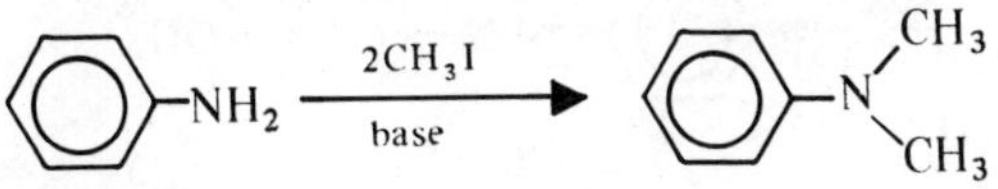

$$\xrightarrow[base]{2CH_3I}$$

(f)

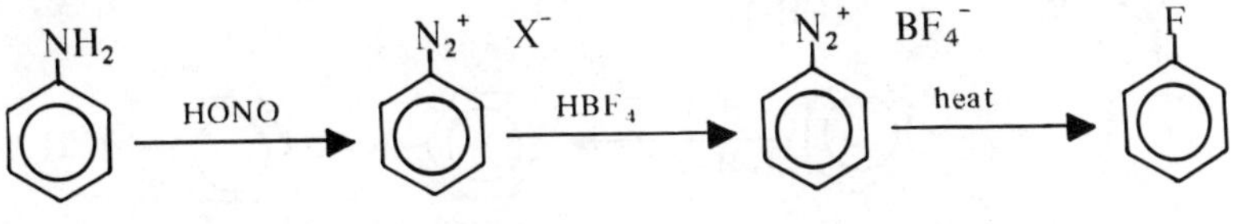

$$\xrightarrow{HONO}$$

$$\xrightarrow{HBF_4}$$

$$\xrightarrow{heat}$$

(g) 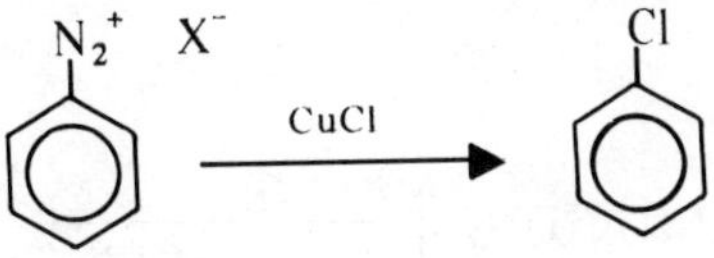 (from part f)

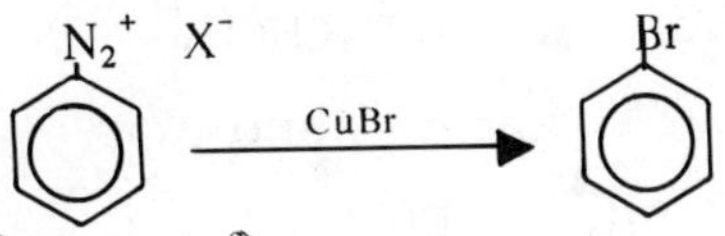

$$\xrightarrow{CuCl}$$

(h) (from part f)

$$\xrightarrow{CuBr}$$

(i) (from part f)

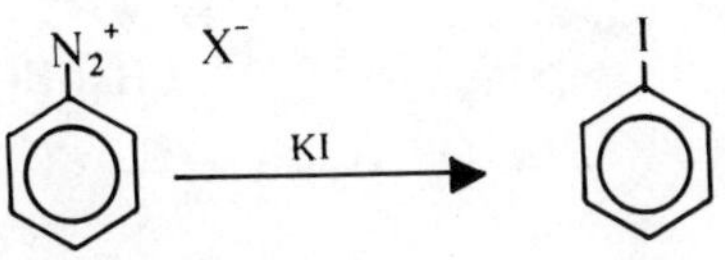

$$\xrightarrow{KI}$$

(j)

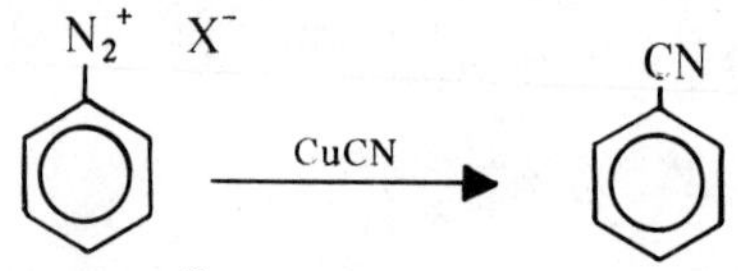
N_2^+ X^-
CuCN
CN

(from part f)

(k)

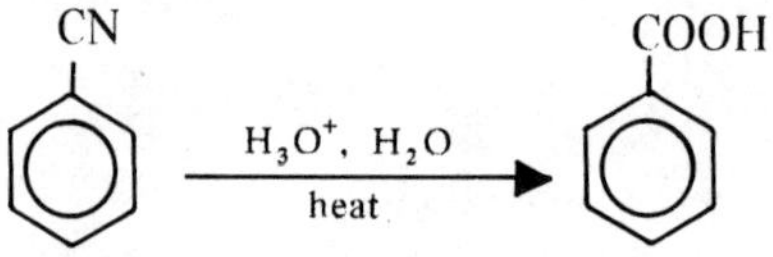
CN
H_3O^+, H_2O
heat
COOH

(from part j)

(l)

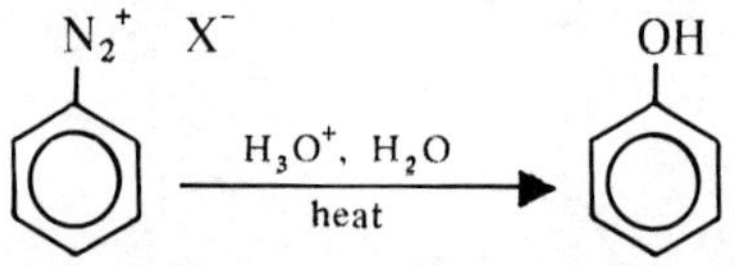
N_2^+ X^-
H_3O^+, H_2O
heat
OH

(from part f)

(m)

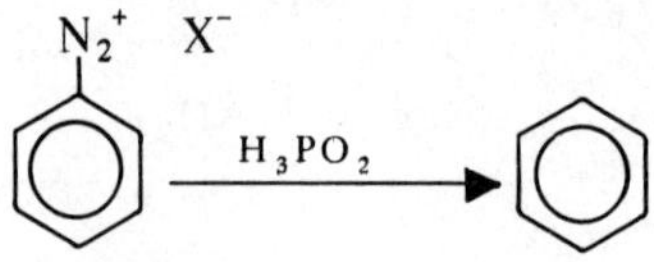
N_2^+ X^-
H_3PO_2

(from part f)

(n)

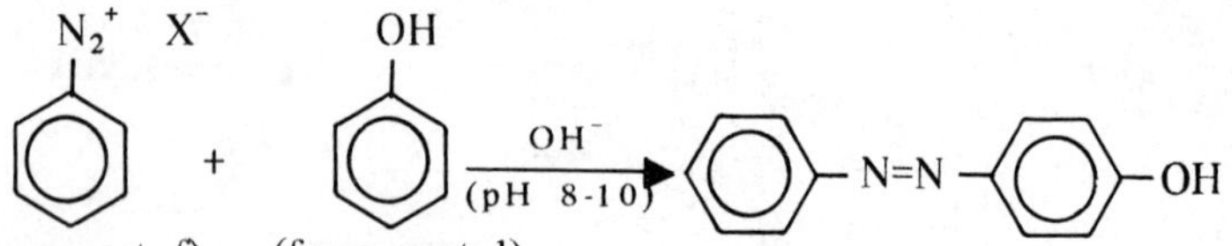
N_2^+ X^-
+
OH
OH^-
(pH 8-10)
N=N
OH

(from part f) (from part l)

(o)

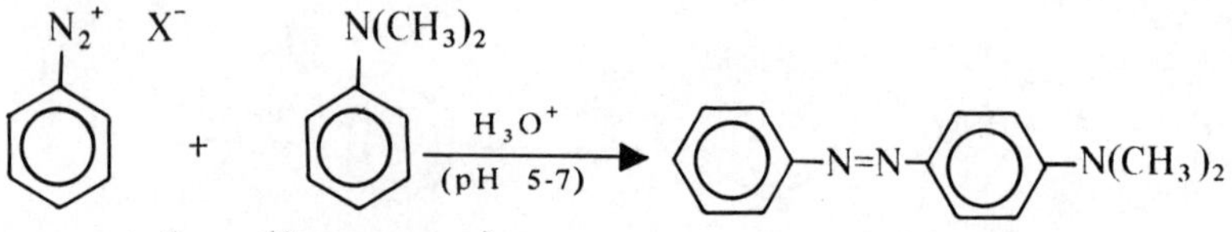
N_2^+ X^-
+
$N(CH_3)_2$
H_3O^+
(pH 5-7)
N=N
$N(CH_3)_2$

(from part f) (from part e)

21.31

(a)

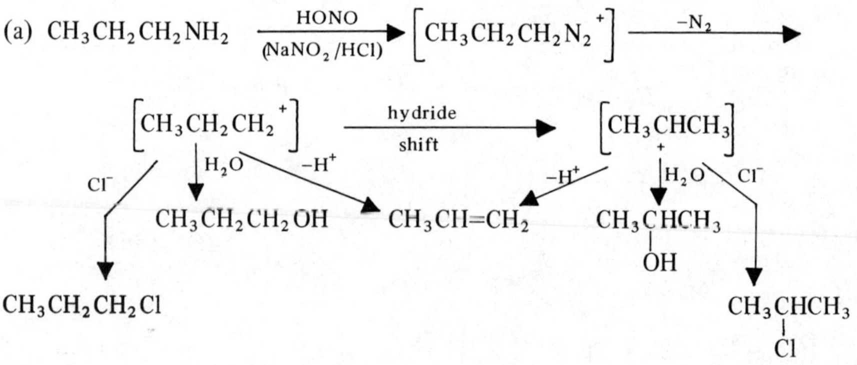
$CH_3CH_2CH_2NH_2$
HONO
($NaNO_2$/HCl)
$[CH_3CH_2CH_2N_2^+]$
$-N_2$
$[CH_3CH_2CH_2^+]$
hydride
shift
$[CH_3\overset{+}{C}HCH_3]$
Cl^-
H_2O
$-H^+$
$-H^+$
H_2O
Cl^-
$CH_3CH_2CH_2OH$
$CH_3CH{=}CH_2$
CH_3CHCH_3
OH
$CH_3CH_2CH_2Cl$
CH_3CHCH_3
Cl

(b) $(CH_3CH_2)_2NH \xrightarrow[\text{(NaNO}_2\text{/HCl)}]{\text{HONO}} (CH_3CH_2)_2N-N=O$

(c) $\xrightarrow[\text{(NaNO}_2\text{/HCl)}]{\text{HONO}}$

(d) $\xrightarrow[\text{(NaNO}_2\text{/HCl)}]{\text{HONO}}$

(e) $CH_3CH_2CH_2-$$-NH_2 \xrightarrow[\text{(NaNO}_2\text{/HCl)}]{\text{HONO, 0-5}^\circ} CH_3CH_2CH_2-$$-N_2^+ \quad Cl^-$

21.32

(a) $CH_3CH_2CH_2NH_2 + C_6H_5SO_2Cl \xrightarrow[H_2O]{KOH} CH_3CH_2CH_2\overset{-}{N}SO_2C_6H_5$
$$\underset{K^+}{}$$
clear solution

$\xrightarrow{H_3O^+} CH_3CH_2CH_2NHSO_2C_6H_5$
precipitate

(b) $(CH_3CH_2CH_2)_2NH + C_6H_5SO_2Cl \xrightarrow[H_2O]{KOH} (CH_3CH_2CH_2)_2NSO_2C_6H_5$
precipitate

$\xrightarrow{H_3O^+}$ No reaction (precipitate remains)

(c) $+ C_6H_5SO_2Cl \xrightarrow[H_2O]{KOH}$
precipitate

$\xrightarrow{H_3O^+}$ No reaction (precipitate remains)

(d) $+ C_6H_5SO_2Cl \xrightarrow[H_2O]{KOH}$ No reaction
(3° amine is insoluble)

$\xrightarrow{H_3O^+}$ $-\overset{+}{N}H(CH_2CH_2CH_3)_2$
3° amine dissolves

(e) $C_3H_7-$$-NH_2 + C_6H_5SO_2Cl \xrightarrow[H_2O]{KOH} C_3H_7-$$-\overset{-}{N}SO_2C_6H_5$
$$\underset{K^+}{} \quad \text{(cont. on p. 244)}$$
clear solution

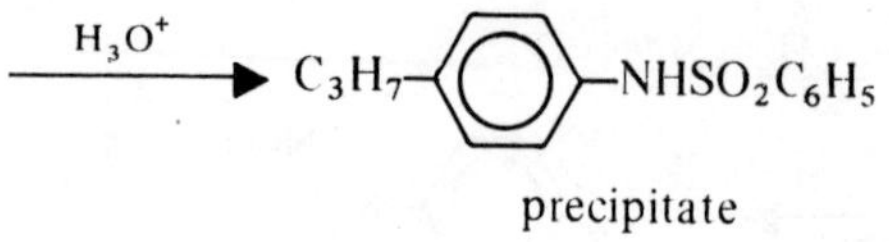

$$\xrightarrow{\text{H}_3\text{O}^+} \text{C}_3\text{H}_7-\text{C}_6\text{H}_4-\text{NHSO}_2\text{C}_6\text{H}_5$$

precipitate

21.33

(a)

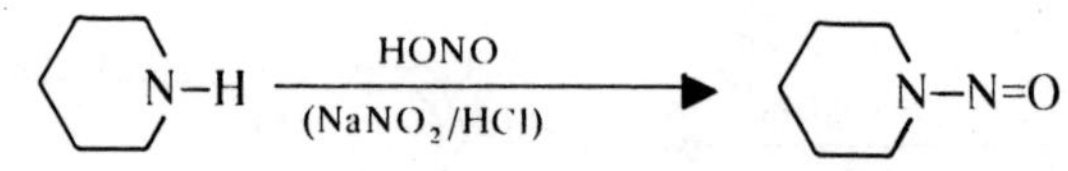

$$\text{(piperidine) N-H} \xrightarrow[\text{(NaNO}_2/\text{HCl)}]{\text{HONO}} \text{N-N=O}$$

(b)

$$\text{N-H} + \text{C}_6\text{H}_5\text{SO}_2\text{Cl} \xrightarrow[\text{H}_2\text{O}]{\text{KOH}} \text{N-SO}_2\text{C}_6\text{H}_5$$

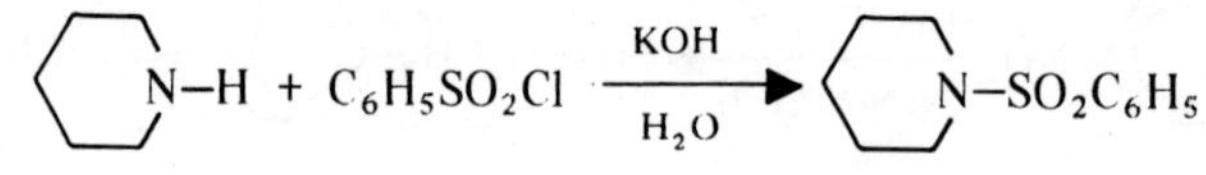

21.34

(a) $2\text{CH}_3\text{CH}_2\text{NH}_2 + \text{C}_6\text{H}_5\text{COCl} \longrightarrow \text{CH}_3\text{CH}_2\text{NHCOC}_6\text{H}_5 + \text{CH}_3\text{CH}_2\text{NH}_3{}^+\text{Cl}^-$

(b) $2\text{CH}_3\text{NH}_2 + (\text{CH}_3\overset{\text{O}}{\overset{\|}{\text{C}}})_2\text{O} \longrightarrow \text{CH}_3\text{NH}\overset{\text{O}}{\overset{\|}{\text{C}}}\text{CH}_3 + \text{CH}_3\overset{+}{\text{N}}\text{H}_3 \ \ \text{CH}_3\overset{\text{O}}{\overset{\|}{\text{C}}}\text{O}^-$

(c)

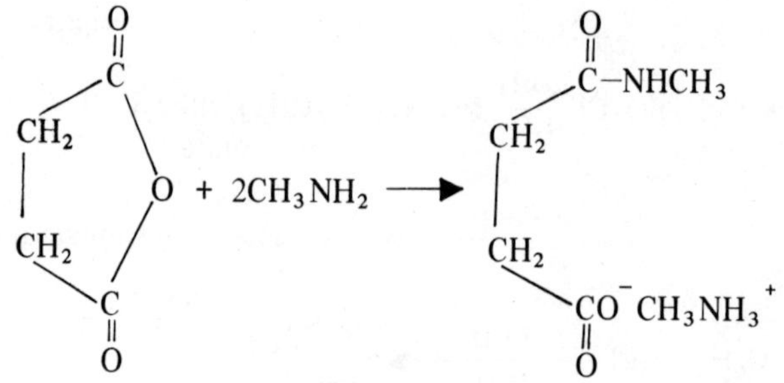

(d) 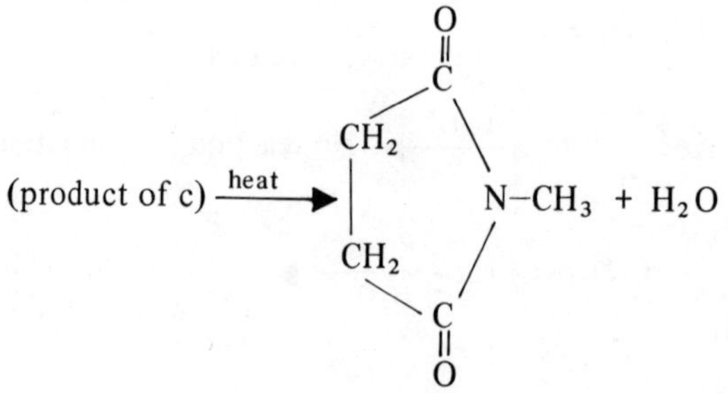 (product of c) $\xrightarrow{\text{heat}}$

$+ \ \text{H}_2\text{O}$

(e)

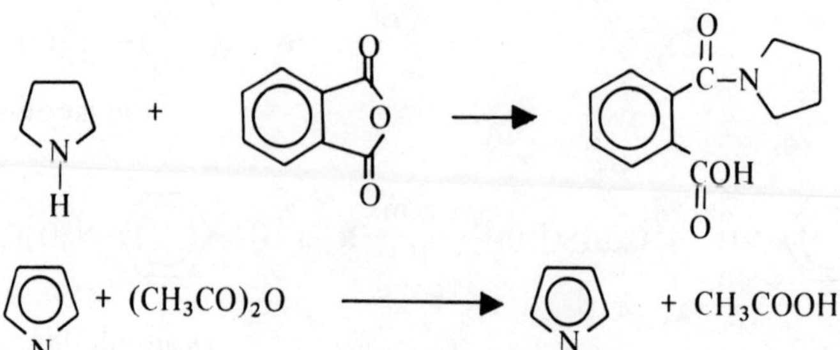

(f)

$+ \ \text{CH}_3\text{COOH}$

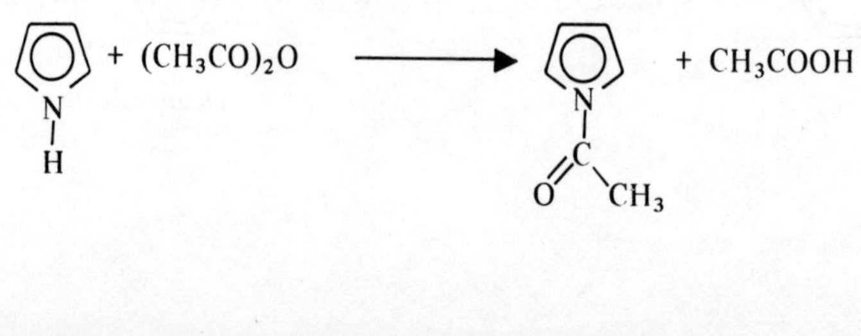

(g) 2 ⬡$-NH_2$ + $CH_3CH_2\overset{\overset{O}{\|}}{C}Cl$ ⟶ ⬡$-NH\overset{\overset{O}{\|}}{C}CH_2CH_3$ + ⬡$-NH_3^+$ Cl^-

(h) $CH_3CH_2-\overset{\overset{\displaystyle CH_2CH_3}{\overset{+}{|}}}{\underset{\underset{\displaystyle CH_2CH_3}{|}}{N}}-CH_2CH_3$ ^-OH ⟶ $CH_2{=}CH_2$ + $(CH_3CH_2)_3N$ + H_2O

(i) [3,5-dinitrobenzene structure with NO_2 groups] + H_2S $\xrightarrow[C_2H_5OH]{NH_3}$ [structure with NO_2 and NH_2]

(j) [4-methylaniline: CH_3 ... NH_2] + $Br_{2\,(excess)}$ $\xrightarrow{H_2O}$ [2,6-dibromo-4-methylaniline with CH_3, Br, Br, NH_2]

21.35

(a) [toluene: CH_3] $\xrightarrow[H_2SO_4]{HNO_3}$ [o-nitrotoluene: CH_3, NO_2] + [p-nitrotoluene: CH_3 ... NO_2]

Separate isomers

[o-nitrotoluene: CH_3, NO_2] $\xrightarrow[(2)\ OH^-]{(1)\ Fe,\ HCl,\ heat}$ [o-toluidine: CH_3, NH_2] $\xrightarrow{HONO}$

[diazonium: CH_3, N_2^+, X^-] $\xrightarrow[heat]{H_3O^+}$ [o-cresol: CH_3, OH]

(b) [m-toluidine: CH_3, NH_2] $\xrightarrow[(2)\ H_3O^+,\ heat]{(1)\ HONO}$ [m-cresol: CH_3, OH]

(from Problem 21.17a)

(c) [p-nitrotoluene: CH_3, NO_2] $\xrightarrow[(2)\ OH^-]{(1)\ Fe,\ HCl,\ heat}$ [p-toluidine: CH_3, NH_2] $\xrightarrow[(2)\ H_3O^+,\ heat]{(1)\ HONO}$ [p-cresol: CH_3, OH]

(from part a)

(d)

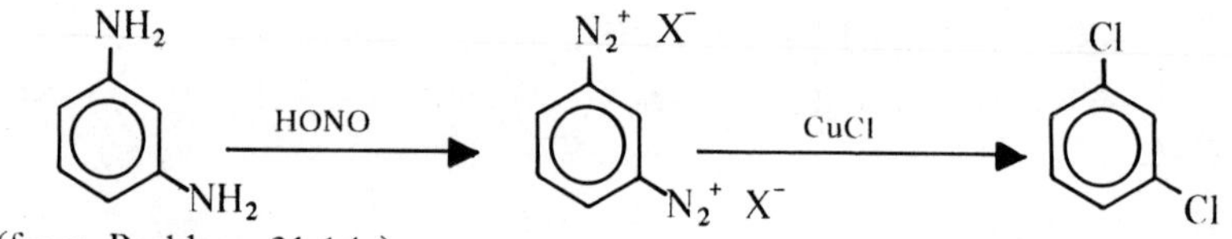

(from Problem 21.14a)

(e)

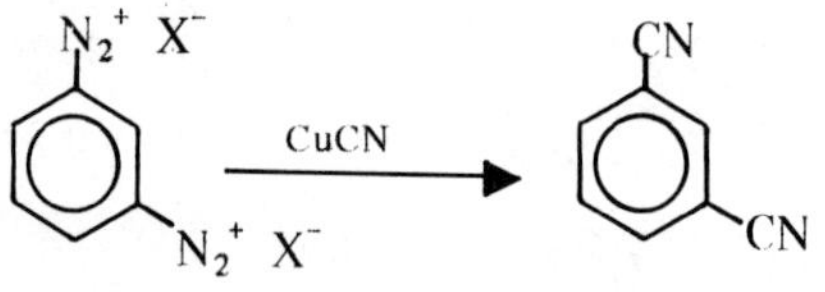

(cf. part d)

(f)

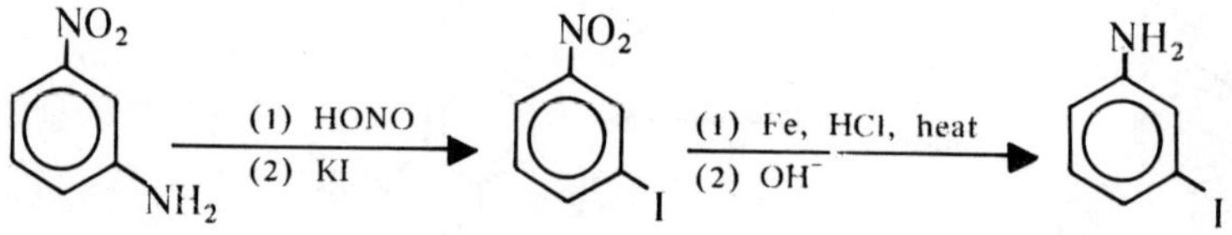

(from Problem 21.14 a)

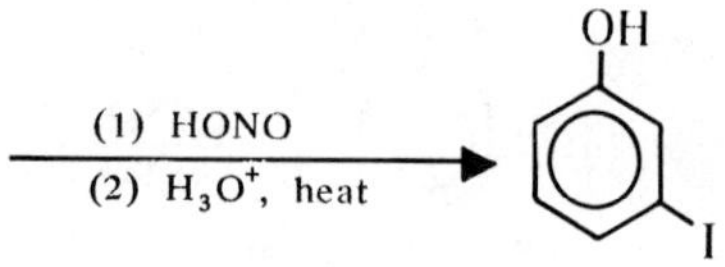

(g)

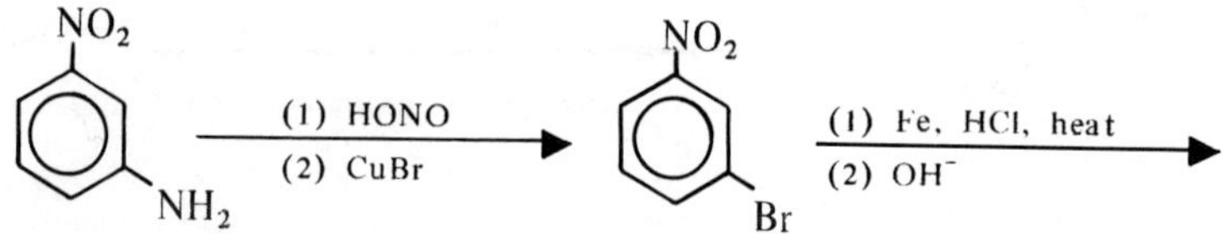

(from Problem
21.14a)

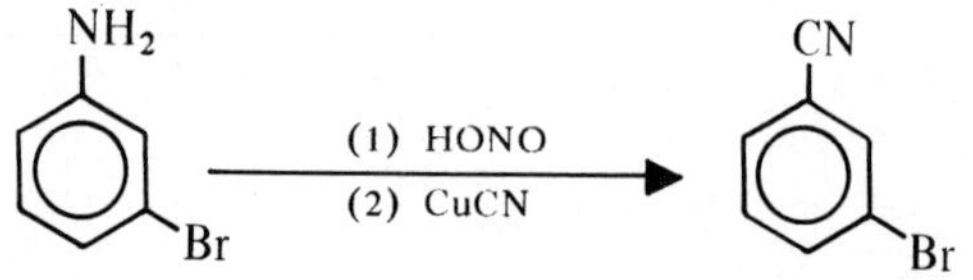

(h)

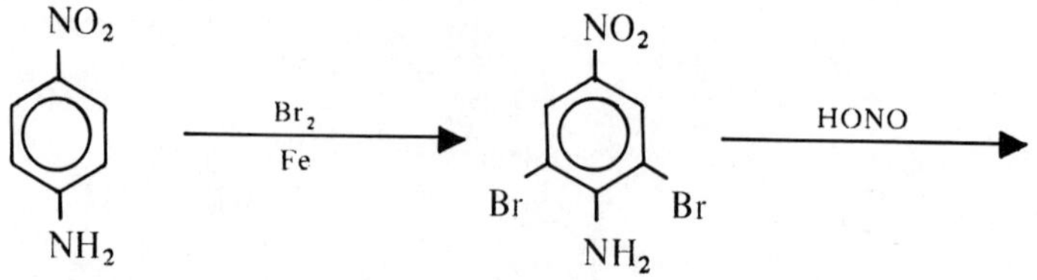

(from Problem 21.14f)

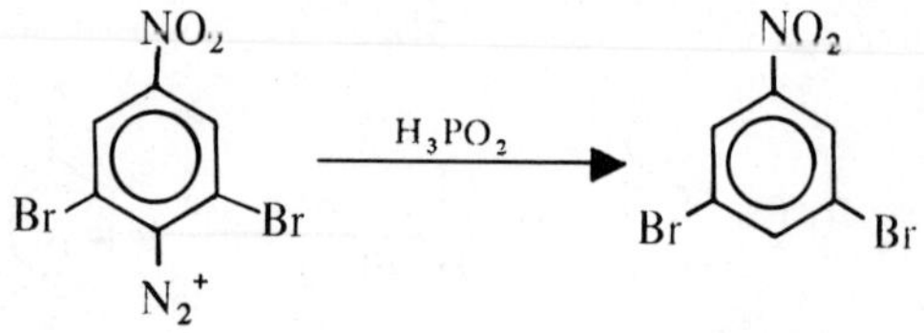

(i)

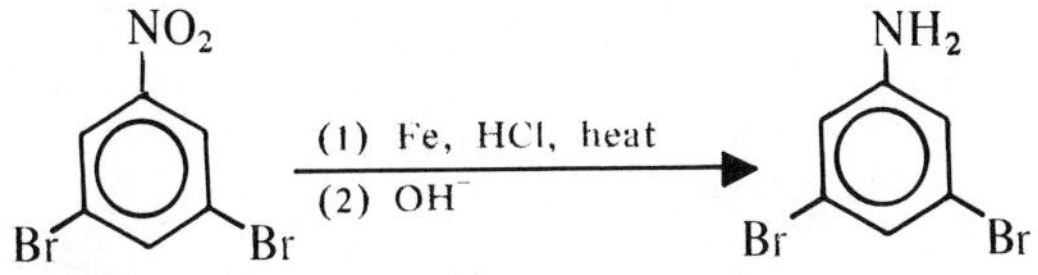

(from part h)

(j)

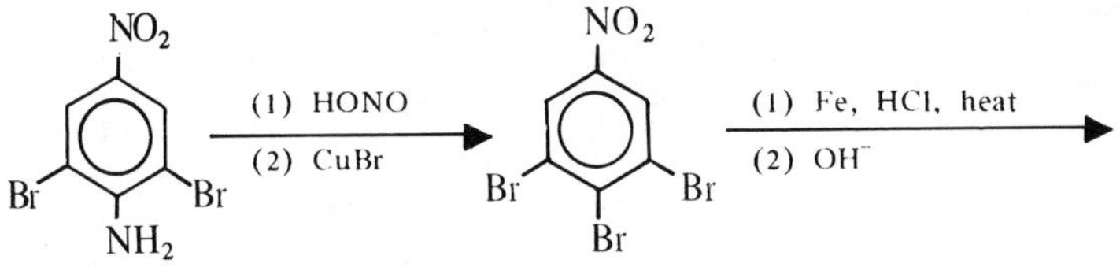

(from part h)

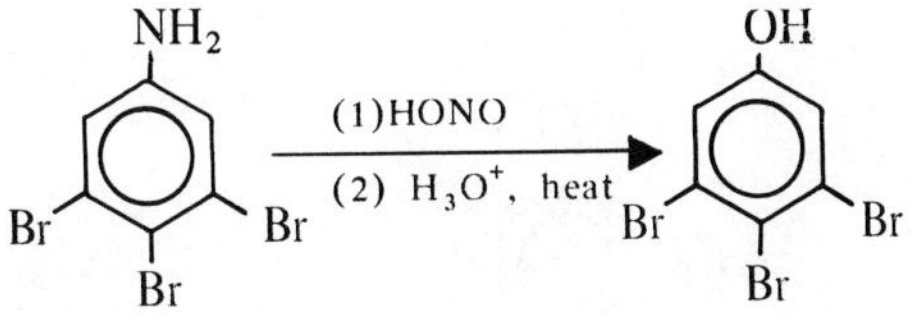

(k)

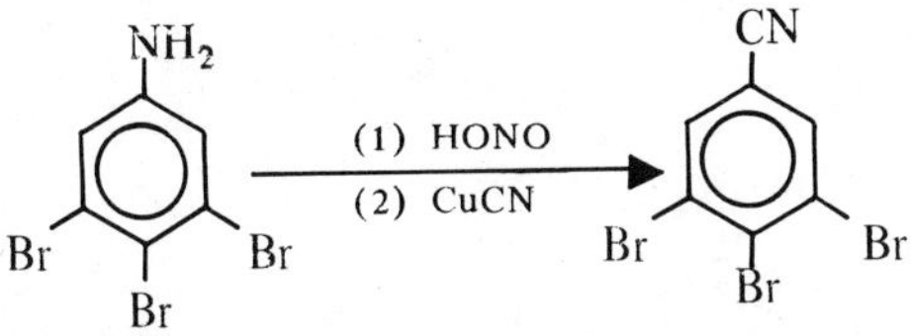

(from part j)

(l)

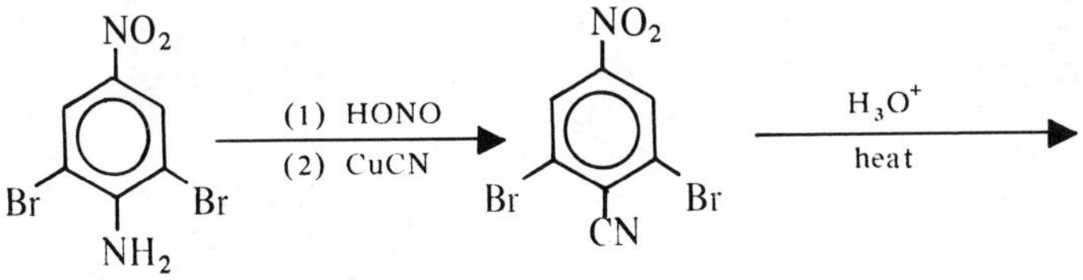

(from part h)

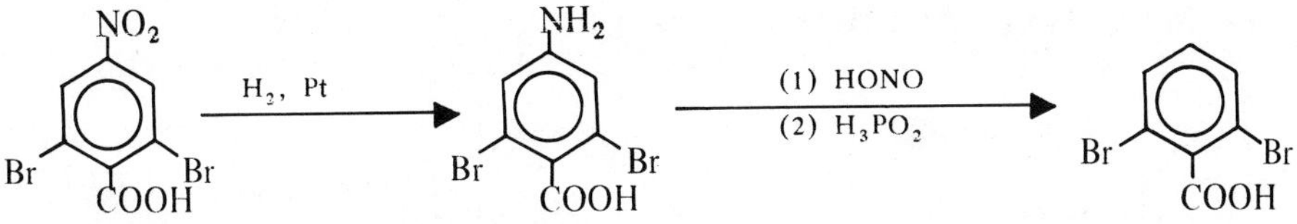

(m)

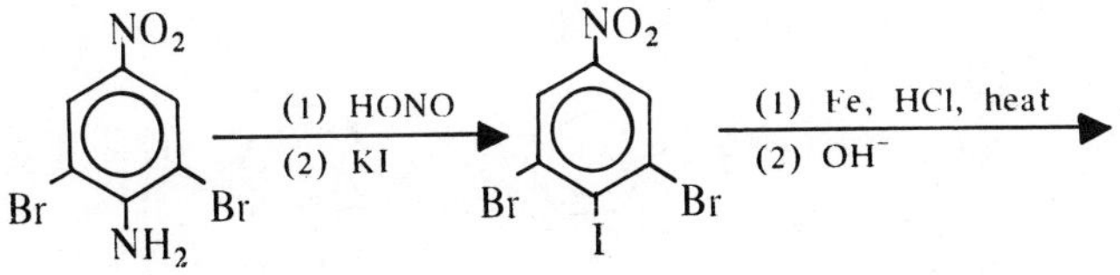

(cont. on p. 248)

(from part h)

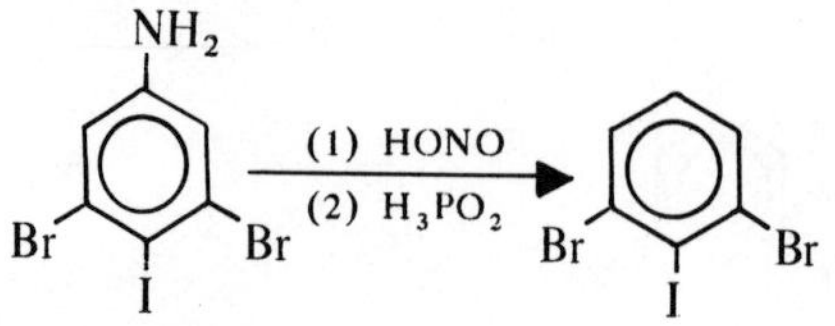

(n)

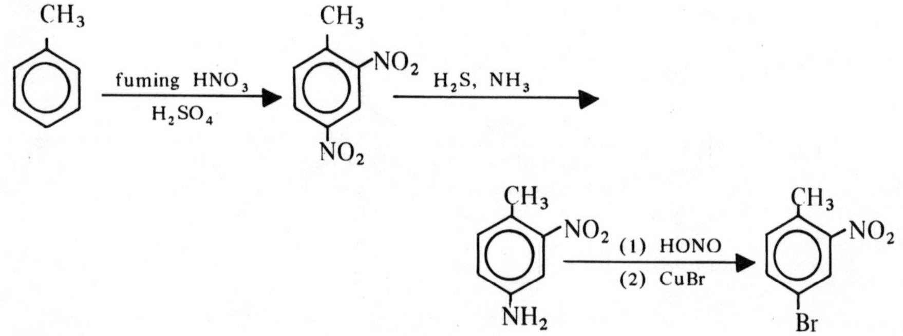

(o) 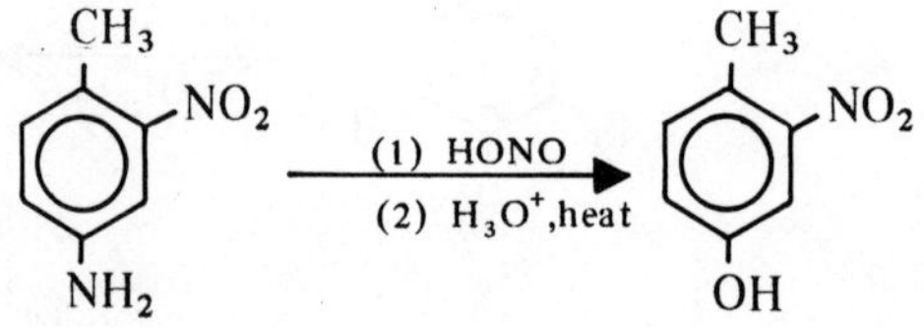 (from part n)

(p) (from part n)

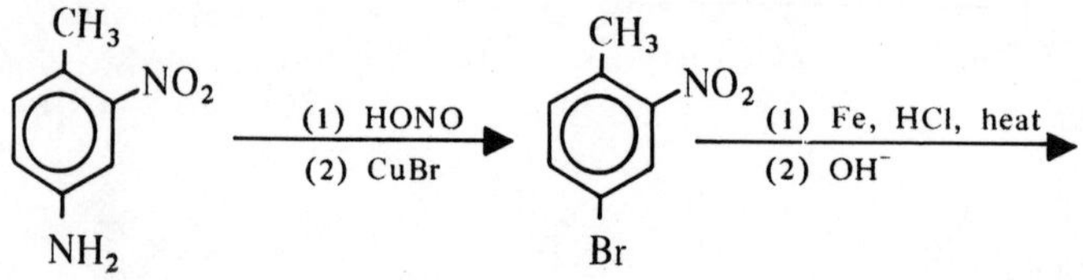

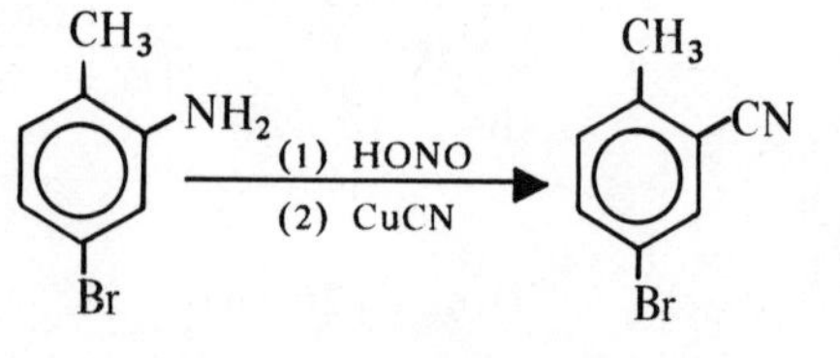

(q) CH₃—⟨⟩—NH₂ →(HONO)→ CH₃—⟨⟩—N₂⁺ X⁻

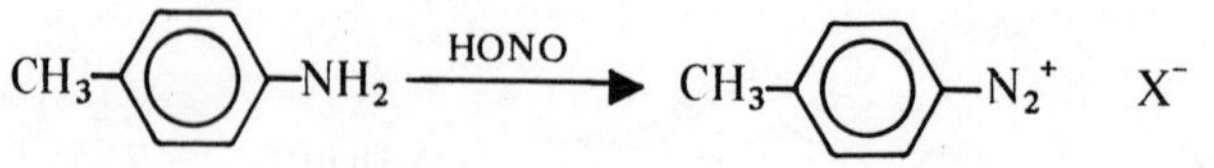

(from part c)

⟨⟩—OH, pH 8-10

CH₃—⟨⟩—N=N—⟨⟩—OH

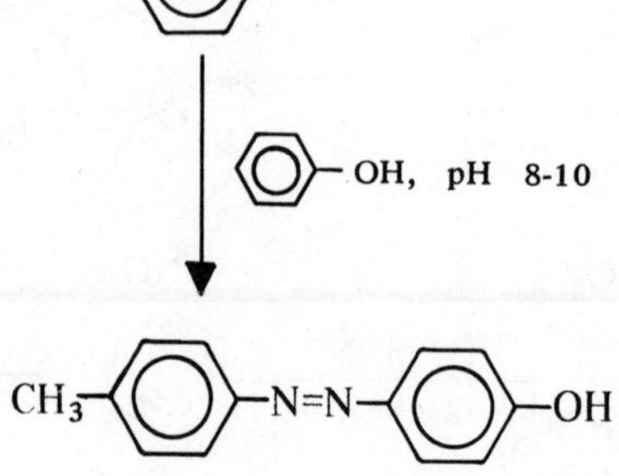

(r)

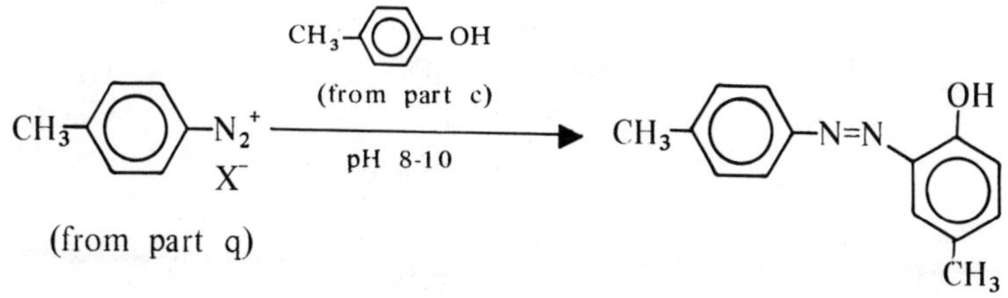

(from part q)

21.36

(a) Benzylamine dissolves in dilute HCl at room temperature,

$$C_6H_5CH_2NH_2 + H_3O^+ + Cl^- \xrightarrow{25°} C_6H_5CH_2\overset{+}{N}H_3Cl^-$$

benzamide does not dissolve:

$$C_6H_5CONH_2 + H_3O^+ + Cl^- \xrightarrow{25°} \text{No reaction}$$

(b) Allylamine reacts with (and decolorizes) bromine in carbon tetrachloride instantly,

$$CH_2{=}CHCH_2NH_2 + Br_2 \xrightarrow{CCl_4} \underset{\underset{Br\ \ Br}{|\ \ |}}{CH_2CHCH_2NH_2}$$

propylamine does not:

$$CH_3CH_2CH_2NH_2 + Br_2 \xrightarrow{CCl_4} \text{No reaction if the mixture is not heated or irradiated.}$$

(c) The Hinsberg test:

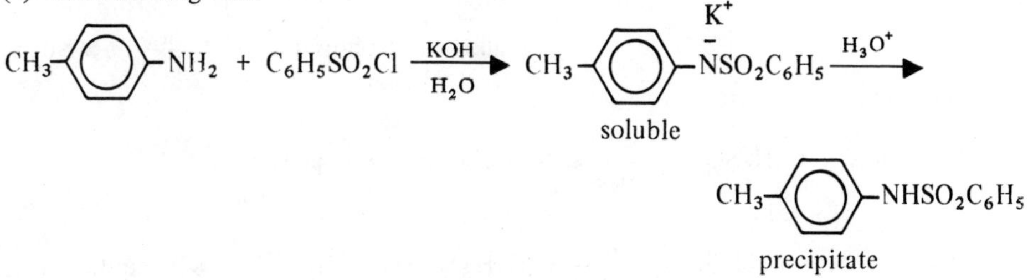

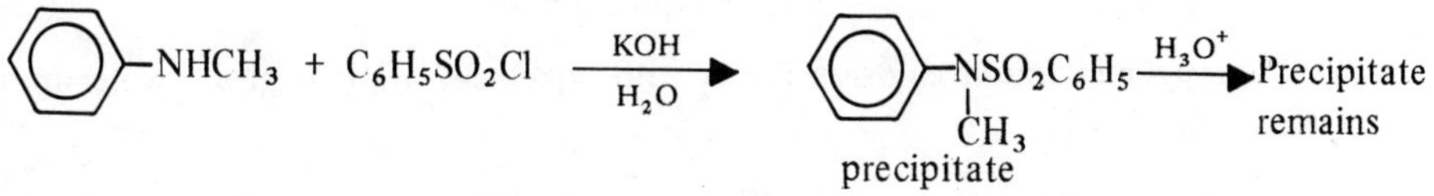

(d) The Hinsberg test:

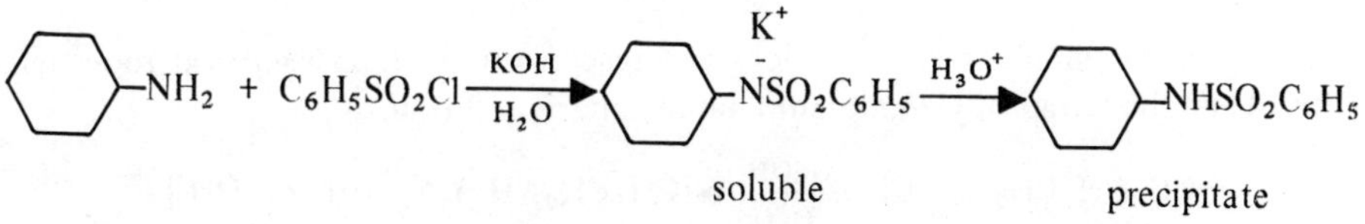

$$\text{(Cyclohexyl)N-H} + C_6H_5SO_2Cl \xrightarrow[H_2O]{KOH} \text{(Cyclohexyl)N-SO}_2C_6H_5 \xrightarrow{H_3O^+} \text{Precipitate remains}$$

precipitate

(e) Pyridine dissolves in dilute HCl,

$$\text{(Pyridine)N} + H_3O^+ + Cl^- \longrightarrow \text{(Pyridine)N}^+\text{-H} \quad Cl^-$$

benzene does not:

$$\text{(benzene)} + H_3O^+ + Cl^- \longrightarrow \text{No reaction}$$

(f) Aniline reacts with nitrous acid at 0-5° to give a stable diazonium salt that couples with 2-naphthol yielding an intensely colored azo compound.

$$\text{(Ph)-NH}_2 \xrightarrow[0-5°]{HONO} \text{(Ph)-N}_2^+ \xrightarrow{2\text{-naphthol}} \text{(Ph)-N=N-(naphthol-HO)}$$

Cyclohexylamine reacts with nitrous acid at 0-5° to yield a highly unstable diazonium salt—one that decomposes so rapidly that the addition of 2-naphthol gives no azo compound.

$$\text{(Cyclohexyl)-NH}_2 \xrightarrow[0-5°]{HONO} \left[\text{(Cyclohexyl)-N}_2^+\right] \xrightarrow{-N_2} \left[\text{(Cyclohexyl)}^+\right] \longrightarrow$$

alkenes, alcohols, etc. $\xrightarrow{2\text{-naphthol}}$ No reaction

(g) The Hinsberg test:

$$(C_2H_5)_3N + C_6H_5SO_2Cl \xrightarrow[H_2O]{KOH} \text{No reaction} \xrightarrow{H_3O^+} (C_2H_5)_3\overset{+}{N}H$$
soluble

$$(C_2H_5)_2NH + C_6H_5SO_2Cl \xrightarrow[H_2O]{KOH} (C_2H_5)_2NSO_2C_6H_5 \xrightarrow{H_3O^+} \text{Precipitate remains}$$
precipitate

(h) Tripropylammonium chloride reacts with aqueous NaOH to give a water insoluble tertiary amine.

$$(CH_3CH_2CH_2)_3\overset{+}{N}H\ Cl^- \xrightarrow[H_2O]{NaOH} (CH_3CH_2CH_2)_3N$$
water soluble ⟶ water insoluble

Tetrapropylammonium chloride does not react with aqueous NaOH (at room temperature) and the tetrapropylammonium ion remains in solution.

$$(CH_3CH_2CH_2)_4\overset{+}{N}\ Cl^- \xrightarrow[H_2O]{NaOH} (CH_3CH_2CH_2)_4\overset{+}{N}\ [Cl^-\ or\ OH^-]$$
water soluble ⟶ water soluble

(i) Tetrapropylammonium chloride dissolves in water to give a neutral solution. Tetrapropylammonium hydroxide dissolves in water to give a strongly basic solution.

21.37
Follow the procedure outlined in the answer to Problem 21.4. Toluene will show the same solubility behavior as benzene.

21.38

(a)

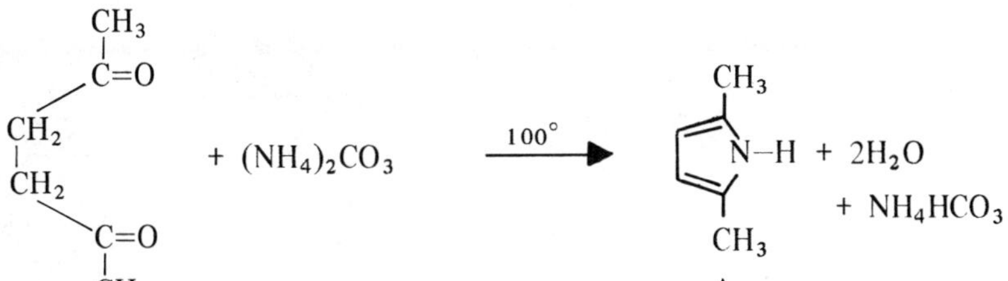

$$+ (NH_4)_2CO_3 \xrightarrow{100^\circ} A + 2H_2O + NH_4HCO_3$$

(b)

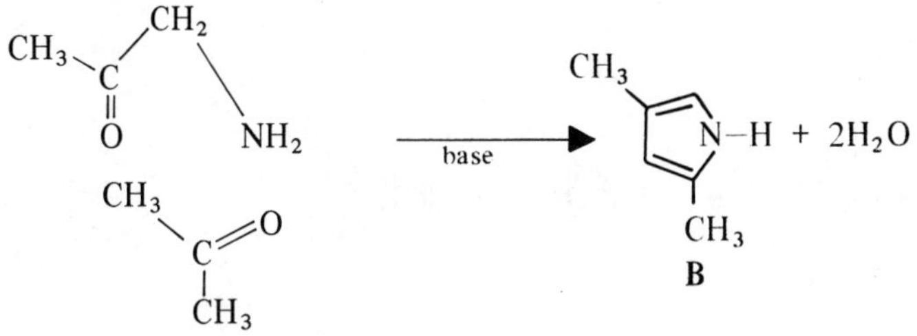

$$\xrightarrow{\text{base}} B + 2H_2O$$

(c)

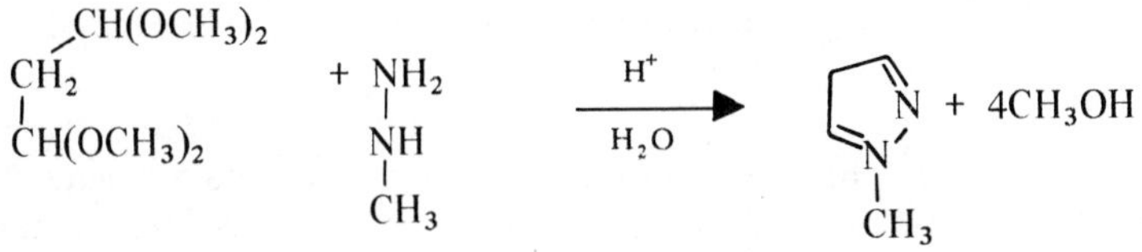

$$+ NH_2 \text{(amine)} \xrightarrow[H_2O]{H^+} C + 4CH_3OH$$

(d)

$$\xrightarrow{} D \xrightarrow{O_2} E$$

(e)

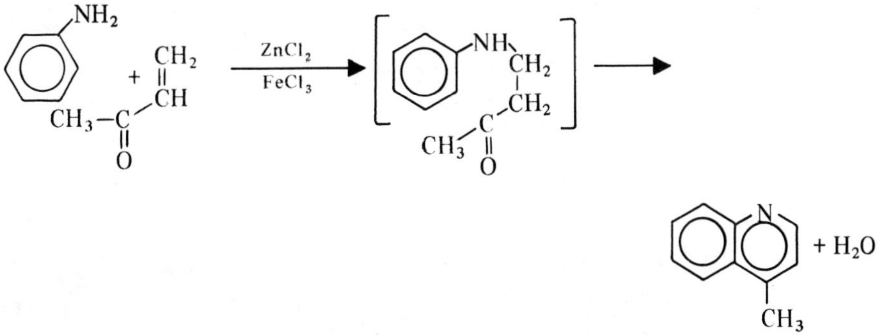

$$\xrightarrow[\text{FeCl}_3]{ZnCl_2} \left[\text{intermediate} \right] \longrightarrow F + H_2O$$

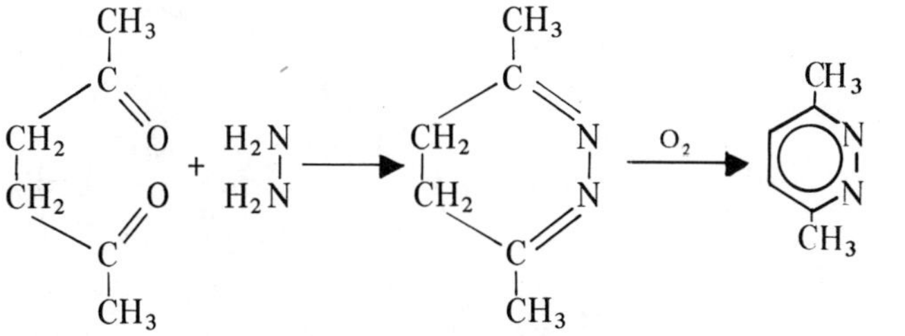

(f)

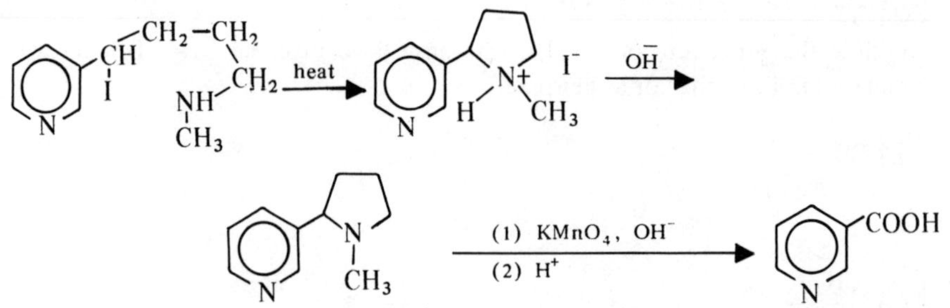

G
Nicotine

H
Nicotinic
acid

21.39

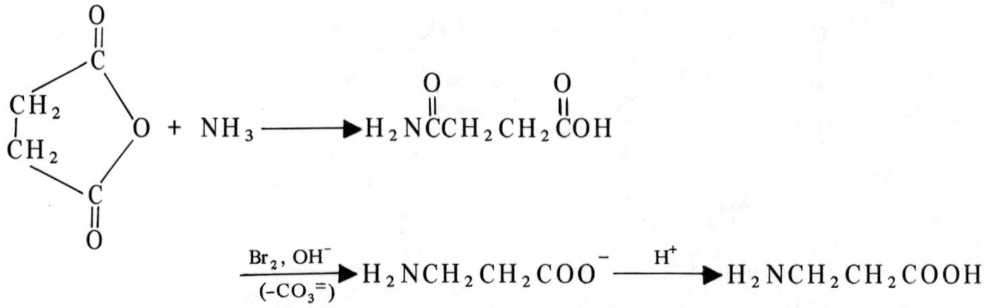

$$Br_2, OH^- \atop (-CO_3^=) \rightarrow H_2NCH_2CH_2COO^- \xrightarrow{H^+} H_2NCH_2CH_2COOH$$

21.40

Compound (b) would probably be inactive because of the meta orientation of the groups. Compound (d) would probably be inactive because the distance that separates the $-NH_2$ from the $-SO_2NH-$ group is too large.

21.41

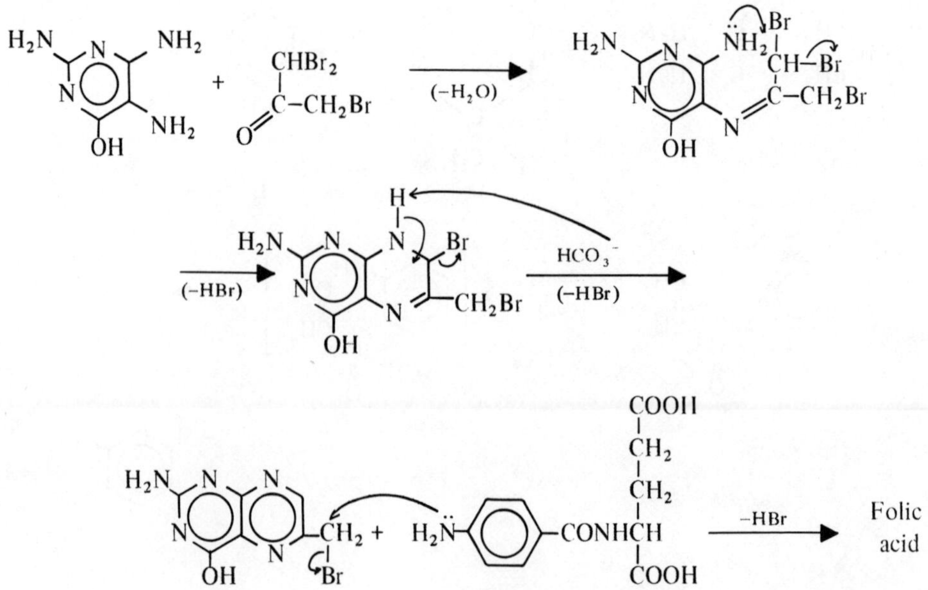

Folic
acid

21.42

The results of the Hinsberg test indicate that compound W is a tertiary amine. The pmr spectrum provides evidence for the following:

(1) Two different C_6H_5- groups (one absorbing at $\delta 7.2$ and one at $\delta 6.7$.)
(2) A CH_3CH_2- group (the quartet at $\delta 3.3$ and the triplet at $\delta 1.2$).
(3) An unsplit $-CH_2-$ group (the singlet at $\delta 4.4$).

There is only one reasonable way to put all of this together,

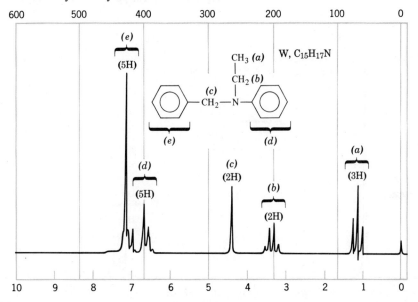

Thus W is N-benzyl-N-ethylaniline.

FIG. 21.3. The pmr spectrum of W (problem 21.42). (Spectrum courtesy of Aldrich Chemical Co.)

21.43

Compound X is benzyl bromide, $C_6H_5CH_2Br$. This is the only structure consistent with the pmr and infrared data. (The mono-substituted benzene ring is strongly indicated by the (5H), $\delta 7.3$ pmr absorption and is confirmed by the peaks at 690 cm^{-1} and 770 cm^{-1} in the infrared spectrum.)

Compound Y, therefore must be phenylacetonitrile, $C_6H_5CH_2CN$, and Z must be 2-phenylethylamine, $C_6H_5CH_2CH_2NH_2$.

$$\underset{\substack{X \\ (C_7H_7Br)}}{\boxed{\bigcirc}-CH_2Br} \xrightarrow{NaCN} \underset{\substack{Y \\ (C_8H_7N)}}{\boxed{\bigcirc}-CH_2CN} \xrightarrow{LiAlH_4} \underset{\substack{Z \\ (C_8H_{11}N)}}{\boxed{\bigcirc}-CH_2CH_2NH_2}$$

Interpretations of the infrared and pmr spectra of Z are given in Fig. 21.4.

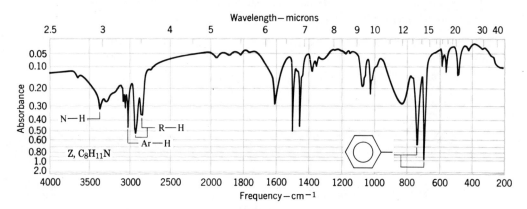

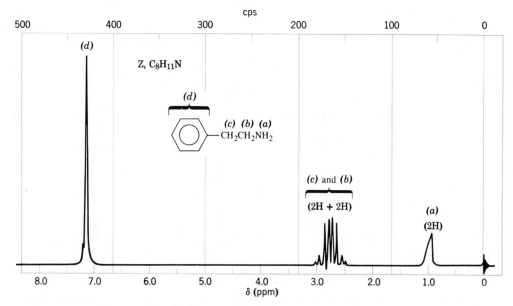

FIG. 21.4. Infrared and pmr spectra for compound Z, problem 21.43. (Spectra courtesy of Sadtler Inc.)

21.44

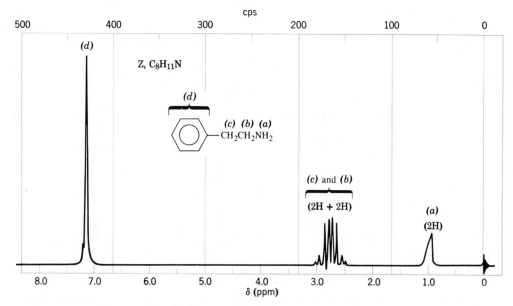

21.45

$$CH_3\overset{O}{\overset{\|}{C}}CH_2\overset{O}{\overset{\|}{C}}OC_2H_5 \ + \ C_6H_5NHNH_2 \ \xrightarrow{\text{heat}} \ CH_3C\text{---}CH_2\overset{O}{\overset{\|}{C}}OC_2H_5$$

with the $\underset{\underset{C_6H_5}{|}}{\underset{NH}{\overset{}{N}}}$ group

$\xrightarrow[\text{heating}]{\text{Further}}$ (pyrazolone ring structure) $\xrightarrow{CH_3OSO_2OCH_3}$ (methylated intermediate) $^-OSO_2OCH_3$ $\xrightarrow{NaOH}$ Antipyrine

22 SPECIAL TOPICS III

22.1

(a)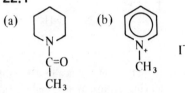

(b) pyridine with N+ CH₃, I⁻

(c) benzene ring with C(=O)-N (pyrrolidine) and C(-OH)=O

(d) pyrrolidinium with N+ CH₃ CH₃, I⁻

(e) $CH_2 = CHCH_2CH_2\underset{\underset{CH_3}{|}}{N} - CH_3$

22.2

(a) The cyclopentadienide ion (cf. p. 424).

(b) The pyrrole anion is a resonance hybrid of the following structures:

The imidazole anion is a hybrid of these:

22.3

A mechanism involving a "pyridyne" intermediate would involve a net loss (of 50%) of the deuterium label.

2-Pyridyne

Since in the actual experiment there was no loss of deuterium this mechanism was disallowed.

The mechanism given on page 861 would not be expected to result in a loss of deuterium, thus it is consistent with the labeling experiment.

22.4

When pyridine undergoes nucleophilic substitution, the leaving group is a hydride ion—an ion that is a strong base and, consequently, a poor leaving group. With 2-halopyridines, on the other hand, the leaving groups are halide ions—ions that are weak bases and thus good leaving groups.

22.5

(a) The first step is similar to a crossed-Claisen condensation:

(b) This step involves hydrolysis of an amide (lactam) and can be carried out with either acid or base. Here we use acid.

(c) This step is the decarboxylation of a substituted malonic acid; it requires only the application of heat and takes place during the acid hydrolysis of step (b).

(d) This is the reduction of a ketone to a secondary alcohol. A variety of reducing agents can be used, sodium borohydride, for example.

(e) Here we convert the secondary alcohol to an alkyl bromide with hydrogen bromide; this also gives a hydrobromide salt of the aliphatic amine.

$$\xrightarrow[\text{heat}]{\text{HBr}}$$

(pyridine ring)—$\overset{\overset{\displaystyle Br}{|}}{C}HCH_2CH_2CH_2\overset{\overset{\displaystyle H}{|}}{\underset{\underset{\displaystyle H}{|}}{N^+}}-CH_3 \quad Br^-$

(f) Treating the salt with base produces the secondary amine; it then acts as a nucleophile and attacks the carbon bearing the bromine. This leads to the formation of a five-membered ring and (±) nicotine.

$$\xrightarrow[\text{(−HBr)}]{\text{base}}$$
(pyridine ring)—$\overset{\overset{\displaystyle Br}{|}}{C}HCH_2CH_2CH_2\overset{\overset{\displaystyle H}{|}}{N}-CH_3$
$$\xrightarrow[\text{(−HBr)}]{\text{base}}$$
(nicotine structure)

(±) nicotine

22.6

(a) The chiral carbon adjacent to the ester carbonyl group is racemized by base (probably through the formation of an anion that can undergo inversion of configuration, cf. Problem 18.24).

(b)

(tropine ester structure with N–CH₃, O, O=C, C₆H₅, H, CH₂OH)

22.7

(a)

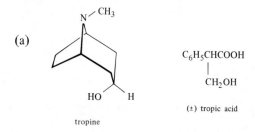

tropine

$C_6H_5CHCOOH$
|
CH_2OH

(±) tropic acid

(b) Tropine is a meso compound; it has a plane of symmetry that passes through the $>$CHOH group, the $>$NCH$_3$ group, and between the two −CH$_2$− groups of the five-membered ring.

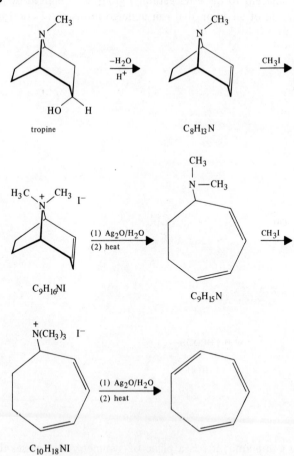

22.9
One possible sequence of steps is the following:

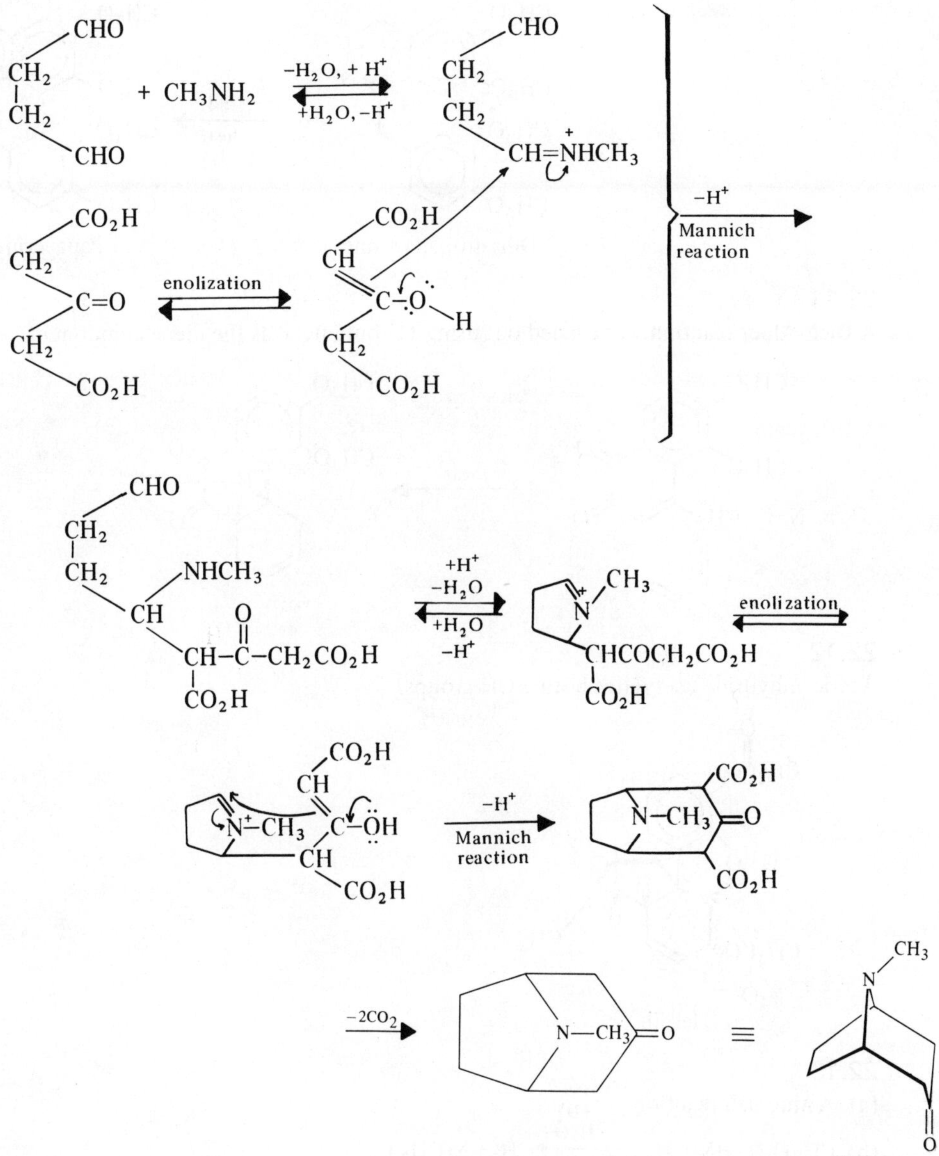

CHO
CH$_2$
CH$_2$
CHO
$+$ CH$_3$NH$_2$ $\xrightarrow[\text{+H}_2\text{O, -H}^+]{\text{-H}_2\text{O, + H}^+}$ CHO
CH$_2$
CH$_2$
CH=$\overset{+}{\text{N}}$HCH$_3$

CO$_2$H
CH$_2$
C=O
CH$_2$
CO$_2$H $\xrightarrow{\text{enolization}}$ CO$_2$H
CH
C–$\overset{..}{\text{O}}$–H
CH$_2$
CO$_2$H

$\xrightarrow[\text{Mannich reaction}]{\text{-H}^+}$

CHO
CH$_2$
CH$_2$
CH NHCH$_3$
CH–C–CH$_2$CO$_2$H
CO$_2$H $\xrightarrow[\text{+H}_2\text{O}]{\substack{\text{+H}^+ \\ \text{-H}_2\text{O} \\ \text{-H}^+}}$ $\overset{+}{\text{N}}$–CH$_3$
CHCOCH$_2$CO$_2$H
CO$_2$H $\xrightarrow{\text{enolization}}$

CO$_2$H
CH
$\overset{+}{\text{N}}$–CH$_3$ C–$\overset{..}{\text{O}}$H
CH
CO$_2$H $\xrightarrow[\text{Mannich reaction}]{\text{-H}^+}$ N–CH$_3$
CO$_2$H
O
CO$_2$H

$\xrightarrow{\text{-2CO}_2}$ N—CH$_3$ O $\equiv$ CH$_3$
N
O
tropinone

22.10

CH$_3$O
CH$_3$O
NH$_2$
$+$ COCl
CH$_3$O
CH$_3$O $\xrightarrow{\text{OH}^-}$ CH$_3$O
CH$_3$O
O
N
H
CH$_3$O
CH$_3$O $\xrightarrow[\substack{\text{heat} \\ (-\text{H}_2\text{O})}]{\text{P}_2\text{O}_5}$

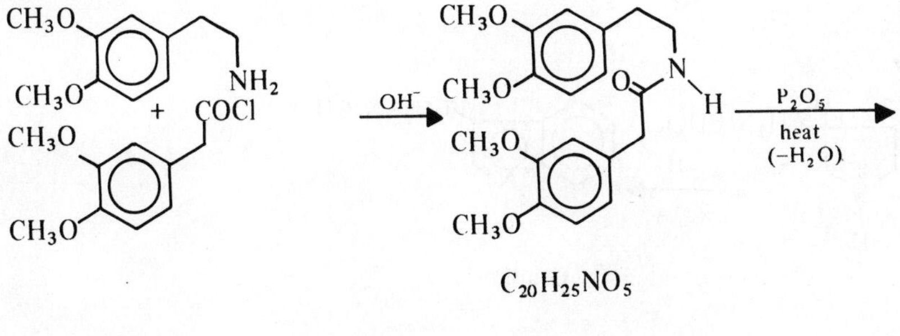

C$_{20}$H$_{25}$NO$_5$

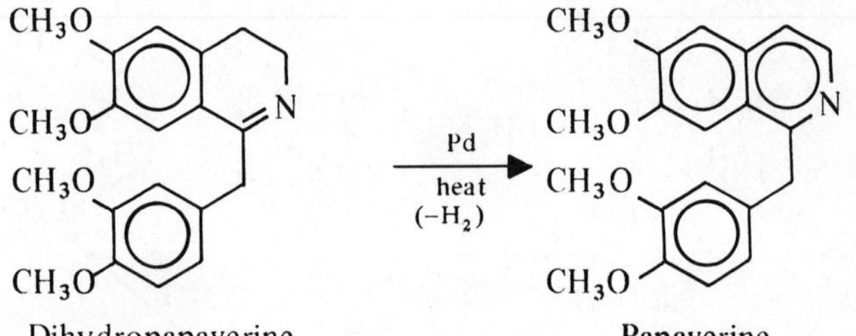

Dihydropapaverine   Papaverine

22.11

A Diels-Alder reaction was carried out using 1,3-butadiene as the diene component.

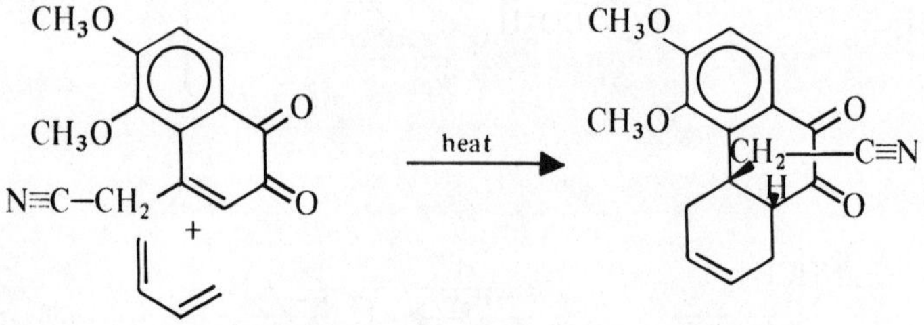

22.12

Acetic anhydride acetylates both —OH groups.

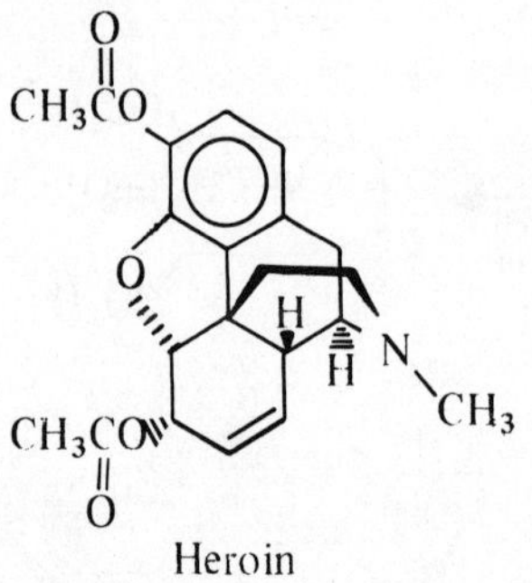

Heroin

22.13

(a) A Mannich reaction.

(b) $CH_2O + HN(CH_3)_2 \underset{\overset{-H_2O}{}}{\overset{+H^+}{\rightleftarrows}} CH_2\overset{+}{=}N(CH_3)_2$

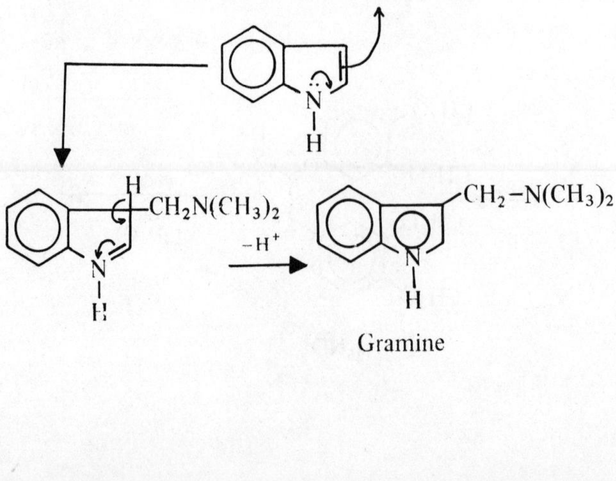

Gramine

22.14

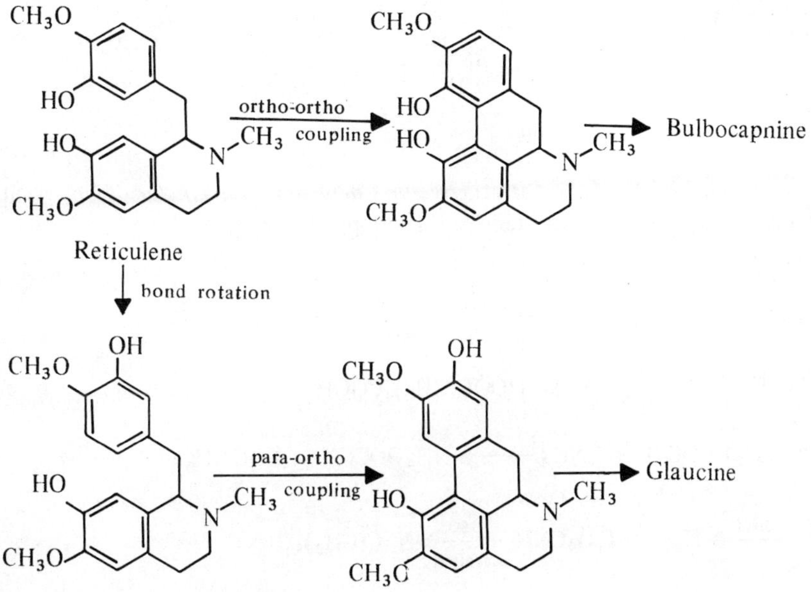

Reticulene

↓ bond rotation

22.15

Yes, because according to the pathway given on pages 871-872, carbons 1 and 3 of papaverine arise from the α-carbons of two molecules of tyrosine.

(b) Yes.

(c) Methylation of the four phenolic hydroxyl groups and dehydrogenation of the nitrogen-containing ring.

22.16

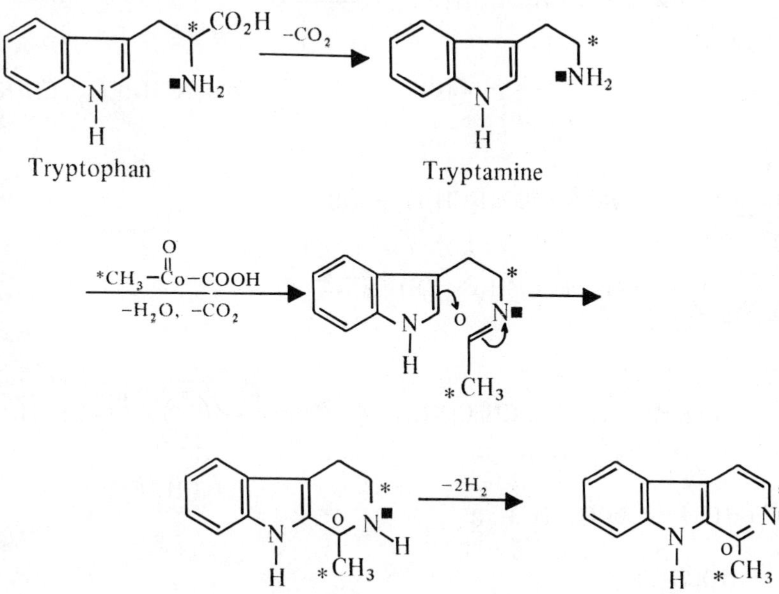

Tryptophan   Tryptamine

(cont. on p. 262)

$$\xrightarrow[\text{methylation}]{\text{hydroxylation}}$$

Harmine

Cf. T. A. Geissman and D. H. G. Crout, *Organic Chemistry of Secondary Plant Metabolism*, Freeman, Cooper & Co., San Francisco, 1969, pp. 473-474.

22.17

(a) $\xrightarrow[\text{H}_2\text{Cr}_2\text{O}_7]{(O)}$ HOOC(CH$_2$)$_4$COOH

(b) HOOC(CH$_2$)$_4$COOH + 2NH$_3$ $\longrightarrow$ NH$_4$OOC(CH$_2$)$_4$COONH$_4$

$\xrightarrow{\text{heat}}$ H$_2$N$\overset{\text{O}}{\overset{\|}{\text{C}}}$(CH$_2$)$_4$$\overset{\text{O}}{\overset{\|}{\text{C}}}NH_2$ $\xrightarrow[\text{catalyst}]{350°}$ N≡C(CH$_2$)$_4$C≡N

$\xrightarrow[\text{catalyst}]{4\text{H}_2}$ H$_2$NCH$_2$(CH$_2$)$_4$CH$_2$NH$_2$

(c) CH$_2$=CH–CH=CH$_2$ $\xrightarrow{\text{Cl}_2}$ ClCH$_2$CH=CHCH$_2$Cl $\xrightarrow{2\text{NaCN}}$

N≡CCH$_2$CH=CHCH$_2$C≡N $\xrightarrow[\text{Ni}]{\text{H}_2}$ N≡C(CH$_2$)$_4$C≡N

$\xrightarrow[\text{catalyst}]{4\text{H}_2}$ H$_2$NCH$_2$(CH$_2$)$_4$CH$_2$NH$_2$

(d) $\xrightarrow{2\text{HCl}}$ ClCH$_2$CH$_2$CH$_2$CH$_2$Cl $\xrightarrow{2\text{NaCN}}$

N≡C(CH$_2$)$_4$C≡N $\xrightarrow[\text{catalyst}]{4\text{H}_2}$ H$_2$NCH$_2$(CH$_2$)$_4$CH$_2$NH$_2$

22.18

(a) HOCH$_2$CH$_2$OH + :B $\rightleftharpoons$ HOCH$_2$CH$_2$O$^-$ + HB

ROC$\overset{\text{O}}{\overset{\|}{}}$—(benzene ring)—$\overset{\text{O}}{\overset{\|}{\text{C}}}OCH_3$ + $^-$OCH$_2$CH$_2$OH $\rightleftharpoons$

ROC$\overset{\text{O}}{\overset{\|}{}}$—(benzene ring)—$\overset{\text{O}^-}{\underset{\text{OCH}_3}{\overset{|}{\text{C}}}}$—OCH$_2CH_2$OH $\rightleftharpoons$ ROC$\overset{\text{O}}{\overset{\|}{}}$—(benzene ring)—$\overset{\text{O}}{\overset{\|}{\text{C}}}OCH_2CH_2$OH

+ CH$_3$O$^-$

CH$_3$O$^-$ + HB $\rightleftharpoons$ CH$_3$OH + :B$^-$

R = CH$_3$– or HOCH$_2$CH$_2$–

(b)

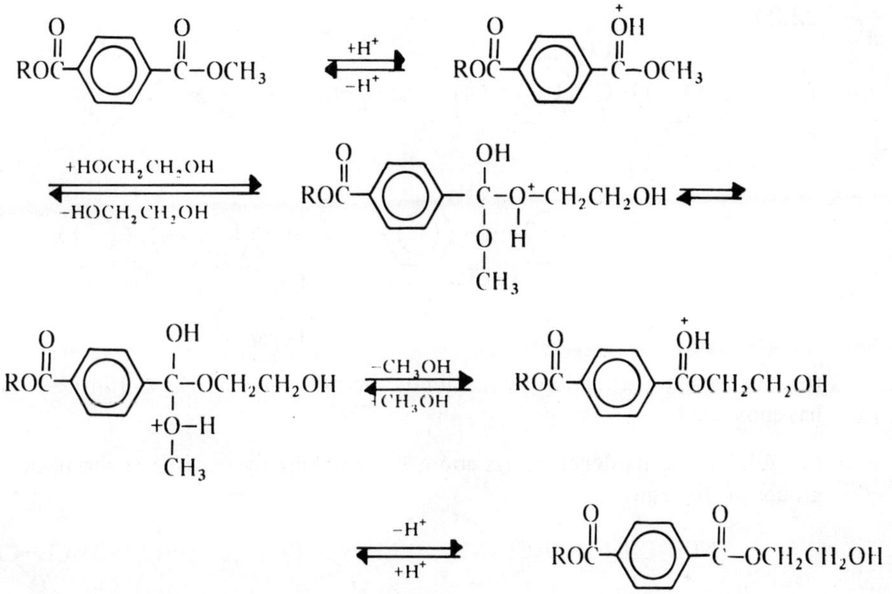

R = CH$_3$- or HOCH$_2$CH$_2$-

22.19

(a)

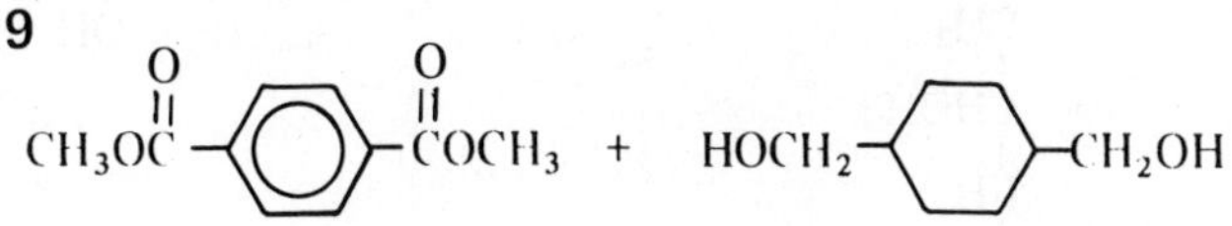

(b) By high-pressure catalytic hydrogenation

22.20

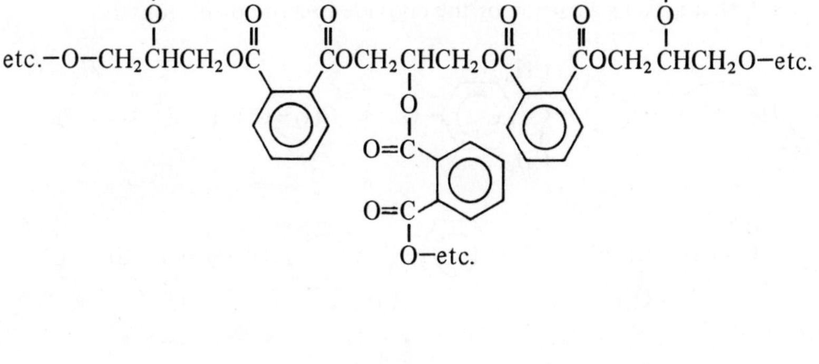

22.21

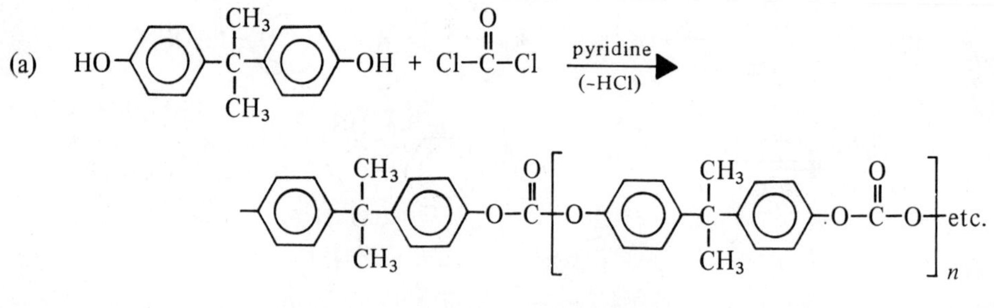

(a)

Lexan

(b) The excess of epichlorohydrin limits the molecular weight and insures that the resin has epoxy ends.

(c) Adding the hardener brings about cross linking by reacting at the terminal epoxide groups of the resin:

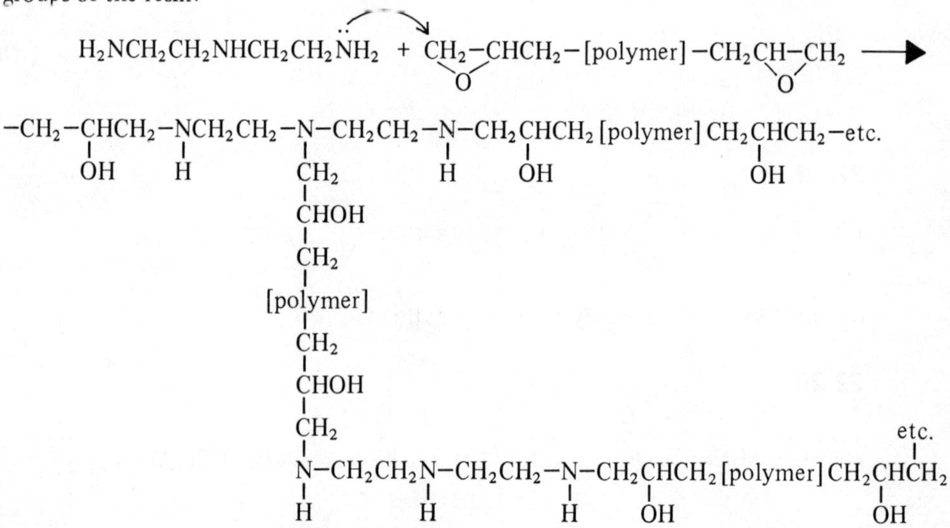

$$H_2NCH_2CH_2NHCH_2CH_2NH_2 \;+\; CH_2-CHCH_2-[\text{polymer}]-CH_2CH-CH_2 \longrightarrow$$

22.22

(a) The resin is probably formed in the following way. Base converts the bisphenol A to a phenoxide ion that attacks a carbon of the epoxide ring of epichlorohydrin:

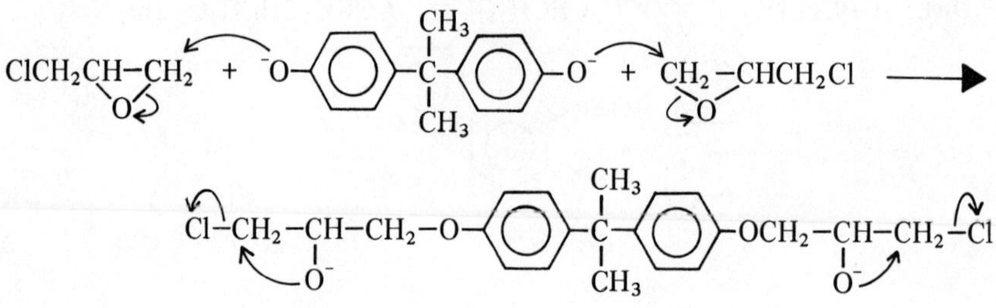

$$\xrightarrow{-2\,Cl^-} \quad \underset{O}{CH_2-CHCH_2O}\!\!-\!\!\bigcirc\!\!-\!\!\overset{CH_3}{\underset{CH_3}{C}}\!\!-\!\!\bigcirc\!\!-\!\!OCH_2CH-CH_2$$

$$\underset{CH_3}{\overset{CH_3}{^-O}}\!\!-\!\!\bigcirc\!\!-\!\!\overset{CH_3}{\underset{CH_3}{C}}\!\!-\!\!\bigcirc\!\!-\!\!O^- \xrightarrow{} \text{then} \xrightarrow{\underset{O}{CH_2-CHCH_2Cl}}$$

$$\underset{O}{CH_2-CHCH_2}\!\!-\!\!\left[O\!\!-\!\!\bigcirc\!\!-\!\!\overset{CH_3}{\underset{CH_3}{C}}\!\!-\!\!\bigcirc\!\!-\!\!OCH_2\underset{OH}{CHCH_2}\right]_n\!\!-\!\!O\!\!-\!\!\bigcirc\!\!-\!\!\overset{CH_3}{\underset{CH_3}{C}}\!\!-\!\!\bigcirc\!\!-\!\!\underset{O}{OCH_2CH-CH_2}$$

23
SPECIAL TOPICS IV LIPIDS

23.1

(a) $CH_3(CH_2)_{16}COOH + C_2H_5OH \underset{}{\overset{H^+}{\rightleftharpoons}} CH_3(CH_2)_{16}COOC_2H_5 + H_2O$

$CH_3(CH_2)_{16}COOH \xrightarrow{SOCl_2} CH_3(CH_2)_{16}COCl \xrightarrow{C_2H_5OH} CH_3(CH_2)_{16}COOC_2H_5$

(b) $CH_3(CH_2)_{16}COCl \xrightarrow{(CH_3)_3COH} CH_3(CH_2)_{16}COOC(CH_3)_3$

(c) $CH_3(CH_2)_{16}COCl \xrightarrow{NH_3} CH_3(CH_2)_{16}CONH_2$

(d) $CH_3(CH_2)_{16}COCl \xrightarrow{(CH_3)_2NH} CH_3(CH_2)_{16}CON(CH_3)_2$

(e) $CH_3(CH_2)_{16}CONH_2 \xrightarrow{LiAlH_4} CH_3(CH_2)_{16}CH_2NH_2$

(f) $CH_3(CH_2)_{16}CONH_2 \xrightarrow{Br_2, OH^-} CH_3(CH_2)_{15}CH_2NH_2$

(g) $CH_3(CH_2)_{16}COCl \xrightarrow{LiAlH[OC(CH_3)_3]_3} CH_3(CH_2)_{16}CHO$

(h) $CH_3(CH_2)_{16}COOC_2H_5 \xrightarrow{H_2, Ni} CH_3(CH_2)_{16}CH_2OH \longrightarrow$

$CH_3(CH_2)_{16}COCl \longrightarrow$

$CH_3(CH_2)_{16}COOCH_2(CH_2)_{16}CH_3$

(i) $CH_3(CH_2)_{16}COOH \xrightarrow[(2)\ H_2O]{(1)\ LiAlH_4} CH_3(CH_2)_{16}CH_2OH$

$CH_3(CH_2)_{16}COOC_2H_5 \xrightarrow{H_2, Ni} CH_3(CH_2)_{16}CH_2OH$

(j) $CH_3(CH_2)_{16}COCl + (CH_3)_2Cd \longrightarrow CH_3(CH_2)_{16}COCH_3$

$\qquad\qquad$ or $(CH_3)_2CuLi$

(k) $CH_3(CH_2)_{16}CH_2OH \xrightarrow{PBr_3} CH_3(CH_2)_{16}CH_2Br$

(l) $CH_3(CH_2)_{16}CH_2Br \xrightarrow[(2)\ H^+,\ H_2O,\ heat]{(1)\ NaCN} CH_3(CH_2)_{16}CH_2COOH$

23.2

(a) $CH_3(CH_2)_{11}CH_2COOH \xrightarrow{Br_2,\ P} CH_3(CH_2)_{11}\underset{\underset{Br}{|}}{C}HCOOH$

(b) $CH_3(CH_2)_{11}\underset{\underset{Br}{|}}{C}HCOOH \xrightarrow[\text{(2) } H^+]{\text{(1) } OH^-, \text{ heat}} CH_3(CH_2)_{11}\underset{\underset{OH}{|}}{C}HCOOH$

(c) $CH_3(CH_2)_{11}\underset{\underset{Br}{|}}{C}HCOOH \xrightarrow[\text{(2) } H^+]{\text{NaCN}} CH_3(CH_2)_{11}\underset{\underset{CN}{|}}{C}HCOOH$

(d) $CH_3(CH_2)_{11}\underset{\underset{Br}{|}}{C}HCOOH \xrightarrow[\text{(2) } H^+]{\text{NH}_3 \text{ (excess)}} CH_3(CH_2)_{11}\underset{\underset{NH_2}{|}}{C}HCOOH \text{ or } CH_3(CH_2)_{11}\underset{\underset{{}^+NH_3}{|}}{C}HCOO^-$

23.3

(a) $CH_3(CH_2)_5CH=CH(CH_2)_7COOH \xrightarrow{I_2} CH_3(CH_2)_5CHICHI(CH_2)_7COOH$

(b) $CH_3(CH_2)_5CH=CH(CH_2)_7COOH \xrightarrow{H_2, \text{ Ni}} CH_3(CH_2)_{14}COOH$

(c) $CH_3(CH_2)_5CH=CH(CH_2)_7COOH \xrightarrow{KMnO_4} CH_3(CH_2)_5CHOHCHOH(CH_2)_7COOH$

(d) $CH_3(CH_2)_5CH=CH(CH_2)_7COOH \xrightarrow{HCl} CH_3(CH_2)_7CH_2CHCl(CH_2)_7COOH$
$+$
$CH_3(CH_2)_7CHClCH_2(CH_2)_7COOH$

23.4

(a) There are two sets of enantiomers, giving a total of four stereoisomers

(b)

(see next page)   (see next page)

(from previous page)   (from previous page)

(±)–*threo*–9, 10–dibromohexadecanoic acids

Formation of a bromonium ion at the other face of palmitoleic acid gives a result such that the *threo*-enantiomers are obtained as a racemic modification.

23.5

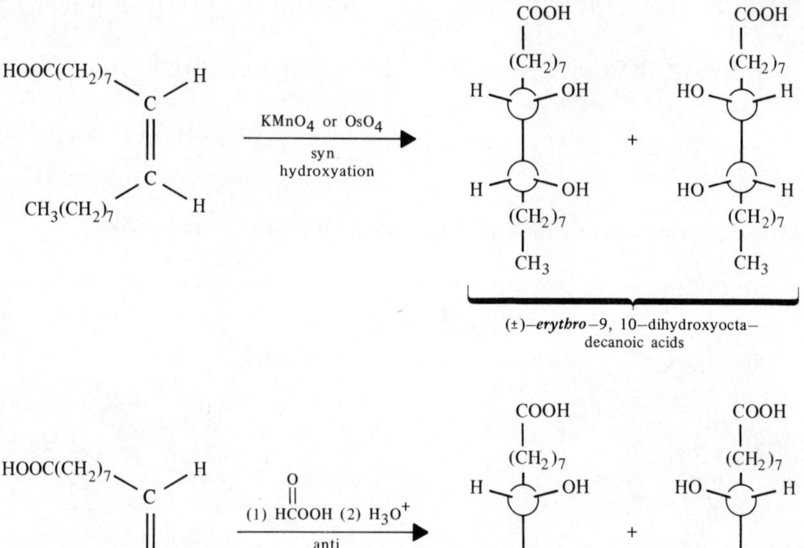

(±)–*erythro*–9, 10–dihydroxyocta–
decanoic acids

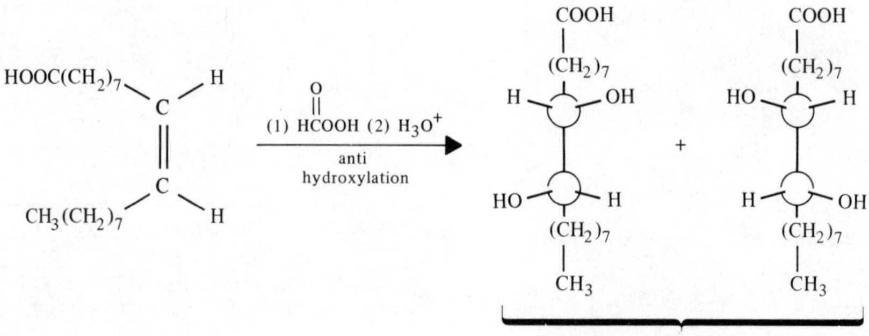

(±)–*threo*–9, 10–dihydroxyocta–
decanoic acids

The designations *erythro*- and *threo*- come from the names of the sugars called *erythrose* and *threose;* see page 935.

23.6

Elaidic acid is *trans*-9-octadecenoic acid:

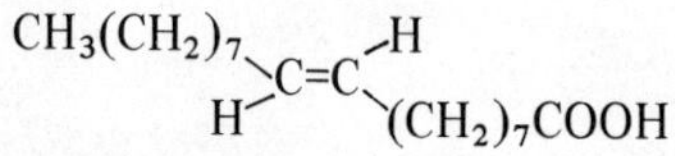

It is formed by the isomerization of oleic acid.

23.7

(a)

$$CH_3(CH_2)_9 \diagdown C=C \diagup (CH_2)_7COOH \quad \text{and} \quad CH_3(CH_2)_9 \diagdown C=C \diagup H$$
$$H \diagup \qquad \diagdown H \qquad\qquad\qquad H \diagup \qquad \diagdown (CH_2)_7COOH$$

(b) Infrared spectroscopy

(c) A peak in the 675-730 cm^{-1} region would indicate that the double bond is *cis*; a peak in the 960-975 cm^{-1} region would indicate that it is *trans*.

23.8

$$CH_3(CH_2)_5C{\equiv}CH + NaNH_2 \xrightarrow[NH_3]{} CH_3(CH_2)_5C{\equiv}CNa$$
$$\mathbf{A}$$

$$\xrightarrow{ICH_2(CH_2)_7CH_2Cl} CH_3(CH_2)_5C{\equiv}CCH_2(CH_2)_7CH_2Cl \xrightarrow{NaCN}$$
$$\mathbf{B}$$

$$CH_3(CH_2)_5C{\equiv}CCH_2(CH_2)_7CH_2CN \xrightarrow{KOH,\ H_2O} CH_3(CH_2)_5C{\equiv}CCH_2(CH_2)_7CH_2COOK$$
$$\mathbf{C} \qquad\qquad\qquad\qquad\qquad\qquad\qquad\qquad \mathbf{D}$$

$$\xrightarrow{H_3O^+} CH_3(CH_2)_5C{\equiv}CCH_2(CH_2)_7CH_2COOH \xrightarrow{H_2,\ Pd\text{-}BaSO_4}$$
$$\mathbf{E}$$

$$CH_3(CH_2)_5 \diagdown C=C \diagup CH_2(CH_2)_7CH_2COOH$$
$$H \diagup \qquad \diagdown H$$
Vaccenic acid

23.9

$$FCH_2(CH_2)_6CH_2Br + HC{\equiv}CNa \longrightarrow FCH_2(CH_2)_6CH_2C{\equiv}CH$$
$$\mathbf{F}$$

$$\xrightarrow[(2)\ I(CH_2)_7Cl]{(1)\ NaNH_2} FCH_2(CH_2)_6CH_2C{\equiv}C(CH_2)_7Cl \xrightarrow[DMSO]{NaCN}$$
$$\mathbf{G}$$

$$FCH_2(CH_2)_6CH_2C{\equiv}C(CH_2)_7CN \xrightarrow[(2)\ H^+]{(1)\ KOH} FCH_2(CH_2)_6CH_2C{\equiv}C(CH_2)_7COOH$$
$$\mathbf{H} \qquad\qquad\qquad\qquad\qquad\qquad\qquad\qquad \mathbf{I}$$

$$\xrightarrow[Pd\text{-}BaSO_4]{H_2} FCH_2(CH_2)_6CH_2 \diagdown C=C \diagup (CH_2)_7COOH$$
$$H \diagup \qquad \diagdown H$$

23.10

(a)
$$\begin{array}{l} CH_2OH \\ | \\ CHOH \\ | \\ CH_2OH \end{array} + R\overset{O}{\overset{\|}{C}}OH + R'\overset{O}{\overset{\|}{C}}OH + H_3PO_4 + HOCH_2CH_2\overset{+}{N}(CH_3)_3\ \ X^-$$

(b)
$$\begin{array}{l} CH_2OH \\ | \\ CHOH \\ | \\ CH_2OH \end{array} + R\overset{O}{\overset{\|}{C}}OH + R'\overset{O}{\overset{\|}{C}}OH + H_3PO_4 + HOCH_2CH_2NH_2$$

(c) $\underset{\displaystyle CH_2OH}{\overset{\displaystyle CH_2OH}{\underset{\displaystyle |}{\overset{\displaystyle |}{CHOH}}}}$ + $CH_3(CH_2)_nCH_2\overset{O}{\overset{\|}{CH}}$ + $R^1\overset{O}{\overset{\|}{COH}}$ + H_3PO_4

$+ \ HOCH_2CH_2\overset{+}{N}(CH_3)_3 \ X^-$

23.11

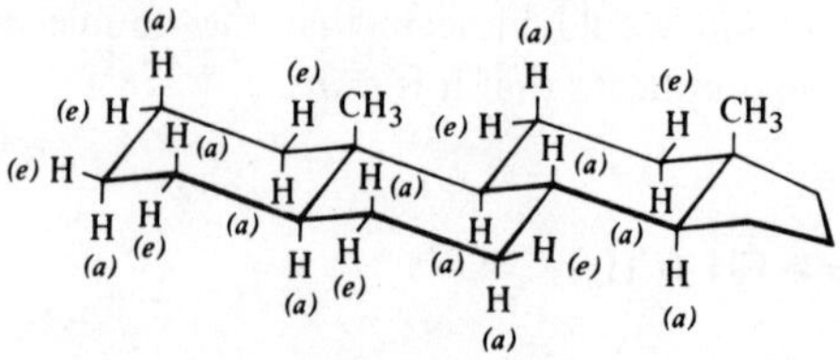

5α–series

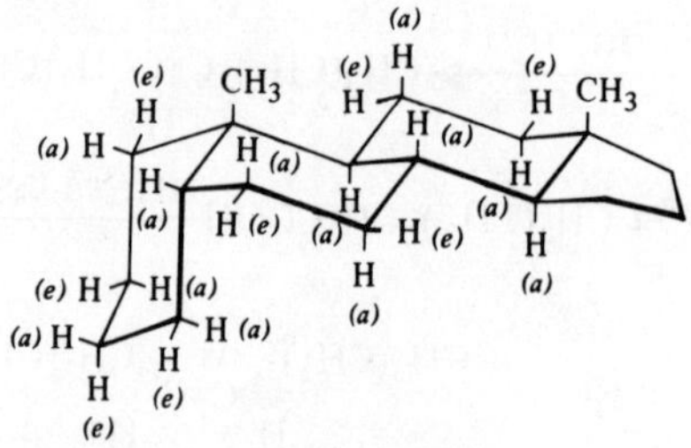

5β–series

23.12

(a)

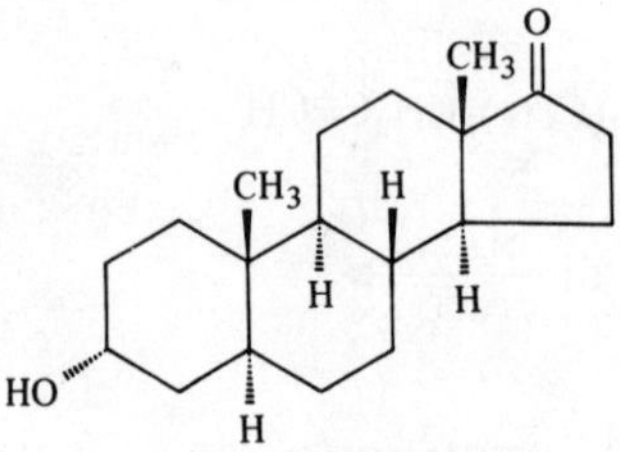

androstan–3α– ol –17–one
(androsterone)

(b)

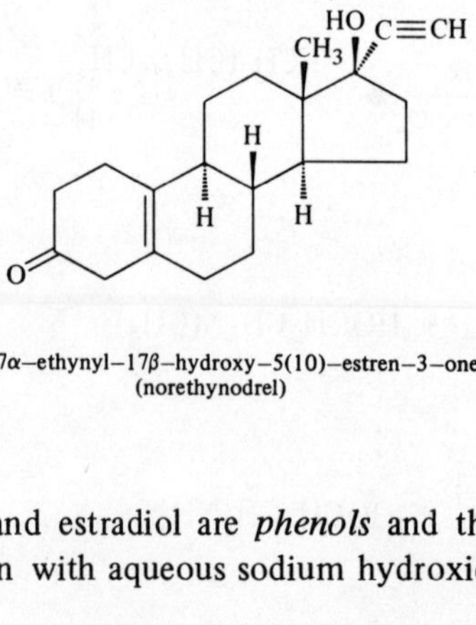

17α–ethynyl–17β–hydroxy–5(10)–estren–3–one
(norethynodrel)

23.13

Estrone and estradiol are *phenols* and thus are soluble in aqueous sodium hydroxide. Extraction with aqueous sodium hydroxide separates the estrogens from the androgens.

23.14

(a)

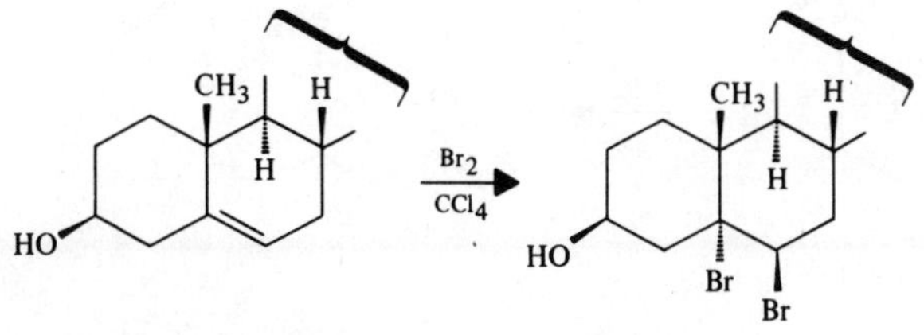

cholesterol   5α, 6β−dibromocholestan−3β− ol

(b)

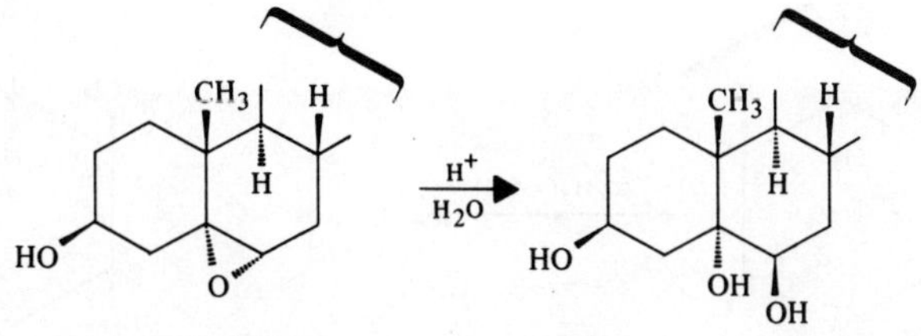

5α, 6α−oxidocholestan−3β− ol
(prepared by epoxidation of
cholesterol, cf. p.902)

cholestan−3β, 5α, 6β− triol

(c)

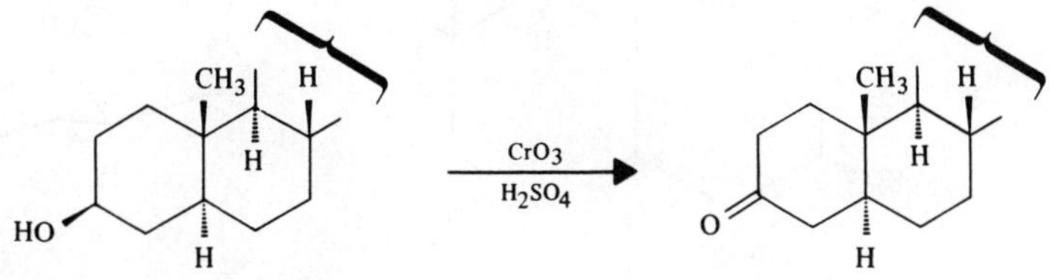

5α−cholestan−3β− ol
(prepared by hydrogenation
of cholesterol, ef. p. 902)

5α−cholestan−3−one

(d)

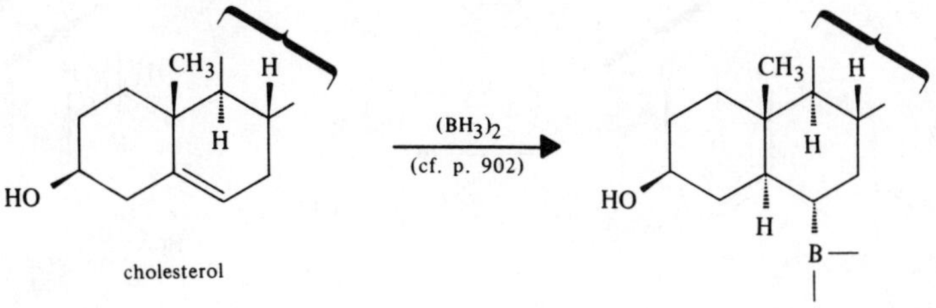

cholesterol

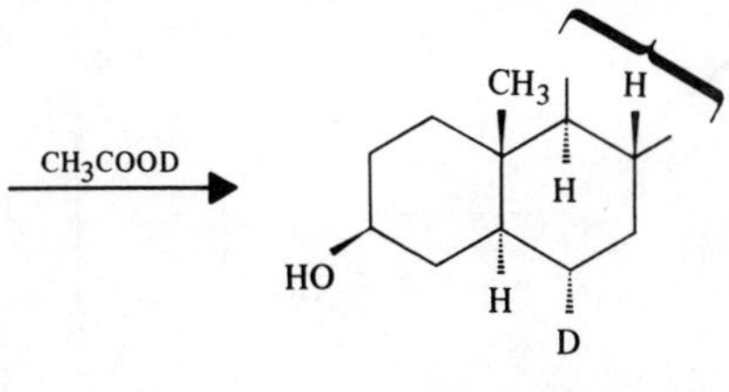

6α−deuterio−5α−cholestan−3β− ol

(e)

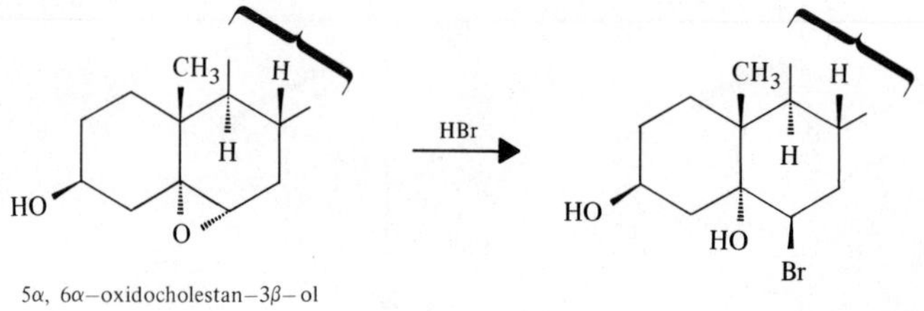

5α, 6α—oxidocholestan—3β—ol

HBr

6β—bromocholestan—3β, 5α—diol

23.15

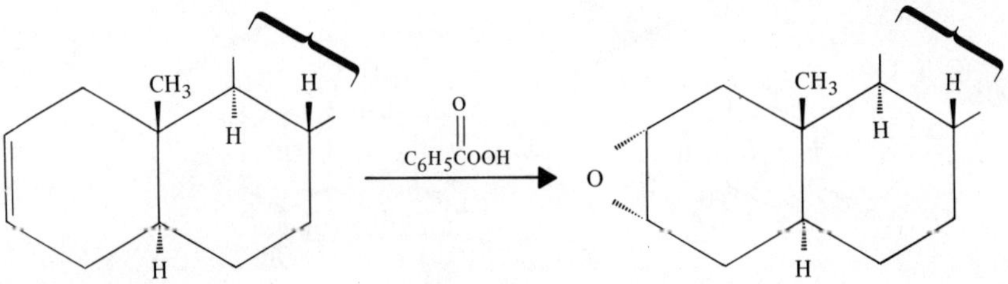

5α—cholest—2—ene

C_6H_5COOH

A

HBr

≡

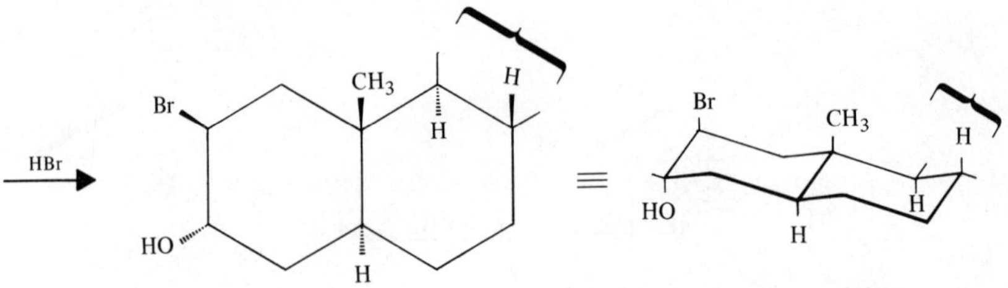

B

Here we find that epoxidation takes place at the less hindered α face (cf. p. 902). Ring opening by HBr takes place in an *anti* fashion to give a product with diaxial substituents.

23.16

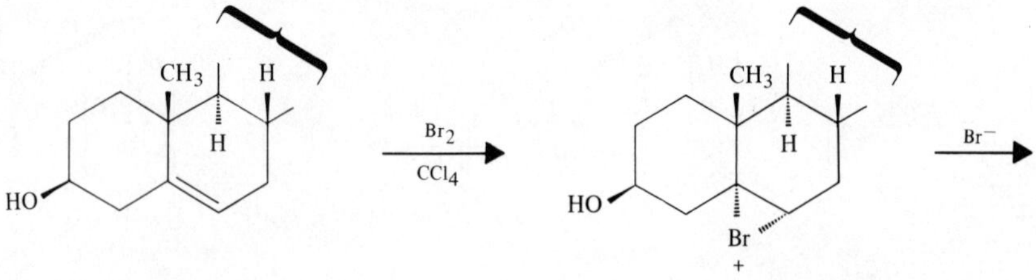

cholesterol

Br_2
CCl_4

5α, 6α—bromonium ion

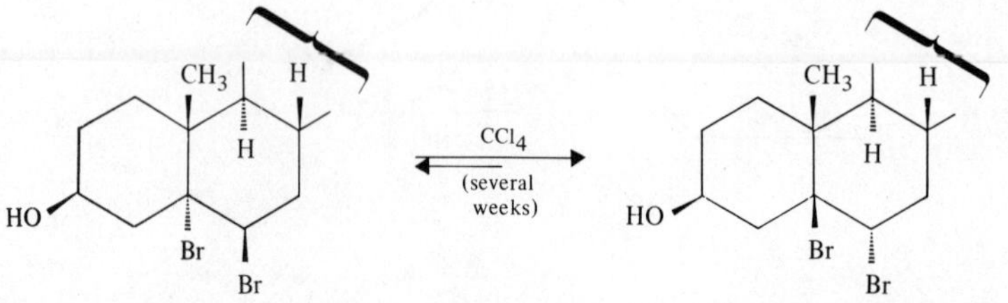

Br^-

+

5α, 6β—dibromocholestan—3β—ol

CCl_4
(several weeks)

5β, 6α—dibromocholestan—3β—ol

Here formation of the bromonium ion takes place preferentially at the α face. Ring opening takes place in an *anti* manner (with a bromide ion attacking the 6-position from above) to give, initially, the 5α,6β-dibromo compound. The 5α,6β-dibromocholestan-3β-ol, however is a *diaxial* dibromide and is, therefore, unstable. It isomerizes to the 5β, 6α-dibromocholestan-3β-ol (below)—a compound in which the bromine substituents are both equatorial.

5α, 6β–dibromide
(bromines are diaxial)

5β, 6α–dibromide
(bromines are diequatorial)

This isomerization does not result from a simple "flipping" of the cylohexane rings; it requires an inversion of configuration at carbons 5 and 6. One mechanism that has been proposed for the isomerization involves the formation of a "bromonium-bromide" ion pair:

diaxial   ion pair   diequatorial

23.17

(a) $CH_2=CH-CH=CH_2$

(b) OH⁻ (Removal of the α-hydrogen allows isomerization to the more stable compound with a *trans* ring junction.)

(c) $LiAlH_4$

(d) H_3O^+ and heat. (Hydrolysis of the enol ether is followed by dehydration of one alcohol group.)

(e) $HCOOC_2H_5$, C_2H_5ONa

(f) OsO_4

(g) $CH_3\overset{O}{\overset{\|}{C}}CH_3$, H^+

(h) H_2, Pd catalyst

(i) H_3O^+, H_2O

(j) Base and heat (This is an aldol condensation.)

(k) and (l) Na_2CrO_4, CH_3COOH to oxidize the aldehyde to an acid, followed by esterification.

(m) H_2 and Pt (Hydrogen addition takes place from the less hindered α-face of the molecule.)

(n), (o), (p) NaBH$_4$ to reduce the keto group; OH$^-$, H$_2$O to hydrolyze the ester; and acetic anhydride to esterify the OH at the 3-position.

(q) and (r) SOCl$_2$ to make the acid chloride, followed by treatment with (CH$_3$)$_2$Cd.

(s) CH$_3$CHCH$_2$CH$_2$CH$_2$MgBr, followed by H$_3$O$^+$.

with a CH$_3$ group on the CH.

(t), (u), (v) acetic acid and heat to dehydrate the 3° alcohol; followed by acetic anhydride to acetylate the 2° alcohol; followed by H$_2$, Pt to hydrogenate the double bond.

24

CARBOHYDRATES

24.1

(a) Two,
$$\begin{array}{c} CHO \\ | \\ {}^*CHOH \\ | \\ {}^*CHOH \\ | \\ CH_2OH \end{array}$$

(b) Two,
$$\begin{array}{c} CH_2OH \\ | \\ C=O \\ | \\ {}^*CHOH \\ | \\ {}^*CHOH \\ | \\ CH_2OH \end{array}$$

(c) There would be four stereoisomers (two sets of enantiomers) with each general structure: $2^2 = 4$.

24.2

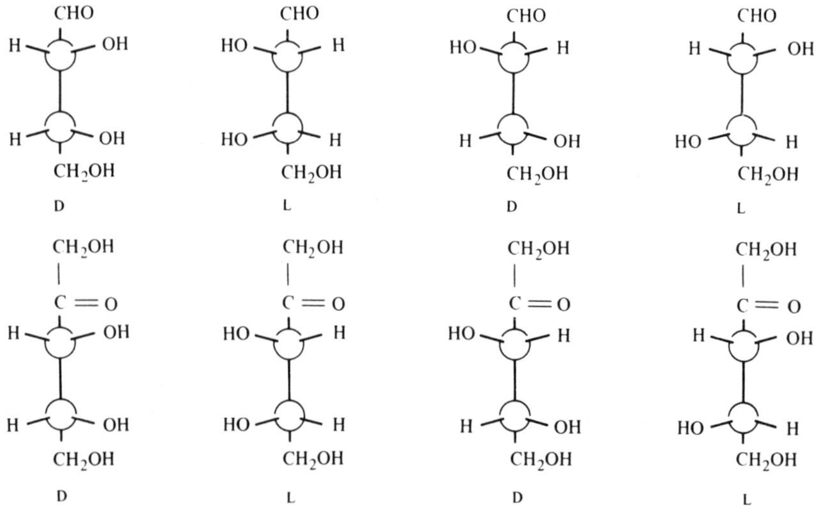

CHO · H—OH · H—OH · CH₂OH — D

CHO · HO—H · HO—H · CH₂OH — L

CHO · HO—H · H—OH · CH₂OH — D

CHO · H—OH · HO—H · CH₂OH — L

CH₂OH · C=O · H—OH · H—OH · CH₂OH — D

CH₂OH · C=O · HO—H · HO—H · CH₂OH — L

CH₂OH · C=O · HO—H · H—OH · CH₂OH — D

CH₂OH · C=O · H—OH · HO—H · CH₂OH — L

24.3

Since glycosides are acetals they undergo hydrolysis in aqueous acid to form cyclic hemiacetals that then undergo mutarotation.

24.4

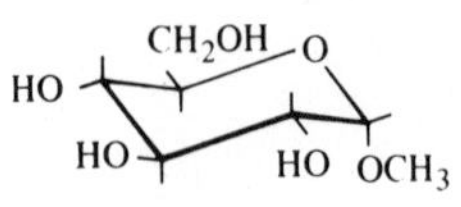

methyl–α D glucopyranoside

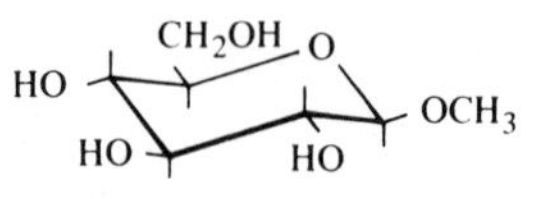

methyl–β D glucopyranoside

24.5

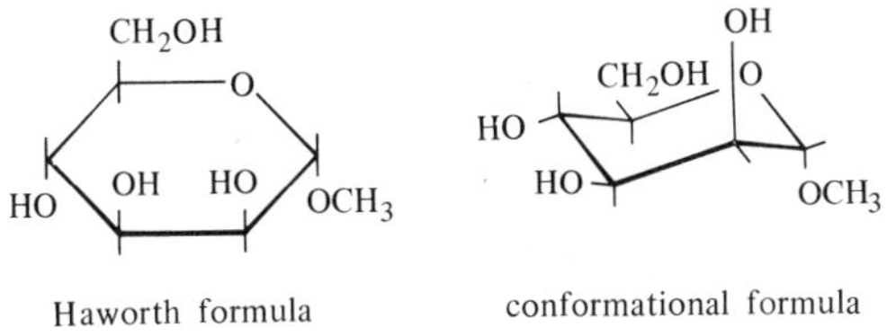

CH$_2$OH

HO   OH   HO   OCH$_3$

Haworth formula

OH

CH$_2$OH   O

HO

HO   OCH$_3$

conformational formula

methyl–α–D–mannopyranoside

24.6

α-D-glucopyranose will give a positive test with Benedict's or Tollens' solution because it is a cyclic hemiacetal. Methyl-α-D-glucopyranoside, because it is a cyclic acetal, will not.

24.7

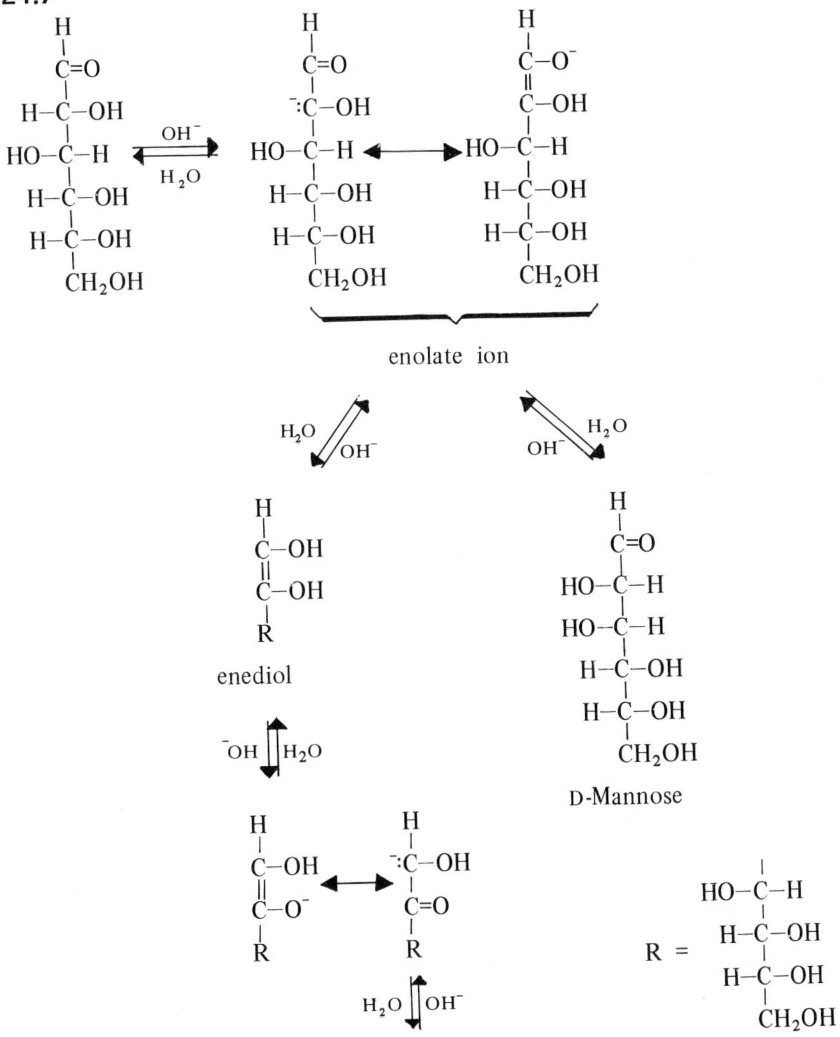

H
C=O
H–C–OH
HO–C–H
H–C–OH
H–C–OH
CH$_2$OH

$\xrightleftharpoons[\text{H}_2\text{O}]{\text{OH}^-}$

H
C=O
$^-$:C–OH
HO–C–H
H–C–OH
H–C–OH
CH$_2$OH

H
C–O$^-$
C–OH
HO–C–H
H–C–OH
H–C–OH
CH$_2$OH

enolate ion

H$_2$O / OH$^-$

OH$^-$ / H$_2$O

H
C–OH
C–OH
R

enediol

H
C=O
HO–C–H
HO–C–H
H–C–OH
H–C–OH
CH$_2$OH

D-Mannose

$^-$OH ‖ H$_2$O

H
C–OH
C–O$^-$
R

H
$^-$:C–OH
C=O
R

H$_2$O ‖ OH$^-$

R =

HO–C–H
H–C–OH
H–C–OH
CH$_2$OH

(see next page)

$$
\begin{array}{c}
\text{H} \\
| \\
\text{H–C–OH} \\
| \\
\text{C=O} \\
| \\
\text{HO–C–H} \\
| \\
\text{H–C–OH} \\
| \\
\text{H–C–OH} \\
| \\
\text{CH}_2\text{OH}
\end{array}
$$

D-Fructose

$$
\begin{array}{c}
\text{H} \\
| \\
\text{C=O} \\
| \\
\text{H–C–OH} \\
| \\
\text{HO–C–H} \\
| \\
\text{H–C–OH} \\
| \\
\text{H–C–OH} \\
| \\
\text{CH}_2\text{OH}
\end{array}
\quad
\underset{\text{H}_2\text{O}}{\overset{\text{OH}^-}{\rightleftharpoons}}
\quad
\begin{array}{c}
\text{H} \\
| \\
\text{C=O} \\
| \\
\text{H–C–OH} \\
| \\
\text{}^-\text{O–C–H} \\
| \\
\text{H–C–OH} \\
| \\
\text{H–C–OH} \\
| \\
\text{CH}_2\text{OH}
\end{array}
$$

$$
\begin{array}{c}
\text{H} \\
| \\
\text{C–O}^- \\
\| \\
\text{H–C–OH}
\end{array}
\quad \longleftrightarrow \quad
\begin{array}{c}
\text{H} \\
| \\
\text{C=O} \\
| \\
\text{H–C–OH} \\
\because
\end{array}
\quad + \quad
\begin{array}{c}
\text{O=C–H} \\
| \\
\text{H–C–OH} \\
| \\
\text{H–C–OH} \\
| \\
\text{CH}_2\text{OH}
\end{array}
$$

D-Erythrose

$$
\underset{\text{H}_2\text{O}}{\overset{\text{OH}^-}{\updownarrow}}
$$

$$
\begin{array}{c}
\text{H} \\
| \\
\text{C=O} \\
| \\
\text{H–C–OH} \\
| \\
\text{H}
\end{array}
$$

Glycolic
aldehyde

24.8

$\beta-\text{D}-\text{mannopyranose}$ → $\delta-\text{D}-\text{mannolactone}$

with reagent $\dfrac{\text{Br}_2}{\text{H}_2\text{O}}$

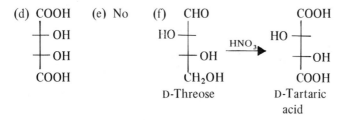

24.9

(a) Yes   (b)   COOH   (c) Yes

HO —
HO —
    — OH
    — OH
    COOH

D-mannaric
acid

(d) COOH   (e) No   (f)   CHO   COOH

  — OH   HO —   HO —
  — OH   — OH $\xrightarrow{\text{HNO}_3}$ — OH
  COOH   CH₂OH   COOH
                    D-Threose   D-Tartaric
                                          acid

(g) The aldaric acid obtained from D-erythrose is *meso*-tartaric acid; the aldaric acid obtained from D-Threose is D-tartaric acid.

24.10

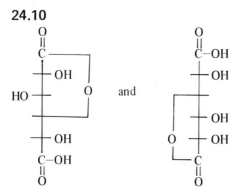

24.11
One way of predicting the products from a periodate oxidation is to place an —OH group
on each carbon at the point where C—C bond cleavage has occurred:

$$
\begin{array}{c}
-\overset{|}{\underset{|}{C}}-OH \\
-\overset{|}{\underset{|}{C}}-OH
\end{array}
\quad \xrightarrow{\text{IO}_4^-} \quad
\begin{array}{c}
-\overset{|}{\underset{|}{C}}-OH \\
OH \\
+ \\
OH \\
-\overset{|}{\underset{|}{C}}-OH
\end{array}
$$

Then if we recall (p. 605) that gem-diols are usually unstable and lose water to produce
carbonyl compounds, we get the following results:

$$
-\overset{|}{\underset{\underset{OH}{|}}{C}}-O-H \longrightarrow -\overset{|}{C}=O + H_2O
$$

$$
-\overset{\overset{OH}{|}}{\underset{|}{C}}-O-H \longrightarrow -C=O + H_2O
$$

Let us apply this procedure to several examples here while we remember that for every
C—C bond that is broken one mole of HIO$_4$ is consumed.

(a)

$$
\begin{array}{c}
CH_3 \\
| \\
H-C-OH \\
---|----- \\
H-C-OH \\
| \\
CH_3
\end{array}
\;+\; HIO_4 \longrightarrow
\begin{array}{c}
CH_3 \\
| \\
H-C-O-H \\
OH \\
+ \\
OH \\
H-C-O-H \\
| \\
CH_3
\end{array}
\;\xrightarrow{-2H_2O}\;
2CH_3\overset{O}{\overset{||}{C}}-H
$$

(b)

$$
\begin{array}{c}
H \\
| \\
H-C-OH \\
--|--- \\
H-C-OH \\
--|--- \\
H-C-OH \\
| \\
CH_3
\end{array}
\;+\; 2HIO_4 \longrightarrow
\begin{array}{c}
H \\
| \\
H-C-O-H \\
OH \\
+ \\
O-H \\
H-C-OH \\
OH \\
+ \\
OH \\
H-C-O-H \\
| \\
CH_3
\end{array}
\;\xrightarrow{-3H_2O}\;
\begin{array}{c}
H \\
| \\
H-C=O \\
+ \\
O \\
|| \\
H-C-OH \\
+ \\
H-C=O \\
| \\
CH_3
\end{array}
$$

(c)

$$
\begin{array}{c}
\text{H} \\
| \\
\text{H–C–OH} \\
\text{- - -|- - -} \\
\text{H–C–OH} \\
| \\
\text{H–C–OCH}_3 \\
| \\
\text{OCH}_3
\end{array}
+ \text{HIO}_4 \longrightarrow
\begin{array}{c}
\text{H} \\
| \\
\text{H–C–OH} \\
| \\
\text{OH} \\
+ \\
\text{OH} \\
| \\
\text{H–C–OH} \\
| \\
\text{H–C–OCH}_3 \\
| \\
\text{OCH}_3
\end{array}
\xrightarrow{-2\text{H}_2\text{O}}
\begin{array}{c}
\text{H} \\
| \\
\text{H–C=O} \\
+ \\
\text{O} \\
|| \\
\text{H–C} \\
| \\
\text{H–C–OCH}_3 \\
| \\
\text{OCH}_3
\end{array}
$$

(d)

$$
\begin{array}{c}
\text{H} \\
| \\
\text{H–C–OH} \\
\text{- -|- - -} \\
\text{H–C–OH} \\
\text{- -|- -} \\
\text{C=O} \\
| \\
\text{CH}_3
\end{array}
+ 2\text{HIO}_4 \longrightarrow
\begin{array}{c}
\text{H} \\
| \\
\text{H–C–OH} \\
| \\
\text{OH} \\
+ \\
\text{OH} \\
| \\
\text{H–C–OH} \\
| \\
\text{OH} \\
+ \\
\text{OH} \\
| \\
\text{C=O} \\
| \\
\text{CH}_3
\end{array}
\xrightarrow{-2\text{H}_2\text{O}}
\begin{array}{c}
\text{H} \\
| \\
\text{H–C=O} \\
+ \\
\text{O} \\
|| \\
\text{H–C–OH} \\
+ \\
\text{O} \\
|| \\
\text{CH}_3\text{COH}
\end{array}
$$

(e)

$$
\begin{array}{c}
\text{CH}_3 \\
| \\
\text{C=O} \\
\text{- -|- -} \\
\text{H–C–OH} \\
\text{- -|- -} \\
\text{C=O} \\
| \\
\text{CH}_3
\end{array}
+ 2\text{HIO}_4 \longrightarrow
\begin{array}{c}
\text{CH}_3 \\
| \\
\text{C=O} \\
| \\
\text{OH} \\
+ \\
\text{OH} \\
| \\
\text{H–C–OH} \\
| \\
\text{OH} \\
+ \\
\text{OH} \\
| \\
\text{C=O} \\
| \\
\text{CH}_3
\end{array}
\xrightarrow{-2\text{H}_2\text{O}}
2\text{CH}_3\overset{\text{O}}{\overset{||}{\text{C}}}\text{OH} + \text{H}\overset{\text{O}}{\overset{||}{\text{C}}}\text{OH}
$$

(f)

$$
\begin{array}{c}
\text{CH}_2 \\
| \quad\;\; \text{H} \\
\text{CH}_2 \quad \overset{|}{\text{C}}\text{–OH} \\
| \;\; \text{- -|- -} \\
\text{CH}_2 \quad \text{C–OH} \\
\quad\quad | \\
\quad\quad \text{H}
\end{array}
+ \text{HIO}_4 \longrightarrow
\begin{array}{c}
\text{CH}_2 \quad \overset{\text{H}}{\overset{|}{\text{C}}}\text{–OH} \\
| \quad\quad\;\; \text{OH} \\
\text{CH}_2 \quad\quad \text{OH} \\
| \quad\quad\;\; | \\
\text{CH}_2 \quad \text{C–OH} \\
\quad\quad | \\
\quad\quad \text{H}
\end{array}
\xrightarrow{-2\text{H}_2\text{O}}
\text{H}\overset{\text{O}}{\overset{||}{\text{C}}}\text{CH}_2\text{CH}_2\text{CH}_2\overset{\text{O}}{\overset{||}{\text{C}}}\text{H}
$$

(g)

$$
\begin{array}{c}
\text{H} \\
\text{H-C-OH} \\
\text{--}\,|\,\text{--} \\
\text{CH}_3\text{-C-OH} \\
\text{CH}_3
\end{array}
\;+\;\text{HIO}_4\;\longrightarrow\;
\begin{array}{c}
\text{H} \\
\text{H-C-OH} \\
\text{OH} \\
+ \\
\text{OH} \\
\text{CH}_3\text{-C-OH} \\
\text{CH}_3
\end{array}
\;\xrightarrow{-2\text{H}_2\text{O}}\;
\begin{array}{c}
\text{H} \\
\text{H-C=O} \\
+ \\
\text{CH}_3\text{-C=O} \\
\text{CH}_3
\end{array}
$$

(h)

$$
\begin{array}{c}
\text{O} \\
\|\\
\text{H-C} \\
\text{--}\,|\,\text{--}\\
\text{H-C-OH}\\
\text{--}\,|\,\text{--}\\
\text{H-C-OH}\\
\text{--}\,|\,\text{--}\\
\text{H-C-OH}\\
\text{H}
\end{array}
\;+\;3\text{HIO}_4\;\longrightarrow\;
\begin{array}{c}
\text{O}\\
\|\\
\text{H-C-OH}\\
+\\
\text{OH}\\
\text{H-C-OH}\\
\text{OH}\\
+\\
\text{OH}\\
\text{H-C-OH}\\
\text{OH}\\
+\\
\text{OH}\\
\text{H-C-OH}\\
\text{H}
\end{array}
\;\xrightarrow{-3\text{H}_2\text{O}}\;
3\overset{\text{O}}{\overset{\|}{\text{HCOH}}} + \overset{\text{O}}{\overset{\|}{\text{HCH}}}
$$

D-erythrose

24.12

Oxidation of an aldohexose and a ketohexose would each require five moles of HIO_4 but would give different results

$$
\begin{array}{c}
\text{CHO}\\
-|---\\
\text{CHOH}\\
-|---\\
\text{CHOH}\\
-|---\\
\text{CHOH}\\
-|---\\
\text{CHOH}\\
-|---\\
\text{CH}_2\text{OH}
\end{array}
\;+\;5\text{HIO}_4\;\longrightarrow\;
\begin{array}{c}
\text{HCOOH}\\
+\\
\text{HCOOH}\\
+\\
\text{HCOOH}\\
+\\
\text{HCOOH}\\
+\\
\text{HCOOH}\\
|\\
\text{HCHO}
\end{array}
\qquad (5\ \text{HCOOH} + \text{HCHO})
$$

Aldohexose

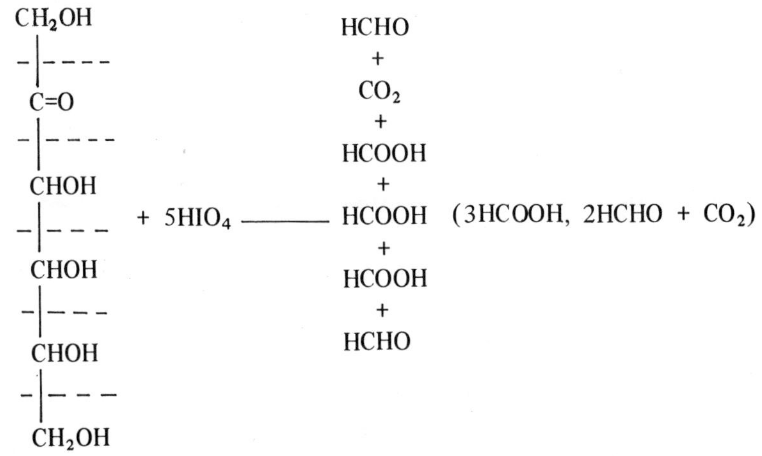

$$
\begin{array}{c}
\text{CH}_2\text{OH} \\
-|---- \\
\text{C}=\text{O} \\
-|--- \\
\text{CHOH} \\
-|--- \\
\text{CHOH} \\
-|-- \\
\text{CHOH} \\
-|-- \\
\text{CH}_2\text{OH}
\end{array}
\;+\;5\text{HIO}_4 \;\longrightarrow\;
\begin{array}{c}
\text{HCHO} \\
+ \\
\text{CO}_2 \\
+ \\
\text{HCOOH} \\
+ \\
\text{HCOOH} \\
+ \\
\text{HCOOH} \\
+ \\
\text{HCHO}
\end{array}
\quad (3\text{HCOOH, } 2\text{HCHO} + \text{CO}_2)
$$

Ketohexose

24.13

(a)

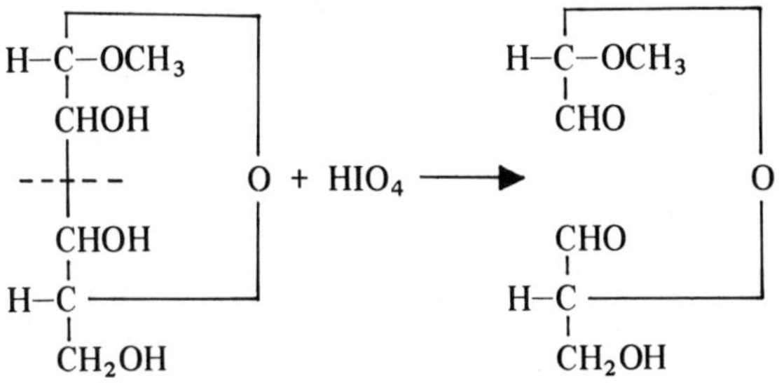

Any methylfuranoside of
the α-D-pentose series

2
Dialdehyde

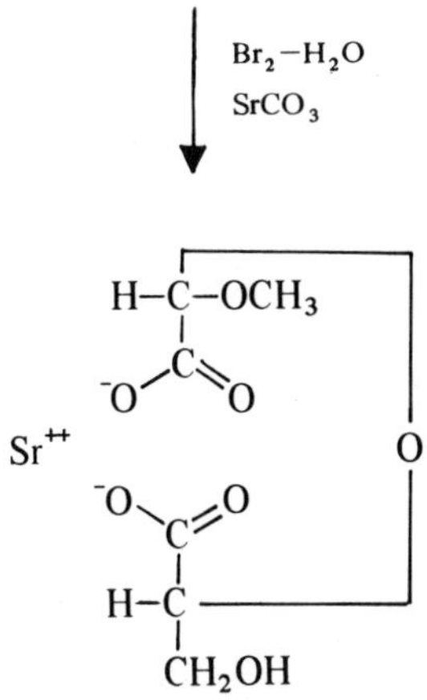

$\text{Br}_2-\text{H}_2\text{O}$
SrCO_3

3

Same strontium salt

(b) Although both compounds yield the same dialdehyde **2** (and Strontium salt **3**), periodate oxidation of a methyl-α-D-pentofuranoside consumes only one mole of HIO_4 and produces no formic acid.

24.14

(a)

```
 H                 COOH
 |                  |
 C=O          H ——— OH
 |                  |
 COOH          CH₂OH

  4                 5

Glyoxylic     D-(-)-Glyceric
  acid             acid
```

(b) This relates the configuration of the highest numbered carbon of the aldose to that of D-(+)-glyceraldehyde, and thus allows us to place the aldose in the D-family.

24.15

(a) Yes, D-glucitol would be optically active; only those alditols (below) whose molecules possess a plane of symmetry would be optically inactive.

(b)

```
 CHO               CH₂OH
  |                 |
  ——OH             ——OH
  |                 |
  ——OH    NaBH₄     ——OH
  |        ——►      |   - - - - - -   plane of symmetry
  ——OH             ——OH
  |                 |
  ——OH             ——OH
  |                 |
 CH₂OH             CH₂OH

                 optically
                 inactive
```

```
 CHO               CH₂OH
  |                 |
  ——OH             ——OH
  |                 |
HO——     NaBH₄   HO——
  |       ——►      |   - - - - -    plane of symmetry
HO——             HO——
  |                 |
  ——OH             ——OH
  |                 |
 CH₂OH             CH₂OH

          optically   inactive
```

24.16

(a)

```
   CH₂OH                    CH=NNHC₆H₅
    |                        |
    C=O                      C=NNHC₆H₅
    |        C₆H₅NHNH₂        |
 HO——        ————————►     HO——
    |                        |
    ——OH                     ——OH
    |                        |
    ——OH                     ——OH
    |                        |
   CH₂OH                    CH₂OH
```

(b) This experiment shows that D-glucose and D-fructose have the same configurations at C-3, C-4, and C-5.

24.17

(a)

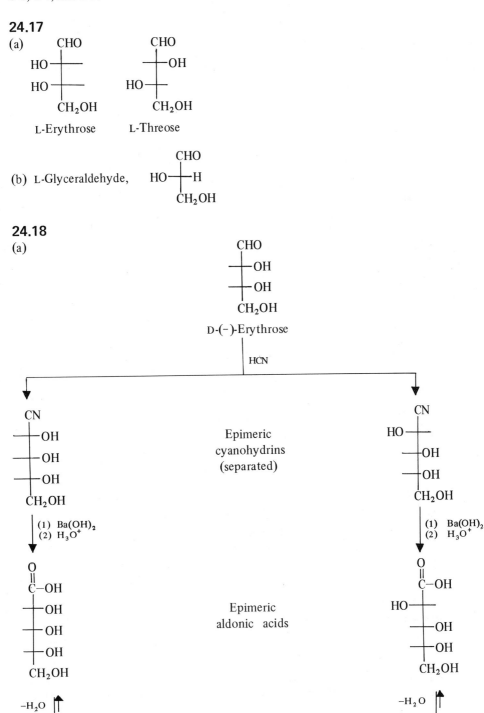

L-Erythrose   L-Threose

(b) L-Glyceraldehyde,

24.18

(a)

D-(−)-Erythrose

HCN

CN

Epimeric
cyanohydrins
(separated)

CN

(1) Ba(OH)$_2$
(2) H$_3$O$^+$

(1) Ba(OH)$_2$
(2) H$_3$O$^+$

Epimeric
aldonic acids

−H$_2$O

−H$_2$O

(see next page)

(see next page)

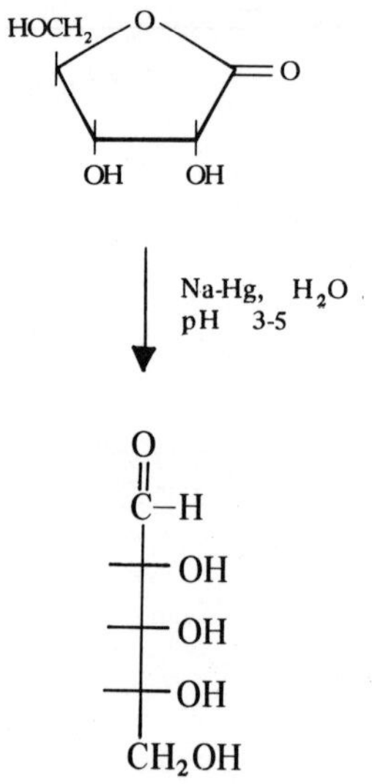

Epimeric
γ-aldonolactones

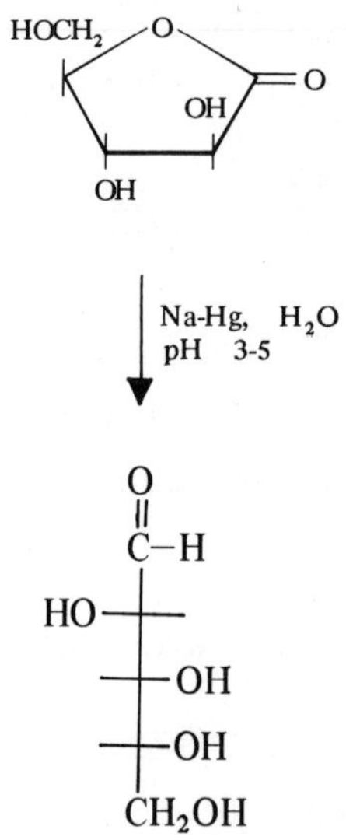

Na-Hg, H₂O
pH 3-5

(b)

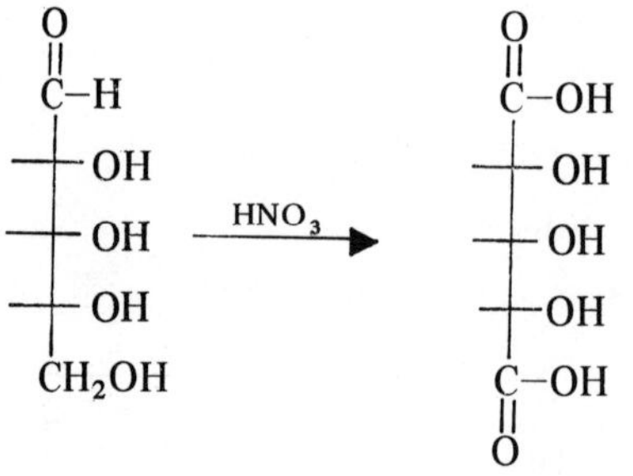

D-(−)-Ribose

optically
inactive

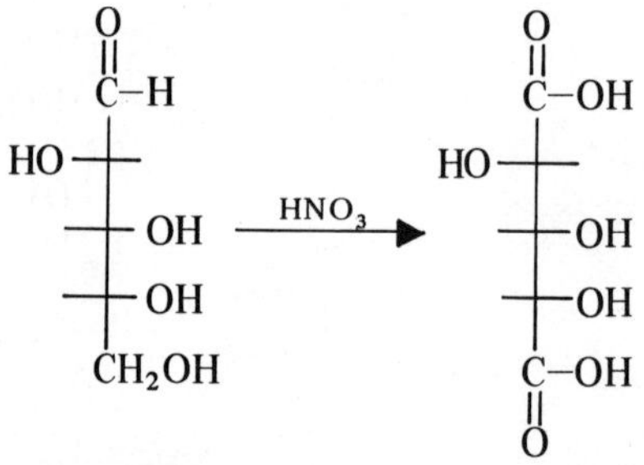

D-(−)-Arabinose

optically
active

24.19

A Kiliani-Fischer synthesis starting with D-(−)-threose would yield **I** and **II**.

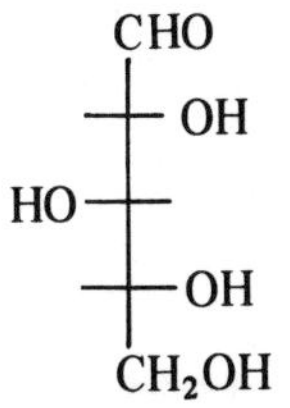

I	II
(D)-(+)-Xylose)	(D-(−)-Lyxose)

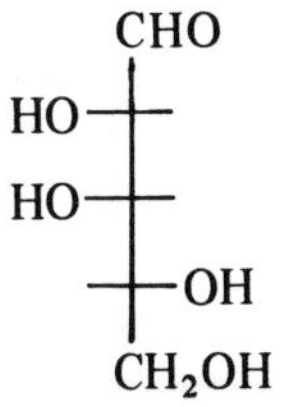

I must be D-(+)-xylose because when oxidized by nitric acid, it yields an optically inactive aldaric acid:

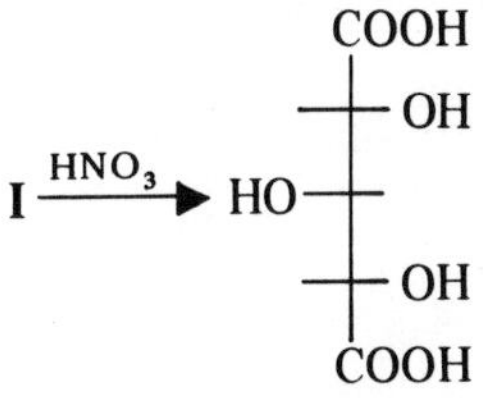

optically
inactive

II must be D-(−)-lyxose because when oxidized by nitric acid it yields an optically active aldaric acid:

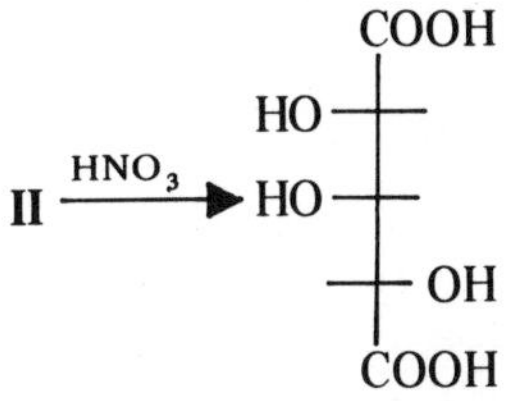

optically
active

24.20

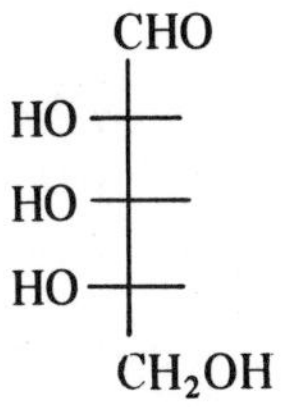

L-(+)-Ribose 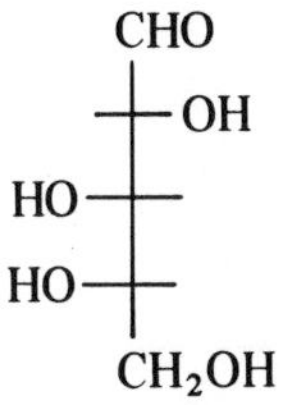   L-(+)-Arabinose 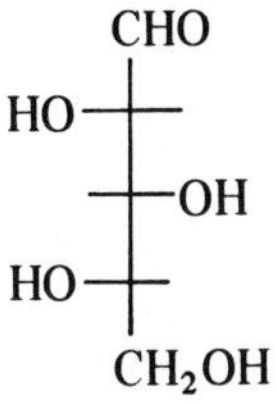   L-(−)-Xylose 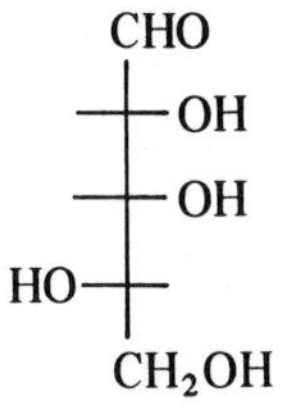   L-(+)-Lyxose

24.21

Since D-(+)-galactose yields an optically inactive aldaric acid it must have either structure

III or structure **IV** below.

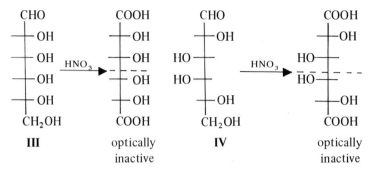

| III | optically inactive | IV | optically inactive |

A Ruff degradation beginning with **III** would yield D-(−)-ribose

$$III \xrightarrow[H_2O]{Br_2} \xrightarrow[Fe_2(SO_4)_3]{H_2O_2}$$

CHO
—OH
—OH
—OH
CH₂OH

D-(−)-Ribose

A Ruff degradation beginning with **IV** would yield D-(−)-lyxose; thus D-(+)-galactose must have structure **IV**.

$$IV \xrightarrow[H_2O]{Br_2} \xrightarrow[Fe_2(SO_4)_3]{H_2O_2}$$

CHO
HO—
HO—
—OH
CH₂OH

D-(−)-Lyxose

24.22

D-(+)-glucose, as shown below.

The other γ-lactone
of D-glucaric acid

24.23

If the methyl glucoside had been a furanoside, hydrolysis of the methylation product would have given:

```
         CHO
          |
          +— OCH₃
          |
 CH₃O ——+
          |
          +— OH
          |
          +— OCH₃
          |
        CH₂CH₃
```

And, oxidation would have given:

```
         CHO                              COOH            COOH       COOH
          |                                |               |          |
          +— OCH₃          HNO₃            +— OCH₃          +—OCH₃  +  +—OCH₃
          |          ————————————▶         |          +      |          |
 CH₃O ——+              CH₃O ——+         COOH          CH₂OCH₃   COOH
          |                                |
          +— OH                          COOH
          |
          +— OCH₃          A dimethoxysuc-    A dimethoxy-  Methoxyma-
          |                 cinic acid         propanoic     lonic acid
        CH₂OCH₃                                 acid
```

$$\downarrow \; -CO_2$$

```
COOH
 |
CH₂OCH₃
```

Methoxyacetic
acid

24.24

(a)
```
 CHO
  |
CHOH
  |
CHOH
  |
CHOH
  |
CH₂OH
```

(b)
```
CH₂OH
  |
 C=O
  |
CHOH
  |
CHOH
  |
CHOH
  |
CH₂OH
```

(c)
```
  CHO
   |
(CHOH)ₙ
   |
HO—⬡—H
   |
 CH₂OH
```
or
```
CH₂OH
  |
 C=O
  |
(CHOH)ₙ
  |
HO—⬡—H
  |
CH₂OH
```

(d)
```
┌ CHOR ──┐
│  |      │
│(CHOH)ₙ  O
│  |      │
└ CH ─────┘
   |
 CH₂OH
```

(e)
```
 COOH
   |
(CHOH)ₙ
   |
 CH₂OH
```

(f)
```
 COOH
   |
(CHOH)ₙ
   |
 COOH
```

(g)

$$\begin{array}{l} \text{O} \\ \| \\ \text{C} \\ | \\ (\text{CHOH})_n \quad \text{O} \\ | \\ \text{CH} \\ | \\ \text{CH}_2\text{OH} \end{array}$$

(h)

$$\begin{array}{l} \text{OH} \\ | \\ \text{CH} \\ | \\ \text{CHOH} \\ | \\ \text{CHOH} \quad \text{O} \\ | \\ \text{CHOH} \\ | \\ \text{CH} \\ | \\ \text{CH}_2\text{OH} \end{array}$$

or

$$\begin{array}{l} \text{CH}_2\text{OH} \\ | \\ \text{CH}\!-\!\!-\!\text{O} \\ \diagup \quad \diagdown \\ \text{CHOH} \quad \text{CHOH} \\ \diagdown \quad \diagup \\ \text{CHOH}\!-\!\text{CHOH} \end{array}$$

(i)

$$\begin{array}{l} \text{OH} \\ | \\ \text{CH} \\ | \\ \text{CHOH} \\ | \quad \text{O} \\ \text{CHOH} \\ | \\ \text{CH} \\ | \\ \text{CHOH} \\ | \\ \text{CH}_2\text{OH} \end{array}$$

or

$$\begin{array}{l} \text{CH}_2\text{OH} \\ | \\ \text{CHOH}\,\text{O} \\ | \\ \text{CH} \qquad \text{CHOH} \\ \diagdown \quad \diagup \\ \text{CHOH}\!-\!\text{CHOH} \end{array}$$

(j) Any sugar that has a free aldehyde or ketone group or one that exists as a cyclic hemiacetal or hemiketal. Examples are:

$$\begin{array}{l} \text{CHO} \\ | \\ (\text{CHOH})_n \\ | \\ \text{CHOH} \\ | \\ \text{CH}_2\text{OH} \end{array} \rightleftharpoons \begin{array}{l} \text{OH} \\ | \\ \text{CH} \\ | \\ (\text{CHOH})_n \quad \text{O} \\ | \\ \text{CH} \\ | \\ \text{CH}_2\text{OH} \end{array}$$

or

$$\begin{array}{l} \text{CH}_2\text{OH} \\ | \\ \text{C}\!=\!\text{O} \\ | \\ (\text{CHOH})_n \\ | \\ \text{CHOH} \\ | \\ \text{CH}_2\text{OH} \end{array} \rightleftharpoons \begin{array}{l} \text{CH}_2\text{OH} \\ | \\ \text{C}\!-\!\text{OH} \\ | \\ (\text{CHOH})_n \quad \text{O} \\ | \\ \text{CH} \\ | \\ \text{CH}_2\text{OH} \end{array}$$

(k)

$$\begin{array}{l} \text{CH}_2\text{OH} \\ | \\ \text{CH}\!-\!\!-\!\text{O} \\ \diagup \quad \diagdown \\ \text{CHOH} \quad \text{CHOR} \\ \diagdown \quad \diagup \\ \text{CHOH}\!-\!\text{CHOH} \end{array}$$

(l)

$$\begin{array}{l} \text{CH}_2\text{OH} \\ | \\ \text{CHOH} \\ | \quad \text{O} \\ \text{CH} \qquad \text{CHOR} \\ \diagdown \quad \diagup \\ \text{CHOH}\!-\!\text{CHOH} \end{array}$$

(m) Any two aldoses that differ only in configuration at C-2. (See also page 932 for a broader definition.) D-Erythrose and D-threose are examples.

$$\begin{array}{l} \text{CHO} \\ \!\!-\!\!\text{OH} \\ \!\!-\!\!\text{OH} \\ \text{CH}_2\text{OH} \end{array} \qquad \begin{array}{l} \text{CHO} \\ \text{HO}\!\!-\!\! \\ \!\!-\!\!\text{OH} \\ \text{CH}_2\text{OH} \end{array}$$

D-Erythrose   D-Threose

(n) Cyclic sugars that differ only in the configuration of C-1. Examples are:

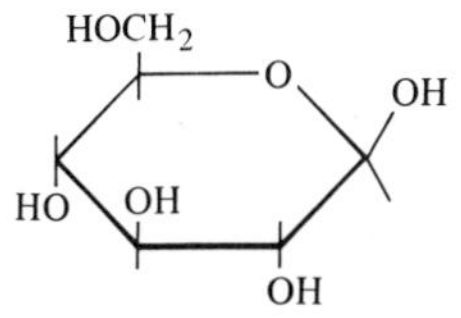

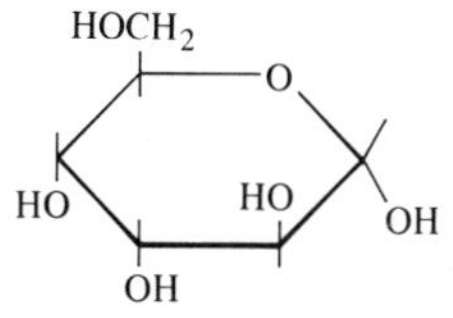

(o) $CH=NNHC_6H_5$

$C=NNHC_6H_5$

$(CHOH)_n$

CH_2OH

(p) Maltose is an example:

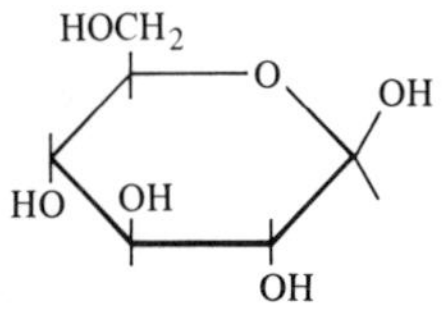

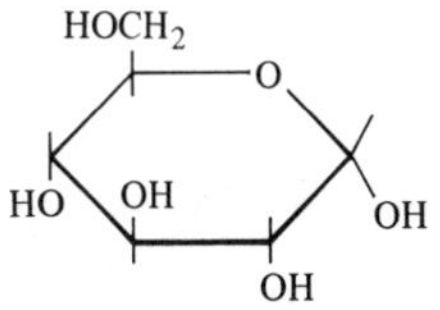

(q) Amylose is an example:

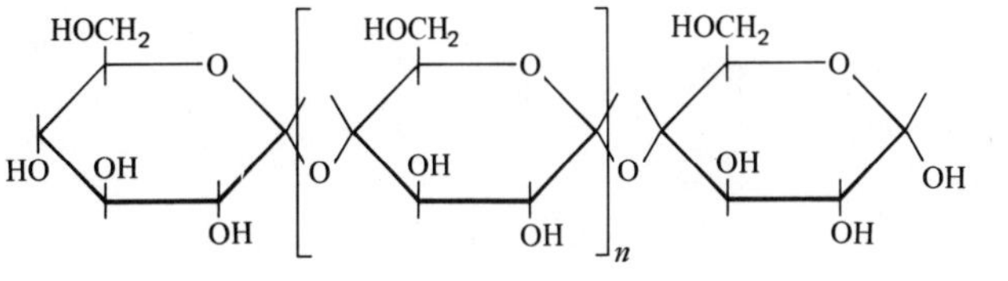

(r) Any sugar in which all potential carbonyl groups are present as acetals or ketals (i.e., as glycosides). Sucrose (p. 939) is an example of a nonreducing disaccharide; the methyl-D-glucopyranosides (p. 922) are examples of nonreducing monosaccharides.

24.25

(a)

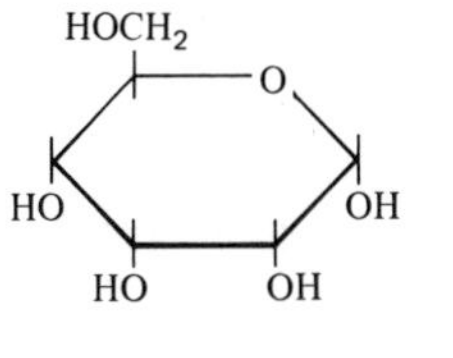

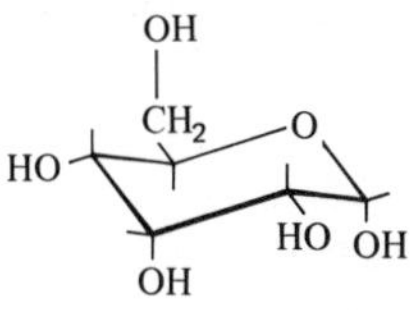

(b)

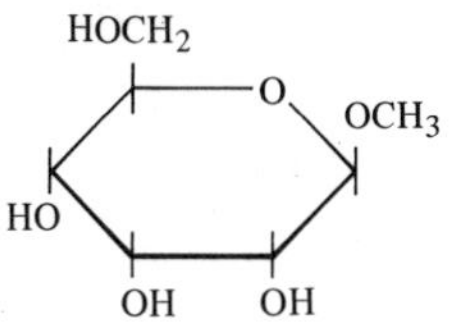

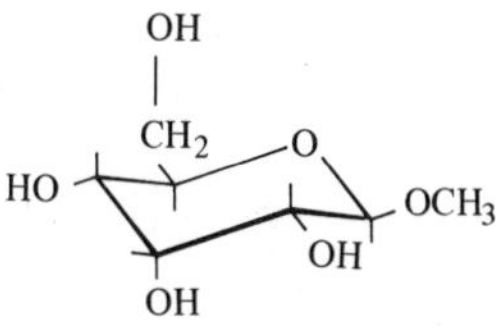

(c)

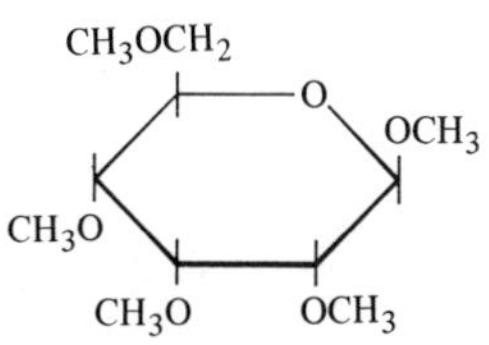

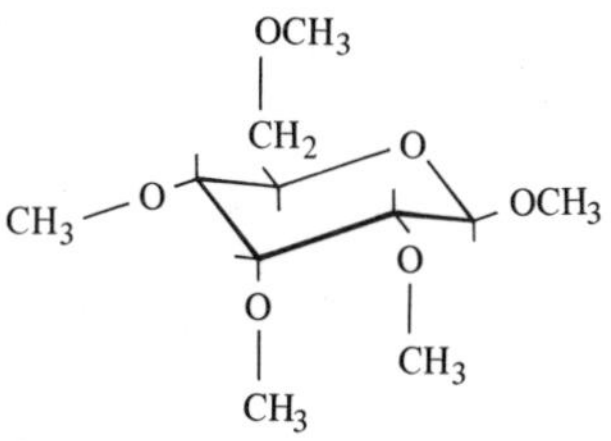

24.26

A methyl ribofuranoside would consume only one mole of HIO_4; a methyl ribopyranoside would consume two moles of HIO_4 and would also produce one mole of formic acid.

24.27

One anomer of D-mannose is dextrorotatory ($[\alpha]_D = +29.3°$), the other is levorotatory ($[\alpha]_D = -17.0°$).

24.28

The microorganism selectively oxidizes the —CHOH group of D-glucitol that corresponds to C-5 of D-glucose.

24.29

L-Gulose and L-idose would yield the same phenylosazone as L-sorbose.

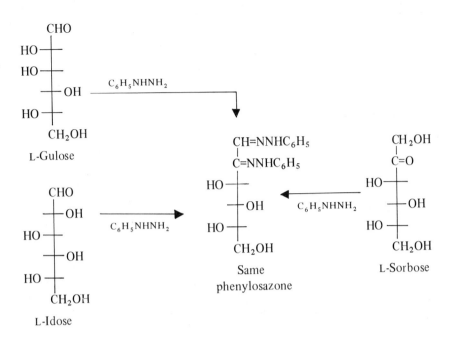

24.30

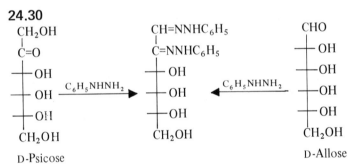

24.31

A is D-altrose, **B** is D-talose, and **C** is D-galactose:

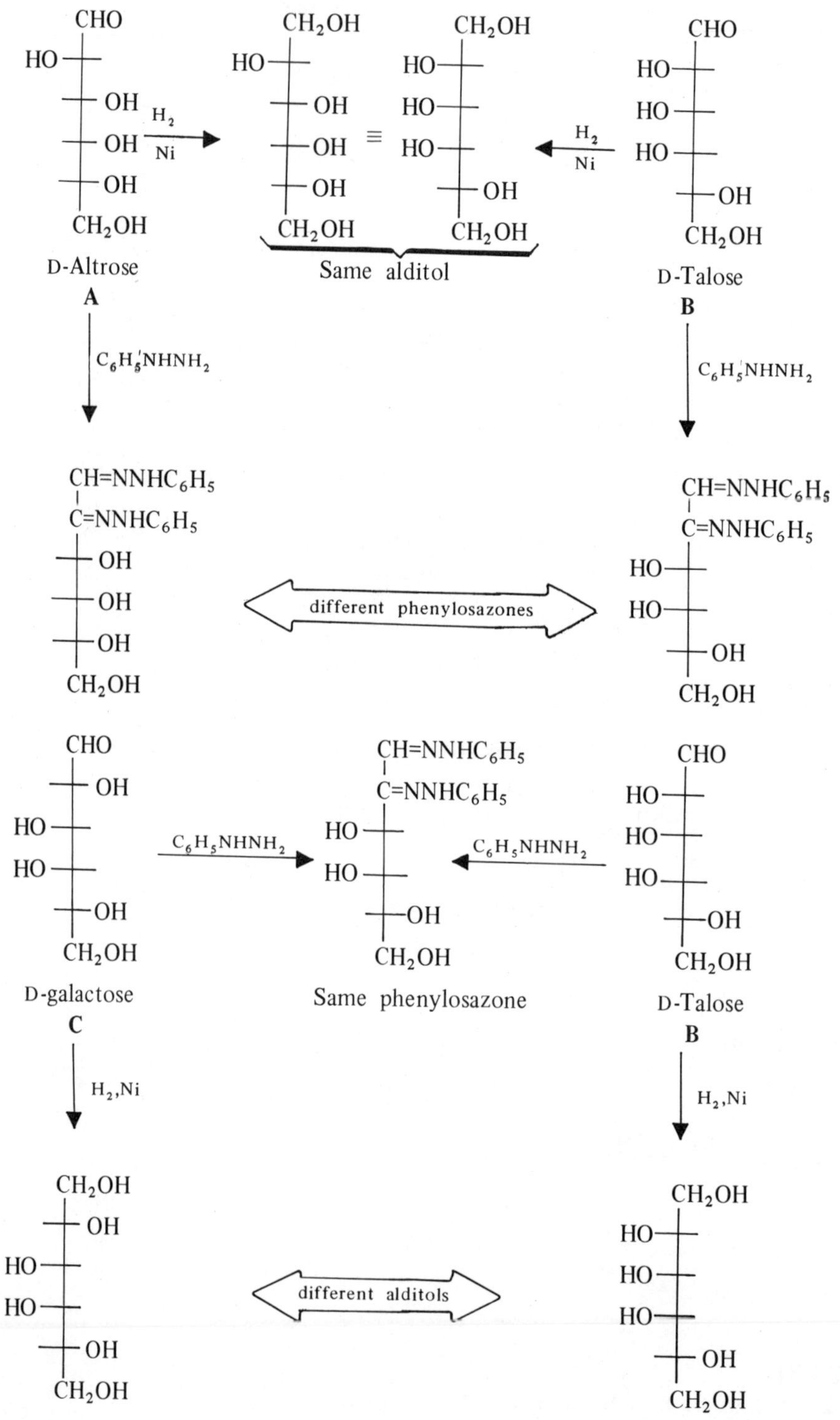

D-Altrose **A** + H_2/Ni → Same alditol ← H_2/Ni + D-Talose **B**

D-Altrose **A** + $C_6H_5NHNH_2$ → (phenylosazone)

different phenylosazones

D-Talose **B** + $C_6H_5NHNH_2$ → (phenylosazone)

D-galactose **C** + $C_6H_5NHNH_2$ → Same phenylosazone ← $C_6H_5NHNH_2$ + D-Talose **B**

D-galactose **C** + H_2,Ni → (alditol)

different alditols

D-Talose **B** + H_2,Ni → (alditol)

(Note: If we had designated D-talose as **A**, and D-altrose as **B**, then **C** is D-allose)

24.32

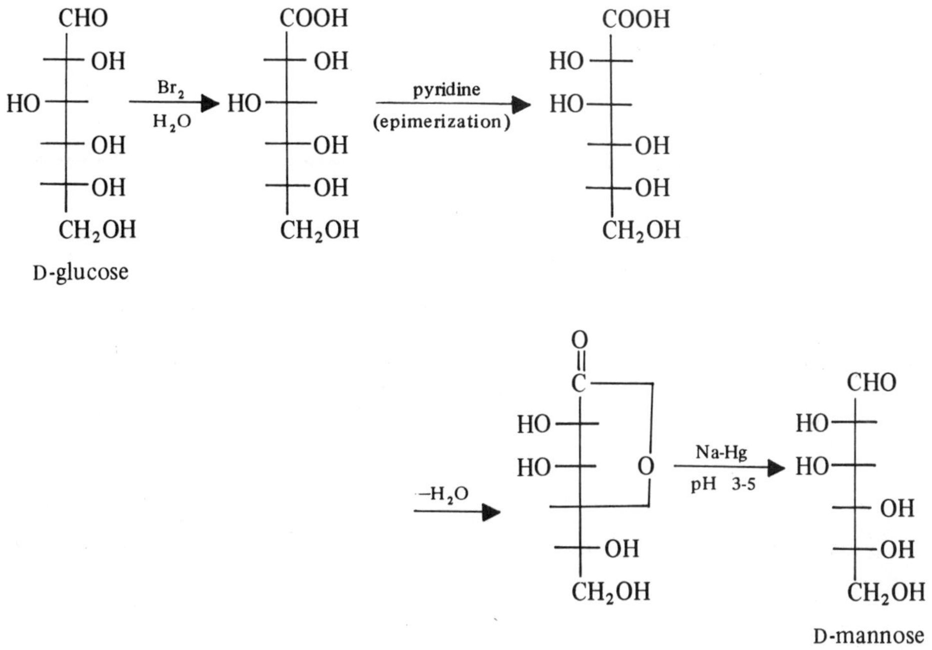

D-glucose

D-mannose

24.33

The conformation of D-idopyranose with four equatorial —OH groups and an axial —CH$_2$OH group is more stable than the one with four axial —OH groups and an equatorial —CH$_2$OH group.

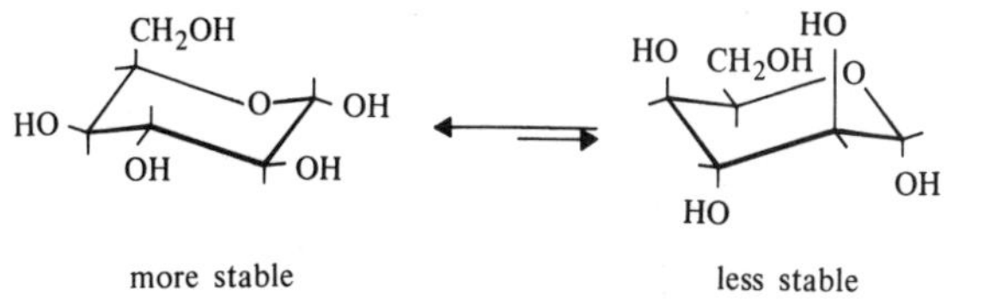

more stable   less stable

4 equatorial —OH groups 4 axial —OH groups
1 axial —CH$_2$OH   1 equatorial —CH$_2$OH

24.34

(a) The anhydro sugar is formed when the axial —CH$_2$OH group reacts with C-1 to form a cyclic acetal.

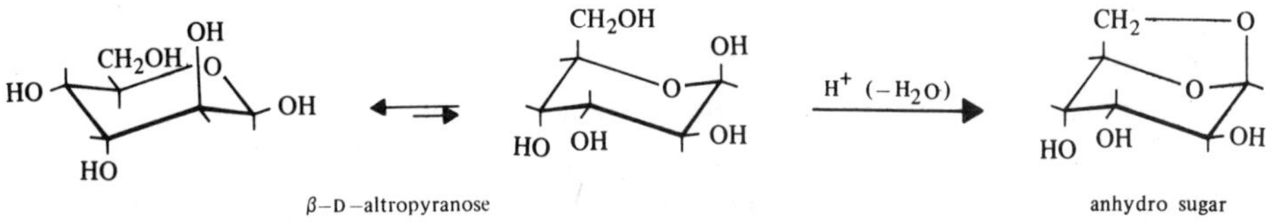

β–D–altropyranose   anhydro sugar

Because the anhydro sugar is an acetal (i.e., an internal glycoside), it is a non-reducing sugar.

Methylation followed by acid hydrolysis converts the anhydro sugar to 2,3,4-tri-O-methyl-D-altrose:

2, 3, 4–tri–O–methyl–D–altrose

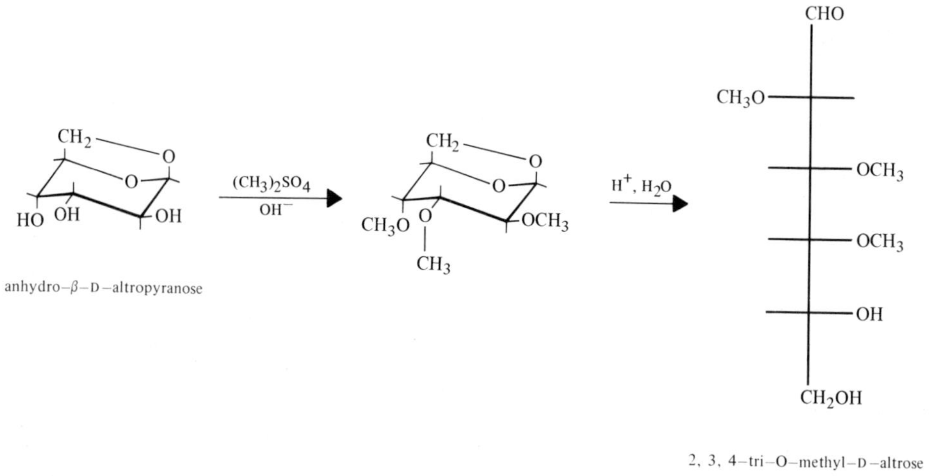

(b) Formation of an anhydro sugar requires that the monosaccharide adopt a chair conformation with the —CH$_2$OH group axial. With β-D-altropyranose this requires that two —OH groups be axial as well. With β-D-glucopyranose, however, it requires that all four —OH groups become axial, and thus that the molecule adopt a very unstable conformation:

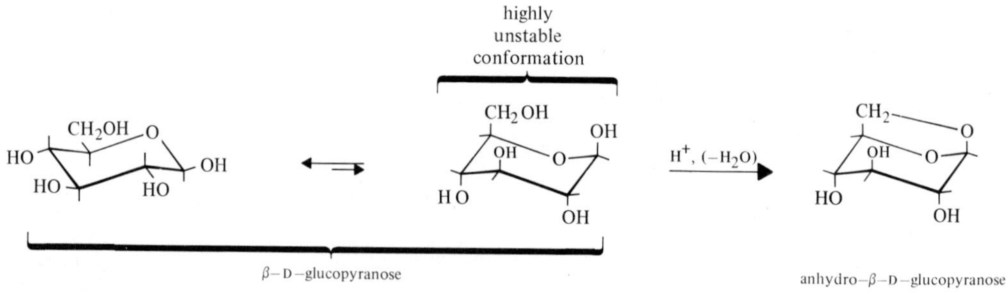

24.35

The initial step in mutarotation—ring opening of the cyclic hemiacetal—requires that an acid donate a proton to the ring oxygen and that a base remove a proton from the anomeric hydroxyl. 2-Hydroxypyridine has, in close proximity, acidic and basic groups that can accomplish both of these tasks.

β−D−glucopyranose aldehyde form α−D−glucopyranose

24.36

1. The molecular formula and the results of acid hydrolysis show that lactose is a disaccharide composed of D-glucose and D-galactose. The fact that lactose is hydrolyzed by a *β-galactosidase* indicates that galactose is present as a glycoside and that the glycosidic linkage is *beta* to the galactose ring.

2. That lactose is a reducing sugar, forms a phenylosazone, and undergoes mutarotation indicates that one ring (presumably that of D-glucose) is present as a hemiacetal and thus is capable of existing to a limited extent as an aldehyde.

3. This experiment confirms that the D-glucose unit is present as a cyclic hemiacetal and that the D-galactose unit is present as a cyclic glycoside.

4. That 2,3,4,6-tetra-O-methyl-D-galactose is obtained in this experiment indicates (by virture of the free −OH at C-5) that the galactose ring of lactose is present as a pyranoside. That the methylated gluconolactone obtained from this experiment has a free −OH group at C-4 indicates that the C-4 oxygen of the glucose unit is connected in a glycosidic linkage to the galactose unit.

     Now only the size of the glucose ring remains in question and the answer to this is provided by experiment 5.

5. That methylation of lactose and subsequent hydrolysis gives 2,3,6-tri-O-methyl-D-glucose—that it gives a methylated glucose derivative with a free −OH at C-4 and C-5— demonstrates that the glucose ring is present as a pyranose. (We know already that the oxygen at C-4 is connected in a glycosidic linkage to the galactose unit; thus a free −OH at C-5 indicates that the C-5 oxygen is a part of the hemiacetal group of the glucose unit and that the ring is six-membered.)

24.37

Melibiose has the following structure:

6-O-(α-D-galactopyranosyl)-D-glucopyranose
We arrive at this conclusion from the data given:

1. That melibiose is a reducing sugar, that it undergoes mutarotation and forms a phenylosazone indicates that one monosaccharide is present as a cyclic hemiacetal.

2. That acid hydrolysis gives D-galactose and D-glucose indicates that melibiose is a disaccharide composed of one D galactose unit and one D-glucose unit. That melibiose is hydrolyzed by an α-galactosidase suggests that melibiose is an α-D-galactosyl-D-glucose.

3. Oxidation of melibiose to melibionic acid and subsequent hydrolysis to give D-galactose and D-gluconic acid confirms that the glucose unit is present as a cyclic hemiacetal and that the galactose unit is present as a glycoside. (Had the reverse been true, this experiment would have yielded D-glucose and D-galactonic acid.)

  Methylation and hydrolysis of melibionic acid produces 2,3,4,6-tetra-O-methyl-D-galactose and 2,3,4,5-tetra-O-methyl-D-gluconic acid. Formation of the first product—a galactose derivative with a free —OH at C-5—demonstrates that the galactose ring is six-membered; formation of the second product—a gluconic acid derivative with a free —OH at C-6—demonstrates that the oxygen at C-6 of the glucose unit is joined in a glycosidic linkage to the galactose unit.

4. That methylation and hydrolysis of melibiose gives a glucose derivative (2,3,4-tri-O-methyl-D-glucose) with free —OH groups at C-5 and C-6 shows that the glucose ring is also six-membered. Melibiose is, therefore, 6-O-(α-D-galactopyranosyl)-D-glucopyranose.

24.38
Trehalose has the following structure:

α-D-glucopyranosyl-D-glucopyranoside
We arrive at this structure in the following way:

1. Acid hydrolysis shows that trehalose is a disaccharide consisting only of D-glucose units.

2. Hydrolysis by α-glucosidases and not by β-glucosidases shows that the glycosidic linkages are *alpha*.

3. That trehalose is a non-reducing sugar, that it does not form phenylosazone, and that it does not react with bromine water indicate that no hemiacetal groups are present. This means that C-1 of one glucose unit and C-1 of the other must be joined in a glycosidic linkage. Fact 2 (above) indicates that this linkage is *alpha* to each ring.

4. That methylation of trehalose followed by hydrolysis yields only 2,3,4,6-tetra-O-methyl-D-glucose demonstrates that both rings are six-membered.

24.39

(a) Tollens' reagent or Benedict's reagent will give a positive test with D-glucose but will give no reaction with D-glucitol.

(b) D-Glucaric acid will give an acidic aqueous solution that can be detected with blue litmus paper. D-Glucitol will give a neutral aqueous solution.

(c) D-Glucose will be oxidized by bromine water and the red brown color of bromine will disappear. D-Fructose will not be oxidized by bromine water since it does not contain an aldehyde group.

(d) Nitric acid oxidation will produce an *optically active* aldaric acid from D-glucose but an *optically inactive* aldaric acid will result from D-galactose.

(e) Maltose is a reducing sugar and will give a positive test with Tollens' or Benedict's solution. Sucrose is a nonreducing sugar and will not react.

(f) Maltose will give a positive Tollens' or Benedict's test; maltonic acid will not.

(g) 2,3,4,6-Tetra-O-methyl-β-D-glucopyranose will give a positive test with Tollens' or Benedict's solution; methyl β-D-glucopyranoside will not.

(h) Periodic acid will react with methyl α-D-ribofuranoside because it has hydroxyl groups on adjacent carbons. Methyl 2-deoxy-α-D-ribofuranoside will not react.

24.40

That the Schardinger dextrins are nonreducing shows that they have no free aldehyde or hemiacetal groups. This strongly suggests the presence of a *cyclic* structure. That methylation and subsequent hydrolysis yields only 2,3,6-tri-O-methyl-D-glucose indicates that the glycosidic linkages all involve C-1 of one glucose unit and C-4 of the next. That α-glucosidases cause hydrolysis of the glycosidic linkages indicates that they are α-glycosidic linkages. Thus we are led to the following general structure.

n = 3, 4, or 5

Note: Schardinger dextrins are extremely interesting compounds. They are able to form complexes with a wide variety of compounds by incorporating these compounds in the cavity in the middle of the cyclic dextrin structure. Complex formation takes place, however, only when the cyclic dextrin and the guest molecule are the right size. Anthracene molecules, for example, will fit into the cavity of a cyclic dextrin with eight glucose units but will not fit into one with seven. For more information about these fascinating compounds, see R. J. Bergeron, "Cycloamyloses," *J. Chem. Educ.*, **54**, 204 (1977).

24.41

Isomaltose has the following structure:

6−*O*−(α−D−glucopyranosyl)−D−glucopyranose

(1) The acid and enzymic hydrolysis experiments tell us that isomaltose has two glucose units linked by an α-linkage.

(2) That isomaltose is a reducing sugar indicates that one glucose unit is present as a cyclic hemiacetal.

(3) Methylation of isomaltonic acid followed by hydrolysis gives us information about the size of the nonreducing pyranoside ring and about its point of attachment to the reducing ring. The formation of the first product (2,3,4,6-tetra-*O*-methyl-D-glucose)—a compound with an −OH at C-5—tells us that the nonreducing ring is present as a pyranoside. The formation of 2,3,4,5-tetra-*O*-methyl-D-gluconic acid—a compound with an −OH at C-6—shows that the nonreducing ring is linked to C-6 of the reducing ring.

(4) Methylation of maltose itself tells the size of the reducing ring. That 2,3,4-tri-*O*-methyl-D-glucose is formed shows that the reducing ring is also six membered; we know this because of the free −OH at C-5.

24.42

Stachyose has the following structure:

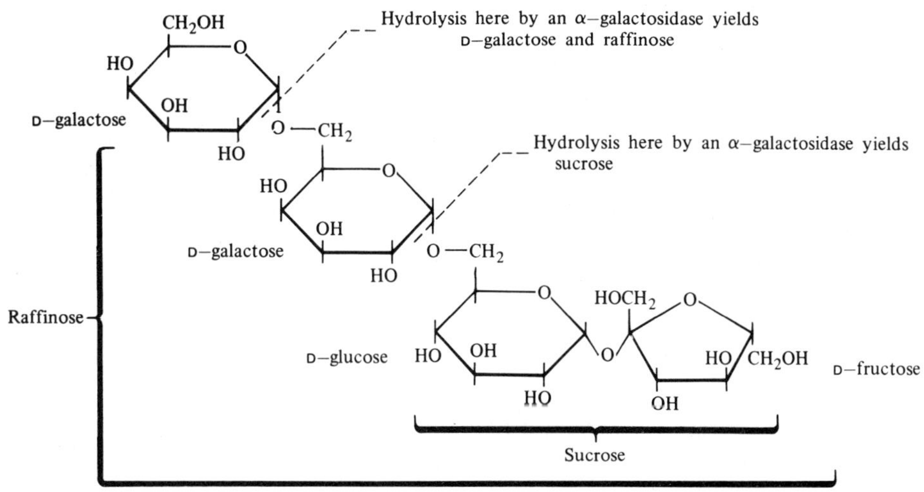

Raffinose has the following structure:

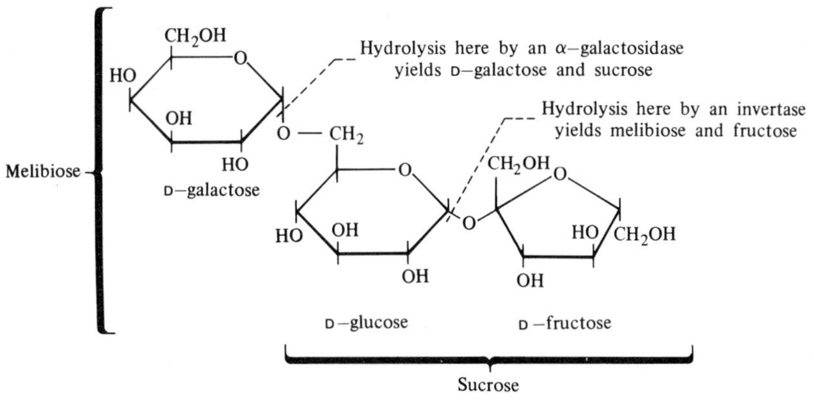

The enzymic hydrolyses (as indicated above) give the basic structure of stachyose and raffinose. The only remaining question is the ring size of the first galactose unit of stachyose. That methylation of stachyose and subsequent hydrolysis yields 2,3,4,6-tetra-O-methyl-D-galactose establishes that this ring is a pyranoside.

24.43

Arbutin has the following structure.

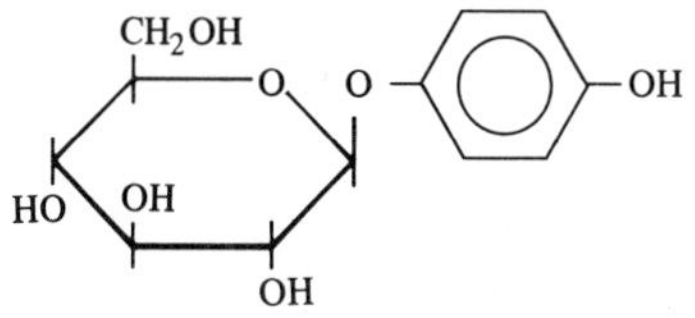

p—hydroxyphenyl—β—D—glucopyranoside

Compounds **X**, **Y**, and **Z** are hydroquinone, *p*-methoxyphenol, and *p*-dimethoxybenzene respectively.

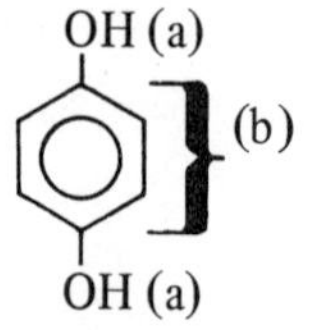

(a) Singlet δ7.9 (2H)
(b) Singlet δ6.8 (4H)

X
Hydroquinone

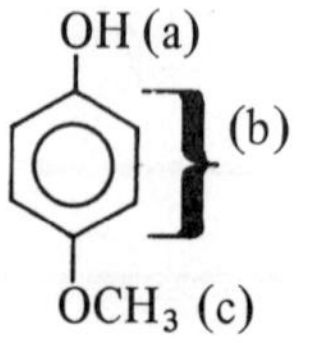

(a) Singlet δ4.8 (1H)
(b) Multiplet δ6.8 (4H)
(c) Singlet δ3.9 (3H)

Y
p-Methoxyphenol

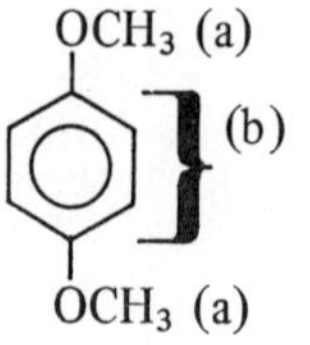

(a) Singlet δ3.75 (6H)
(b) Singlet δ6.8 (4H)

Z
p-Dimethoxybenzene

The reactions that take place are the following:

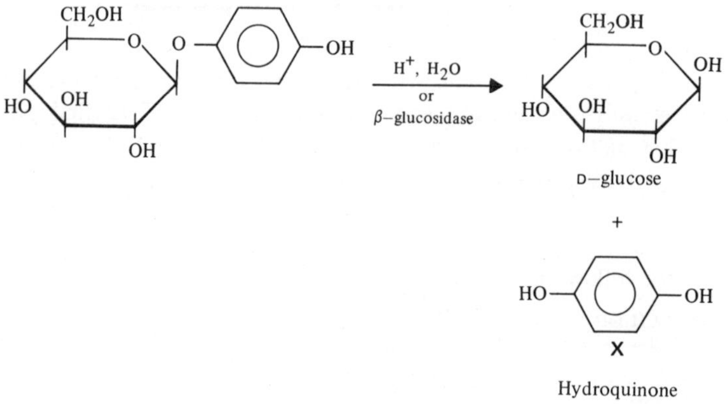

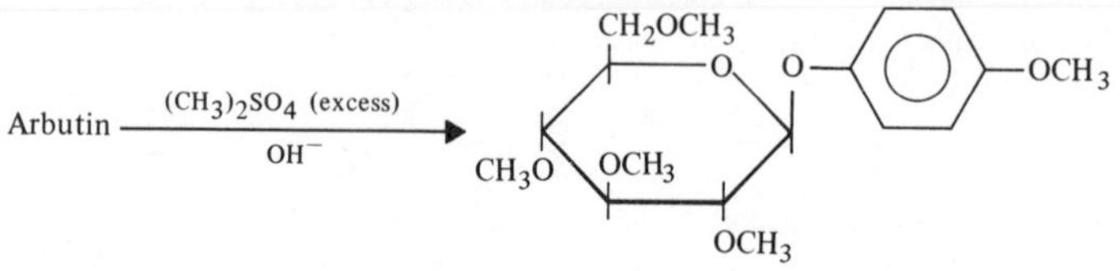

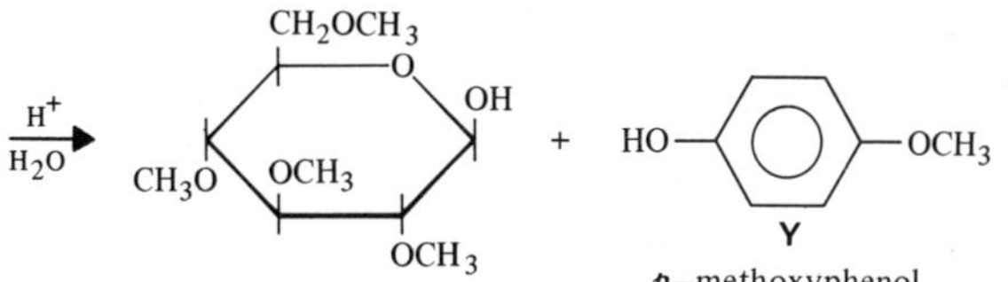

2,3,4,6–tetra–*O*–methyl
D–glucose

p–methoxyphenol **Y**

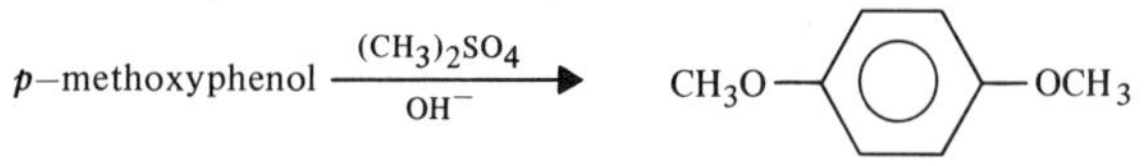

p–dimethoxybenzene **Z**

24.44

(a) and (b) Two molecules of acetone react with four hydroxyl groups of D-glucose to yield a compound (below) containing two cyclic ketal linkages. Reaction with *cis* hydroxyl groups is preferred in reactions like this. Thus D-glucose reacts with acetone preferentially in the furanose form because reaction in the pyranose form would require the formation of a cyclic ketal from the *trans* hydroxyl groups at C-3 and C-4. This would introduce greater strain into the product.

W-50

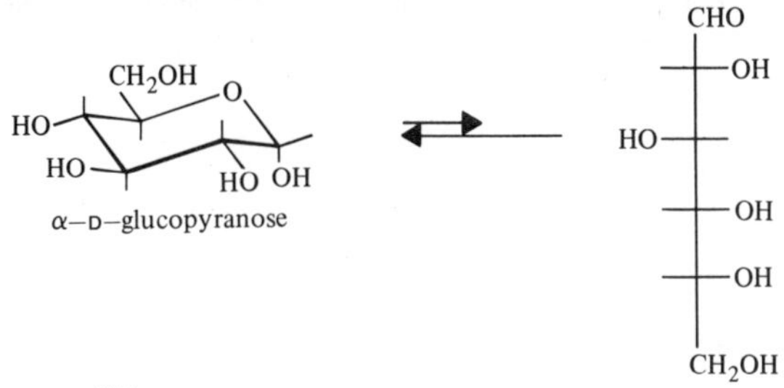

α–D–glucopyranose

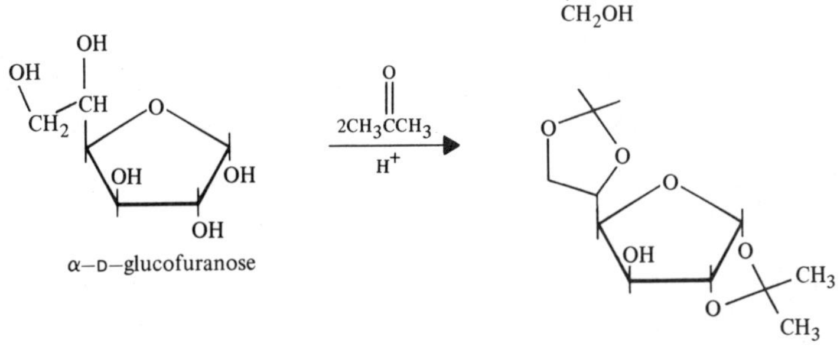

α–D–glucofuranose

"Diacetone glucose"

D-Galactose reacts similarly, but it can react in the pyranose form because the hydroxyl groups at C-3 and C-4 are *cis* (as are those at C-1 and C-2).

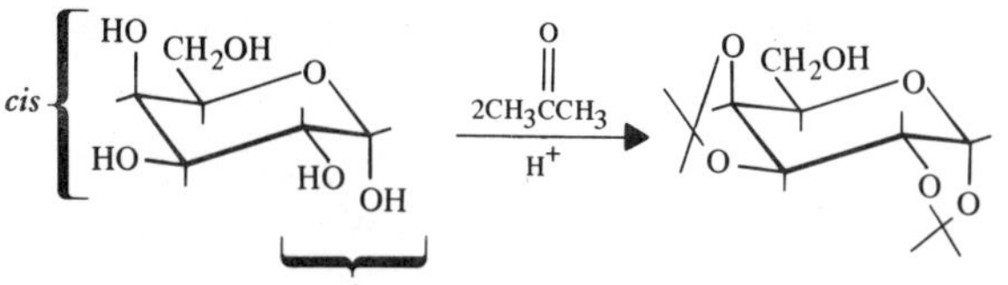

α–D–galactopyranose

25 AMINO ACIDS AND PROTEINS

25.1

(a) $\overset{+}{H_3N}CH_2CH_2CH_2CH_2\underset{\underset{+NH_3}{|}}{C}HCOOH$

(b) $H_2NCH_2CH_2CH_2CH_2\underset{\underset{NH_2}{|}}{C}HCOO^-$

(c) The α-amino group is less basic than the ϵ-amino group because of the proximity of the electron-withdrawing carboxylate group.

25.2

(a) $HOOCCH_2CH_2\underset{\underset{+NH_3}{|}}{C}HCOOH$

(b) $^-OOCCH_2CH_2\underset{\underset{NH_2}{|}}{C}HCOO^-$

(c) $HOOCCH_2CH_2\underset{\underset{+NH_3}{|}}{C}HCOO^-$ predominates at the isoelectric point rather than $^-OOCCH_2-$

$\underset{\underset{+NH_3}{|}}{C}H_2CHCOOH$ because of the acid-strengthening inductive effect of the α-amino group.

(d) Since glutamic acid is a dicarboxylic acid, acid must be added (i.e., the pH must be made lower) to suppress the ionization of the second carboxyl group and thus achieve the isoelectric point. Glutamine, with only one carboxyl group, is similar to glycine or phenylalanine and has its isoelectric point at a higher pH.

25.3

(a) $C_6H_5CONHCH(CO_2C_2H_5)_2 \xrightarrow[C_6H_5CH_2Br]{NaOCH_2CH_3}$

$C_6H_5CONH\underset{\underset{\underset{C_6H_5}{|}}{CH_2}}{C}(CO_2C_2H_5)_2 \xrightarrow[\text{heat}]{HBr} \left[H_3\overset{+}{N}-\underset{\underset{\underset{C_6H_5}{|}}{CH_2}}{\overset{\overset{COOH}{|}}{C}}-COO^- \right] \xrightarrow{-CO_2} C_6H_5CH_2\underset{\underset{+NH_3}{|}}{C}HCOO^-$

Phenylalanine

296

(b) $C_6H_5CONHCH(CO_2C_2H_5)_2 \xrightarrow[\text{BrCH}_2\text{CO}_2\text{C}_2\text{H}_5]{\text{NaOCH}_2\text{CH}_3}$

$$C_6H_5CONHC(CO_2C_2H_5)_2 \xrightarrow{\text{HBr}} \left[\begin{array}{c} \text{COOH} \\ | \\ H_3\overset{+}{N}-\overset{|}{C}-COO^- \\ | \\ CH_2 \\ | \\ COOH \end{array} \right]$$

with side chain:
$\begin{array}{c} | \\ CH_2 \\ | \\ CO_2C_2H_5 \end{array}$

$$\xrightarrow{-CO_2} \begin{array}{c} HOOCCH_2CHCOO^- \\ | \\ \overset{+}{N}H_3 \end{array}$$

Aspartic acid

(c) $C_6H_5CONHCH(CO_2C_2H_5)_2 \xrightarrow[\text{(CH}_3)_2\text{CHBr}]{\text{NaOCH}_2\text{CH}_3}$

$$C_6H_5CONHC(CO_2C_2H_5)_2 \xrightarrow[\text{heat}]{\text{HBr}} \left[\begin{array}{c} \text{COOH} \\ | \\ H_3\overset{+}{N}-\overset{|}{C}-COO^- \\ | \\ CHCH_3 \\ | \\ CH_3 \end{array} \right]$$

with side chain:
$\begin{array}{c} | \\ CHCH_3 \\ | \\ CH_3 \end{array}$

$$\xrightarrow{-CO_2} \begin{array}{c} CH_3CH-CHCOO^- \\ | \quad\quad | \\ CH_3 \quad \overset{+}{N}H_3 \end{array}$$

Valine

25.4

(a) $NK \; + \; BrCH(CO_2C_2H_5)_2 \longrightarrow$

$N{-}CH(CO_2C_2H_5)_2 \xrightarrow[\text{(CH}_3)_2\text{CHCH}_2\text{Br}]{\text{NaOCH}_2\text{CH}_3}$

$\xrightarrow[\text{heat}]{\text{NaOH}}$

$$\xrightarrow[\text{Heat}]{\text{HCl}} \quad \begin{array}{c} (CH_3)_2CHCH_2CHCOO^- \\ | \\ \overset{+}{N}H_3 \end{array} \; + \; CO_2 \; + \; \text{(phthalic acid structure)}$$

Leucine

(b)

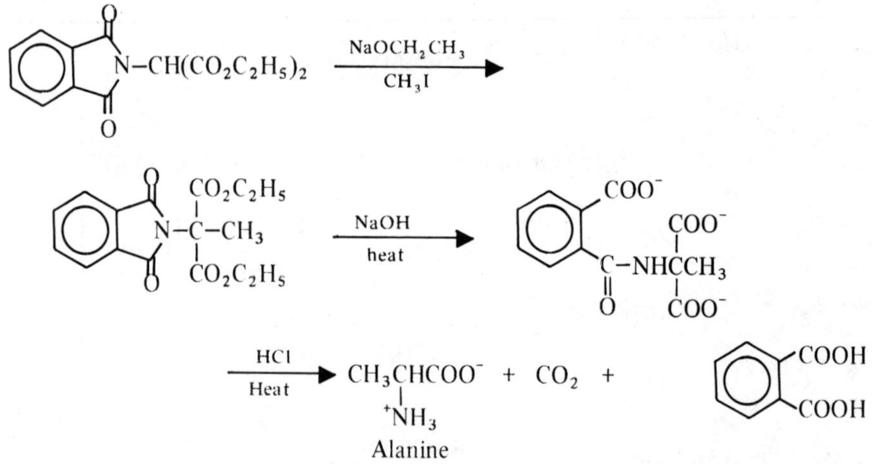

$$\xrightarrow[\text{Heat}]{\text{HCl}} \quad \underset{\underset{^+NH_3}{|}}{CH_3CHCOO^-} + CO_2 + \quad$$

Alanine

(c)

$$\xrightarrow[\text{heat}]{\text{HCl}} \quad \underset{\underset{^+NH_3}{|}}{C_6H_5CH_2CHCOO^-} + CO_2 + \quad$$

Phenylalanine

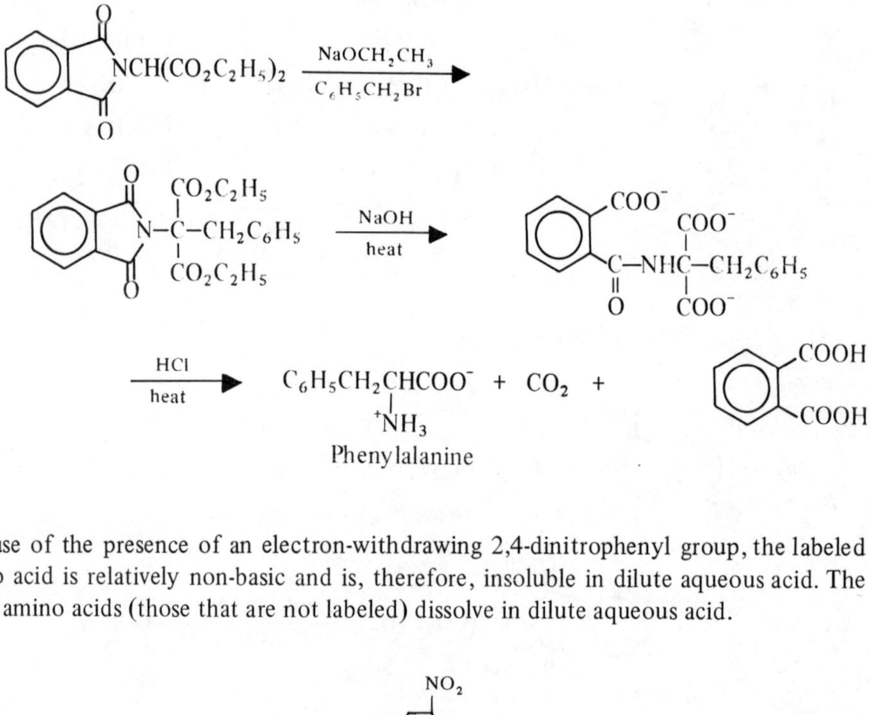

25.5

Because of the presence of an electron-withdrawing 2,4-dinitrophenyl group, the labeled amino acid is relatively non-basic and is, therefore, insoluble in dilute aqueous acid. The other amino acids (those that are not labeled) dissolve in dilute aqueous acid.

25.6

(a)

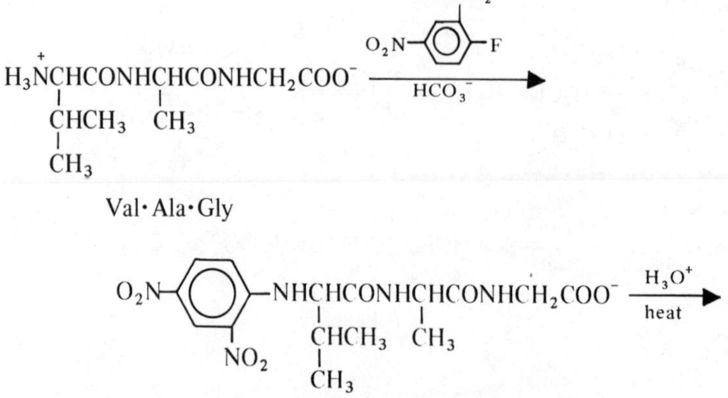

Val·Ala·Gly

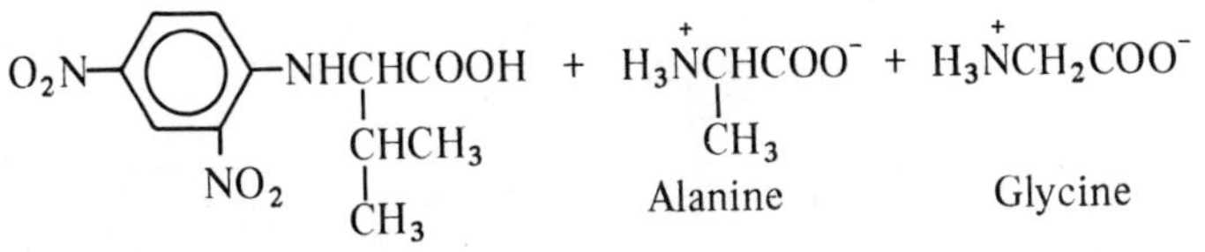

$$O_2N-\langle\rangle-NHCHCOOH \;+\; H_3\overset{+}{N}CHCOO^- \;+\; H_3\overset{+}{N}CH_2COO^-$$

with side chains:
- under first: CHCH₃, NO₂, CH₃
- under second: CH₃ — Alanine
- third: Glycine

Labeled valine
(separate and identify)

(b) $O_2N-\langle\rangle-NHCHCOOH \;+\; O_2N-\langle\rangle-NHCH_2CH_2CH_2CH_2CHCOO^-$

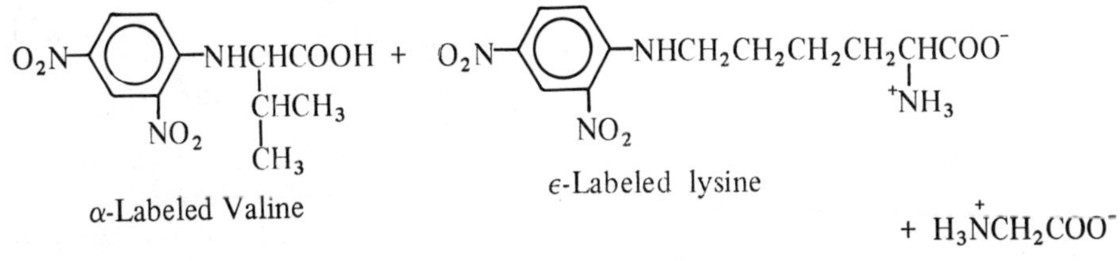

α-Labeled Valine (side chains NO₂, CHCH₃, CH₃)

ε-Labeled lysine (NO₂; $^+NH_3$)

$+\; H_3\overset{+}{N}CH_2COO^-$

Glycine

25.7

$$\langle\rangle-N=C=S \;+\; H_2\ddot{N}-CHCO-NHCHCO-NHCHCOO^- \xrightarrow{\;OH^-\;}$$

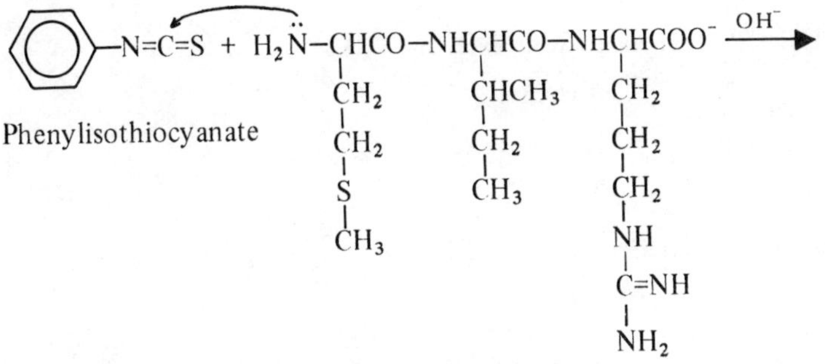

Phenylisothiocyanate

side chains:
CH₂ / CH₂ / S / CH₃ ;
CHCH₃ / CH₂ / CH₃ ;
CH₂ / CH₂ / CH₂ / NH / C=NH / NH₂

Met·Ile·Arg

$$\langle\rangle-NH-\overset{\overset{S}{\|}}{C}-NH-CHCO-NHCHCO-NHCHCOO^- \xrightarrow{\;H^+\;}$$

side chains:
CH₂ / CH₂ / S / CH₃ ;
CHCH₃ / CH₂ / CH₃ ;
CH₂ / CH₂ / CH₂ / NH / C=NH / NH₂

$$\langle\rangle-N\overset{\overset{S}{\|}}{\underset{}{C}}NH \quad + \quad H_2NCHCO-NHCHCOO^-$$

(ring: N–C(=S)–NH and C–CH with O and CH₂CH₂SCH₃)

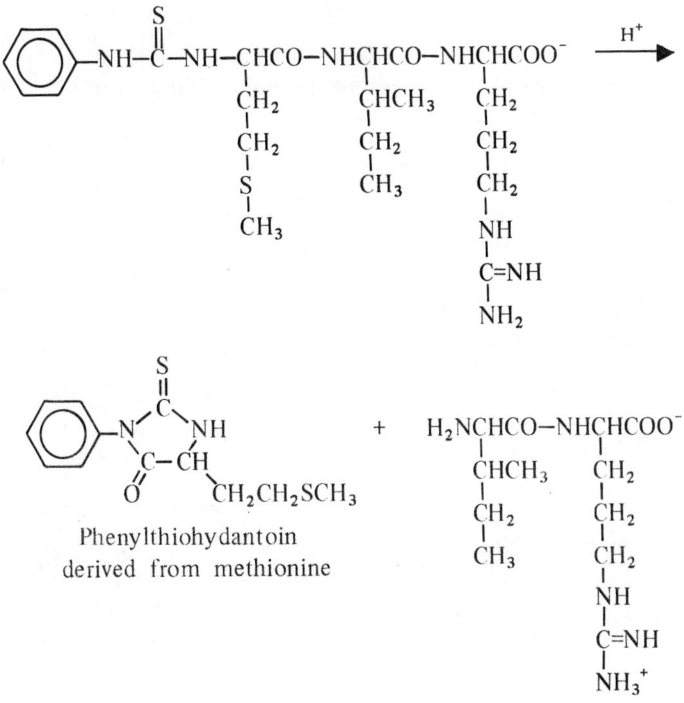

Phenylthiohydantoin
derived from methionine

side chains:
CHCH₃ / CH₂ / CH₃ ;
CH₂ / CH₂ / CH₂ / NH / C=NH / NH_3^+

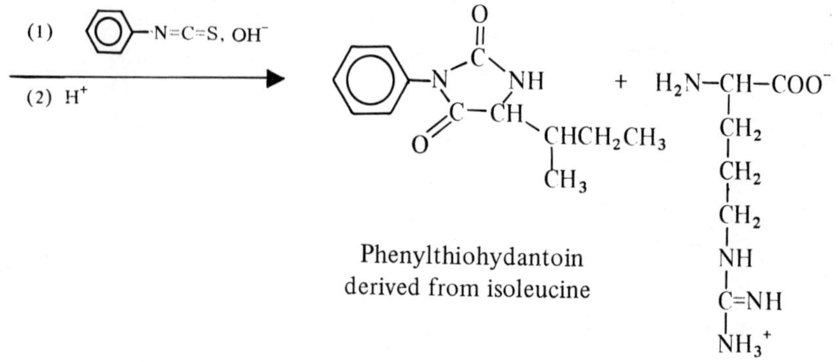

Phenylthiohydantoin
derived from isoleucine

25.8

(a) Two structures are possible with the sequence Glu·Cys·Gly. Glutamic acid may be linked to cysteine through its α-carboxyl group,

$$\underset{\underset{^+NH_3}{|}}{HOOCCH_2CH_2CHCO}-\underset{\underset{CH_2SH}{|}}{NHCHCO}-NHCH_2COO^-$$

or through its γ-carboxyl group,

$$\underset{\underset{COO^-}{|}}{H_3\overset{+}{N}CHCH_2CH_2CO}-\underset{\underset{CH_2SH}{|}}{NHCHCO}-NHCH_2COO^-$$

(b) This shows that the second structure above is correct, that in glutathione the γ-carboxyl group is linked to cysteine.

25.9

We look for points of overlap to determine the amino acid sequence in each case.

(a)   Ser · Thr
            Thr · Hyp
   <u>Pro · Ser</u>
   Pro · Ser · Thr · Hyp

(b)   Ala · Cys
            Cys · Arg
                  Arg · Val
   <u>Leu · Ala</u>
   Leu · Ala · Cys · Arg · Val

25.10

Sodium in liquid ammonia brings about reductive cleavage of the disulfide linkage of oxytocin to two thiol groups, then air oxidizes the two thiol groups back to a disulfide linkage:

See also pages 623-624 in the text.

25.11

$$\overset{+}{H_3}NCH_2COO^- + (CH_3)_3COCN_3 \xrightarrow[25°]{OH^-}$$

Glycine   *tert*-Butoxy-
carbonyl azide

$$(CH_3)_3C-OCNHCH_2COOH \xrightarrow[(2)\ ClCO_2C_2H_5]{(1)\ (C_2H_5)_3N}$$
Boc-Gly

$$(CH_3)_3COCNHCH_2COCOC_2H_5 \xrightarrow[(-CO_2,\ -C_2H_5OH)]{Valine}$$
Mixed anhydride

with Valine: $\overset{+}{H_3}NCHCOO^-$, $CHCH_3$, CH_3

$$(CH_3)_3COCNHCH_2CNHCHCOOH \xrightarrow[(2)\ ClCO_2C_2H_5]{(1)\ (C_2H_5)_3N}$$
Boc-Gly·Val   CHCH₃ CH₃

$$(CH_3)_3COCNHCH_2CNHCHCOCOC_2H_5 \xrightarrow{Alanine}$$
Mixed anhydride   CHCH₃ CH₃

with Alanine: $\overset{+}{H_3}NCHCOO^-$, CH_3

$$(CH_3)_3COCNHCH_2CNHCHCNHCHCOOH \xrightarrow[25°]{CF_3COOH \atop CH_3COOH}$$
Boc-Gly·Val·Ala   CHCH₃ CH₃

(cont. on p. 302)

$$(CH_3)_2C=CH_2 \ + \ CO_2 \ + \ H_3\overset{+}{N}CH_2\overset{O}{\overset{\|}{C}}NH\overset{}{\underset{\underset{\underset{CH_3}{|}}{\underset{CHCH_3}{|}}}{CH}}\overset{O}{\overset{\|}{C}}NHCHCOO^-$$

$$\underset{\underset{CH_3}{|}}{}$$

Gly·Val·Ala

25.12

(a) $2C_6H_5CH_2O\overset{O}{\overset{\|}{C}}Cl \ + \ H_2NCH_2CH_2CH_2CH_2\underset{\underset{NH_2}{|}}{CH}COO^- \xrightarrow[25°]{OH^-}$

Benzyl chloro-
carbonate   Lysine

$C_6H_5CH_2O\overset{O}{\overset{\|}{C}}NHCH_2CH_2CH_2CH_2\underset{\underset{\underset{C_6H_5CH_2O\overset{}{\underset{}{C}}=O}{|}}{NH}}{CH}COOH \xrightarrow[\text{(2) ClCOOC}_2\text{H}_5]{\text{(1) (C}_2\text{H}_5)_3\text{N}}$

$C_6H_5CH_2O\overset{O}{\overset{\|}{C}}NHCH_2CH_2CH_2CH_2\underset{\underset{\underset{C_6H_5CH_2O\overset{}{\underset{}{C}}=O}{|}}{NH}}{CH}\overset{O}{\overset{\|}{C}}\overset{O}{\overset{\|}{C}}OC_2H_5 \xrightarrow[(-CO_2, \ -C_2H_5OH)]{\underset{\underset{\underset{CH_3 \ NH_3^+}{}}{}}{CH_3CH_2CH-CHCOO^-}}$

$C_6H_5CH_2O\overset{O}{\overset{\|}{C}}NHCH_2CH_2CH_2CH_2\underset{\underset{\underset{C_6H_5CH_2O\overset{}{\underset{}{C}}=O}{|}}{NH}}{CH}\overset{O}{\overset{\|}{C}}NH\underset{\underset{\underset{\underset{CH_3}{|}}{\underset{CH_2}{|}}}{CHCH_3}}{CH}COO^- \xrightarrow[\substack{CH_3COOH \\ cold}]{HBr}$

$$2C_6H_5CH_2Br \ + \ 2CO_2 \ + \ H_3\overset{+}{N}CH_2CH_2CH_2CH_2\underset{\underset{NH_2}{|}}{CH}\overset{O}{\overset{\|}{C}}NHCHCOO^-$$

Lys·Ile $\qquad\qquad \underset{\underset{\underset{CH_3}{|}}{\underset{CH_2}{|}}}{CHCH_3}$

(b) $3C_6H_5CH_2O\overset{O}{\overset{\|}{C}}Cl \ + \ H_2N\overset{NH}{\overset{\|}{C}}NHCH_2CH_2CH_2\underset{\underset{NH_2}{|}}{CH}COO^- \xrightarrow[25°]{OH^-}$

$C_6H_5CH_2O\overset{O}{\overset{\|}{C}}NH\overset{NH}{\overset{\|}{C}}NCH_2CH_2CH_2\underset{\underset{\underset{C_6H_5CH_2O}{|}}{\underset{C=O}{|}}}{CH}COOH \xrightarrow[\text{(2) ClCOOC}_2\text{H}_5]{\text{(1) (C}_2\text{H}_5)_3\text{N}}$

$\underset{\underset{C_6H_5CH_2O}{}}{\underset{\underset{C=O}{|}}{}}$

$$
\underset{\substack{\text{C}_6\text{H}_5\text{CH}_2\text{OCNHCNCH}_2\text{CH}_2\text{CH}_2\text{CHCOCOC}_2\text{H}_5}}{\overset{\overset{\displaystyle O}{\|}\;\overset{\displaystyle NH}{\|}\qquad\qquad\qquad\overset{\displaystyle O}{\|}\;\overset{\displaystyle O}{\|}}{}}
\xrightarrow[\;(-\text{CO}_2,\;-\text{C}_2\text{H}_5\text{OH})\;]{\overset{\text{CH}_3\text{CHCOO}^-}{\underset{^+\text{NH}_3}{|}}}
$$

with substituents:

C=O below, attached to C₆H₅CH₂O ; and NH, C=O, C₆H₅CH₂O

$$
\text{C}_6\text{H}_5\text{CH}_2\overset{\overset{\displaystyle O}{\|}}{\text{OCNHC}}\overset{\overset{\displaystyle NH}{\|}}{\text{NCH}_2}\text{CH}_2\text{CH}_2\overset{}{\text{CHC}}\overset{\overset{\displaystyle O}{\|}}{\text{NHCHCOOH}}
\xrightarrow[\substack{\text{CH}_3\text{COOH}\\ \text{cold}}]{\text{HBr}}
$$

with C=O, C₆H₅CH₂O below; NH, C=O, C₆H₅CH₂O ; CH₃

$$
3\text{C}_6\text{H}_5\text{CH}_2\text{Br} + 3\text{CO}_2 + {}^+\text{H}_3\text{N}\overset{\overset{\displaystyle NH}{\|}}{\text{C}}\text{NHCH}_2\text{CH}_2\text{CH}_2\overset{\underset{\displaystyle NH_2}{|}}{\text{CH}}\text{CONHCH}\underset{\underset{\displaystyle CH_3}{|}}{}\text{COO}^-
$$

Arg·Ala

25.13
The weakness of the benzyl-oxygen bond (p. 969) allows these groups to be removed by catalytic hydrogenolysis.

25.14
(a) An electrophilic aromatic substitution reaction:

$$
\left(\text{CH}_2\text{CH}\right)_{\overline{n}} + \text{CH}_3\text{OCH}_2\text{Cl} \xrightarrow{\text{BF}_3} \left(\text{CH}_2\text{CH}\right)_{\overline{n}} + \text{CH}_3\text{OH}
$$

(with benzene ring attached; product ring para-substituted with CH₂Cl)

(b) The linkage between the resin and the polypeptide is a benzylic ester. It is cleaved by HBr in CF₃COOH at room temperature because the carbocation that is formed initially is the relatively stable, benzylic cation.

25.15

$$
\bigcirc\!\!-\text{CH}_2\text{Cl} + \text{HO}\overset{\overset{\displaystyle O}{\|}}{\text{C}}\text{CHNH}\overset{\overset{\displaystyle O}{\|}}{\text{C}}\text{OC}(\text{CH}_3)_3
$$
(with CH₃ below)

$\Big\downarrow$ base

1 Add Boc·Ala

$$
\bigcirc\!\!-\text{CH}_2\text{O}\overset{\overset{\displaystyle O}{\|}}{\text{C}}\text{CHNH}\overset{\overset{\displaystyle O}{\|}}{\text{C}}\text{OC}(\text{CH}_3)_3
$$
(with CH₃ below)

2 Purify by washing

$\Big\downarrow$ CF₃COOH, CH₂Cl₂

3 Remove protecting group

$\downarrow$

(cont. on p. 304)

$$\text{)}-CH_2O\overset{\overset{\displaystyle O}{\|}}{C}\overset{\underset{\displaystyle CH_3}{|}}{C}HNH_2$$

4 Purify by washing

$$HO\overset{\overset{\displaystyle O}{\|}}{C}\overset{\underset{\displaystyle CH_2C_6H_5}{|}}{C}HNH\overset{\overset{\displaystyle O}{\|}}{C}OC(CH_3)_3$$
and
dicyclohexylcarbodiimide

5 Add Boc·Phe

$$\text{)}-CH_2O\overset{\overset{\displaystyle O}{\|}}{C}\overset{\underset{\displaystyle CH_3}{|}}{C}HNH\overset{\overset{\displaystyle O}{\|}}{C}\overset{\underset{\underset{\displaystyle C_6H_5}{|}}{\underset{\displaystyle CH_2}{|}}}{C}HNH\overset{\overset{\displaystyle O}{\|}}{C}OC(CH_3)_3$$

6 Purify by washing

CF_3COOH, CH_2Cl_2

7 Remove protecting group

$$\text{)}-CH_2O\overset{\overset{\displaystyle O}{\|}}{C}\overset{\underset{\displaystyle CH_3}{|}}{C}HNH\overset{\overset{\displaystyle O}{\|}}{C}\overset{\underset{\underset{\displaystyle C_6H_5}{|}}{\underset{\displaystyle CH_2}{|}}}{C}HNH_2$$

8 Purify by washing

$$HO\overset{\overset{\displaystyle O}{\|}}{C}\overset{\underset{\underset{\displaystyle O=COC(CH_3)_3}{|}}{\underset{\displaystyle NH}{|}}}{C}HCH_2CH_2CH_2CH_2NH\overset{\overset{\displaystyle O}{\|}}{C}OC(CH_3)_3$$
and
dicyclohexylcarbodiimide

9 Add Protected Lys

$$\text{)}-CH_2O\overset{\overset{\displaystyle O}{\|}}{C}\overset{\underset{\displaystyle CH_3}{|}}{C}HNH\overset{\overset{\displaystyle O}{\|}}{C}\overset{\underset{\underset{\displaystyle C_6H_5}{|}}{\underset{\displaystyle CH_2}{|}}}{C}HNH\overset{\overset{\displaystyle O}{\|}}{C}\overset{\underset{\displaystyle CH_2\cdots}{|}}{C}HNH\overset{\overset{\displaystyle O}{\|}}{C}OC(CH_3)_3$$

with side chain: CH_2—CH_2—CH_2—CH_2—$NHCOC(CH_3)_3$ (with C=O)

10 Purify by washing

CF_3COOH, CH_2Cl_2

11 Remove protecting groups

$$\text{⟩-CH}_2\text{O}\overset{\overset{\text{O}}{\|}}{\text{C}}\text{CHNH}\overset{\overset{\text{O}}{\|}}{\text{C}}\text{CHNH}\overset{\overset{\text{O}}{\|}}{\text{C}}\text{CHNH}_2$$

with side chains:
- CH_3
- CH_2 — C_6H_5
- CH_2 — CH_2 — CH_2 — CH_2 — NH_2

12 Purify by washing

$\downarrow$ HBr, CF_3COOH

13 Detach tripeptide

$$\text{⟩-CH}_2\text{Br} + \text{}^-\text{O}\overset{\overset{\text{O}}{\|}}{\text{C}}\text{CHNH}\overset{\overset{\text{O}}{\|}}{\text{C}}\text{CHNH}\overset{\overset{\text{O}}{\|}}{\text{C}}\text{CHNH}_2$$

with side chains:
- CH_3
- CH_2 — C_6H_5
- CH_2 — CH_2 — CH_2 — CH_2 — NH_3^+

14 Isolate product

Ala·Phe·Lys

25.16

(a) Isoleucine, threonine, hydroxyproline, and cystine.

(b)

$$\begin{array}{c} COO^- \\ H_3\overset{+}{N}\!-\!\!\!\mid\!-H \\ CH_3\!-\!\!\!\mid\!-H \\ CH_2 \\ CH_3 \end{array} \quad\text{and}\quad \begin{array}{c} COO^- \\ H_3\overset{+}{N}\!-\!\!\!\mid\!-H \\ H\!-\!\!\!\mid\!-CH_3 \\ CH_2 \\ CH_3 \end{array}$$

$$\begin{array}{c} COO^- \\ H_3\overset{+}{N}\!-\!\!\!\mid\!-H \\ H\!-\!\!\!\mid\!-OH \\ CH_3 \end{array} \quad\text{and}\quad \begin{array}{c} COO^- \\ H_3\overset{+}{N}\!-\!\!\!\mid\!-H \\ HO\!-\!\!\!\mid\!-H \\ CH_3 \end{array}$$

$$\begin{array}{c} COO^- \\ H_2\overset{+}{N}\!-\!\!\!\!-\!\!\!\!-H \\ CH_2 \quad CH_2 \\ \diagdown\; H\; \diagup \\ OH \end{array} \quad\text{and}\quad \begin{array}{c} COO^- \\ H_2\overset{+}{N}\!-\!\!\!\!-\!\!\!\!-H \\ CH_2 \quad CH_2 \\ \diagdown\; OH\; \diagup \\ H \end{array}$$

(With cystine, both chiral carbons are α-carbons, thus according to the problem, both must have the L-configuration, and no isomers of this type can be written.)

(c) Diastereomers

25.17
(a) Alanine

$$CH_3\underset{\overset{|}{^+NH_3}}{CH}COO^- + HONO \longrightarrow CH_3\underset{\overset{|}{OH}}{CH}COOH + N_2$$

(b) Proline and hydroxyproline. All of the other amino acids have at least one primary amino group.

(c) HO—⟨ ⟩—CH₂CHCOO⁻ with Br substituents and NH₃⁺

$$(c)\ HO-\underset{Br}{\overset{Br}{\bigcirc}}-CH_2\underset{\overset{|}{NH_3^+}}{CH}COO^-$$

$$(d)\ \bigcirc-CH_2\underset{\overset{|}{NH_3^+}}{CH}COOC_2H_5$$

(e) $CH_3\underset{\overset{|}{NH}}{CH}COO^-$
  $\underset{\overset{|}{C_6H_5}}{C}{=}O$

25.18
(a)

$$^+H_3N-\underset{\overset{|}{CH_2OH}}{\overset{|}{\overset{COO^-}{|}}}-H \xrightarrow[CH_3OH]{HCl} {}^+H_3N-\underset{\overset{|}{CH_2OH}}{\overset{|}{\overset{COOCH_3}{|}}}-H \quad Cl^-$$

(−)-Serine

A
$(C_4H_{10}ClNO_3)$

$$\xrightarrow{PCl_5} {}^+H_3N-\underset{\overset{|}{CH_2Cl}}{\overset{|}{\overset{COOCH_3}{|}}}-H \quad Cl^- \xrightarrow[(2)\ OH^-]{(1)\ H_3O^+,\ H_2O,\ heat}$$

B
$(C_4H_9Cl_2NO_2)$

$$^+H_3N-\underset{\overset{|}{CH_2Cl}}{\overset{|}{\overset{COO^-}{|}}}-H \xrightarrow[dil.\ H_3O^+]{Na-Hg} {}^+H_3N-\underset{\overset{|}{CH_3}}{\overset{|}{\overset{COO^-}{|}}}-H$$

C
$(C_3H_6ClNO_2)$

L-(+)-Alanine

(b)

$$B \xrightarrow{OH^-} H_2N \overset{COOCH_3}{\underset{CH_2Cl}{\rule{0pt}{1em}|\!\!-\!\!|\, H}} \xrightarrow{NaSH} H_2N \overset{COOCH_3}{\underset{CH_2SH}{\rule{0pt}{1em}|\!\!-\!\!|\, H}}$$

$$\begin{array}{cc} \mathbf{D} & \mathbf{E} \\ (C_4H_8ClNO_2) & (C_4H_9NO_2S) \end{array}$$

$$\xrightarrow[\text{(2) } OH^-]{\text{(1) } H_3O^+,\ H_2O,\ \text{heat}} {}^+H_3N \overset{COO^-}{\underset{CH_2SH}{\rule{0pt}{1em}|\!\!-\!\!|\, H}}$$

L-(+)-Cysteine

(c)

$$H_3\overset{+}{N} \overset{COO^-}{\underset{\underset{\displaystyle O}{\overset{\displaystyle \|}{CH_2CNH_2}}}{\rule{0pt}{1em}|\!\!-\!\!|\, H}} \xrightarrow{NaOBr,\ OH^-} H_2N \overset{COO^-}{\underset{CH_2NH_2}{\rule{0pt}{1em}|\!\!-\!\!|\, H}}$$

$$\begin{array}{cc} \text{L-Asparagine} & \mathbf{F} \\ & (C_3H_7N_2O_2) \end{array}$$

$${}^+H_3N \overset{COO^-}{\underset{CH_2Cl}{\rule{0pt}{1em}|\!\!-\!\!|\, H}} \longrightarrow NH_3$$

C
(from part A)

25.19

(a) $CH_3\overset{O}{\overset{\|}{C}}NHCH(CO_2C_2H_5)_2 + CH_2=CH-C\equiv N \xrightarrow[C_2H_5OH]{NaOC_2H_5}$

$$CH_3\overset{O}{\overset{\|}{C}}NH-\overset{CO_2C_2H_5}{\underset{CO_2C_2H_5}{\overset{|}{\underset{|}{C}}}}-CH_2CH_2C\equiv N \xrightarrow[\text{reflux}]{\text{conc HCl}}$$

G

$$HOOCCH_2CH_2\overset{}{\underset{NH_3^+}{\overset{|}{CHCOO^-}}} + CH_3COOH + 2C_2H_5OH + NH_4^+$$
$$+ CO_2$$

D L-Glutamic acid

(b)

$$CH_3\overset{O}{\overset{\|}{C}}NH-\overset{CO_2C_2H_5}{\underset{CO_2C_2H_5}{\overset{|}{\underset{|}{C}}}}-CH_2CH_2C\equiv N \xrightarrow[68°,\ 1000\ psi]{H_2(Ni)}$$

$$\left[CH_3\overset{O}{\overset{\|}{C}}NH-\overset{CO_2C_2H_5}{\underset{CO_2C_2H_5}{\overset{|}{\underset{|}{C}}}}-CH_2CH_2CH_2NH_2 \right] \xrightarrow{-C_2H_5OH}$$

H (cont. on p. 308)

$$\xrightarrow[\text{reflux}]{\text{conc. HCl}} \overset{+}{H_3}NCH_2CH_2CH_2CHCOO^- + CH_3COOH + CO_2 + C_2H_5OH$$

$$\underset{Cl^-}{}\underset{NH_3^+}{|}$$

DL-Ornithine hydrochloride

25.20

(a) $C_6H_5CH_2\overset{\overset{O}{\|}}{CH} \xrightarrow[\text{HCN}]{NH_3} C_6H_5CH_2\underset{\underset{NH_2}{|}}{CH}C\equiv N \xrightarrow{H_3O^+} C_6H_5CH_2\underset{\underset{NH_3^+}{|}}{CH}COO^-$

Phenyl acetaldehyde   DL-Phenylalanine

(b) $CH_3SH + CH_2=CH-\overset{\overset{O}{\|}}{CH} \xrightarrow{\text{base}} CH_3SCH_2CH_2\overset{\overset{O}{\|}}{CH}$

$\xrightarrow[\text{HCN}]{NH_3} CH_3SCH_2CH_2\underset{\underset{NH_2}{|}}{CH}C\equiv N \xrightarrow{H_3O^+} CH_3SCH_2CH_2\underset{\underset{NH_3^+}{|}}{CH}COO^-$

DL-Methionine

25.21

$C_6H_5CH_2\underset{\underset{NH_3^+}{|}}{CH}COO^- + HOOCCH_2CH_2\overset{\overset{O}{\|}}{C}COOH \underset{\longleftarrow}{\overset{\text{transaminase}}{\longrightarrow}}$

Phenylalanine   α-Ketoglutaric acid

$C_6H_5CH_2\overset{\overset{O}{\|}}{C}COOH + HOOCCH_2CH_2\underset{\underset{NH_3^+}{|}}{CH}COO^-$

Phenylpyruvic acid   Glutamic acid

Then:

$HOOCCH_2CH_2\underset{\underset{NH_3^+}{|}}{CH}COO^- + HOOCCH_2\overset{\overset{O}{\|}}{C}COOH \underset{\longleftarrow}{\overset{\text{transaminase}}{\longrightarrow}}$

Glutamic acid   Oxaloacetic acid

$HOOCCH_2CH_2\overset{\overset{O}{\|}}{C}COOH + HOOCCH_2\underset{\underset{NH_3^+}{|}}{CH}COO^-$

α-Ketoglutaric acid   Aspartic acid

This amounts to:

Phenylalanine + α-Ketoglutaric acid $\rightleftharpoons$ Phenylpyruvic acid + Glutamic acid

Glutamic acid + Oxaloacetic acid $\rightleftharpoons$ α-Ketoglutaric acid + Aspartic acid

Net: Phenylalanine + Oxaloacetic acid $\rightleftharpoons$ Phenylpyruvic acid + Aspartic acid

25.22

We look for points of overlap:

```
                              Phe  ·  Ser
                  Pro  ·  Gly  ·  Phe
         Pro  ·  Pro                    Ser  ·  Pro  ·  Phe
Arg  ·  Pro                                    Phe  ·  Arg
Arg  ·  Pro  ·  Pro  ·  Gly  ·  Phe  ·  Ser  ·  Pro  ·  Phe  ·  Arg
```

Bradykinin

25.23

1. This shows that valine is the N-terminal amino acid and that valine is attached to leucine. (Lysine labeled at the ϵ-amino group is to be expected if lysine is not the N-terminal amino acid and if it is linked in the polypeptide through its α-amino group.)

2. This shows that alanine is the C-terminal amino acid and that it is linked to glutamic acid.

At this point, then, we have the following information about the structure of the heptapeptide.

Val · Leu (Ala, Lys, Phe) Glu · Ala

the sequence here is
unknown

3. (a) This shows that the dipeptide, **A**, is

Leu · Lys

(b) The carboxypeptidase reaction shows that the C-terminal amino acid of the tripeptide, **B**, is glutamic acid; the DNP labeling experiment shows that the N-terminal amino acid is phenylalanine. Thus the tripeptide **B** is:

Phe · Ala · Glu

Putting these pieces together in the only way possible, we arrive at the following amino acid sequence for the heptapeptide.

```
Val  ·  Leu
         Leu  ·  Lys
                       Phe  ·  Ala  ·  Glu
                                    Glu  ·  Ala
Val  ·  Leu  ·  Lys  ·  Phe  ·  Ala  ·  Glu  ·  Ala
```

25.24

At pH 2-3 the γ-carboxyl groups of polyglutamic acid are uncharged (they are present as —COOH groups). At pH 5 the γ-carboxyl groups ionize and become negatively charged (they become γ-COO$^-$ groups). The repulsive forces between these negatively charged groups cause an unwinding of the α-helix and the formation of a random coil.

25.25

The observation that the pmr spectrum taken at room temperature shows two different signals for the methyl groups suggests that they are in different environments. This would be true if rotation about the carbon-nitrogen bond was not taking place.

$$\delta 8.05 \; H \underset{O}{\overset{\diagdown}{\underset{\diagup}{C-N}}} \overset{CH_3 \; \delta 2.95}{\underset{CH_3 \; \delta 2.80}{}}$$

We assign the $\delta 2.80$ signal to the methyl group that is on the same side as the electronegative oxygen.

The fact that the methyl signals appear as doublets (and that the formyl signal is a multiplet) indicates that long-range coupling is taking place between the methyl protons and the formyl proton.

That the two doublets are not simply the result of spin-spin coupling is indicated by the observation that the distance that separates one doublet from the other changes when the applied magnetic field strength is lowered. [Remember (pp 511 and 517) the magnitude of a chemical shift is proportional to the strength of the applied magnetic field while the magnitude of a coupling constant is not.]

That raising the temperature (to 111°) causes the doublets to coalesce into a single signal indicates that at higher temperatures the molecules have enough energy to surmount the energy barrier of the carbon-nitrogen bond. Above 111°, rotation is taking place so rapidly that the spectrometer is unable to discriminate between the two methyl groups.

26 SPECIAL TOPICS V NUCLEIC ACIDS: PROTEIN SYNTHESIS

26.1

Adenine:

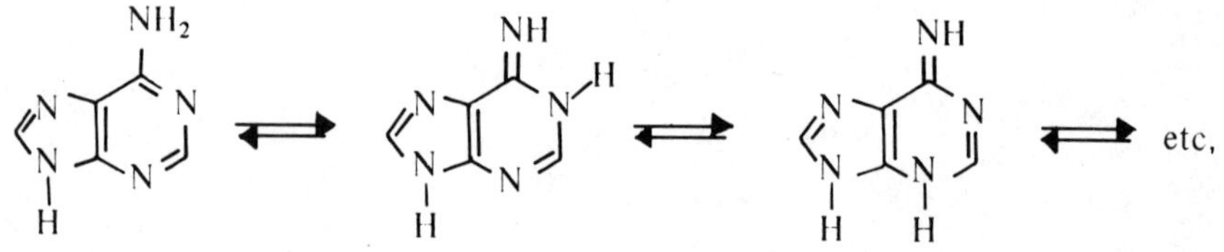

Guanine:

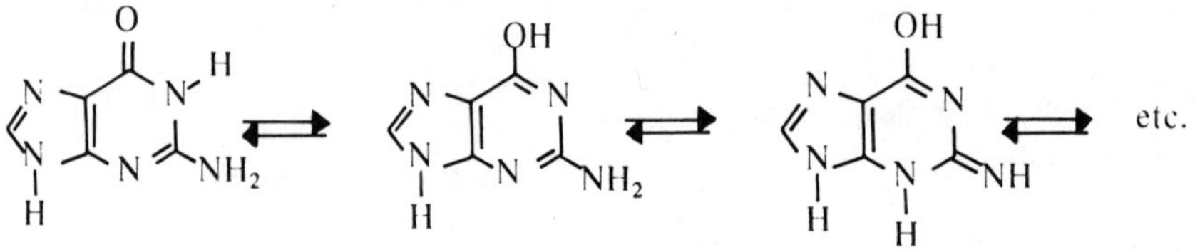

Cytosine:

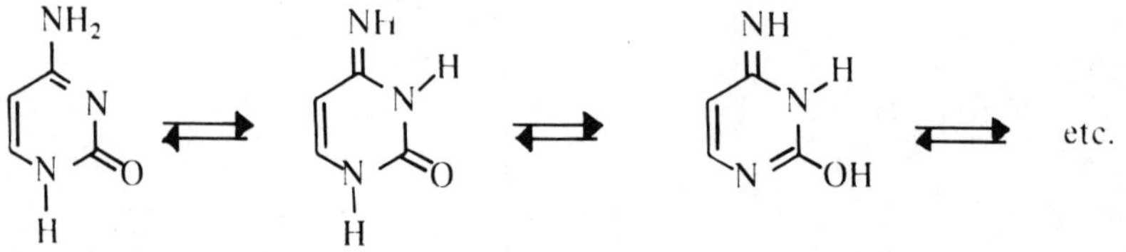

Thymine (R = CH_3) or Uracil (R = H):

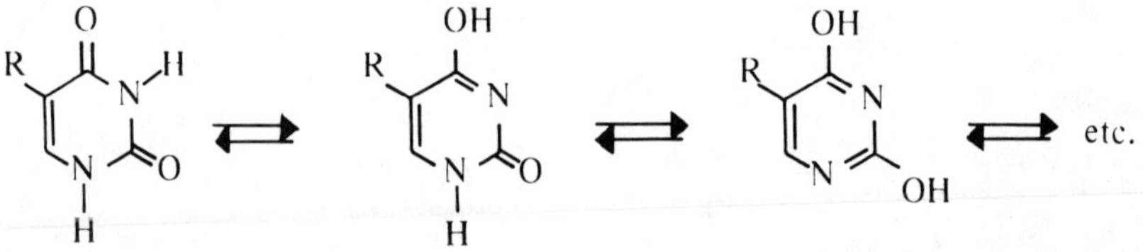

26.2

(a) The nucleosides have an N-glycosidic linkage that (like an O-glycosidic linkage) is rapidly hydrolyzed by aqueous acid but one that is stable in aqueous base.

(b)

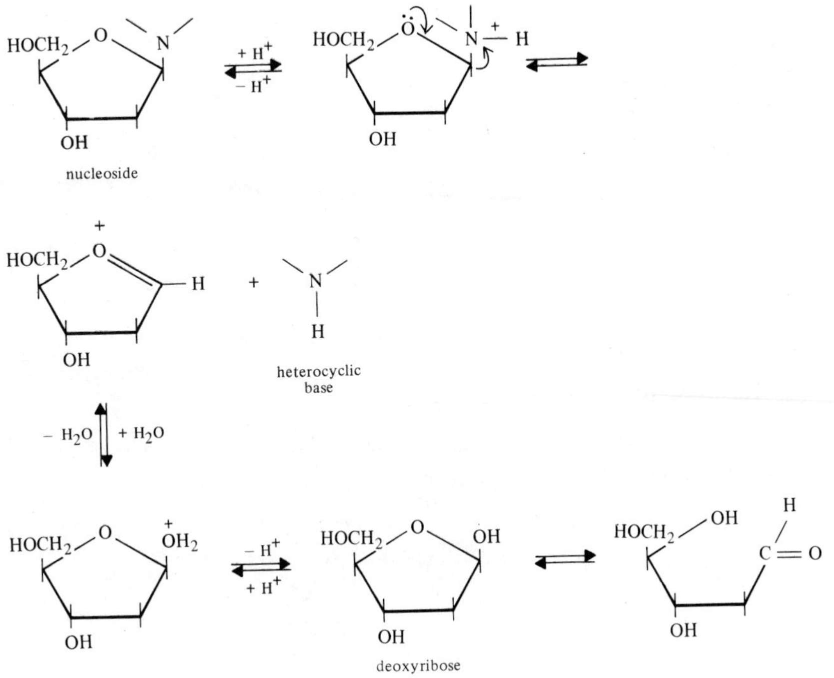

nucleoside

heterocyclic base

$- H_2O$ $+ H_2O$

deoxyribose

26.3

The reaction appears to take place through an S_N2 mechanism. Attack occurs preferentially at the primary 5'-carbon rather than at the secondary 3'-carbon.

26.4

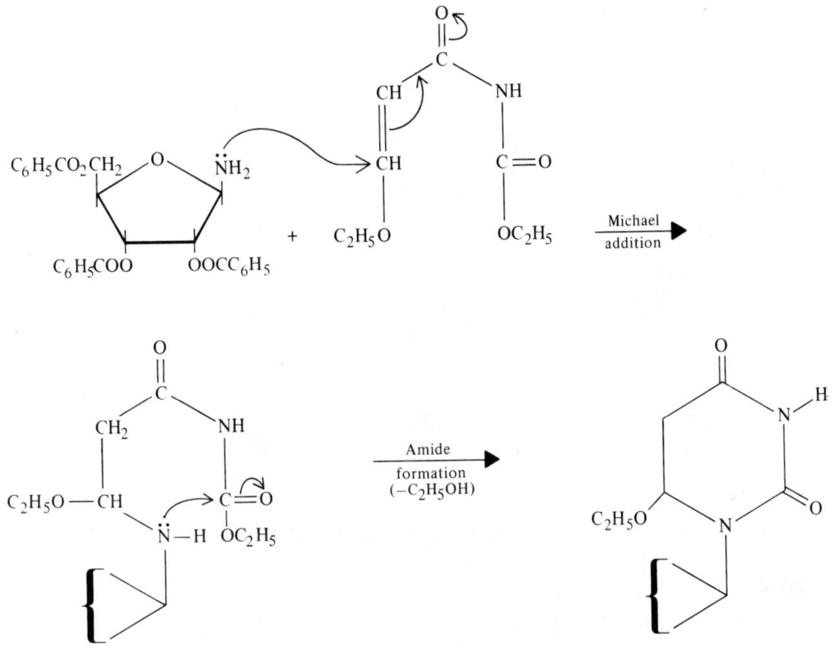

26.5

(a) The isopropylidene group is a cyclic ketal.

(b) It can be installed by treating the nucleoside with acetone and a trace of acid and by simultaneously removing the water that is produced.

26.6

(a) 6×10^9 base pairs $\times \dfrac{34\text{Å}}{10\text{ base pairs}} \times \dfrac{10^{-10}\text{ meters}}{\text{Å}} \cong 2$ meters

(b) $6 \times 10^{-12} \dfrac{\text{g}}{\text{ovum}} \times 3 \times 10^9$ ova $= 1.8 \times 10^{-2}\text{g}$

26.7

(a)

Lactim form   Thymine
of guanine

(b) Thymine would pair with adenine and thus adenine would be introduced into the complementary strand where guanine should occur.

26.8

(a) A diazonium salt and a heterocyclic analog of a phenol.

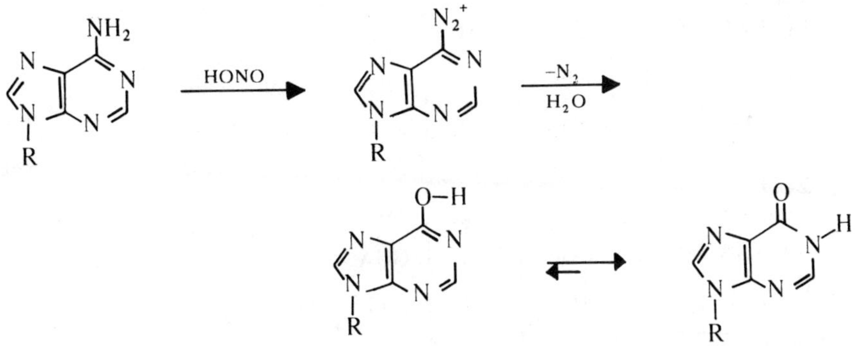
HONO
$-N_2$
H_2O

Hypoxanthine
nucleotide

(b)

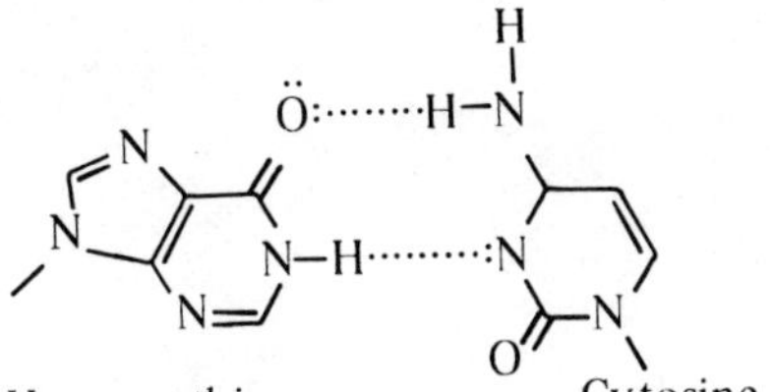

Hypoxanthine Cytosine

(c) Original double strand

First replication

Second replication

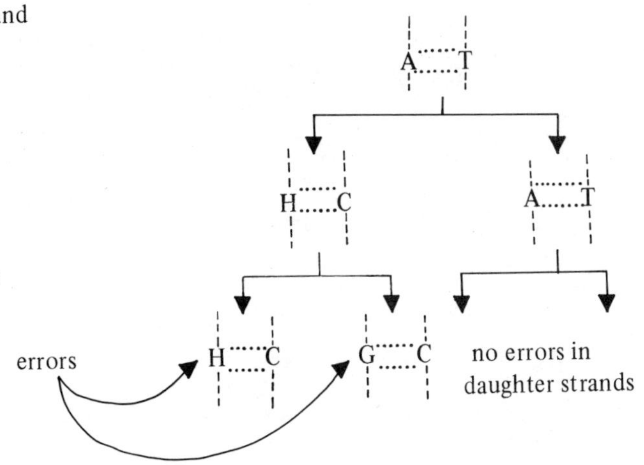
A T
H C
A T
errors
H C
G C
no errors in
daughter strands

26.9

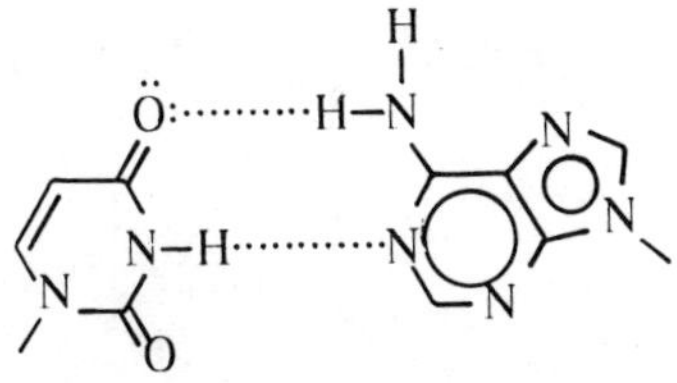

Uracil
(in mRNA)

Adenine
(in DNA)

26.10

(a)	UGG	GGG	UUU	UAC	AGC	mRNA
(b)	Tyr	Gly	Phe	Tyr	Ser	Amino acids
(c)	ACC	CCC	AAA	AUG	UCG	Anticodons

26.11

	Arg ·	Ile ·	Cys ·	Tyr ·	Val	Amino acids
(a)	AGA	AUA	UGC	UGG	GUA	mRNA
(b)	TCT	TAT	ACG	ACC	CAT	DNA
(c)	UCU	UAU	ACG	ACC	CAU	anticodons

26.12

A change from C–T–T to C–A–T or a change from C–T–C to C–A–C.

27

APPENDIX

A.1

The compound is methane, CH_4. The molecular ion is at m/e 16. (This happens also to be the base peak.)

$$H-\underset{\underset{H}{|}}{\overset{\overset{H}{|}}{C}}-H + e^- \longrightarrow H-\underset{\underset{H}{|}}{\overset{\overset{H}{|}}{\overset{+}{C}\cdot}}H + 2e^-$$

$$m/e\ 16$$
$$M^{\ddot{+}}$$

The peaks at m/e 15, 14, 13, and 12 are caused by successive losses of hydrogen atoms.

$$H-\underset{\underset{H}{|}}{\overset{\overset{H}{|}}{\overset{+}{C}\cdot}}H \longrightarrow H-\underset{\underset{H}{|}}{\overset{\overset{H}{|}}{\overset{+}{C}}} + H\cdot$$

$$m/e\ 15$$

$$H-\underset{\underset{H}{|}}{\overset{\overset{H}{|}}{\overset{+}{C}}} \longrightarrow H-\underset{\underset{\cdot}{}}{\overset{\overset{H}{|}}{\overset{+}{C}}} + H\cdot$$

$$m/e\ 14$$

$$H-\underset{\underset{\cdot}{}}{\overset{\overset{H}{|}}{\overset{+}{C}}} \longrightarrow H-C\overset{+}{:} + H\cdot$$

$$m/e\ 13$$

$$H-C\overset{+}{:} \longrightarrow \cdot C\overset{+}{:} + H\cdot$$

$$m/e\ 12$$

The small peak at m/e 17 ($M^{\ddot{+}}$ + 1) comes mainly from methane molecules that contain ^{13}C.

$$H-{}^{13}\underset{\underset{H}{|}}{\overset{\overset{H}{|}}{C}}-H + e^- \longrightarrow H-{}^{13}\underset{\underset{H}{|}}{\overset{\overset{H}{|}}{\overset{+}{C}\cdot}}H + 2e^-$$

$$m/e\ 17$$
$$(M^{\ddot{+}} + 1)$$

A.2

The compound is water.

$$H-\ddot{O}-H + e^- \longrightarrow H-\overset{+}{\underset{\cdot\cdot}{\ddot{O}}}-H + 2e^-$$
$$m/e\ 18$$
$$(M^{\ddagger})$$

$$H-\overset{+}{\underset{\cdot\cdot}{\ddot{O}}}-H \longrightarrow H-\ddot{O}^+ + H\cdot$$
$$m/e\ 17$$

$$H-\ddot{O}^+ \longrightarrow \cdot\ddot{O}^+ + H\cdot$$
$$m/e\ 16$$

The peaks at m/e 19 and m/e 20 are due (primarily) to naturally occurring oxygen isotopes.

$$H-^{17}\ddot{O}-H + e^- \longrightarrow H-^{17}\overset{+}{\ddot{O}}-H + 2e^-$$
$$m/e\ 19$$
$$(M^{\ddagger} + 1)$$

$$H-^{18}\ddot{O}-H + e^- \longrightarrow H-^{18}\overset{+}{\ddot{O}}-H + 2e^-$$
$$m/e\ 20$$
$$(M^{\ddagger} + 2)$$

A.3

The compound is methyl fluoride, CH_3F.

$$CH_3-F + e^- \longrightarrow [CH_3F]^{\ddagger} + 2e^-$$
$$m/e\ 34$$
$$(M^{\ddagger})$$

$$[CH_3F]^{\ddagger} \longrightarrow [CH_2F]^+ + H\cdot$$
$$m/e\ 33$$

$$[CH_2F]^+ \longrightarrow [CHF]^{\ddagger} + H\cdot$$
$$m/e\ 32$$

$$[CHF]^{\ddagger} \longrightarrow [CF]^+ + H\cdot$$
$$m/e\ 31$$

$$[CH_3F]^{\ddagger} \longrightarrow [F]^+ + CH_3\cdot$$
$$m/e\ 19$$

$$[CH_3F]^{\ddagger} \longrightarrow [CH_3]^+ + F\cdot$$
$$m/e\ 15$$

$$[CH_3]^+ \longrightarrow [CH_2]^{\ddagger} + H\cdot$$
$$m/e\ 14$$

A.4

The $M^{\ddagger} + 1$ peak should be approximately 0.43% of the molecular ion.

$$M^{\ddagger} + 1 \cong 0.38 + 3(0.016) \cong 0.43\%$$
$$\text{(from } {}^{15}\text{N)} \text{ (from } {}^{2}\text{H)}$$

This correlates very well with the m/e 18 peak in the spectrum of ammonia (Fig. A.3).

A.5

The $M^{\ddagger} + 1$ peak should be approximately 0.07% that of the molecular ion.

$$M^{\ddagger} + 1 \cong 0.04 + 2(0.016) \cong 0.07\%$$
$$\text{(from } {}^{17}\text{O)} \text{ (from } {}^{2}\text{H)}$$

In the actual spectrum (Fig. A.5) the peak is 0.06%.
The $M^{\ddagger} + 2$ peak should be approximately 0.20%

$$M^{\ddagger} + 2 \cong 0.20$$
$$\text{(from } {}^{18}\text{O)}$$

This agrees with a measured value of 0.2%.

A.6

(a)

(b) Only the first three. (The peak at 1730 cm^{-1} is due to a C=O group.)

A.7

Recalculating the intensities to base on $M^{\ddagger}$

Peak	m/e	% of Base Peak	% of $M^{\ddagger}$
$M^{\ddagger}$	73	86.1	100
$M^{\ddagger} + 1$	74	3.2	3.72
$M^{\ddagger} + 2$	75	0.2	0.23

These best fit the formula C_3H_7NO.

A.8

(a) There is a large natural abundance of the $M^{\ddagger} + 2$ isotope.

(b) The $M^{\ddagger} + 2$ peak due to $CH_3{}^{37}Cl$ (at m/e 52) should be almost one third (32.5%) as large as the $M^{\ddagger}$ peak at m/e 50.

(c) The peaks due to $CH_3{}^{79}Br$ and $CH_3{}^{81}Br$ (at m/e 94 and m/e 96 respectively) should be of nearly equal intensity.

(d) That the $M^{\ddagger}$ and $M^{\ddagger} + 2$ peaks are of nearly equal intensity tells us that the compound contains bromine. C_3H_7Br is therefore a likely molecular formula.

$$
\begin{array}{ll}
C_3 = 36 & C_3 = 36 \\
H_7 = 7 & H_7 = 7 \\
{}^{79}Br = \underline{79} & {}^{81}Br = \underline{81} \\
m/e = 122 & m/e = 124
\end{array}
$$

A.9

(a) A *tert*-butyl cation, $(CH_3)_3C^+$.

$$
\left[\begin{array}{c} CH_3 \\ | \\ CH_3-C-CH_3 \\ | \\ CH_3 \end{array}\right]^{\ddagger} \longrightarrow \begin{array}{c} CH_3 \\ | \\ CH_3-C^+ \\ | \\ CH_3 \end{array} + CH_3\cdot
$$

$$m/e\ 57$$

A.10

A peak at $M^{\ddagger} - 15$ involves the loss of a methyl radical and the formation of a 1° carbocation.

$$
\begin{array}{c} CH_3 \\ | \\ [CH_3CH_2CHCH_2CH_3] \end{array}^{\ddagger} \longrightarrow \begin{array}{c} CH_3 \\ | \\ CH_3CH_2CHCH_2{}^+ \end{array} + CH_3\cdot
$$

$$
\begin{array}{ccc} M^{\ddagger} & \qquad & M^{\ddagger} - 15 \end{array}
$$

A peak at $M^{\ddagger} - 29$ arises from the loss of an ethyl radical and the formation of a 2° carbocation.

$$
\begin{array}{c} CH_3 \\ | \\ [CH_3CH_2CHCH_2CH_3] \end{array}^{\ddagger} \longrightarrow \begin{array}{c} CH_3 \\ | \\ CH_3CH_2CH^+ \end{array} + CH_3CH_2\cdot
$$

$$
\begin{array}{ccc} M^{\ddagger} & \qquad & M^{\ddagger} - 29 \end{array}
$$

Since a 2° carbocation is more stable, the peak at $M^{\ddagger} - 29$ is more intense.

A.11

Both peaks arise from allylic fragmentations

$$
{}^+CH_2-\overset{\cdot}{C}H-CH_2-\underset{\underset{\displaystyle CH_3}{|}}{C}HCH_2CH_3 \longrightarrow \overset{\cdot}{C}H_2-CH=CH_2 + {}^+\underset{\underset{\displaystyle CH_3}{|}}{C}HCH_2CH_3
$$

Allyl radical $m/e\ 57$

$$CH_2 \overset{+\cdot}{\frown} CH - CH_2 \overset{\frown}{:} CHCH_2CH_3 \longrightarrow \overset{+}{C}H_2 - CH = CH_2 \; + \; \cdot CHCH_2CH_3$$

with CH_3 below the first structure and CH_3 below the product $\cdot CHCH_2CH_3$.

m/e 41
Allyl cation

A.12

(a) Alcohols undergo rapid cleavage of a carbon-carbon bond next to oxygen because this leads to a resonance-stabilized cation.

$$1° \text{ alcohol } R\overset{\frown}{:}CH_2 - \overset{+\cdot}{O}H \xrightarrow{-R\cdot} CH_2 = \overset{+}{O}H \longleftrightarrow \overset{+}{C}H_2 - \overset{..}{O}H$$

$$2° \text{ alcohol } R - \overset{R}{\underset{}{CH}} - \overset{+\cdot}{O}H \xrightarrow{-R\cdot} RCH = \overset{+}{O}H \longleftrightarrow R\overset{+}{C}H - \overset{..}{O}H$$

$$3° \text{ alcohol } R - \overset{R}{\underset{R}{C}} - \overset{+\cdot}{O}H \xrightarrow{-R\cdot} R - \overset{}{\underset{R}{C}} = \overset{+}{O}H \longleftrightarrow R - \overset{+}{\underset{R}{C}} - \overset{..}{O}H$$

The cation obtained from a tertiary alcohol is the most stable (because of the electron-releasing R groups).

(b) Primary alcohols give a peak at *m/e* 31 due to $CH_2 = \overset{+}{O}H$.

(c) Secondary alcohols give peaks at *m/e* 45, 59, 73, and so forth, because ions like the following are produced.

$$CH_3CH = \overset{+}{O}H \qquad CH_3CH_2CH = \overset{+}{O}H \qquad CH_3CH_2CH_2CH = \overset{+}{O}H$$

m/e 45   *m/e* 59   *m/e* 73

(d) Tertiary alcohols give peaks at *m/e* 59, 73, 87, and so forth, because ions like the following are produced.

$$CH_3\overset{+}{C}=\overset{}{O}H \qquad CH_3CH_2\overset{+}{C}=\overset{}{O}H \qquad CH_3CH_2CH_2\overset{+}{C}=\overset{}{O}H$$

with CH_3 below each central carbon.

m/e 59   *m/e* 73   *m/e* 87

A.13

The spectrum given in Fig. A.12 is that of isopropyl butyl ether. The main clues are the peaks at *m/e* 101 and *m/e* 73 due to the following fragmentations.

$$\left[\begin{matrix} CH_3 \\ | \\ CH_3 - CH - OCH_2CH_2CH_2CH_3 \end{matrix} \right]^{\ddagger} \xrightarrow{-CH_3\cdot} CH_3CH = \overset{+}{O}CH_2CH_2CH_2CH_3$$

m/e 101

$$\left[\begin{matrix} CH_3 \\ | \\ CH_3CH - O - CH_2CH_2CH_2CH_3 \end{matrix} \right]^{\ddagger} \xrightarrow{-CH_3CH_2CH_2\cdot} CH_3\overset{CH_3}{\underset{|}{C}}HO^+ = CH_2$$

m/e 73

Propyl butyl ether (Fig. A.13) has no peak at m/e 101 but has a peak at m/e 87 instead.

$$\left[CH_3CH_2CH_2-O-CH_2CH_2CH_2CH_3\right]^{\ddagger} \xrightarrow{-CH_3CH_2\cdot}$$

$$CH_2=\overset{+}{O}CH_2CH_2CH_2CH_3$$
$$m/e\ 87$$

Propyl butyl ether also has a peak at m/e 73.

$$\left[CH_3CH_2CH_2-O-CH_2CH_2CH_2CH_3\right]^{\ddagger} \xrightarrow{-CH_3CH_2CH_2\cdot}$$

$$CH_3CH_2CH_2-\overset{+}{O}=CH_2$$
$$m/e\ 73$$

[Although it does not help us decide it is interesting to notice that both spectra have intense peaks at m/e 43 and m/e 57 corresponding to propyl (or isopropyl) and butyl cations formed by carbon-oxygen bond cleavage.]

A.14

    (a) $[C_6H_6O]^{\ddagger}$ m/e 94   Zero nitrogens

    (b) $[C_2H_5Cl]^{\ddagger}$ m/e 64   Zero nitrogens

    (c) $[C_6H_{12}]^{\ddagger}$ m/e 84   Zero nitrogens

    (d) $[C_3H_9N]^{\ddagger}$ m/e 59   One nitrogen

    (e) $[CN_2H_6]^{\ddagger}$ m/e 48   Two nitrogens

    (f) $[C_2H_5NO]^{\ddagger}$ m/e 59   One nitrogen

A.15

The compound is butanal. The peak at m/e 44 arises from a McLafferty rearrangement.

$$\left[H-C\overset{O}{\underset{CH_2-CH_2}{\diagdown}}\overset{H}{\underset{CH_2}{\diagup}}\right]^{\ddagger} \longrightarrow \left[H-C\overset{O-H}{\underset{CH_2}{\diagdown}}\right]^{\ddagger} + \overset{CH_2}{\underset{CH_2}{\|}}$$

$$m/e\ 72 \qquad\qquad\qquad m/e\ 44$$
$$M^{\ddagger} \qquad\qquad\qquad\qquad (M^{\ddagger}-28)$$

The peak at m/e 29 arises from a fragmentation producing an acylium ion.

$$\overset{H}{\underset{\underset{\underset{CH_3}{\overset{|}{CH_2}}}{\overset{|}{CH_2}}}{\diagdown}}C=\overset{+}{\underset{\cdot}{O}}{}^{\ddagger} \longrightarrow H-C\equiv\overset{+}{O} + CH_3CH_2CH_2\cdot$$

$$m/e\ 29$$

A.16

The ion, $CH_2=\overset{+}{N}H_2$, produced by the following fragmentation.

$$R\!:\!CH_2\!-\!\overset{+}{N}H_2 \xrightarrow{-R\cdot} CH_2\!=\!\overset{+}{N}H_2 \longleftrightarrow \overset{+}{C}H_2\!-\!\overset{..}{N}H_2$$

$$m/e \; 30$$

A.17

Compound **A** is *tert*-butylamine. Our first clue is the molecular ion at m/e 73 (an odd-numbered mass unit) indicating the presence of an odd number of nitrogen atoms. The base peak at m/e 58 is our second important clue. It arises from the following fragmentation.

$$\left[\begin{array}{c} CH_3 \\ | \\ CH_3\!-\!\overset{|}{C}\!-\!NH_2 \\ | \\ CH_3 \end{array} \right]^{\ddot{+}} \xrightarrow{-CH_3\cdot} \begin{array}{c} CH_3\!-\!\overset{+}{C}\!=\!NH_2 \\ | \\ CH_3 \end{array}$$

$$m/e \; 58$$

The pmr spectrum confirms the structure

$$\begin{array}{cc} (a) & (b) \\ (CH_3)_3C\!-\!NH_2 \end{array}$$

(a) Singlet $\delta 1.2$ (9H)

(b) Singlet $\delta 1.3$ (2H)

A.18

The compound is 2-methyl-2-butanol. Although the molecular ion is not discernible, we are given that it is at m/e 88. This gives us the molecular weight of **B** and rules out the possibility of a structure with an odd number of nitrogen atoms.

The infrared absorption (3200-3600 cm^{-1}) suggests the presence of an $-OH$ group.

Two important peaks in the mass spectrum are the intense peaks at m/e 59 and m/e 73. These correspond to fragmentation reactions that produce resonance-stabilized oxonium ions and strongly suggest that we have a tertiary alcohol [see problem A.12, part (d)].

$$\left[\begin{array}{c} CH_3 \\ | \\ CH_3CH_2\overset{|}{C}\!-\!OH \\ | \\ CH_3 \end{array} \right]^{\ddot{+}} \xrightarrow{-CH_3CH_2\cdot} \begin{array}{c} CH_3 \\ | \\ \overset{+}{C}\!=\!OH \\ | \\ CH_3 \end{array}$$

$$m/e \; 59$$

$$\xrightarrow{-CH_3\cdot} \begin{array}{c} CH_3CH_2C\!=\!\overset{+}{O}H \\ | \\ CH_3 \end{array}$$

$$m/e \; 73$$

The peak at m/e 70 corresponds to the loss of a molecule of water from the molecular ion and the peak at m/e 55 probably arises from a subsequent allylic cleavage

$$\left[CH_3CH_2\overset{\underset{\displaystyle CH_3}{|}}{\underset{\underset{\displaystyle CH_3}{|}}{C}}-OH\right]^{\ddot{+}}$$

$$\xrightarrow{-H_2O}\left[\underset{m/e\ 70}{CH_3CH=\overset{\overset{\displaystyle CH_3}{|}}{C}-CH_3}\right]^{\ddot{+}}$$

$$\xrightarrow{-H_2O}\left[\underset{m/e\ 70}{CH_3CH_2\overset{\overset{\displaystyle CH_3}{|}}{C}=CH_2}\right]^{\ddot{+}}$$

$$\xrightarrow{-CH_3\cdot}\underset{m/e\ 55}{\overset{+}{CH_2}-\overset{\overset{\displaystyle CH_3}{|}}{C}=CH_2}$$

The pmr spectrum of **B** confirms that it is 2-methyl-2-butanol

$$\underset{(d)}{\overset{(a)\quad(b)\quad\overset{(c)}{\overset{\displaystyle CH_3}{|}}}{CH_3-CH_2-\overset{|}{\underset{|}{C}}-CH_3(c)}}\\ OH$$

(a) Triplet, $\delta 0.9$ (3H)

(b) Quartet, $\delta 1.6$ (2H)

(c) and (d) Overlapping singlets, $\delta 1.1$ (7H)

A. 19

Compound **C** is 3-methyl-1-butanol. Here, (because the compound is a primary alcohol) the molecular ion (m/e 88) is small but discernible. Again, the even-numbered mass of the molecular ion rules out a compound with an odd number of nitrogen atoms and the infrared absorption suggests the presence of an $-OH$ group.

An important indication that **C** is a primary alcohol is the peak at m/e 31 corresponding to the following fragmentation [see also problem A.12, part (b)].

$$\left[\underset{m/e\ 88}{CH_3-\overset{\overset{\displaystyle CH_3}{|}}{CH}CH_2CH_2OH}\right]^{\ddot{+}}\xrightarrow{-CH_3\overset{\overset{\displaystyle CH_3}{|}}{CH}CH_2\cdot}\underset{m/e\ 31}{CH_2=\overset{+}{O}H}$$

The peak at m/e 70 ($M^{\ddot{+}}-18$) corresponds to the loss of water from the molecular ion.

$$\left[\begin{array}{c} \text{CH}_3 \\ | \\ \text{CH}_3\text{CHCH}_2\text{CH}_2\text{OH} \end{array}\right]^{\ddagger} \xrightarrow{-\text{H}_2\text{O}} \left[\begin{array}{c} \text{CH}_3 \\ | \\ \text{CH}_3\text{CHCH}=\text{CH}_2 \end{array}\right]^{\ddagger}$$

$$m/e\ 88 \qquad\qquad\qquad m/e\ 70$$

The peak at m/e 55 probably comes from a subsequent allylic cleavage.

$$\left[\begin{array}{c} \text{CH}_3 \\ | \\ \text{CH}_3\text{CHCH}=\text{CH}_2 \end{array}\right]^{\ddagger} \xrightarrow{-\text{CH}_3\cdot} \text{CH}_3\overset{+}{\text{C}}\text{HCH}=\text{CH}_2$$

$$m/e\ 70 \qquad\qquad\qquad m/e\ 55$$

The pmr spectrum is consistent with this structure. We can make the following assignments.

$$\begin{array}{c} \text{(a)} \\ \text{CH}_3 \\ \overset{\text{(a)}}{}\ | \\ \text{CH}_3-\text{CH}-\text{CH}_2-\text{CH}_2\text{OH} \\ \text{(b)}\ \ \text{(c)}\quad\text{(d)}\ \ \text{(e)} \end{array}$$

(a) Doublet, $\delta 0.9$

(b) and (c) Multiplet $\delta 1.5$

(d) Triplet $\delta 3.7$

(e) Singlet $\delta 2.2$

A.20

The compound is 2-pentanone. The infrared absorption at $1710\ \text{cm}^{-1}$ strongly indicates the presence of a carbonyl group. In the mass spectrum the molecular ion peak at m/e 86 gives us the molecular weight and rules out structures with an odd number of nitrogens. A possible formula is $C_5H_{10}O$.

The peaks at m/e 71 and m/e 43 correspond to $M^{\ddagger} - 15$ and $M^{\ddagger} - 43$. Fragmentations of 2-pentanone would produce acylium ions with these mass numbers.

$$\left[\begin{array}{c} \text{O} \\ \| \\ \text{CH}_3\text{CCH}_2\text{CH}_2\text{CH}_3 \end{array}\right]^{\ddagger} \Bigg\langle \begin{array}{l} \xrightarrow{-\text{CH}_3\cdot} \overset{+}{\text{O}}{\equiv}\text{CCH}_2\text{CH}_2\text{CH}_3 \\ \qquad\qquad\qquad m/e\ 71 \\ \\ \xrightarrow{-\text{CH}_3\text{CH}_2\text{CH}_2\cdot} \text{CH}_3\text{C}{\equiv}\overset{+}{\text{O}} \\ \qquad\qquad\qquad m/e\ 43 \end{array}$$

$$m/e\ 86$$

The peak at m/e 58 ($M^{\ddagger} - 28$) comes from a McLafferty rearrangement.

$$\left[\begin{array}{c} \text{CH}_3-\text{C} \overset{\text{O}}{\underset{\text{CH}_2-\text{CH}_2}{\diagdown}}\overset{\text{H}}{\underset{}{\diagdown}}\text{CH}_2 \end{array}\right]^{\ddagger} \longrightarrow \left[\begin{array}{c} \text{CH}_3-\text{C} \overset{\text{OH}}{\underset{\text{CH}_2}{\diagdown}} \end{array}\right]^{\ddagger} + \begin{array}{c} \text{CH}_2 \\ \| \\ \text{CH}_2 \end{array}$$

$$m/e\ 58$$

The pmr spectrum confirms our structure.

(a) (b) (c) (d)
$$CH_3\overset{\overset{\textstyle O}{\|}}{C}CH_2CH_2CH_3$$

(a) Singlet, δ 2.2

(b) Triplet, δ 2.4

(c) Multiplet, δ 1.6

(d) Triplet, δ 0.9

A.21

The compound is bromobenzene. That the compound contains bromine is indicated by the $M^{\ddag}$ and $M^{\ddag} + 2$ peaks of nearly equal intensity at m/e 156 and m/e 158. The peak at m/e 77 (the base peak) strongly suggests the presence of a benzene ring.

m/e 77

Putting these facts together with the molecular weight (156) leads us to only one logical conclusion.

A.22

Compound **F** is 2-bromoethyl phenyl ether, $C_6H_5OCH_2CH_2Br$. The presence of bromine is indicated by the $M^{\ddag}$ and $M^{\ddag} + 2$ peaks of nearly equal intensity at m/e 200 and m/e 202. The presence of a benzene ring is indicated by the peak at m/e 77.

m/e 77

If we subtract the masses of these two large fragments (i.e., Br = 79 and C_6H_5 = 77) from the mass of the molecular ion (200) we are left with a mass difference of 44. One way that we can account for this mass difference is with C_2H_4O.

$$
\begin{aligned}
C_6H_5 &= 77 \\
Br &= 79 \\
C_2H_4O &= \underline{44} \\
MW &= 200
\end{aligned}
$$

Thus we can tentatively assume that the compound has a molecular formula C_8H_9BrO.

The complex multiplet (5H) at $\delta 7.0$ in the pmr spectrum confirms our assignment of a benzene ring, C_6H_5-. The triplets at $\delta 2.5$ and at $\delta 4.2$ suggest mutually coupled $-CH_2-$ groups. Thus we are led to the following as a possible structure.

The peak in the mass spectrum at m/e 107 strongly suggests that this assignment is correct.

The peak at m/e 94 comes from the following rearrangement.

m/e 94

The small peaks of nearly equal intensity at m/e 93 and m/e 95 come from $CH_2=Br^+$ fragments in which the bromines are ^{79}Br and ^{81}Br.